QUEUEING SYSTEMS

VOLUME II: COMPUTER APPLICATIONS

*"As gold which he cannot spend will make no man rich,
so knowledge which he cannot apply will make no man wise."*

Samuel Johnson: *The Idler No. 84*

QUEUEING SYSTEMS

VOLUME II:
COMPUTER APPLICATIONS

Leonard Kleinrock

Professor
Computer Science Department
School of Engineering and Applied Science
University of California, Los Angeles

A Wiley-Interscience Publication

JOHN WILEY AND SONS
New York • Chichester • Brisbane • Toronto • Singapore

Library of Congress Cataloging in Publication Data (*Revised*)

Kleinrock, Leonard.
    Queueing systems.

    "A Wiley-Interscience publication."
    CONTENTS: v. 1. Theory.—v. 2. Computer applications.
    1. Queueing theory.   I. Title.

T57.9.K6   519.8'2   74-9846
ISBN 0-471-49111-X (v. 2)

Printed in the United States of America

20 19 18 17

TO STELLA

Preface

Recently, I made the mistake of flying across the country in a Boeing 747. As a queueing systems analyst, I should have known better! As soon as I arrived at the airport, I immediately realized my error, for there was a mob of passengers waiting to be checked in. This was clearly a rush-hour situation for the airline, and the peak load was saturating the service facility. I, of course, had two extremely heavy suitcases filled with notes on queueing systems (what else?) and so I could not morally avoid this queue. The situation was a multiple-server multiple-queue system with clearly unequal rates of service; however, once invested in a (particularly slow) queue, I could not afford to risk giving up my position. After clearing the check-in procedure, I then found my way to the departure lounge where an enormous queue had been formed in a snakelike fashion awaiting seat assignments and boarding passes. This was a two-server system with a common queue that also was unbelievably over-loaded. The snaked queue convoluted itself in such a way that its tail was immediately adjacent to its head, so, as you might imagine, considerable cheating took place as passengers bypassed the queue completely; there was absolutely no control of this effect except for a few angry passengers (however, those at the head of the queue who were in the only position to stop this cheating really did not care since they were ready to be served at this point). Aboard the aircraft, of course, things went from bad to worse as the overworked crew provided a comedy of errors and a tragedy of frustrations for all of us. Would you believe that one of the meals was served in buffet fashion so that the entire passenger population was asked to stand up and file past a collection of randomly placed delicacies in the lounge at the front of the aircraft? The ingenuity of this last queue was too much for me. They arranged matters so that first those passengers on the port side of the aircraft queued up in the port aisle, worked their way to the front, received (fought for) the delicacies, returned via the star-board aisle, and then found it quite impossible to reach their seats because of the port aisle queue. A similar delight later developed for the passengers on the starboard side. I was foolish enough to suggest the obvious improvement whereby port-side passengers would queue up on the starboard aisle, leaving their own aisle clear for return to their seats, but was quickly informed that such a suggestion "could not possibly work!" And so it continued up to and including baggage recovery (you can imagine the fun).

That travel adventure is just one of many similar situations that all of us have encountered. As systems analysts, we have a moral and personal obligation to study these real-life systems and provide some relief from their aggravations even if they do not lend themselves easily to analysis. We should be able to develop models for the physical systems we interact with, produce an analysis that describes the behavior of these systems, and then find methods for improving their behavior. In so doing, the first task of creating the mathematical model is perhaps the most difficult, since it is here that we make the most compromises in describing the reality of the physical systems. In the second (analytical) stage, we look for tools to aid us in solving these systems, and in the third we look for design methodologies. For many of the technological systems of today, we find that tools from queueing theory are useful in describing the systems' behavior. Unfortunately, these tools are not nearly as sharp as we would like, and so we often are faced with the necessity of using tools that are not wholly adequate for the job. This is a common problem whenever one deals in applications of a theory, and it is our purpose in this book to study the difficulty and the success in the application of queueing systems to various fields, and in particular to the field of computer systems.

The class of systems that generally lends itself to queueing analysis is the (huge) one in which customers compete for access to a limited (i.e., finite-capacity) resource. In fact, many of today's significant problems can be reduced to the problem of *resource allocation* and *resource sharing*. Throughout this book, we will encounter such problems, and we will emphasize the need for resource sharing in computer systems for purposes of efficiency. For example, imagine that a pair of users requires the use of a communication channel. The classical approach to satisfying this requirement is to provide a channel for their use as long as that need continues (and to charge them for the full cost of this channel). It has long been recognized that such allocation of scarce resources is extremely wasteful, as witnessed by their low utilization. Rather than provide channels on a user-pair basis, we much prefer to provide a large number of users with a single high-speed channel that can be shared in some fashion; this then allows us to take advantage of the powerful "large-number laws" which state that, with very high probability, the demand at any instant will be very nearly equal to the sum of the average demands of that user population. In this way, the required channel capacity to support the user traffic may be considerably less than in the unshared case of dedicated channels. This approach has been used to great effect for many years now in a number of different contexts; for example, in the use of graded channels in the telephone industry, in the introduction of asynchronous time-division multiplexing, in the implementation of time-shared and multiaccess computer systems, and in the packet-switching concepts introduced in the 1960s and recently implemented in the AR-PANET. The essential observation is that the full-time allocation of a fraction of the channel to each user-pair is highly inefficient compared to the part-time use of the full capacity of the channel (this is precisely the notion of time-sharing). We gain this efficient sharing particularly when the traffic consists of rapid but short *bursts* of data. The classical schemes of synchronous time-division multiplexing and frequency-division multiplexing are examples of the inefficient partitioning of channels for bursty data sources. As soon as we introduce the notion of a shared channel in a packet-switching mode, then we must be prepared to resolve conflicts that arise when more than one demand is simultaneously placed upon the channel. There are two obvious solutions to this problem: the first is to "throw out" or "lose" any demands made while the channel is in use; and the second is to form a queue of conflicting demands and serve them in some order as the channel becomes free. The latter approach is that taken in the ARPANET since storage may be provided economically at the points of conflict. The former approach is taken in the ALOHA system, which uses packet switching with radio channels; in this system, in fact, *all simultaneous* demands made on the channel are lost. These concepts are discussed in Chapters 5 and 6, and represent some of the more successful modern applications of queueing theory.

Our purpose in this book, then, is twofold: first, to modify the tools of queueing theory in a way that permits them to be applied to real-world problems; and second, to make an extensive application of these tools to various and important modern-day computer systems.* Indeed, one of the fastest-growing industries and one of the most advanced technologies of today is that of computer systems. Sad to say, the theory is badly lagging behind the practice in this field, and we have not yet come up with an acceptable measure of computer power. Thus we find a situation in which one grasps for analytical tools in attempting to evaluate performance of computer systems. Fortunately, in the early 1960s it was found that queueing theory was an effective tool for studying the throughput, response, and other measures of performance for computer systems.

* This twofold purpose is the reason why Chapter 2 (Bounds, Inequalities and Approximations) was placed here in Volume II rather than in Volume I (Theory).

Since that time, the queueing theory literature and the computer applications literature have virtually exploded with analytic models for computer systems, and in this book we focus on some of these recent and successful applications. Indeed, for those readers who have invested considerable time, effort, and even discomfort in developing a set of queueing-theoretic tools, this text represents one payoff for that effort. Now, finally, we have the opportunity to use these tools in real-life applications. We have chosen computer applications in this book principally since these applications are the most recent and most successful for the theory of queues (and also because, in no small way, I personally have been deeply involved in the application of the theory to these cases). In fact, the use of queueing theory to analyze resource allocation and job flow through computer systems is perhaps the *only* method available to computer scientists in understanding the behavior of the complex interconnection of their systems.

This book was written over a period of many years while being used as class notes at the graduate level in the Computer Science Department of the School of Engineering and Applied Science at the University of California, Los Angeles. Much of this material has come to life in the course of this development and in the participation of UCLA as a modeling and analysis center for research in computer systems. A vast number of important problems have been identified as a result of these applications and many opportunities exist for the clever and capable analyst.

This book is at the first-year graduate level and naturally follows material such as that contained in my companion volume, *Queueing Systems, Volume I: Theory*. In order that this stand on its own as a self-contained book, we begin with a *queueing theory primer* in Chapter 1 in which we review the fundamental material in queueing systems analysis. Those readers who have been exposed to a first course in queueing systems will easily manage the material in this book. On the other hand, for those readers who have not been through a formal course in queueing systems, this book is well within their reach if they accept the material presented in Chapter 1 as a point of departure for the balance of the book.

Chapter 2 is the bridge that permits us to pass from the abstract tools of queueing theory to the real world of applied results. In this chapter, we acknowledge the weaknesses and shortcomings of the formal theory in its ability to describe the real-world applications that we encounter; consequently, we look for engineering guidelines in the form of *bounds, inequalities, and approximations* that are permissible and valid in various situations. Following that, we discuss the interesting field of *priority queueing* in Chapter 3. We look upon this chapter as a further bridge between theory and applications, since it is through priority queueing that we are able to model the behavior of *computer time-sharing and multiaccess systems*, which form the subject material for Chapter 4. In this, our first real applications chapter, we emphasize computer systems in isolation that handle demands from a large collection of competing users. We look for throughput and response time as well as utilization of resources. The major portion of this chapter is devoted to a particular class of algorithms known as processor-sharing algorithms, since they are singularly well suited to queueing analysis and they capture the essence of more difficult and more complex algorithms seen in real scheduling problems. Chapter 5 addresses itself to the *analysis and design of computer-communication networks*. In Chapter 6 we discuss *simulation, measurement, flow control, and traps in computer-communication networks*. These networks represent perhaps the fastest-growing field in the young computer industry itself. In the Epilogue of Volume I, we promised only one chapter in Volume II on computer networks; however, in the months that separated the preparation of these two volumes, the field has expanded so quickly that we have been forced to meet the challenge of this material with an additional chapter. If the reader glances at the list of references for these last two chapters, he will see that the majority are drawn from the last three or four years—a tell-tale indicator indeed; as a result, it is virtually impossible to produce a truly up-to-date chapter in a printed volume such as this. Perhaps the best solution is to keep an updated on-line file that is accessible through a computer network! Chapter 5 is devoted to developing methods of analysis and design for computer-communication networks; it identifies many unsolved, important problems. A specific existing network, the ARPANET, is used throughout as an example to guide the reader through the motivation and evaluation of the various techniques developed. The chapter closes with the consideration of some of the newer packet-switching concepts, namely, data transmission using satellite and ground radio communications. Chapter 6 points to the operational and measurement aspects of networks, drawing heavily from the ARPANET. On page xiv we show the precedence structure among the chapters of this book.

A glossary at the end of the book summarizes the commonly used symbols and notation. Each chapter contains its own list of references keyed alphabetically to the author and year; for example, [KLEI 75] would refer to the companion Volume I. All equations of importance have been marked with a symbol ▬, and these are included in the summary of important equations (the equations in Chapter 1 are exceptions to this in that the symbol is omitted, since *all* of them are

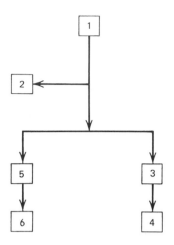

important). Each chapter (except the first) includes a set of exercises that in some cases extends the material in the chapter; the reader is urged to work them out.

We view this as a "living book" in that its five principal chapters (omitting the review in Chapter 1) must, by their very nature, evolve and change in coming years. It is in these fields that we expect improvements to occur; in that sense, this book will undoubtedly pass through other editions as the need for updating demands it.

Acknowledgement and thanks are due to many of my friends, colleagues, and students who provided the encouragement and environment at the UCLA Computer Science Department which made this work possible. Most of my research results contained in Chapters 3, 4, 5, and 6 represent areas that the United States Department of Defense Advanced Research Projects Agency funded; for this support, and for making it possible for me to participate in some of the most advanced computer systems and networks ever developed, I am happy to express my thanks, especially to Lawrence G. Roberts and Robert E. Kahn. During 1971–1972, I was fortunate enough to be a John Simon Guggenheim Fellow, and for this I gratefully thank that Foundation. A book such as this requires constant evaluation in the harsh testing stage of the graduate classroom; I thank the hundreds of graduate students who have helped me in that regard and also am happy to acknowledge the special help offered by Arne Nilsson, Mario Gerla, Faroula Kamoun, Parviz Kermani, Simon Lam, William Naylor, Louis Nelson, Holger Opderbeck, Michel Scholl, Fouad Tobagi, David Wong, and Johnny Wong. To my research and administrative support staff, I owe a great debt of thanks. Most of the typing was done by Charlotte LaRoche to whom I am deeply indebted;

Carol Mason and Barbara Warren were responsible for handling my unending revisions, and they were aided by Jean D'Fucci, Jean Dubinsky, George Ann Hornor, Cathy Pfennig, and Gloria Roy. To Diana Skocypec and Lynn Johnson fell the tedious task of proofreading the garrulous galleys and ponderous pages; my admiration and thanks goes to them. To my children, I offer my regret at having taken time for this which might have been spent with them, and my thanks for their indulgence. To my parents, I give my gratitude for the principles they instilled in me and for the opportunities they afforded me. To Stella, my wife, I owe much for her unending patience, encouragement, and understanding, without which I could hardly have completed this work.

LEONARD KLEINROCK

Los Angeles, California
January 1976

CONTENTS

VOLUME I

PART I: PRELIMINARIES

PART II: ELEMENTARY QUEUEING THEORY

QUEUEING SYSTEMS

VOLUME II: COMPUTER APPLICATIONS

I

A Queueing Theory Primer

In this chapter we summarize the important results to which one is exposed in a first course on queueing theory. This material is drawn from the companion volume [KLEI 75] in which will be found a list of results that key the reader to the location where each result is derived. Our purpose is to lay the foundation for the remainder of the book, which is devoted to the application of this theory in real-world situations; these applications require sound judgment and experience in formulating models as well as in developing operational formulas (exact or approximate) that may be used for analysis and design of systems. We give a rather complete review here (by stating—*not* deriving—results) so that this material will form a self-contained body of results, to be used in later chapters.

Consider any system that has a capacity C, the maximum rate at which it can perform work. Assume that R represents the average rate at which work is demanded from this system. One fundamental law of nature states that if $R < C$ then the system can "handle" the demands placed upon it, whereas if $R > C$ then the system capacity is insufficient and all the unpleasant and catastrophic effects of saturation will be experienced. However, even when $R < C$ we still experience a different set of un-pleasantnesses that come about because of the *irregularity* of the demands. For example, consider the corner telephone booth, which on the average can handle the load demanded of it. Suppose now that two people approach that telephone booth almost simultaneously; it is clear that only one of the two can obtain service at a given time and the other must wait in a queue until that one is finished. Such queues arise from two sources: the first is the unscheduled arrival times of the customers; the second is the random demand (duration of service) that each customer requires of the system. The characterization of these two unpredictable quantities (the arrival times and the service times) and the evaluation of their effect on queueing phenomena form the essence of queueing theory. In the following section we introduce some of the usual notation for queueing systems and then we proceed to summarize the major results for various systems.

1

1.1. NOTATION

Here we introduce only that notation required for the statement of results in this chapter. A more complete listing is given in the glossary at the end of the book.

We let C_n denote the nth customer to arrive at a queueing facility. The important random variables to associate with C_n are

$$\tau_n = \text{arrival time for } C_n \tag{1.1}$$

$$t_n = \tau_n - \tau_{n-1} = \text{interarrival time between } C_n \text{ and } C_{n-1} \tag{1.2}$$

$$x_n = \text{service time for } C_n \tag{1.3}$$

It is the sequence of random variables $\{t_n\}$ and $\{x_n\}$ that really "drives" the queueing system. All these random variables are selected independently of each other, and so we define the two generic random variables

$$\tilde{t} = \text{interarrival time} \tag{1.4}$$

$$\tilde{x} = \text{service time} \tag{1.5}$$

Associated with each is a probability distribution function (PDF), that is,

$$A(t) = P[\tilde{t} \le t] \tag{1.6}$$

$$B(x) = P[\tilde{x} \le x] \tag{1.7}$$

and the related probability density function (pdf), namely,

$$a(t) = \frac{dA(t)}{dt} \tag{1.8}$$

$$b(x) = \frac{dB(x)}{dx} \tag{1.9}$$

In this last definition for the pdf we permit the use of impulse functions as discussed, for example, in Volume I of this text. The moments associated with these random variables are denoted by

$$E[\tilde{t}] = \bar{t} = \frac{1}{\lambda} \tag{1.10}$$

$$E[(\tilde{t})^k] = \overline{t^k} \tag{1.11}$$

$$E[\tilde{x}] = \bar{x} = \frac{1}{\mu} \tag{1.12}$$

$$E[(\tilde{x})^k] = \overline{x^k} \tag{1.13}$$

where the symbol μ is often reserved only for the case of exponentially distributed service times. Furthermore, we need the Laplace transform associated with these pdf's, namely

$$E[e^{-s\tilde{t}}] = A^*(s) \qquad (1.14)$$

$$E[e^{-s\tilde{x}}] = B^*(s) \qquad (1.15)$$

The integral representation of this transform [say for $a(t)$] is simply

$$A^*(s) = \int_{0^-}^{\infty} a(t)e^{-st}\, dt \qquad (1.16)$$

A key use of this transform is its moment generating property; for example, the moments $\overline{t^k}$ may be generated from $A^*(s)$ through the relationship

$$\left. \frac{d^k A^*(s)}{ds^k} \right|_{s=0} = (-1)^k \overline{t^k} \qquad (1.17)$$

We often denote the kth derivative of a function $f(t)$ evaluated at $t = t_0$ by

$$\left. \frac{d^k f(t)}{dt^k} \right|_{t=t_0} = f^{(k)}(t_0) \qquad (1.18)$$

Thus Eq. (1.17) may be written as $A^{*(k)}(0) = (-1)^k \overline{t^k}$.

Both $\tilde{t}$ and $\tilde{x}$ are the input random variables to the queueing system; now we must define some of the important *performance* variables, namely, the number of customers in the system, the waiting time per customer, and the total time that a customer spends in the system, that is,

$$N(t) = \text{number of customers in system at time } t \qquad (1.19)$$

$$w_n = \text{waiting time (in queue) for } C_n \qquad (1.20)$$

$$s_n = \text{system time (queue plus service) for } C_n \qquad (1.21)$$

The corresponding limiting random variables (after the system has been in operation a long time) for a stable queue are N, $\tilde{w}$, and $\tilde{s}$. As with $\tilde{t}$ and $\tilde{x}$ we may define the PDF, the pdf, the first moment, and the appropriate transform for N, $\tilde{w}$, and $\tilde{s}$ as follows:

$$P[N \le k] \qquad W(y) = P[\tilde{w} \le y] \qquad S(y) = P[\tilde{s} \le y]$$

$$P[N = k] \qquad w(y) = \frac{dW(y)}{dy} \qquad s(y) = \frac{dS(y)}{dy}$$

$$E[N] = \bar{N} \qquad E[\tilde{w}] = W \qquad E[\tilde{s}] = T$$

$$E[z^N] = Q(z) \qquad E[e^{-s\tilde{w}}] = W^*(s) \qquad E[e^{-s\tilde{s}}] = S^*(s)$$

The study of queues naturally breaks into three cases: elementary queueing theory, intermediate queueing theory, and advanced queueing theory. What distinguishes these three cases are the assumptions regarding $a(t)$ and $b(x)$. In order to name the different kinds of systems we wish to discuss, a rather simple shorthand notation is used for describing queues. This involves a three-component description, A/B/m, which denotes an m-server queueing system where A and B "describe" the interarrival time distribution and service time distribution, respectively. A and B take on values from the following set of symbols, which are meant to remind the reader which distributions they refer to:

$$M = \text{exponential (i.e., Markovian)}$$

$$E_r = r\text{-stage Erlangian}$$

$$H_R = R\text{-stage Hyperexponential}$$

$$D = \text{Deterministic}$$

$$G = \text{General}$$

Specifically, if one of these symbols were used in place of B then it would refer to the following pdf $(x \geq 0)$:

$$M: \qquad b(x) = \mu e^{-\mu x} \qquad\qquad\qquad\qquad (1.22)$$

$$E_r: \qquad b(x) = \frac{r\mu(r\mu x)^{r-1}e^{-r\mu x}}{(r-1)!} \qquad\qquad (1.23)$$

$$H_R: \qquad b(x) = \sum_{i=1}^{R} \alpha_i \mu_i e^{-\mu_i x} \quad \left(\sum_{i=1}^{R} \alpha_i = 1\right) \qquad (\alpha_i \geq 0) \qquad (1.24)$$

$$D: \qquad b(x) = u_0\left(x - \frac{1}{\mu}\right) \qquad\qquad\qquad (1.25)$$

$$G: \qquad b(x) \text{ is arbitrary}$$

where in the next to last expression $u_0(x - 1/\mu)$ refers to a unit impulse occurring at the position $x = 1/\mu$. Any distribution is permitted when G is assumed. Occasionally we add one or two more items to our three-component description in order to describe the system's storage capacity (denoted by K) or the size of the customer population (denoted by M), and these will be commented on appropriately when used (otherwise they are assumed to be infinite). The simplest interesting system we consider in this chapter is the M/M/1 queue in which we have exponential interarrival times, exponential service times, and a single server (see Section 1.4). The most complicated system we consider in this chapter is G/G/1 in which the exponential distributions are replaced by arbitrary distributions (see Sections 1.2 and 1.10). In this review the majority of our results apply only to the first-come–first-serve queueing discipline; in Chapter 3, we study the effect of other queueing disciplines. Let us now proceed with our summary of results.

1.2. GENERAL RESULTS

Perhaps the most important system parameter for G/G/1 is the *utilization factor* ρ, defined as the product of the average arrival rate of customers to the system times the average service time each requires, that is,

$$\rho = \lambda \bar{x} \tag{1.26}$$

This quantity gives the fraction of time that the single server is busy and is also equal to the ratio of the rate at which work arrives to the system divided by the capacity of the system to do work, that is, R/C as discussed earlier.* In the multiple-server system G/G/m the corresponding definition is

$$\rho = \frac{\lambda \bar{x}}{m} \tag{1.27}$$

which also is equal to R/C and may be interpreted as the expected fraction of busy servers when each server has the same distribution of service time; more generally, ρ is the expected fraction of the system's capacity that is in use. In all cases a stable system (one that yields finite average delays and queue lengths) is one for which

$$0 \le \rho < 1 \tag{1.28}$$

and we note that the case $\rho = 1$ is not permitted (except in the very special situation of a D/D/m queue). As we shall see, the closer ρ approaches unity, the larger are the queues and the waiting times; it is this quantity that essentially reflects the way in which the system performance varies with the *average* system load.

The average time in system is simply related to the average service time and the average waiting time through the fundamental equation

$$T = \bar{x} + W \tag{1.29}$$

and it is the quantity W that reflects the price we must pay for sharing a given resource (server) with other customers. Whereas ρ is the most important system parameter, it is fair to say that one of the more famous

---

* On the average, λ customers arrive per second and each brings $\bar{x}$ sec of work for the system; thus $R = \lambda \bar{x}$. The (single-server) system can perform 1 sec of work per second of elapsed time, and so $C = 1$.

formulas from queueing is *Little's result*, which relates the average number in the system to the average arrival rate and the average time spent in that system, namely,

$$\bar{N} = \lambda T \tag{1.30}$$

This result enters most of the calculations we make in this book and is extremely general in its application. The corresponding result for number and time in *queue* is simply given by

$$\bar{N}_q = \lambda W \tag{1.31}$$

where $\bar{N}_q$ is merely the average queue size. Furthermore, it is true in G/G/m that these quantities are related by*

$$\bar{N}_q = \bar{N} - m\rho \tag{1.32}$$

We have already given one fundamental law that applies to queueing systems, namely that $R < C$ in order for the system to be stable. A second common and general law of nature also finds its way into our analyses; it relates the rate at which accumulation within a system occurs as a function of the input and output rates to and from that system. In particular, if we let E_k denote the system state in which k customers are present and if we let

$$P_k(t) = P[N(t) = k] \tag{1.33}$$

which is merely the probability that the system state at time t is E_k, then, loosely stated, we have

$$\frac{dP_k(t)}{dt} = [\text{flow rate of probability into } E_k \text{ at time } t]$$

$$-[\text{flow rate of probability out of } E_k \text{ at time } t] \tag{1.34}$$

Equation (1.34) will allow us to write down time-dependent relationships among the system probabilities in a straightforward fashion. Now consider a stable system, for which the probability $P_k(t)$ has a limiting value (as $t \to \infty$) which we denote by p_k, (this represents the fraction of time that the system will contain k customers in the steady state). If the interarrival times are exponentially distributed (that is, they form a Poisson arrival process), then the equilibrium probability, r_k, that an arriving customer finds k in the system upon his arrival will in fact equal the long-run probability of there being k customers in the system, that is $p_k = r_k$. On the other hand, if we denote by d_k the equilibrium probability that a departure leaves behind k customers in the system, then $d_k = r_k$ if

* This follows from $T = \bar{x} + W$ and Little's result.

the system state $N(t)$ is permitted to change by at most one at any time. Thus, if we have unit state changes and Poisson arrivals, then we have the situation in which $p_k = r_k = d_k$.

1.3. MARKOV, BIRTH–DEATH, AND POISSON PROCESSES

Before we proceed to discuss the results for elementary queueing systems it is convenient to list some of the well-known results for some simple and important random processes that form the foundation for the queueing results we shall quote.

We begin with discrete-state discrete-time Markov processes such that X_n denotes the discrete value of the (random) process at its nth step. The defining condition for such a Markov chain is

$$P[X_n = j \mid X_{n-1} = i_{n-1}, \ldots, X_1 = i_1] = P[X_n = j \mid X_{n-1} = i_{n-1}] \quad (1.35)$$

This is merely an expression of the fact that the present state completely summarizes all of the pertinent past history so far as that history affects the future of the process. If we let

$$\pi_i^{(n)} = P[X_n = i] \quad (1.36)$$

and denote the vector of these probabilities by

$$\boldsymbol{\pi}^{(n)} = [\pi_0^{(n)}, \pi_1^{(n)}, \ldots] \quad (1.37)$$

and moreover if we denote the one-step transition probabilities for homogeneous Markov chains by

$$p_{ij} = P[X_n = j \mid X_{n-1} = i] \quad (1.38)$$

and collect these into a square matrix denoted by $\mathbf{P} = (p_{ij})$, then we have the basic results for the time-dependent probabilities of this Markov process, namely,

$$\boldsymbol{\pi}^{(n)} = \boldsymbol{\pi}^{(n-1)}\mathbf{P} \quad (1.39)$$

$$\boldsymbol{\pi}^{(n)} = \boldsymbol{\pi}^{(0)}\mathbf{P}^n \quad (1.40)$$

The sequence $\mathbf{P}^n$ ($n = 0, 1, 2, \ldots$) is equal to the inverse z-transform of the matrix $[\mathbf{I} - z\mathbf{P}]^{-1}$, where $\mathbf{I}$ represents the identity matrix and -1 refers to the matrix inverse. The more useful steady-state behavior of these probabilities may be found by solving the equation

$$\boldsymbol{\pi} = \boldsymbol{\pi}\mathbf{P} \quad (1.41)$$

along with the condition that

$$\sum_{i=0}^{\infty} \pi_i = 1 \tag{1.42}$$

where we have used the notation $\pi_i = \lim_{n \to \infty} \pi_i^{(n)}$. Finally, we comment that the time the process spends in any state is geometrically distributed (an inherent property of all Markov processes); this distribution is, of course the only discrete memoryless distribution.

Let us now consider the case of a discrete-state continuous-time homogeneous Markov process $X(t)$; here we have a defining property much as we did in Eq. (1.35). The time the process spends in any state is exponentially distributed for all continuous-time Markov processes; this is the only (continuous) memoryless distribution, and it is this property that makes the analysis simple. We now define the transition probabilities as

$$p_{ij}(t) = P[X(s+t) = j \mid X(s) = i] \tag{1.43}$$

The matrix of these transition probabilities will be denoted by $\mathbf{H}(t)$, and in terms of this matrix we may express the Chapman–Kolmogorov equations as

$$\mathbf{H}(t) = \mathbf{H}(t-s)\mathbf{H}(s) \tag{1.44}$$

In a real sense $\mathbf{H}(t)$ corresponds to $\mathbf{P}^n$ and that which corresponds to $\mathbf{P}$ itself is $\mathbf{H}(\Delta t)$ (namely the transition probabilities over an infinitesimal interval). Of more use is the matrix $\mathbf{Q} = [q_{ij}]$, referred to as the infinitesimal generator of the process; it is defined by

$$\mathbf{Q} = \lim_{\Delta t \to 0} \frac{\mathbf{H}(\Delta t) - \mathbf{I}}{\Delta t} \tag{1.45}$$

In terms of this matrix we may then express the time-dependent behavior of our Markov process by the equation

$$\frac{d\mathbf{H}(t)}{dt} = \mathbf{H}(t)\mathbf{Q} \tag{1.46}$$

whose solution is

$$\mathbf{H}(t) = e^{\mathbf{Q}t} \tag{1.47}$$

The steady-state behavior of this process, namely, the stable probabilities $\boldsymbol{\pi}$, are given through the basic equation

$$\boldsymbol{\pi}\mathbf{Q} = 0 \tag{1.48}$$

along with the normalizing equation (1.42). We have occasion to discuss the discrete-state continuous-time and continuous-state continuous-time processes in Chapter 2 below. (A more complete summary for Markov chains is given in tabular form in the summary of results in Volume I.)

Perhaps the most fundamental random process we encounter in queueing theory is the Poisson process that describes a collection of arrivals for which the interarrival times are independent and exponentially distributed with a mean interarrival time $\bar{t} = 1/\lambda$. In particular, the probability $P_k(t)$ of k arrivals in an interval whose duration is t sec is given by

$$P_k(t) = \frac{(\lambda t)^k}{k!} e^{-\lambda t} \qquad (1.49)$$

The average number of arrivals during this interval is merely

$$\bar{N}(t) = \lambda t \qquad (1.50)$$

and the variance is given by

$$\sigma^2_{N(t)} = \lambda t \qquad (1.51)$$

We note that the mean and variance for this process are identical. The z-transform for this process is simply given by

$$E[z^{N(t)}] = e^{\lambda t(z-1)} \qquad (1.52)$$

The assumption of an exponential interarrival time means, of course,

$$a(t) = \lambda e^{-\lambda t} \qquad t \geq 0 \qquad (1.53)$$

which, we repeat, is the memoryless distribution. Here, the mean and variance are, respectively, $\bar{t} = 1/\lambda$ and $\sigma^2 = 1/\lambda^2$.

Among the class of continuous-time Markov processes there is the special case of birth–death processes in which the system state changes by at most one (up or down) in any infinitesimal interval. In such cases we talk about the birth rate λ_k, which is the average rate of births when the system contains k customers, and also of the death rate μ_k, which is the average rate at which deaths occur when the population is of size k. The time-dependent behavior for such a system is essentially given in Eq. (1.47). The equilibrium behavior as defined in Eq. (1.48) takes on an especially simple form for this class of birth–death processes whose solution is given as follows (here we use the more usual notation p_k rather than π_k to denote the probability of having k customers in the system):

$$p_k = p_0 \prod_{i=0}^{k-1} \frac{\lambda_i}{\mu_{i+1}} \qquad (1.54)$$

with the constant p_0 being evaluated through

$$p_0 = \frac{1}{1 + \sum_{k=1}^{\infty} \prod_{i=0}^{k-1} \lambda_i/\mu_{i+1}} \qquad (1.55)$$

The application of this equilibrium solution leads us directly to the class of elementary queueing systems which we discuss in the next three sections.

1.4. THE M/M/1 QUEUE

The M/M/1 queue is the simplest interesting queueing system we present. It is the classic example and the analytical techniques required are rather elementary. Whereas these *techniques* do not carry over into more complex queueing systems, the *behavior* of M/M/1 is in many ways similar to that observed in the more complex cases.

Since this system has a Poisson input (with an average arrival rate λ) and makes unit step changes (single service and single arrivals), then $p_k = r_k = d_k$. (Recall that the average service time is $\bar{x} = 1/\mu$.) This distribution is given by

$$p_k = (1-\rho)\rho^k \qquad (1.56)$$

and so we immediately find that the average number in the system is given by

$$\bar{N} = \frac{\rho}{1-\rho} \qquad (1.57)$$

with variance

$$\sigma_N^2 = \frac{\rho}{(1-\rho)^2} \qquad (1.58)$$

Using Little's result and Eq. (1.32), we may immediately write down the two basic performance expressions for average delays in M/M/1:

$$W = \frac{\rho/\mu}{1-\rho} \qquad (1.59)$$

$$T = \frac{1/\mu}{1-\rho} \qquad (1.60)$$

The terms $\bar{N}$, W, and T all demonstrate the same common behavior as regards the utilization factor ρ; namely, they all behave inversely with respect to the quantity $(1-\rho)$. This effect is dominant for M/M/1 as well as for most common queueing systems, and in Figure 1.1 we show the average time in system as a function of the utilization factor. Thus as ρ approaches unity from below, these average delays and queue sizes grow without bound! This is true of essentially every queueing system one will encounter and shows the extreme price that must be paid if one is interested in running the system close to its capacity ($\rho = 1$).

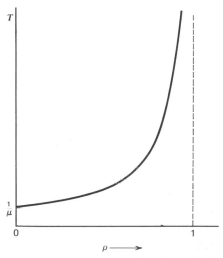

Figure 1.1 Average delay as a function of ρ for M/M/1.

As for the distributions, we have already seen that there is a geometrically distributed number of customers in the system and we now give the waiting time and system time pdf's along with the corresponding PDF's for the case of first-come–first-serve (FCFS):

$$w(y) = (1-\rho)u_0(y) + \lambda(1-\rho)e^{-\mu(1-\rho)y} \qquad y \geq 0 \qquad (1.61)$$

[where $u_0(y)$ is the unit impulse (Dirac delta) function],

$$W(y) = 1 - \rho e^{-\mu(1-\rho)y} \qquad y \geq 0 \qquad (1.62)$$

$$s(y) = \mu(1-\rho)e^{-\mu(1-\rho)y} \qquad y \geq 0 \qquad (1.63)$$

$$S(y) = 1 - e^{-\mu(1-\rho)y} \qquad y > 0 \qquad (1.64)$$

With the exception of the accumulation of probability at the origin for the waiting time, we note that these are all exponential in nature. The idle period I (the interval of time from the departure of a customer who leaves the system empty until the next arrival) and the interdeparture time D (the time between successive departures) are also both exponentially distributed with the parameter λ:

$$P[I \leq y] = P[D \leq y] = 1 - e^{-\lambda y} \qquad y \geq 0 \qquad (1.65)$$

The busy period (the interval of time between successive idle periods) has a pdf denoted by $g(y)$ given in terms of the modified Bessel function of the first kind as

$$g(y) = \frac{1}{y\sqrt{\rho}} e^{-(\lambda+\mu)y} I_1(2y\sqrt{\lambda\mu}) \qquad (1.66)$$

The probability f_n that n customers are served during a busy period is given by

$$f_n = \frac{1}{n}\binom{2n-2}{n-1}\rho^{n-1}(1+\rho)^{1-2n} \tag{1.67}$$

Two simple extensions for the M/M/1 system are easily described. First, there is the case of bulk arrivals where with probability g_k a group of k customers arrives at each arrival instant from the Poisson process; we then define the generating function for this distribution as usual by $G(z)=\sum_{k=0}^{\infty} g_k z^k$ with which we may then give the generating function for the number of customers in this bulk arrival M/M/1 system,* namely,

$$Q(z) = \frac{\mu(1-\rho)(1-z)}{\mu(1-z)-\lambda z[1-G(z)]} \tag{1.68}$$

The second generalization is a bulk service system in which a free server will take up to, but no more than, r customers and serve them collectively (as if they were a single customer) with an exponentially distributed service time. The probability of finding k customers in this system is given by

$$p_k = \left(1-\frac{1}{z_0}\right)\left(\frac{1}{z_0}\right)^k \qquad k = 0, 1, 2, \ldots \tag{1.69}$$

where z_0 is that unique root lying outside the unit disk, that is, $|z_0|>1$, for the equation

$$r\rho z^{r+1}-(1+r\rho)z^r+1=0 \tag{1.70}$$

and where, as usual, $\rho = \lambda/r\mu$.

A final generalization, which we will use in Chapter 4, involves the case of an M/M/1 system with a finite number of customers, namely M, that behave in the following way. A customer is either in the system (waiting for or being served) or outside the system and arriving; the interval from the time he leaves the system until he returns once again is exponentially distributed with mean $1/\lambda$. This case gives the following expression for the probability for finding k customers in the system:

$$p_k = \frac{[M!/(M-k)!](\lambda/\mu)^k}{\sum_{i=0}^{M} [M!/(M-i)!](\lambda/\mu)^i} \tag{1.71}$$

* That is, recall $Q(z)=E[z^N]$, not to be confused with the infinitesimal generator $\mathbf{Q}$ defined in Eq. (1.45).

So much for the classic M/M/1 system. In the next section, we retain the Markovian assumptions but consider the case of multiple servers.

1.5. THE M/M/m QUEUEING SYSTEM

We now consider the generalization to the case of m servers. A single queue forms in front of this collection of m servers and the customer at the head of the queue will be handled by the first available server. As usual, λ is the arrival rate and $1/\mu$ is the average service time, with $\rho = \lambda/m\mu$. The equilibrium probability of finding k customers in the system is given by

$$p_k = \begin{cases} p_0 \dfrac{(m\rho)^k}{k!} & k \leq m \\[2mm] p_0 \dfrac{(\rho)^k m^m}{m!} & k \geq m \end{cases} \tag{1.72}$$

where

$$p_0 = \left[\sum_{k=0}^{m-1} \frac{(m\rho)^k}{k!} + \frac{(m\rho)^m}{m!\,(1-\rho)} \right]^{-1} \tag{1.73}$$

A. K. Erlang, the father of queueing theory, considered this system as one model for the behavior of telephone systems early in this century [BROC 48]. Identified with his name is the Erlang-C formula, which gives the probability that an arriving customer must wait for a server; his expression is given by p_m from Eq. (1.72). Extensive tables of this quantity are available in the many books dealing with telephony [TELE 70].

Further results for M/M/m may be found in Section 1.9, which discusses G/M/m. Specifically W and $W(y)$ are given in Eqs. (1.113) and (1.114), respectively, where for M/M/m we have simply that $\sigma = \rho$.

Erlang considered a second model for telephone systems that is the same as M/M/m but permits no customers to wait; that is, it is a loss system with at most m customers present at any one time. In this case, the probability of finding k customers in the system is given by

$$p_k = \frac{(\lambda/\mu)^k/k!}{\sum\limits_{i=0}^{m} (\lambda/\mu)^i/i!} \tag{1.74}$$

for the range $0 \leq k \leq m$. The important quantity of interest here is the probability that a customer upon arrival to the system will find no empty servers and will therefore be "lost;" this is referred to as the Erlang-B formula or as Erlang's Loss Formula (also commonly tabulated) and is given simply by p_m from Eq. (1.74).

1.6. MARKOVIAN QUEUEING NETWORKS

Before leaving the comfortable world of exponential distributions, we wish to discuss another class of results that applies to networks of queues in which customers move from one queueing facility to another in some random fashion until they depart from the system at various points. Specifically, we consider an N-node network in which the ith node consists of a single queue served by m_i servers, each of which has an exponentially distributed service time of mean $1/\mu_i$. The ith node receives from outside the network a sequence of arrivals from an independent Poisson source at an average rate of γ_i customers per second. When a customer completes service at the ith node he will proceed next to the jth node with probability r_{ij}; thus he becomes an "internal" arrival to the jth node. On the other hand, upon leaving the ith node a customer will depart from the entire network with probability $1 - \sum_{j=1}^{N} r_{ij}$. We define the total arrival rate to the ith node to be, on the average, λ_i customers per second, and this consists both of external and internal arrivals. The set of defining equations for λ_i is given by

$$\lambda_i = \gamma_i + \sum_{j=1}^{N} \lambda_j r_{ji} \qquad (1.75)$$

A large measure of independence exists among the nodes in such a network, as may be seen from the expression given below for the joint distribution of finding k_1 customers in the first node, k_2 customers in the second node, and so on:

$$p(k_1, k_2, \ldots, k_N) = p_1(k_1) p_2(k_2) \ldots p_N(k_N) \qquad (1.76)$$

The factoring of this joint distribution exposes the independence. In Chapters 4 and 5 we are delighted to take advantage of this independence. In particular, each factor in this last expression, say $p_i(k_i)$, is merely the solution to an *isolated* M/M/m_i queueing facility operating by itself with an input rate λ_i; the solution for $p_i(k_i)$ is given in Eq. (1.72).

Another class of Markovian queueing networks consists of those networks in which customers are permitted neither to leave nor to enter. In particular, we assume that K customers are placed (trapped) within a network similar to the one described above and that they move around from node to node, but no departures from any node are permitted; that is, $1 - \sum_{j=1}^{N} r_{ij} = 0$ for all i. These closed networks have the following solution for the joint distribution of finding customers in various nodes:

$$p(k_1, k_2, \ldots, k_N) = \frac{1}{G(K)} \prod_{i=1}^{N} \frac{x_i^{k_i}}{\beta_i(k_i)} \qquad (1.77)$$

where the set of numbers $\{x_i\}$ must satisfy the following linear equations [similar to Eq. (1.75) with $\gamma_i = 0$]:

$$\mu_i x_i = \sum_{j=1}^{N} \mu_j x_j r_{ji} \qquad i = 1, 2, \ldots, N \tag{1.78}$$

and where

$$G(K) = \sum_{k \in A} \prod_{i=1}^{N} \frac{x_i^{k_i}}{\beta_i(k_i)} \tag{1.79}$$

where $\mathbf{k} = (k_1, k_2, \ldots, k_N)$ and A is that set of vectors $\mathbf{k}$ for which $k_1 + k_2 + \cdots + k_N = K$ and where

$$\beta_i(k_i) = \begin{cases} k_i! & k_i \leq m_i \\ m_i! \, m_i^{k_i - m_i} & k_i \geq m_i \end{cases} \tag{1.80}$$

These open and closed networks will be developed further in Chapter 4.

1.7. THE M/G/1 QUEUE

In this and the following two sections we study systems that fall in the domain of intermediate queueing theory. This classification refers to those systems in which we permit either (but not both) the interarrival time or the service time to be nonexponentially distributed; the case when both these random variables are nonexponential forms part of advanced queueing theory which we discuss in Section 1.10. For the M/G/1 system we cannot give explicit distributions for the number in system or for the time in system as we did for the M/M/1 system [specifically, see Eqs. (1.56) and (1.64) above]. Rather, we find expressions for the transforms of these distributions.

The M/G/1 system is characterized by a Poisson arrival process at a mean rate of λ arrivals per second and with an arbitrary or general service time distribution of form $B(x)$ with a mean service time of $\bar{x}$ sec and with kth moment equal to $\overline{x^k}$. Due to the Poisson arrival process and due to the fact that the number in the system changes by at most one, we again have $p_k = r_k = d_k$.

The basic (difference) equation describing the relationship among random variables for this first-come–first-serve M/G/1 system is

$$q_{n+1} = \begin{cases} q_n - 1 + v_{n+1} & \text{if } q_n > 0 \\ v_{n+1} & \text{if } q_n = 0 \end{cases} \tag{1.81}$$

where q_n is the number of customers left behind by the departure of customer C_n, and v_n is the number of customers who enter during his service time (x_n). The sequence $\{q_n\}$ forms a (discrete-state continuous-time) Markov chain. The entire transient and equilibrium behavior for the system is contained in this equation, and from it we may derive most of our results for M/G/1.

By far the most well-known result for the M/G/1 system is the Pollaczek–Khinchin (P-K) mean value formula, which gives the following compact expression for the (equilibrium) average waiting time in the queue:

$$W = \frac{\lambda \overline{x^2}/2}{(1-\rho)} \tag{1.82}$$

The numerator term, denoted by $W_0 = \lambda \overline{x^2}/2$, is, in fact, equal to the expected time that a newly arriving customer must spend in the queue while that customer (if any) which he finds in service completes his remaining required service time.* From this formula one may easily calculate T using Eq. (1.29); combining that result with the results quoted in Eqs. (1.31) and (1.32) we easily come up with the P-K mean-value formula for number in system as

$$\bar{N} = \rho + \frac{\lambda^2 \overline{x^2}/2}{1-\rho} \tag{1.83}$$

---

* This quantity is related to the concept of *residual life*, which we will use in this book. To elaborate, let us consider the sequence of instants located on the real-time axis such that the set of distances between adjacent points is a set of independent, identically distributed random variables whose density we shall denote by $f(x)$ (that is, we are dealing with a renewal process). Let m_n denote the nth moment of these interval lengths. Let us now select a point along the time axis at random; the interval in which this point falls will be referred to as the "sampled" interval. The length of the sampled interval is known as the *lifetime* of the interval, the time from the start of the sampled interval to this point is known as the *age* of the interval, and the distance from this selected point until the end of the sampled interval is known as the *residual life* of the interval. We are concerned with the statistics of the residual life. The pdf for residual life is given by $\hat{f}(x) = [1 - F(x)]/(m_1)$ and the Laplace transform of this density is given by $\hat{F}^*(s) = [1 - F^*(s)]/(sm_1)$; the notation here is that $F(x) = \int_0^x f(y)\,dy$ and $F^*(s)$ is the Laplace transform associated with the pdf $f(x)$. Perhaps the most significant statistic is the *mean residual life*, given by $m_2/2m_1$; that is, the expected value of the remaining length of the interval is merely the second moment over twice the first moment of the interval lengths themselves. Also, the pdf for the lifetime of the sampled interval is $xf(x)/m_1$.

The last quantity we wish to describe is the probability that the length of an interval (or that the value of any random variable) lies between x and $x + dx$ given that it exceeds x; dividing this probability by dx, we have a quantity referred to as the *failure rate* of the random variable, given by $f(x)/[1 - F(x)]$, where f and F refer to the pdf and the PDF of the random variable itself.

One sees that W_0 is merely the mean residual life of a service time (i.e., the average remaining service time) $(\overline{x^2}/2\bar{x})$ times the probability $(\rho = \lambda\bar{x})$ that, in fact, someone is occupying the service facility.

As mentioned above, the best we can do regarding the distributions of the various performance measures is to give the transforms associated with these random variables. Specifically, then, we recall the definition of the z-transform for the distribution p_k to be $Q(z) = \sum_{k=0}^{\infty} p_k z^k$ and find that it is given through

$$Q(z) = B^*(\lambda - \lambda z) \frac{(1-\rho)(1-z)}{B^*(\lambda - \lambda z) - z} \qquad (1.84)$$

where $B^*(\lambda - \lambda z)$ is the Laplace transform of the service time density $b(x)$ evaluated at the point $s = \lambda - \lambda z$. This last is referred to as the P-K transform equation for the number in system, and from it we easily derive Eq. (1.83).† The Laplace transform of the waiting time pdf is merely

$$W^*(s) = \frac{s(1-\rho)}{s - \lambda + \lambda B^*(s)} \qquad (1.85)$$

and for the time in system we have

$$S^*(s) = B^*(s) \frac{s(1-\rho)}{s - \lambda + \lambda B^*(s)} \qquad (1.86)$$

These last two equations are also referred to as P-K transform equations. Due to the independence of service times, we see that Eq. (1.86) is related to Eq. (1.85) through the obvious relationship $S^*(s) = B^*(s)W^*(s)$, that is, the transform for the pdf of the sum of two independent random variables is equal to the product of the transforms of the pdf of each separately. From Eq. (1.85) we easily obtain W in Eq. (1.82) by differentiation as usual; similarly, the second moment (and therefore, the variance of the waiting time, denoted by $\sigma_{\tilde{w}}^2$) may be obtained to give

$$\sigma_{\tilde{w}}^2 = W^2 + \frac{\lambda \overline{x^3}}{3(1-\rho)} \qquad (1.87)$$

Because of the Poisson arrival process, one immediately finds that the idle time I is distributed exponentially, that is,

$$P[I \le y] = 1 - e^{-\lambda y} \qquad (1.88)$$

The busy-period duration has a pdf whose transform $G^*(s)$ is given through the functional equation

$$G^*(s) = B^*(s + \lambda - \lambda G^*(s)) \qquad (1.89)$$

---

† For the case of bulk arrivals as discussed in introducing Eq. (1.68) above, the M/G/1 system gives an expression for $Q(z)$ identical to that in Eq. (1.84), except that $B^*(s)$ is evaluated at the point $s = \lambda - \lambda G(z)$ rather than as above; $G(z)$ is as given for Eq. (1.68).

which, in general, cannot be solved. However, we may determine various moments of the busy period through the moment-generating properties of this transform, and so, for example, g_1 (the mean duration of the busy period) and σ_g^2 (the variance of this duration) are given by

$$g_1 = \frac{\bar{x}}{1-\rho} \tag{1.90}$$

$$\sigma_g^2 = \frac{\sigma_b^2 + \rho(\bar{x})^2}{(1-\rho)^3} \tag{1.91}$$

where σ_b^2 is the variance of the service time. Similarly, the z-transform for the number served during the busy period, which we denote by $F(z)$, is given functionally by

$$F(z) = zB^*[\lambda - \lambda F(z)] \tag{1.92}$$

with mean and variance for this number given respectively by

$$h_1 = \frac{1}{1-\rho} \tag{1.93}$$

$$\sigma_h^2 = \frac{\rho(1-\rho) + \lambda^2 \overline{x^2}}{(1-\rho)^3} \tag{1.94}$$

An important stochastic process, which we have so far neglected, is the unfinished work $U(t)$ in the system at time t. This is a Markov process whose value represents the time required to empty the system of all customers present at time t, assuming that no new customers enter the system after time t; that is, $U(t)$ is the system backlog expressed in time units.

For a first-come–first-serve system, the unfinished work also represents the waiting time of an arrival *if* it were to enter at time t, and so $U(t)$ is sometimes referred to as the "virtual" waiting time; in the case of a first-come–first-serve system with Poisson arrivals (M/G/1), the unfinished work has the same statistics as the true waiting time for arrivals. We shall deal with this function in numerous places throughout the balance of this book. For the moment we wish to quote two important results regarding its distribution. For this purpose we define

$$F(w, t) = P[U(t) \le w] \tag{1.95}$$

and we may then cite the well-known Takács integrodifferential equation, namely,

$$\frac{\partial F(w, t)}{\partial t} = \frac{\partial F(w, t)}{\partial w} - \lambda F(w, t) + \lambda \int_{x=0}^{w} B(w-x) \, d_x F(x, t) \tag{1.96}$$

which defines the transient behavior of the unfinished work distribution. Defining the double Laplace transform $F^{**}(r, s)$ for $F(w, t)$, where r carries out the transform in the w-domain and s in the t-domain, we have the following transform equation for this time-dependent behavior:

$$F^{**}(r, s) = \frac{(r/\eta)e^{-\eta w_0} - e^{-rw_0}}{\lambda B^*(r) - \lambda + r - s} \tag{1.97}$$

Here η is the unique root (for r) of the equation $s - r + \lambda - \lambda B^*(r) = 0$ in the region $\text{Re}(s) > 0$, $\text{Re}(r) > 0$, and w_0 is the initial value of the unfinished work at time 0, that is, $U(0) = w_0$. We make use of these transient results in Chapter 2.

Much more can be said about the M/G/1 system, but for purposes of this primer we have said enough. In the natural order of things we should next consider the system M/G/m, but unfortunately there are very few substantive results that can be given for this system. On the other hand, the limiting case for the M/G/∞ system is itself in some ways a trivial system since no queueing ever takes place; indeed, a very lovely result for the number of busy servers (that is the number of customers in the system) is given simply by

$$p_k = \frac{\rho^k}{k!} e^{-\rho} \tag{1.98}$$

We note that this result is independent of the form for $B(x)$, depending only upon its first moment. Similarly we can immediately write down that $T = \bar{x}$ and $s(y) = b(y)$.

It is possible to interpret some of the above transforms as probabilities using the *method of collective marks*. The concept is to assume that each entering customer is "marked" independently with probability $(1 - z)$. Then we may interpret the generating function $P(z, t) = E[z^{N(t)}]$ for an arrival process [e.g., for Poisson arrivals, $P(z, t) = e^{\lambda t(z-1)}$] as being equal to the probability that no customers arriving in $(0, t)$ are marked. Similarly, consider any interval whose duration is given by a random variable X whose pdf has a Laplace transform, say, $X^*(s)$; if we further consider an independent Poisson arrival process (at mean rate λ) and ask for the probability P that no arrivals are marked that enter during the interval X, then $P = X^*(\lambda - \lambda z)$. Again consider an interval and an independent Poisson process as above; let us think of the epochs generated by the Poisson process as "catastrophes." If we ask for the probability Q that no catastrophes occur in the random interval, then $Q = X^*(\lambda)$. Thus we are able to give interesting probabilistic interpretations for many of the basic transform expressions that we encounter in queueing theory.

1.8. THE G/M/1 QUEUE

The G/M/1 system is in fact the "dual" of the M/G/1 system. Surprisingly, G/M/1 yields to analysis more easily than M/G/1 and so we can quote distributions directly. The system, of course, corresponds to the case of an arbitrary interarrival time whose PDF is given by $A(t)$ and with pdf $a(t)$ the transform of which is denoted by $A^*(s)$; service times are distributed exponentially with mean $1/\mu$.

The basic recurrence relation that governs the behavior of G/M/1 (and also G/M/m), similar to that for M/G/1 given in Eq. (1.81), is

$$q'_{n+1} = q'_n + 1 - v'_{n+1} \tag{1.99}$$

where q'_n is the number of customers found in the system by C_n and v'_{n+1} is the number of customers served between the arrival of C_n and C_{n+1}. The sequence $\{q'_n\}$ forms a Markov chain. Many of the G/M/m results follow from this equation.

All our results are expressed in terms of a root σ that is the unique root in the range $0 \le \sigma < 1$ of the functional equation

$$\sigma = A^*(\mu - \mu\sigma) \tag{1.100}$$

Once σ is evaluated, the following results are immediately available. The distribution for the number of customers found in the system by a new arrival is given by

$$r_k = (1 - \sigma)\sigma^k \qquad k = 0, 1, 2, \ldots \tag{1.101}$$

The PDF for waiting time is given by

$$W(y) = 1 - \sigma e^{-\mu(1-\sigma)y} \qquad y \ge 0 \tag{1.102}$$

and the mean waiting time is

$$W = \frac{\sigma}{\mu(1 - \sigma)} \tag{1.103}$$

It is remarkable that the waiting times are exponentially distributed, independent of the form of the interarrival time distribution (except insofar as it affects the value for σ).

1.9. THE G/M/m QUEUE

In contrast to the M/G/m system, we find that the G/M/m system does in fact yield to analysis, the results for which we quote in this section. The G/M/m system, of course, has arbitrarily distributed interarrival times and a single queue served first-come–first-serve by m servers, each of which has an exponentially distributed service time of mean $1/\mu$. As with the system G/M/1, σ is a key parameter and in this case it is found as the unique solution in the range $0 \le \sigma < 1$ for the equation

$$\sigma = A^*(m\mu - m\mu\sigma) \tag{1.104}$$

We have that the distribution of queue size found by a new arrival, conditioned by the fact that this arrival must queue, is given by

$$P[\text{queue size} = n \mid \text{arrival queues}] = (1-\sigma)\sigma^n \qquad n \ge 0 \tag{1.105}$$

We note here as with the G/M/1 system that the queue size is geometrically distributed. As earlier, we define r_k as the probability that a newly arriving customer finds k in the system ahead of him; in terms of these probabilities we define

$$R_k = \begin{cases} r_k/J & 0 \le k \le m-2 \\ \sigma^{k-m+1} & m-2 < k \end{cases} \tag{1.106}$$

We must evaluate J and the $m-1$ terms R_k for $0 \le k \le m-2$. The equation for J is given by

$$J = \frac{1}{[1/(1-\sigma)] + \sum_{k=0}^{m-2} R_k} \tag{1.107}$$

and the values for the terms R_k are given through the set of equations

$$R_{k-1} = \frac{R_k - \sum_{i=k}^{m-2} R_i p_{ik} - \sum_{i=m-1}^{\infty} \sigma^{i+1-m} p_{ik}}{p_{k-1,k}} \tag{1.108}$$

where the transition probabilities p_{ij} are nontrivial and are calculated through the following four equations, depending on the range of the subscripts i and j:

$$p_{ij} = 0 \qquad j > i+1 \tag{1.109}$$

$$p_{ij} = \int_0^\infty \binom{i+1}{j}[1-e^{-\mu t}]^{i+1-j} e^{-\mu tj} \, dA(t) \qquad j \le i+1 \le m \tag{1.110}$$

$$\beta_n = p_{i,i+1-n} = \int_{t=0}^\infty \frac{(m\mu t)^n}{n!} e^{-m\mu t} \, dA(t) \qquad 0 \le n \le i+1-m, \, m < i \tag{1.111}$$

$$p_{ij} = \int_0^\infty \binom{m}{j} e^{-j\mu t}\left[\int_0^t \frac{(m\mu y)^{i-m}}{(i-m)!}(e^{-\mu y}-e^{-\mu t})^{m-j} m\mu \, dy\right] dA(t) \qquad j < m < i+1 \tag{1.112}$$

(Who said it would be easy!) Once these constants are evaluated we may then calculate the average waiting time as

$$W = \frac{J\sigma}{m\mu(1-\sigma)^2} \tag{1.113}$$

The PDF of the waiting time is given through

$$W(y) = 1 - \frac{\sigma e^{-m\mu(1-\sigma)y}}{1+(1-\sigma)\sum_{k=0}^{m-2} R_k} \qquad y \geq 0 \tag{1.114}$$

Whereas these last two equations require the calculation of difficult constants, the waiting time pdf conditioned on the fact that the customer must queue is simply given by

$$w(y \mid \text{arrival queues}) = (1-\sigma)m\mu e^{-m\mu(1-\sigma)y} \qquad y \geq 0 \tag{1.115}$$

This only requires the calculation of σ. Note that even for the G/M/m system we have an exponentially distributed conditional waiting time.

1.10.   THE G/G/1 QUEUE

Advanced queueing theory deals with the system G/G/1 and things beyond (for example, G/G/m, about which we can say so very little—recall that even the system M/G/m confounded us). In this section we give some of the principal well-known results for G/G/1 and describe a method of attack that sometimes yields the required solution or at least some simplified measures of performance. In addition we present a point of view that describes the underlying operations involved in solving the G/G/1 system.

As mentioned in the first section of this chapter, the random variables that drive any queueing system are the interarrival times t_n and the service times x_n. In the general formulation of the G/G/1 system, we find that these random variables do not appear separately in the solution but in fact always appear as a difference; thus we are led to consider a new random variable associated with the nth customer C_n, namely,

$$u_n = x_n - t_{n+1} \tag{1.116}$$

This random variable represents the difference between the amount of work (x_n) that C_n demands of the system and the "breathing space" (t_{n+1}), or time, between the arrival of this demand and the arrival of the next demand by C_{n+1}; hopefully this difference will be negative on the average so that there will be more breathing space than load on the system. In fact if we take the average of Eq. (1.116) we find

$$E[u_n] = \bar{t}(\rho - 1) \tag{1.117}$$

which, first of all, is independent of n (as we expected) and, second, will have a negative mean value so long as $\rho < 1$; this is no different than requiring that $R < C$ if our system is to be stable. Associated with the random variable u_n, whose generic form we now write as $\tilde{u}$, we have its PDF $C(u)$, its pdf $c(u)$ and the Laplace transform of this pdf, which we denote by $C^*(s)$. Expressing these last two in terms of the pdf's and Laplace transforms thereof for the interarrival time and service times we have

$$c(u) = \int_{-\infty}^{\infty} b(u+t)a(t)\, dt \tag{1.118}$$

and

$$C^*(s) = A^*(-s)B^*(s) \tag{1.119}$$

The integral in Eq. (1.118) is, of course, the convolution integral between $a(-u)$ and $b(u)$, which we henceforth denote by $c(u) = a(-u) \circledast b(u)$. Thus once we know the interarrival time and service time pdf we also have the pdf for our random variable $\tilde{u}$.

Of basic interest to the G/G/1 system is the behavior of the waiting time w_n for customer C_n. This random variable is related to others in the sequence through the following difference equation, in which we see the basic role played by the random variable u_n:

$$w_{n+1} = \max[0, w_n + u_n] \tag{1.120}$$

This is the key defining equation for G/G/1 [as was Eq. (1.81) for M/G/1 and Eq. (1.99) for G/M/m]. The sequence $\{w_n\}$ forms a (continuous-time continuous-state) Markov process (in fact, it is an imbedded Markov process). The maximum operator shown above is often rewritten in the following fashion: $(x)^+ = \max(0, x)$. In the case of a stable system $(\rho < 1)$ there will exist a limiting random variable representing the equilibrium waiting time, which we denote by $\tilde{w}$. It can be seen from Eq. (1.120) that $\tilde{w}$ must have the same distribution as $(\tilde{w} + \tilde{u})^+$; the pdf that satisfies this condition will be the unique solution for the waiting time pdf. Let us denote the pdf for w_n by $w_n(y)$. The (nonlinear) functional equation that defines this pdf is given through

$$w_{n+1}(y) = \pi(w_n(y) \circledast c(y)) \tag{1.121}$$

where $\circledast$ is the convolution operator and π is a special operator that modifies the pdf of its argument by replacing all of the probability associated with negative values of y (the argument of the pdf) with an impulse at $y = 0$

whose area equals this probability. The pdf $w(y)$ for our limiting random variable $\tilde{w}$ must, from Eq. (1.121), satisfy the following basic equation:

$$w(y) = \pi(w(y) \circledast c(y)) \qquad (1.122)$$

whose solution will be the equilibrium density for the waiting time in G/G/1. Equation (1.122) states that this equilibrium pdf must be such that when it is convolved with $c(y)$ and when the resulting density has all of its probability on the negative half-line moved to an impulse at the origin, then we must have a resulting pdf that is the same as the $w(y)$ with which we began.

Another way to describe the random variable $\tilde{w}$ is through the equation

$$\tilde{w} = \sup_{n \geq 0} U_n \qquad (1.123)$$

where $U_n = u_0 + u_1 + \cdots + u_{n-1}$ ($n \geq 1$) and $U_0 = 0$.

A random variable related to w_n that forms the "other half" for w_n is

$$y_n = -\min[0, w_n + u_n] \qquad (1.124)$$

Thus we see that

$$w_{n+1} - y_n = w_n + u_n \qquad (1.125)$$

Taking expectations of this equation in the limit as $n \to \infty$, we obtain

$$\bar{y} = -\bar{u} \qquad (1.126)$$

Another defining relationship for the waiting time PDF is given by the well-known Lindley's integral equation:

$$W(y) = \begin{cases} \int_{-\infty}^{y} W(y-u) \, dC(u) & y \geq 0 \\ 0 & y < 0 \end{cases} \qquad (1.127)$$

This equation is of the Wiener–Hopf type. We now let $\Phi_+(s)$ denote the Laplace transform for the waiting time PDF $W(y)$; note that this is the transform for the PDF and not for the pdf $w(y)$, whose transform we had previously denoted by $W^*(s)$ and which is related to this new transform through the equation $W^*(s) = s\Phi_+(s)$. We wish to solve for $\Phi_+(s)$. The procedure we are about to describe is formally correct for those G/G/1 systems for which $A^*(s)$ and $B^*(s)$ may be written as rational functions of s. In this case our task is to find a suitable representation of the following form:

$$A^*(-s)B^*(s) - 1 = \frac{\Psi_+(s)}{\Psi_-(s)} \qquad (1.128)$$

where for $\text{Re}(s) > 0$, $\Psi_+(s)$ must be an analytic function of s that contains no zeros in this half-plane; similarly, for $\text{Re}(s) < D$, $\Psi_-(s)$ must be an analytic function of s and be zero-free (where $D > 0$). In addition, we require for $|s|$ approaching infinity that the behavior of $\Psi_+(s)$ should be $\Psi_+(s) \cong s$ for $\text{Re}(s) > 0$ and that the behavior of $\Psi_-(s)$ should be $\Psi_-(s) \cong -s$ for $\text{Re}(s) < D$. Having accomplished this "spectrum factorization" we may write our solution for $\Phi_+(s)$ as

$$\Phi_+(s) = \frac{K}{\Psi_+(s)} \tag{1.129}$$

where the constant K may be evaluated through

$$K = \lim_{s \to 0} \frac{\Psi_+(s)}{s} \tag{1.130}$$

This constant represents the probability that an arriving customer need not queue. We note that once we have found $\Phi_+(s)$ then we have found the transform for the waiting time PDF, which is what we were seeking.

Although we have described a procedure above for calculating the waiting time pdf, we have not been able to extract the properties of this solution and in fact we have not even given an expression for the average waiting time W in the G/G/1 system. Sad to say, this quantity is, in general, unknown! Its value can be expressed, however, in terms of other system variables as follows. For example, the average waiting time is simply the negative sum of the mean residual life of the random variable $\tilde{u}$ and of $\tilde{y}$ (which is the limiting random variable for the sequence y_n); that is,

$$W = -\frac{\overline{u^2}}{2\bar{u}} - \frac{\overline{y^2}}{2\bar{y}} \tag{1.131}$$

It can be shown that the mean residual life for $\tilde{y}$ is exactly equal to the mean residual life for the random variable I, which denotes the length of an idle period in G/G/1; this last observation coupled with the easy evaluation of the first two moments of the random variable $\tilde{u}$ yields the following expression for the mean wait in G/G/1:

$$W = \frac{\sigma_a^2 + \sigma_b^2 + (\bar{t})^2(1-\rho)^2}{2\bar{t}(1-\rho)} - \frac{\overline{I^2}}{2\bar{I}} \tag{1.132}$$

where σ_a^2 and σ_b^2 are, respectively, the variance of the interarrival time and service time. We shall make use of this last formula in evaluating bounds on the mean waiting time in Chapter 2.

We include no exact results for the G/G/m queue, but refer the reader to the approximations and bounds in Chapter 2. An elegant approach to the exact analysis of G/G/m has been given by Kiefer and Wolfowitz [KIEF 55] involving the (usually impossible) task of solving an integral equation (which reduces to Lindley's Integral Equation for G/G/1). More recently, de Smit [DESM 73] has extended the theory due to Pollaczek [POLL 61] for G/G/m and has elaborated upon the G/M/m and G/H$_R$/m queues.

This completes our very rapid summary of the elements of queueing theory. We will need much of this material in the following chapters. It should be clear that a number of important behavioral properties for these queueing systems remain as yet unsolved. Nevertheless we are faced in the real world with applying the tools from queueing theory to solve immediate problems. The balance of this textbook discusses such problems and methods for applying the theory developed. Consequently, we begin with a rather advanced chapter in queueing systems where the goal is *not* to extend the rigorous theory as summarized here but rather to find *effective approximation methods* that permit one to use the theory in a true engineering sense.

REFERENCES

BROC 48   Brockmeyer, E., H. L. Halstrøm, and A. Jensen, "The Life and Works of A. K. Erlang," *Transactions of the Danish Academy of Technology and Science*, **2**, (1948).

DESM 73   de Smit, J. H. A., "Some General Results for Many Server Queues," pp. 153–169 and, "On the Many Server Queue with Exponential Service Times," *Advances in Applied Probability*, **5**, No. 1 (April 1973).

KIEF 55   Kiefer, J., and J. Wolfowitz, "On the Theory of Queues with Many Servers," *Transactions of the American Mathematics Society*, **78**, 1–18 (1955).

KLEI 75   Kleinrock, L., *Queueing Systems, Volume I: Theory*, Wiley-Interscience (New York), 1975.

POLL 61   Pollaczek, F., *Théorie Analytique de Problèms Stochastiques Relatifs à un Groupe de Lignes Teléphonique avec Dispositif d'attente*, Gauthiers-Villars (Paris), 1961.

TELE 70   *Telephone Traffic Theory Tables and Charts, Part 1*, Siemens Altiengesellschaft, Telephone and Switching Division (Munich), 1970.

2

Bounds, Inequalities, and Approximations

An exciting "new" branch of queueing theory is emerging that deals with methods for finding approximate or bounding behaviour for queues.* It is not hard to convince oneself that queueing theory is rather difficult and that exact results are hard to obtain; in fact, *many* of the interesting queueing phenomena have not as yet yielded to exact analysis (and perhaps never will!). Moreover, in those simpler systems where exact results can be obtained, their form is sometimes so complex as to render them ineffectual for practical applications.

If one examines why we study queueing theory in the first place, one readily admits that it is to answer questions regarding real queues in the real world. The mathematical structures we have created in attempting to describe these real situations are merely idealized fictions, and one must not become enamoured with them for their own sake if one is really interested in practical answers. We must face the fact that authentic queueing problems seldom satisfy the assumptions made throughout most of the literature available on queueing theory: stationarity is rare, independence occurs only occasionally, and ergodicity is not only unlikely but is also impossible to establish with measurements over a finite time! Therefore if our mathematical models are so crude, we should be willing to accept much less than an exact solution to the systems of equations they give rise to; rather, we should be happy to accept approximate solutions to these "approximate" mathematical models and hope that such solutions provide information about the behavior of real-world queues. Even more important is the search for "robust" qualitative behavior of queues which provides "rules of thumb" for estimating the

---

* Perhaps the first approximations used in queueing theory date back to Erlang himself through his introduction of the method of stages (see [KLEI 75], Section 4.2); he tried to approximate the underlying distributions of a queueing system with tractable analytic functions. The reader is referred to [SYSK 62] and to the elucidation of Erlang's work and era in [BROC 48] for some of the historical flavor.

27

behavior of complex systems. An excellent example of successful robust models is given in Chapter 4, where we apply Markovian queueing networks to multiaccess computer system modelling; Buzen [BUZE 74] discusses the way in which system structure can be used to generate simple robust models. A second example is the use of diffusion approximations in a variety of applications (see Sections 2.9, 4.13, and 4.14). A third example is described in Chapter 5 in which a robust model is developed for computer network delay. All three of these examples demonstrate the success in the use of simple (often Markovian) models to predict behavior of rather complicated real-world systems. Issues such as these are addressed in this chapter and we emphasize that this approach to the study of queues is relatively new and potentially highly rewarding.

The chapter is organized as follows. We begin by establishing a robust approximation for the distribution of queueing time in the G/G/1 heavy-traffic case ($\rho \rightarrow 1$ from below). This approximation lurks just beneath the surface of many of the results we have already seen. A tight upper bound on the average wait W is then established from first principles (good for $0 \leq \rho < 1$); lower bounds for W are more difficult to come by and certain available results are presented. (It is sad but true that even W cannot be expressed exactly in terms of the simple system parameters for the G/G/1 queue!) We also give a bound on the tail of the waiting time distribution in Section 2.4. Most of the results in the first four sections were inspired by the work of Kingman [KING 61, 62a, 62b, 64, 70] and pursued by Marshall [MARS 68c], Brummelle [BRUM 71, 73], and others [SUZU 70, MARC 74]. A simple discrete approximation for the G/G/1 system is then presented in Section 2.5 using techniques from elementary queueing theory. Next we make a few remarks concerning bounds on W for G/G/m. At this point in the chapter we abandon our former approach of attempting to find approximate solutions to the given system equations and take the point of view that we will approximate the stochastic processes themselves (that is, the arrival and departure processes). We begin with a "first-order" approximation whereby stochastic processes are replaced simply by their average values (perhaps time-dependent) and this leads us to the *fluid approximation* for queues. Next, we study a "second-order" approximation in which a stochastic process is represented both by its mean and its variance, and this gives us the *diffusion approximation* for queues. This diffusion approximation refines the fluid approximation by describing the time-dependent processes with means given by the fluid approximation but with a normal (Gaussian) distribution describing the fluctuations about that (possibly time-varying) mean. In the case of stable queues ($\rho < 1$) the limit of this diffusion approximation as $t \rightarrow \infty$ is in fact the heavy-traffic approximation of Section 2.1! These are related because the diffusion approximation assumes that the queue never empties, and this is just the kind of approximation made in the heavy-traffic case. Following this, a careful discussion of the diffusion approximation for the M/G/1 queue is given. These methods are then applied to give approximate solutions to the "rush-hour" behavior so common in practical life. The diffusion approximation methods have been studied by various workers including Newell [NEWE 65, 68, 71], Gaver [GAVE 68], Iglehart and Whitt [IGLE 69], McNeil [McNE 73] and Kobayashi [KOBA 74a, 74b].

2.1. THE HEAVY-TRAFFIC APPROXIMATION

In this section we study the behavior of the system G/G/1 in the *heavy-traffic case* [KING 62a]. This is the case where $\rho \cong 1$ (but remains strictly less than one, preserving stability). We establish the central result for heavy-traffic theory, which states that the waiting time distribution is, as an approximation, exponentially distributed with the mean wait given by $(\sigma_a^2 + \sigma_b^2)/2(1-\rho)\bar{t}$. This is a remarkable result and it pervades most of our approximation methods. (It is valid when the denominator is small compared to the square root of the numerator.)

Our point of departure in establishing the central result is Eq. (1.128) repeated below:

$$A^*(-s)B^*(s) - 1 = \frac{\Psi_+(s)}{\Psi_-(s)} \tag{2.1}$$

We will examine this expression in the case $\rho \cong 1$. Let us begin by considering the Taylor series expansion for $B^*(s)$ and $A^*(-s)$ as follows:

$$B^*(s) = \sum_{k=0}^{\infty} \frac{s^k}{k!} B^{*(k)}(0) \tag{2.2}$$

However, from Eq. (1.17) we know that $B^{*(k)}(0) = (-1)^k \overline{x^k}$; using this and considering $B^*(s)$ near the origin ($s \to 0$), we have†

$$B^*(s) = 1 - \bar{x}s + \frac{\overline{x^2}s^2}{2!} + o(s^2) \tag{2.3}$$

Similarly, we have

$$A^*(-s) = 1 + \bar{t}s + \frac{\overline{t^2}s^2}{2!} + o(s^2) \tag{2.4}$$

† As usual, the notation $o(x)$ denotes any function which goes to zero faster than x, that is, $\lim_{x \to 0} [o(x)/x] = 0$.

Now, since we are considering the heavy-traffic case, we recognize that our interest must lie in the distribution of *large* waiting times. Recall that the waiting time distribution, $W(y)$, has a Laplace transform $\Phi_+(s)$, whereas the density, $w(y)$, has a transform $W^*(s) = s\Phi_+(s)$. It can be seen that the behavior of $W(y)$ for large values of y is governed by that pole (singularity) of $\Phi_+(s)$ which has the smallest Re (s) in absolute value; this follows since the decay rate of each exponential term in $w(y)$ or $W(y)$ is inversely related to the (negative) real part of the pole associated with that term. The expression in Eq. (2.1) has a zero at $s = 0$ and in fact, has an additional zero near $s = 0$ for the heavy-traffic case; as we shall see, this additional zero forms the pole of $\Phi_+(s)$ that governs the behavior of large waiting times (this is merely the final-value theorem for Laplace transforms).[†] Let us find this nearby zero (which has a small but negative real part). Using the expansions in Eqs. (2.3) and (2.4) we have

$$A^*(-s)B^*(s) - 1 = \left(1 - \bar{x}s + \frac{\overline{x^2}s^2}{2}\right)\left(1 + \bar{t}s + \frac{\overline{t^2}s^2}{2}\right) - 1 + o(s^2)$$

$$= 1 + s(\bar{t} - \bar{x}) + s^2\left(\frac{\overline{x^2}}{2} + \frac{\overline{t^2}}{2} - \bar{x}\bar{t}\right) - 1 + o(s^2)$$

$$= s\left[\bar{t} - \bar{x} + s\left(\frac{\overline{x^2}}{2} + \frac{\overline{t^2}}{2} - \bar{x}\bar{t}\right)\right] + o(s^2) \qquad (2.5)$$

From this last we clearly see the root at $s = 0$. Solving for the second root in the vicinity of $s = 0$ we first note that

$$\frac{\overline{x^2}}{2} + \frac{\overline{t^2}}{2} - \bar{x}\bar{t} = \frac{\sigma_b^2 + \sigma_a^2}{2} + \frac{(\bar{x} - \bar{t})^2}{2} \qquad (2.6)$$

Since $\rho \cong 1$, we choose to assume at this point that the last term on the right-hand side of Eq. (2.6) (the squared difference of the first moments) is negligible compared to the first term in that equation (the sum of the variances). Using Eq. (2.6) under this approximation and dropping $o(s^2)$ (since we are examining the vicinity of the origin), we may then solve Eq. (2.5) for our second root (which we denote by s_0) as

$$\bar{t} - \bar{x} + s_0\frac{\sigma_b^2 + \sigma_a^2}{2} \cong 0$$

which yields

$$s_0 \cong -\frac{2\bar{t}(1 - \rho)}{\sigma_a^2 + \sigma_b^2} \qquad (2.7)$$

[†] One can already see the exponential approximation emerging from this single critical pole.

Clearly $s_0 < 0$. Note from this and Eq. (2.5) that s_0 is (approximately) the inverse of the mean residual life (see footnote on p. 16) of the random variable $\tilde{u} = \tilde{x} - \tilde{t}$. Thus, as an approximation near the origin, we have

$$A^*(-s)B^*(s) - 1 \cong s(s - s_0)\frac{(\sigma_a^2 + \sigma_b^2)}{2}$$

Returning to our direct argument now, when s is near the origin we may then use the expression in Eq. (2.1) and arrive at the approximation

$$\Psi_+(s) \cong s(s - s_0)C \tag{2.8}$$

where $C = \Psi_-(0)[\sigma_a^2 + \sigma_b^2]/2$. In order to proceed to our solution for $\Phi_+(s)$, we see from Eq. (1.129) that we must evaluate the constant K; this we do by using Eq. (1.130) as follows:

$$K = \lim_{s \to 0} (s - s_0)C = -s_0 C$$

which then gives from Eq. (1.129)

$$\Phi_+(s) \cong \frac{-s_0}{s(s - s_0)}$$

(The unknown constant C cancels!) Making a partial fraction expansion we have

$$\Phi_+(s) \cong \frac{1}{s} - \frac{1}{s - s_0}$$

Finally, using the expression for s_0, this inverts to give

$$W(y) \cong 1 - \exp\left(-\frac{2\bar{t}(1 - \rho)}{\sigma_a^2 + \sigma_b^2}y\right) \qquad \blacksquare \quad (2.9)$$

This last gives us an approximation for the distribution of waiting time in the vicinity of large waiting times for $\rho \cong 1$. The factor s_0 is given specifically through Eq. (2.7). We note that the average wait W is given by $(-1/s_0)$ and so

$$W \cong \frac{(\sigma_a^2 + \sigma_b^2)}{2(1 - \rho)\bar{t}} \qquad \blacksquare \quad (2.10)$$

Equations (2.9) and (2.10) form the *central results* for heavy-traffic theory as applied to G/G/1. These results are extremely robust and give the general behavior of queues with long waiting times. From Eq. (2.10) we see that the numerator contribution to the average waiting time is due to fluctuations in the arrival and service processes, whereas the denominator (which dominates in the heavy-traffic case) depends only on first moments (in particular, on ρ). The exponential character of these large waiting times is in some sense a central limit theorem for queueing theory and we shall see it appear again in our diffusion approximations below *

2.2. AN UPPER BOUND FOR THE AVERAGE WAIT

The heavy-traffic approximation studied in the previous section leads to an exponential distribution of large waiting times whose mean is given by Eq. (2.10). In this section we are interested not in an approximation, but in a firm upper bound on the average wait W in the system G/G/1.

The following development is simple and is due again to Kingman [KING 62b]. We recall from Section 1.10 that the limiting random variable $\tilde{w}$ must have the same distribution as the random variable $(\tilde{w} + \tilde{u})^+$. Therefore, assuming the following moments exist, we must have

$$E[(\tilde{w})^k] = E\{[(\tilde{w} + \tilde{u})^+]^k\} \tag{2.11}$$

Let us introduce the definition

$$(X)^- \triangleq -\min[0, X] \tag{2.12}$$

Then, recalling that $(X)^+ \triangleq \max[0, X]$ we have the simple relationship

$$X = (X)^+ - (X)^- \tag{2.13}$$

and it must also be true from their definitions that

$$(X)^+(X)^- = 0 \tag{2.14}$$

Squaring Eq. (2.13) and using Eq. (2.14) we then see that

$$X^2 = [(X)^+]^2 + [(X)^-]^2 \tag{2.15}$$

Taking X to be a random variable, we may form expectations in Eq. (2.13) to yield

$$\bar{X} = \overline{(X)^+} - \overline{(X)^-} \tag{2.16}$$

And likewise, from Eq. (2.15) we have

$$\overline{X^2} = \overline{[(X)^+]^2} + \overline{[(X)^-]^2}$$

* Queues in series have also been studied by means of the heavy-traffic approximation [HARR 73]. Again it is shown that the total waiting time is asymptotically distributed in a way depending only on the mean and variance of the interarrival and service time distributions. When all variances are identical, then it is shown that the waiting time distribution is an exponential function of these moments.

Since $\sigma_X^2 = \overline{X^2} - (\bar{X})^2$, we may use the above relationships to yield

$$\sigma_X^2 = \sigma_{(X)^+}^2 + \sigma_{(X)^-}^2 + 2\overline{(X)^+}\,\overline{(X)^-} \qquad (2.17)$$

This last result is true for any random variable X.

Now taking $X = \tilde{w} + \tilde{u}$, we see from Eq. (2.16) that $\bar{X} = \bar{w} + \bar{u}$ is given by

$$\bar{w} + \bar{u} = \overline{(\tilde{w} + \tilde{u})^+} - \overline{(\tilde{w} + \tilde{u})^-} \qquad (2.18)$$

However, from Eq. (2.11) (with $k = 1$) we have $\bar{w} = \overline{(\tilde{w} + \tilde{u})^+}$, and so Eq. (2.18) may be rewritten as*

$$\bar{u} = -\overline{(\tilde{w} + \tilde{u})^-}$$

Furthermore, from Eq. (2.11) we have that

$$\sigma_{\tilde{w}}^2 = \sigma_{(\tilde{w}+\tilde{u})^+}^2 \qquad (2.19)$$

Once again, taking $X = \tilde{w} + \tilde{u}$ we see that the term $\sigma_{(X)^+}^2$ from Eq. (2.17) may be set equal to $\sigma_{\tilde{w}}^2$ due to the relationship in Eq. (2.19). Furthermore, since w_n and u_n are independent, it must be that $\sigma_{(\tilde{w}+\tilde{u})}^2 = \sigma_{\tilde{w}}^2 + \sigma_{\tilde{u}}^2$, and so Eq. (2.17) finally takes the form

$$\sigma_{\tilde{w}}^2 + \sigma_{\tilde{u}}^2 = \sigma_{\tilde{w}}^2 + \sigma_{(X)^-}^2 + 2\overline{(\tilde{w} + \tilde{u})^+}\,\overline{(\tilde{w} + \tilde{u})^-} \qquad (2.20)$$

Regarding the last term in this equation, we have already established that $\overline{(\tilde{w} + \tilde{u})^+} = \bar{w}$ and $\overline{(\tilde{w} + \tilde{u})^-} = -\bar{u}$; using these and canceling the variance of $\tilde{w}$ from both sides of our last equation, we have

$$\sigma_{\tilde{u}}^2 = \sigma_{(X)^-}^2 - 2\bar{w}\bar{u} \qquad (2.21)$$

By definition $\bar{u} = \bar{x} - \bar{t}$ and so, as we have seen many times before, $\bar{u} = \bar{t}(\rho - 1)$; similarly, since $\tilde{x}$ and $\tilde{t}$ are independent, it must be that $\sigma_u^2 = \sigma_t^2 + \sigma_x^2$. However, we already have notation for the variance of interarrival time and variance of service time, namely, σ_a^2 and σ_b^2, respectively. With these observations, and solving for $\bar{w}$ (which, in the past, we have written simply as W) we may rewrite Eq. (2.21) as follows:

$$W = \frac{\sigma_a^2 + \sigma_b^2}{2\bar{t}(1-\rho)} - \frac{\sigma_{(X)^-}^2}{2\bar{t}(1-\rho)}$$

Since variances are always non-negative, we may drop the last term in this equation and thereby create our final upper bound on the average

---

* We point out that the limiting random variable $\bar{y} = \lim_{n\to\infty} y_n$ must have the same distribution as $(\tilde{w} + \tilde{u})^-$, as may be seen from Eq. (1.124).

waiting time:

$$W \le \frac{\sigma_a^2 + \sigma_b^2}{2\bar{t}(1-\rho)}$$ ■ (2.22)

This result is correct for $0 \le \rho < 1$ and improves (is asymptotically sharp) as $\rho \to 1$.* This result is familiar! It is, in fact, the mean waiting time that we obtained in the previous section for the heavy-traffic approximation. What we now see is that the heavy-traffic approximation to the mean wait forms a strict upper bound for the mean wait in any G/G/1 queue. In Section 1.4 we boldly stated that the behavior of the mean waiting time for the queue M/M/1 was typical of most queueing systems in that the dominant behavior is due to a simple pole at $\rho = 1$; we have now confirmed that statement by our basic results in this and the previous section.

Our upper bound is essentially distribution-free in that it depends only on the first two moments of the service and interarrival time; this simplicity is a key virtue since often we are willing to specify only some gross properties of the input (e.g., mean and variance). Unfortunately, this simplicity does not extend to the lower bound, which we discuss next.

2.3. LOWER BOUNDS FOR THE AVERAGE WAIT

The simple upper bound obtained in the last section may also be derived easily (see Exercise 2.6) from Eq. (1.132), which we may express as follows:

$$W = W_U + \frac{1}{2} \bar{t}(1-\rho) - \frac{\overline{I^2}}{2\bar{I}}$$ (2.23)

---

* We note that the upper bound exceeds the known exact mean wait for M/G/1 [as given by the P-K mean value formula in Eq. (1.82)] by $(\bar{x} + \bar{t})/2$, which is less than one interarrival time. Marchal has proposed that the upper bound in Eq. (2.22) be scaled down so that it is *exact* for M/G/1; thus his approximation is

$$W \cong \frac{1 + C_b^2}{(1/\rho)^2 + C_b^2} \left[\frac{\sigma_a^2 + \sigma_b^2}{2\bar{t}(1-\rho)} \right]$$

where C_b, the service time coefficient of variation, is defined as $C_b = \sigma_b/\bar{x}$. Both he [MARC 74] and Gross [GROS 73] consider the effectiveness of this (and other) approximations to W. Their numerical studies show that the fit to G/M/1 is good, so far as percentage error is concerned; for G/G/1 it is fair, degrading with an increase in the coefficient of variation of either the interarrival times or the service times, and improving as ρ increases.

where we have defined W_U to be our upper bound

$$W_U \triangleq \frac{\sigma_a{}^2 + \sigma_b{}^2}{2\bar{t}(1-\rho)} \qquad (2.24)$$

We also had an alternative expression for the mean wait equivalent to that given in Eq. (2.23) and expressed it as Eq. (1.131), which we repeat here:

$$W = -\frac{\overline{u^2}}{2\bar{u}} - \frac{\overline{y^2}}{2\bar{y}} \qquad (2.25)$$

These last two expressions for W form our point of departure in establishing lower bounds on the mean wait in $G/G/1$. It is clear that if we are to obtain such lower bounds, then we must place an upper bound on $\overline{I^2}/2\bar{I}$, which is the mean residual life of the idle time period I. We have already introduced the random variable $\tilde{y} = (\tilde{w} + \tilde{u})^-$ [see Section 1.10, Eqs. (1.124), (2.12) and the footnote on p. 33] whose mean residual life is shown in Exercise 2.6 to be equal to that of the idle time, that is,

$$\frac{\overline{I^2}}{2\bar{I}} = \frac{\overline{y^2}}{2y}$$

and since $\overline{y^2} = \upsilon_{\tilde{y}}{}^2 + (\bar{y})^2$, we see that our main task is to place an upper bound on the variance of $\tilde{y}$; in this endeavor we follow the approach of Kingman [KING 62b]. First recall from Eqs. (1.116) and (1.126) that

$$\bar{t}(1-\rho) = -\bar{u} = \bar{y} \qquad (2.26)$$

Now, since $\tilde{y}$ has the same distribution as $(\tilde{w} + \tilde{u})^-$, and furthermore, since $\tilde{w} \ge 0$, then from a stochastic point of view,* $\tilde{w} + \tilde{u} \ge \tilde{u}$ and $(\tilde{w} + \tilde{u})^- \le (\tilde{u})^-$; thus we may write

$$(\bar{y})^2 = [\overline{(\tilde{w} + \tilde{u})^-}]^2 \le [\overline{(\tilde{u})^-}]^2$$

Finally,

$$\overline{y^2} \le \overline{[(\tilde{u})^-]^2}$$

Using this last and taking advantage of Eq. (2.15) (with $X = \tilde{u}$) we may also write

$$\overline{y^2} \le \overline{(\tilde{u})^2} - \overline{[(\tilde{u})^+]^2}$$

---

* To say that a random variable X_1 is stochastically smaller than X_2 means that $P[X_1 \le x] \ge P[X_2 \le x]$.

which upon application of Eq. (2.26) yields

$$\frac{\overline{[(\bar{u})^+]^2}}{2\bar{y}} \leq -\frac{\overline{u^2}}{2\bar{u}} - \frac{\overline{y^2}}{2\bar{y}} \tag{2.27}$$

So, finally, we substitute back into Eq. (2.25) and establish the following lower bound on the average waiting time:*

$$W_K \triangleq \frac{\overline{[(\bar{u})^+]^2}}{2\bar{t}(1-\rho)} \leq W \qquad \blacksquare \tag{2.28}$$

This is the first of our lower bounds. We note that it depends on much more than just the first two moments of our input process. This is not an especially tight bound, and in order to do better, we must place conditions on our arrival process, as we shall see later.

Marshall [MARS 68a, b] has established a lower bound on W different from that given in Eq. (2.28). This new bound is an improvement over the other (W_K) in the light-traffic case, and the converse is true in the heavy-traffic case. To establish this new bound, our point of departure is once again the basic relationship

$$w_{n+1} = \max[0, w_n + u_n]$$

This piecewise linear expression takes on the value zero whenever $u_n < -w_n$; therefore, if we condition on the event $w_n = y \geq 0$, then any calculation for the expected value of w_{n+1} need only consider the range for which $u_n \geq -y$, and in this range it must be true that $w_{n+1} = y + u_n$. We may therefore form the conditional expectation on w_{n+1} as

$$E[w_{n+1} \mid w_n = y] = \int_{u=-y}^{\infty} (y+u) \, dP[u_n \leq u] \tag{2.29}$$

Recall that $P[u_n \leq u] = C(u)$. Integrating by parts, it is easy to show that the integral in the following equation is identical to that in Eq. (2.29), namely,

$$E[w_{n+1} \mid w_n = y] = \int_{u=-y}^{\infty} [1 - C(u)] \, du \tag{2.30}$$

this being good for all $y \geq 0$. It is convenient to define $g(y)$ as the integral above, that is,

$$g(y) \triangleq \int_{-y}^{\infty} [1 - C(u)] \, du$$

---

* We use the notation W_K for this lower bound since it is due to Kingman.

Now let us show that $g(y)$ is convex.* We observe that the PDF $C(u)$ is nondecreasing with u (for all u in the range $-\infty \le u \le \infty$), and so $C(-u)$ is nonincreasing with u; therefore, $1 - C(-u)$ is nondecreasing with u. We also have

$$\frac{dg(y)}{dy} = 1 - C(-y) \tag{2.31}$$

Due to the property for $1 - C(-u)$, we see that $dg(y)/dy$ is nondecreasing with y; thus $g(y)$ is convex.

Let us now proceed with the calculation of W. We define $W_n(y) \triangleq P|w_n \le y|$. Unconditioning Eq. (2.30), we then have

$$E[w_{n+1}] = \int_0^\infty E[w_{n+1} \mid w_n = y]\, dW_n(y)$$

$$= \int_0^\infty \int_{-y}^\infty [1 - C(u)]\, du\, dW_n(y)$$

$$= \int_0^\infty g(y)\, dW_n(y)$$

Thus

$$E[w_{n+1}] = E[g(w_n)] \tag{2.32}$$

where the expectation on the right-hand side of this equation is with respect to the distribution of the random variable w_n. However, we have already shown that $g(y)$ is a convex function of its argument. Thus we may apply Jensen's inequality, which states, for any convex function g of a random variable X, that we must have

$$E[g(X)] \ge g(E[X]) \tag{2.33}$$

From Eqs. (2.32) and (2.33) we therefore have

$$E[w_{n+1}] \ge g(E[w_n])$$

If we allow $n \to \infty$ we obtain

$$W \ge g(W) \tag{2.34}$$

Let us now consider the equation $y = g(y)$, that is,

$$y = \int_{-y}^\infty [1 - C(u)]\, du \tag{2.35}$$

---

* That is, for $y_1 \le y_2$ and $0 \le \alpha \le 1$, $g(y)$ will be shown to have the following property:

$$g(\alpha x_1 + (1 - \alpha)x_2) \le \alpha g(x_1) + (1 - \alpha)g(x_2)$$

This is equivalent to requiring that $dg(y)/dy$ be nondecreasing.

where $y \geq 0$. We are interested in the value of y that satisfies this equation since, as we shall see, this value will be our lower bound, which we denote by W_M (the subscript reminds us that it is due to Marshall). We may rewrite Eq. (2.35) as

$$y = \int_{-y}^{0} [1 - C(u)]\, du + g(0) \tag{2.36}$$

for $y \geq 0$. In Figure 2.1 we show y and g(y) versus y. Note that $g^{(1)}(0) = 1 - C(0^-) = P[u_n \geq 0] \geq 0$. We see that a solution to Eq. (2.35) will be obtained if and only if the two curves shown in Figure 2.1 intersect; of course this point of intersection is W_M. Let us next show that these curves cross *exactly once* (for $y \geq 0$) and therefore W_M is unique. We note from Eq. (2.36) that for $g(0) = 0$, $y = 0 = W_M$ will be a solution (and if W_M is to be our lower bound on W, then this value is useless). Moreover, if $g(0) > 0$, then the two curves will cross if and only if for sufficiently large y we have

$$y > g(y)$$

$$= g(0) + \int_{-y}^{0} [1 - C(u)]\, du$$

$$= g(0) + y - \int_{-y}^{0} C(u)\, du$$

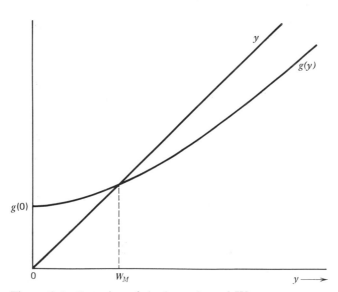

Figure 2.1 Location of the lower bound W_M.

This last condition reduces to

$$g(0) < \int_{-y}^{0} C(u)\,du \tag{2.37}$$

Now, as $y \to \infty$, this last integral is simply

$$\int_{-\infty}^{0} C(u)\,du = -E[\min(0, u_n)] = E[(u_n)^-]$$

Furthermore, $g(0)$ may be written as

$$g(0) = \int_{0}^{\infty} [1 - C(u)]\,du$$

$$= E[\max(0, u_n)] = E[(u_n)^+]$$

Thus

$$E[\tilde{u}] = E[(u_n)^+] - E[(u_n)^-]$$

$$= g(0) - \int_{-\infty}^{0} C(u)\,du$$

But $E[\tilde{u}] = -\bar{t}(1 - \rho)$, and so as $y \to \infty$, the condition in Eq. (2.37) is equivalent to the condition

$$\bar{t}(\rho - 1) < 0$$

or

$$\rho < 1 \tag{2.38}$$

The condition expressed in inequality (2.38) is the condition that guarantees that both curves cross and thereby guarantees a (nontrivial) solution to Eq. (2.35); however, inequality (2.38) is our usual condition for stable queueing systems! Moreover, since we have just shown (for $\rho < 1$) that

$$\lim_{y \to \infty}(y - g(y)) > 0$$

and since $g(y)$ is a convex function, these curves will cross exactly once [if they crossed more than once, then by the convexity of $g(y)$ they would cross exactly twice, and the last inequality above would have to be reversed—a contradiction]. However, one might inquire whether these two curves can coincide over some interval $(a \le y < b)$, say. This would require that $dg(y)/dy = 1$ over this interval. However, due to Eq. (2.31) our assumptions would then also require that $1 - C(-y) = 1$ over this region. But the function $1 - C(-y) \le 1$ and is nondecreasing with y, thereby requiring that $dg(y)/dy = 1$ over the *entire* range $(a \le y)$. Thus we come to the conclusion that if the curves coincide over any finite interval, then they must coincide over a semi-infinite interval and will never separate from one another; however, condition (2.38) guarantees that they will separate. We have thus arrived at a contradiction, thereby removing the possibility of the two curves coinciding over any finite range.

Thus, for $\rho < 1$, W_M is the *unique* solution to Eq. (2.35). Now, if $W_M = 0 = g(0)$, then clearly $W \geq W_M$. On the other hand, if $W_M > 0$, then $g(0) > 0$ as shown above, and therefore for all $0 \leq y < W_M$ we have that

$$y < g(0) + \int_{-y}^{0} [1 - C(u)] \, du \qquad (2.39)$$

due to the uniqueness arguments given above (see also Figure 2.1). Suppose now that $W < W_M$; in this case we would then be able to write

$$W < g(W) \qquad (2.40)$$

since W would fall in the range for which Eq. (2.39) holds. However, Eq. (2.40) directly contradicts Eq. (2.34), and therefore we conclude that

$$W_M \leq W \qquad \blacksquare \qquad (2.41)$$

which finally establishes the lower bound we were seeking. The value for the lower bound W_M is given as the unique solution to Eq. (2.35). Comparing this calculation with that required for Kingman's lower bound W_K, we see that they both require nontrivial computations.

We comment here that in Exercise 2.7 we show by methods similar to those described above that upper and lower bounds on the variance of the waiting time may be given as

$$\sigma_b{}^2 \leq \sigma_{\tilde{w}}{}^2 \leq \sigma_a{}^2 + \sigma_b{}^2 - 2 W_M \bar{t}(1 - \rho) \qquad \blacksquare \qquad (2.42)$$

If we are willing to place some simple constraints on the interarrival time distribution $A(t)$, then we find that we can simplify the lower bound on the average waiting time considerably. These constraints require that we define certain properties of the mean residual life and of the failure rate of distribution functions; the mean residual life and failure rate were defined in the footnote on p. 16. The following definitions are commonly used in reliability theory [BARL 65].

DEFINITION OF γ-MRLA (AND γ-MRLB):   A nondiscrete distribution function F has its mean residual life bounded above (below) by γ [and is then said to be γ-MRLA (γ-MRLB)] if and only if

$$\int_{t}^{\infty} \frac{1 - F(u)}{1 - F(t)} \, du \underset{(\geq)}{\overset{\leq}{}} \gamma \qquad (2.43)$$

for all t and $1 - F(t) > 0$.

DEFINITION OF DMRL (AND IMRL):   A nondiscrete distribution F has decreasing (increasing) mean residual life DMRL (IMRL) if and only if

$$\int_t^\infty \frac{1-F(u)}{1-F(t)}\,du \text{ decreases (increases) with } t \qquad (2.44)$$

for $t \geq 0$ and $1 - F(t) > 0$.

DEFINITION OF IFR (AND DFR):   A nondiscrete distribution function F has increasing (decreasing) failure rate IFR (DFR) if and only if for any $\varepsilon > 0$ we have that

$$\frac{F(t+\varepsilon)-F(t)}{1-F(t)} \text{ increases (decreases) with } t \qquad (2.45)$$

for all $t > 0$ and for $1 - F(t) > 0$.

The first definition merely describes distributions whose mean residual life may be bounded independent of the age of the random variable. The second definition describes distribution functions whose mean residual life behaves monotonically with the age of the random variable. The third definition describes distribution functions whose death rate (failure rate) behaves monotonically with age. It can be shown that

$$\text{IFR} \subset \text{DMRL} \subset \bar{X} - \text{MRLA} \qquad (2.46)$$

where $\subset$ is read as "implies" and $\bar{X}$ is the mean value of the random variable under consideration.

We now wish to apply the notion of the mean residual life and the definitions for this quantity described above for the case of the interarrival distribution $A(t)$ in our queueing system G/G/1. For an interarrival time distribution that is γ-MRLA in the system G/G/1 we use the special notation γ-MRLA/G/1, whereas if $A(t)$ is IFR, then we write IFR/G/1. In Exercise 2.8 we show for the queueing system γ-MRLA/G/1 that

$$\frac{\overline{I^2}}{2\overline{I}} \leq \gamma \qquad (2.47)$$

where I is the random variable describing the idle time as earlier. As we commented previously, as soon as we are able to place an upper bound on the mean residual idle time, as we have just done, then Eq. (2.23) will immediately provide for us a lower bound on the mean wait W. We judiciously choose $\gamma = \bar{t}$, in which case Eqs. (2.23) and (2.47) give

$$W \geq W_U + \frac{1}{2}\bar{t}(1-\rho) - \bar{t}$$

$$= W_U - \frac{1}{2}\bar{t}(1+\rho)$$

Thus, for the queueing system $\bar{t}$-MRLA/G/1, we have the upper and lower bounds on the mean wait given by

$$W_U - \frac{1}{2}\bar{t}(1+\rho) \leq W \leq W_U \qquad \blacksquare \quad (2.48)$$

If we now apply Little's result to this last equation and recall that $\bar{N}_q$ denotes the average number of customers in the queue, then we may bound this quantity as

$$\lambda W_U - \frac{1+\rho}{2} \leq \bar{N}_q \leq \lambda W_U \qquad \blacksquare \quad (2.49)$$

where $\lambda = 1/\bar{t}$ is the average arrival rate of customers to this queue. This last equation gives upper and lower bounds on the expected queue size; note that the difference between these bounds is less than unity!

In Exercise 2.11 we show for the IFR/G/1 queue that

$$\frac{\bar{I^2}}{2\bar{I}} \leq \frac{\bar{t^2}}{2\bar{t}} = \frac{1}{2}\bar{t}(1+C_a^2) \qquad (2.50)$$

where $\bar{t^2}$ and C_a $(=\sigma_a/\bar{t})$ are the second moment and the coefficient of variation, respectively, for the interarrival time. Thus again we have an upper bound on the mean idle time, and so we may apply this to Eq. (2.23) to yield the following lower bound on the mean wait:

$$W \geq W_U + \frac{1}{2}\bar{t}(1-\rho) - \frac{1}{2}\bar{t}(1+C_a^2)$$

$$= W_U - \frac{1}{2}\bar{t}(\rho+C_a^2)$$

Combining this with our upper bound we have that the IFR/G/1 queue has a mean waiting time bounded as follows:

$$W_U - \frac{1}{2}\bar{t}(C_a^2+\rho) \leq W \leq W_U \qquad \blacksquare \quad (2.51)$$

It can easily be shown that any distribution that is IFR must have a coefficient of variation less than unity; therefore, the lower bound in Eq. (2.51) is tighter than the lower bound in Eq. (2.48). This is a reflection of the relationship in Eq. (2.46) which states that the IFR constraint is the strongest among the three. Applying Little's result here we find

$$\lambda W_U - \frac{C_a^2 + \rho}{2} \le \bar{N}_q \le \lambda W_U \qquad \blacksquare \quad (2.52)$$

which again reduces the range of uncertainty for the average queue size to less than one customer. For example, the system D/G/1 which is IFR and for which $C_a^2 = 0$ results in an average queue size that is bounded to within one-half a customer.

Except for these last two cases of $\bar{t}$-MRL A/G/1 and IFR/G/1, the lower bounds we have found in this section are not simply expressed in terms of the first two moments of the interarrival and service time distributions (which was the happy situation with regard to the upper bound of Section 2.2). Marchal [MARC 74] has developed such a lower bound which we now present. Our approach, once again, is to place an upper bound on $\bar{I^2}/2\bar{I}$ in Eq. (2.23), and our point of departure is the expression for y_n given in Eq. (1.124). We have already noted that the limiting form of this random variable, $\tilde{y}$, may be expressed as

$$\tilde{y} = -\min [0, \tilde{w} + \tilde{u}]$$

We have already shown (in Exercise 2.6) that $\bar{I^2}/2\bar{I} = \overline{y^2}/2\bar{y}$, and so we will study the moments of $\tilde{y}$. We may rewrite $\tilde{y}$ as

$$\tilde{y} = \max [0, -\tilde{w} - \tilde{u}]$$
$$= \max [0, \tilde{t} - \tilde{x} - \tilde{w}]$$

Now, since $\tilde{x}$, $\tilde{t}$ and $\tilde{w}$ are all non-negative random variables, then from this last expression, it must be that $\tilde{y}$ is stochastically smaller than $\tilde{t}$. It then follows that $\overline{y^k} \le \overline{t^k}$. Now, since $\overline{t^2} = \sigma_a^2 + (1/\lambda)^2$ we have

$$\overline{y^2} \le \sigma_a^2 + \frac{1}{\lambda^2}$$

But $\bar{y} = (1 - \rho)/\lambda$ and so

$$\frac{\overline{y^2}}{2\bar{y}} = \frac{\bar{I^2}}{2\bar{I}} \le \frac{\lambda(\sigma_a^2 + 1/\lambda^2)}{2(1 - \rho)}$$

Substituting this upper bound into Eq. (2.23), we finally obtain

$$W_U - \frac{\rho(2 - \rho) + C_a^2}{2\lambda(1 - \rho)} \le W \qquad \blacksquare \quad (2.53)$$

This may also be expressed as

$$\frac{\rho^2 C_b^2 + \rho(\rho - 2)}{2\lambda(1 - \rho)} \le W \qquad \blacksquare \quad (2.54)$$

This is the lower bound we were seeking. Note that it is not symmetrical in σ_a^2 and σ_b^2, as is W_U in Eq. (2.22). This bound will be non-negative only for service time coefficients of variation that satisfy $C_b^2 \geq (2 - \rho)/\rho$. The exact value for W for the system M/G/1 exceeds this lower bound by an amount $\bar{x}/(1 - \rho)$; therefore, the bound degrades as ρ increases (but we have seen that the upper bound improves as ρ increases). The main virtue of this bound seems to be its simplicity.

Let us now look for bounds on the waiting time distribution itself rather than on the mean wait.

2.4. BOUNDS ON THE TAIL OF THE WAITING TIME DISTRIBUTION

We recognize that a customer's waiting time is the sum of the service times for all those customers he finds in the queue upon his arrival plus the residual service time for the customer he finds in service. Of course, each of the queued customers' service times is independent and identically distributed, and so we might expect that a result similar to the Chernoff bound [KLEI 75] would perhaps provide an upper and lower bound on the tail of the waiting time distribution. This is indeed the case, and we follow Kingman's approach [KING 70] in establishing these bounds.

Once again we begin with the equation $w_{n+1} = \max[0, w_n + u_n]$. Therefore, for $y > 0$ we may write

$$P[w_{n+1} \geq y] = P[w_n + u_n \geq y]$$

Conditioning this on the value for u_n and recognizing that $P[w_n \geq 0] = 1$, we have

$$P[w_{n+1} \geq y] = \int_{-\infty}^{\infty} P[w_n \geq y - u] \, dC(u)$$

$$= \int_{-\infty}^{y} P[w_n \geq y - u] \, dC(u) + 1 - C(y) \tag{2.55}$$

Now let us consider $C^*(-s) \triangleq E[e^{su_n}]$ where s is taken to be a real (rather than a complex) variable; we recognize that s must lie in a restricted range if this transform is to remain bounded. In particular, if there exists a real s' such that $B^*(-s') \triangleq E[e^{s'\bar{x}}] < \infty$, then a permissible range for s is $0 \leq s \leq s'$. Furthermore, there will be a range in which $C^*(-s) \leq 1$ (for example, in this stable case, $C^*(0) = 1$ and for $s = 0$, $dC^*(-s)/ds = \bar{u} < 0$, thus identifying a neighborhood in this range), and we let s_0 denote the largest value for s such that this remains true. We may thus write the following inequality:

$$e^{-s_0 y} \geq e^{-s_0 y} C^*(-s_0)$$

$$= e^{-s_0 y} \int_{-\infty}^{\infty} e^{s_0 u} \, dC(u)$$

$$= \int_{-\infty}^{\infty} e^{-s_0(y-u)} \, dC(u) \tag{2.56}$$

Now since $s_0 > 0$, for the range $u \geq y$ it must be that $e^{-s_0(y-u)} \geq 1$; thus inequality (2.56) may be extended to

$$e^{-s_0 y} \geq \int_{-\infty}^{y} e^{-s_0(y-u)} \, dC(u) + \int_{y}^{\infty} dC(u)$$

$$= \int_{-\infty}^{y} e^{-s_0(y-u)} \, dC(u) + 1 - C(y) \tag{2.57}$$

so long as $y > 0$.

Let us now assume that w_0 (an initial customer's waiting time) is chosen so that $P[w_0 \geq y] \leq e^{-s_0 y}$; we wish to prove that this hypothesis carries over for all w_n. We prove this by induction, assuming that we have already established its truth up to the nth step, that is $P[w_n \geq y] \leq e^{-s_0 y}$. Then applying this to Eq. (2.55) we have

$$P[w_{n+1} \geq y] \leq \int_{-\infty}^{y} e^{-s_0(y-u)} \, dC(u) + 1 - C(y)$$

But this right-hand side is exactly the expression we bounded in Eq. (2.57), and so we conclude that $P[w_{n+1} \geq y] \leq e^{-s_0 y}$ also, completing the inductive proof. Thus we have established the following exponential bound on the tail of the equilibrium waiting time distribution (by letting $n \to \infty$):

$$P[\tilde{w} \geq y] \leq e^{-s_0 y} \tag{2.58}$$

where, as we stated earlier, s_0 is found from

$$s_0 = \sup \{s > 0 : C^*(-s) \leq 1\}$$

The result given in Eq. (2.58) is, as we had predicted, similar to the form of the Chernoff bound. It is possible also to prove that this tail has a lower bound of a similar form [KING 70], which combines with Eq. (2.58) to give

$$\gamma e^{-s_0 y} \leq 1 - W(y) \leq e^{-s_0 y} \quad \blacksquare \tag{2.59}$$

where we have used our usual notation $W(y) \triangleq P[\tilde{w} \leq y]$ and where γ must satisfy the inequality

$$\gamma \leq \frac{1 - C(y)}{\displaystyle\int_y^\infty e^{-s_0(y-u)}\, dC(u)} \tag{2.60}$$

for all values of $y > 0$; therefore, γ is the smallest value that the ratio in this last equation takes on. From these bounds on the distribution function itself it is trivial to show that the mean wait may also be bounded by

$$\frac{\gamma}{s_0} \leq W \leq \frac{1}{s_0} \tag{2.61}$$

These bounds on W are sometimes sharper than those we considered earlier.

Kobayashi [KOBA 74c] also derives the Kingman upper bound in Eq. (2.58) using Kolmogorov's inequality for submartingales; Ross [ROSS 74] improves on Kingman's upper bound and studies these results for some special cases.

2.5. SOME REMARKS FOR G/G/m

So little is known about the queue G/G/m that any results available for its approximate behavior are extremely worthwhile. Much of the work has been addressed at bounding the mean wait and it is this which we discuss below.

As we know, the appropriate definition for the utilization factor of this system is

$$\rho = \frac{\bar{x}}{m\bar{t}} \tag{2.62}$$

and it has been shown [KIEF 55] that the condition for stability in this case is still

$$\rho < 1$$

Now the most general multiple-server queue that we have so far seen is G/M/m, and from Eq. (1.115) we observed that the conditional pdf for waiting time is exponentially distributed with parameter $m\mu(1-\sigma)$ where $\bar{x} = 1/\mu$; in the heavy-traffic case we expect the unconditional waiting time density to approach this conditional density, and so in that case we may write

$$W \cong \frac{1}{m\mu(1-\sigma)} \qquad \rho \to 1 \tag{2.63}$$

We must solve for the value of σ, which is given as the appropriate root of Eq. (1.104) repeated here:

$$\sigma = A^*(m\mu - m\mu\sigma)$$

Making the change of variable $\alpha = m\mu(1-\sigma)$ the last equation becomes

$$1 - \frac{\alpha\bar{x}}{m} = A^*(\alpha)$$

If we now expand $A^*(\alpha)$ in a power series about the origin as in Eq. (2.4) we have

$$1 - \frac{\alpha\bar{x}}{m} = 1 - \bar{t}\alpha + \frac{\overline{t^2}\alpha^2}{2!} + o(\alpha^2)$$

Since we are considering the heavy-traffic case, we see that $\alpha \ll 1$ [that is, $W \gg \bar{x}$ and $\sigma \cong 1$; see Eq. (2.63)], and so we may neglect the higher-order terms; neglecting $o(\alpha^2)$ and solving for α we have

$$\alpha \cong \frac{2\bar{t}(1-\rho)}{\sigma_a^2 + \bar{t}^2}$$

but since $m\bar{t} \cong \bar{x}$ and since for the exponential service time $\sigma_b^2 = \bar{x}^2$, we may rewrite this last expression as

$$\alpha \cong \frac{2\bar{t}(1-\rho)}{\sigma_a^2 + (1/m^2)\sigma_b^2}$$

We may finally use this result in Eq. (2.63) to give the following[†] as the approximate mean wait in G/M/m as $\rho \to 1$:

$$W \cong \frac{\sigma_a^2 + (1/m^2)\sigma_b^2}{2\bar{t}(1-\rho)} \qquad \blacksquare \quad (2.64)$$

This observation led Kingman [KING 64] to generalize from G/M/m to G/G/m and to suggest (conjecture) for the heavy-traffic approximation for G/G/m that the waiting time should be distributed exponentially with a mean wait given by Eq. (2.64). This conjecture has recently been established by Köllerström [KOLL 74]; thus the Kingman–Köllerström approximation to the waiting time distribution for G/G/m is

$$W(y) \cong 1 - \exp\left(-\frac{2\bar{t}(1-\rho)}{\sigma_a^2 + (\sigma_b^2/m^2)}y\right) \qquad \blacksquare \quad (2.65)$$

The proof of this result uses a G/G/1 approximation to G/G/m in heavy traffic with interarrival times t_n and service times x_n/m. (The Brumelle

---

† Note for $m = 1$, that this approximation for W reduces exactly to Kingman's G/G/1 approximation.

lower bound below also uses this approach.) This G/G/1 approximation was developed earlier by Kiefer and Wolfowitz [KIEF 55]. This heavy-traffic approximation for $W(y)$ also implies that the heavy-traffic approximation for W is as given in Eq. (2.64) for G/G/m.

Suzuki and Yoshida [SUZU 70] have shown that Kingman's conjecture is truly an upper bound for W for $\rho \leq 1/m$. Kingman himself [KING 70] suggests that the approximation is an upper bound for $0 \leq \rho < 1$, but does not prove it, and so far this remains only a conjecture. We state the known *bounds* on W without proof. Kingman [KING 70] derives the following upper bound for the mean wait:

$$W \leq \frac{\sigma_a^2 + (1/m)\sigma_b^2 + [(m-1)/m^2]\bar{x}^2}{2\bar{t}(1-\rho)} \tag{2.66}$$

Brumelle [BRUM 71] also finds this upper bound for G/G/m.

As for the lower bounds on G/G/m, Kingman [KING 70] shows the following:

$$W \geq \frac{2W^*\bar{t} - (\sigma_b^2 + m\sigma_a^2) - [(m-1)/m]\bar{x}^2}{2\bar{x}} \triangleq K_L \tag{2.67}$$

where W^* is the average waiting time in a G/G/1 system with service times $\{x_n\}$ and interarrival times $\{mt_n\}$. Brumelle [BRUM 71] also gives a lower bound in the following form:

$$W \geq \hat{W} - \frac{[(m-1)/m]\overline{x^2}}{2\bar{x}} \triangleq B_L \tag{2.68}$$

where $\hat{W}$ is the average waiting time in a G/G/1 system with service times $\{x_n/m\}$ and interarrival times $\{t_n\}$. Let us compare these last two bounds. If one plots the unfinished work for Kingman's special single-server system (whose average wait is W^* and whose average unfinished work will be denoted by $\bar{U}^*$) and if one also plots the unfinished work for Brumelle's equivalent single-server system (whose average waiting time is given by $\hat{W}$ and whose average unfinished work will be denoted by $\bar{\hat{U}}$) then one readily finds that

$$\overline{U^*} = m\bar{\hat{U}} \tag{2.69}$$

This is easily seen by comparing the two unfinished work functions and recognizing that the average of the unfinished work on a scaled time axis is independent of the scaling. In Chapter 3 below, we show that [see Eq. (3.23)] the average unfinished work, $\bar{U}$, is simply

$$\bar{U} = \rho W + \frac{\overline{x^2}}{2\bar{t}}$$

for G/G/1. If we form $\bar{U}^*$ and $\bar{\bar{U}}$ it is clear from this last equation that

$$W^* = m\hat{W} \tag{2.70}$$

Now if we subtract Kingman's lower bound from Brumelle's lower bound and denote this by $B_L - K_L$ we have

$$B_L - K_L = \hat{W} - \frac{\bar{t}}{\bar{x}} W^* + \frac{m\sigma_a^2 + (1/m)\sigma_b^2}{2\bar{x}}$$

Using Eq. (2.70) we then see

$$B_L - K_L = \hat{W}\left(\frac{\rho - 1}{\rho}\right) + \frac{\sigma_a^2 + (1/m^2)\sigma_b^2}{2\bar{t}\rho}$$

Now the average wait $\hat{W}$ in Brumelle's single-server system clearly has an upper bound given by Eq. (2.22), where the mean service time is $\bar{x}/m$ and the service time variance is σ_b^2/m^2; throughout we maintain the definition for ρ as given in Eq. (2.62). Therefore we may write

$$\hat{W} \le \frac{\sigma_a^2 + (1/m^2)\sigma_b^2}{2\bar{t}(1-\rho)}$$

Using this inequality in the expression for $B_L - K_L$ we immediately have

$$B_L - K_L \ge 0$$

which clearly shows that Brumelle's lower bound is tighter (larger) than Kingman's lower bound.

In summary then the best published bounds for the average wait in G/G/m are[†]

$$\hat{W} - \frac{[(m-1)/m]\overline{x^2}}{2\bar{x}} \le W \le \frac{\sigma_a^2 + (1/m)\sigma_b^2 + [(m-1)/m^2]\overline{x^2}}{2\bar{t}(1-\rho)} \quad \blacksquare \tag{2.71}$$

One sees that these bounds are consistent with the Kingman–Köllerström heavy-traffic approximation of an exponentially distributed waiting time [Eq. (2.65)] with mean given by Eq. (2.64). The term $\hat{W}$ is Brumelle's single-server system to which we may apply any of our earlier bounds; in particular if one is willing to assume more about the interarrival times such as we did in Section 2.3 (for example IFR) then a tighter lower bound may be obtained.

An improvement in the upper bound may be found for the special case G/M/m. Although G/M/m has been solved exactly, as we saw in Section 1.9, we observed there that the solution required the difficult calculation of J and R_k ($k = 0, 1, \ldots, m-2$). Therefore an easily calculated bound

[†]For $\rho \le 1/m$, the upper bound can be tightened by the results of Suzuki and Yoshida mentioned above.

serves a useful purpose. The key result here (due to Brumelle [BRUM 73]) is once again to consider his single-server system G/M/1 with service times $\{x_n/m\}$ and interarrival times $\{t_n\}$; again we denote all variables for this system by a caret. Brumelle shows that $P[w_n > y] < P[\hat{w}_n > y]$, which yields, as a corollary, $W \leq \hat{W}$. To calculate $\hat{W}$, we need deal only with a single-server system, which avoids the calculation of J and R_k; the mean wait $\hat{W}$ is in fact given in Eq. (1.103), which involves finding σ from Eq. (1.100)—a much simpler task. On the other hand, to make the G/M/1 calculation even easier, we may use our earlier result in Eq. (2.22), which is good for any G/G/1 system, to obtain finally for G/M/m

$$W \leq \hat{W} \leq \frac{\sigma_a^2 + (\sigma_b^2/m^2)}{2\bar{t}(1-\rho)} \qquad \blacksquare \quad (2.72)$$

which is an improvement over Eq. (2.71) and which shows that the Kingman–Köllerström heavy-traffic approximation is, in fact, an upper bound to W for G/M/m. In fact, the bounds are rather tight, since we now have shown* that for G/M/m,

$$\hat{W} - \left(\frac{m-1}{m}\right) \bar{x} \leq W \leq \hat{W} \qquad \text{(G/M/m)} \qquad \blacksquare \quad (2.73)$$

Using Little's result, we have

$$\bar{\bar{N}} - \rho(m-1) \leq \bar{N} \leq \bar{\bar{N}} \qquad \text{(G/M/m)} \qquad \blacksquare \quad (2.74)$$

and since $\rho < 1$, we have bounded the average number in system to within $m-1$ of its true value [and this true value happens also to be within $m-1$ of the average number in the equivalent G/M/1 system, i.e., $\bar{\bar{N}} - \bar{N} \leq \rho(m-1)$].

If we now take advantage of Marchal's lower bound for G/G/1 in Eq. (2.54) and use it with Brumelle's lower bound in Eq. (2.68), we arrive at a simple explicit lower bound G/G/m as follows. In particular, we bound $\hat{W}$ by

$$\frac{\rho^2 C_b^2 - \rho(2-\rho)}{2\lambda(1-\rho)} \leq \hat{W}$$

Using this in Eq. (2.68), we get

$$\frac{\rho^2 C_b^2 - \rho(2-\rho)}{2\lambda(1-\rho)} - \frac{[(m-1)/m]\overline{x^2}}{2\bar{x}} \leq W \qquad \blacksquare \quad (2.75)$$

This is a simpler explicit lower bound for G/G/m.

The results of this section only begin to provide some answers for G/G/m; much more work needs to be done in this area.

---

* We note that $\overline{x^2}/2\bar{x} = \bar{x}$ for the exponential service time.

2.6. A DISCRETE APPROXIMATION

So far in this chapter we have handled the complexity of the G/G/1 queue by finding approximations and bounds for the exact solution. Throughout most of the rest of this chapter we take a different point of view: rather than attempt an approximate solution for the original problem, we attempt an exact solution for an approximation of the original problem. That is, we purposefully distort the equations of motion for the given G/G/1 queue and reformulate them in a fashion that permits the system equations to be solved. In this section we discuss a rather crude discrete approximation.

The key to the approach in this section is to alter the input distributions $[A(t)$ and $B(x)]$ in such a way that our basic recurrence relationship [given again in Eq. (2.76) below] permits a direct analytic solution for the distribution of waiting time,

$$w_{n+1} = \max [0, w_n + u_n] \tag{2.76}$$

We observe that the iterative application of this equation is quite straightforward when the interarrival time and service time are both discrete random variables whose only nonzero values occur at the instants $k\tau$ $(k = 0, 1, 2, \ldots)$ where τ is the basic time unit. In such cases one may write down the limit of such recursions to yield a set of linear difference equations that may then be handled by the method of z-transforms [KLEI 75]. We see that this approach requires little more sophistication than that which one uses in elementary queueing theory. If our original random variables are of this discrete nature to begin with, then we have a simple method for giving the *exact* distribution of waiting time. On the other hand, if our given random variables are continuous, then we are faced with an approximation problem; that is, we must approximate the continuous random variables with discrete ones in a fashion that preserves the essence of the solution we seek. Just how one goes about choosing this approximation is as yet basically unstudied and the only recommendation we make at this point is that if one wishes to represent a continuous distribution with a finite set of discontinuities then one should use this approximation to match as many of the moments of the original distribution as possible, working from the first moment and proceeding upwards. We emphasize again, however, that the precision of this approximation has only begun to be studied.

Perhaps the best way to present this method is through an example. We avoid the question of how one should approximate a continuous random variable and assume we begin with discrete interarrival time and service time distributions. Thus we assume, by way of example,

$$A(t) = \begin{cases} 0 & t < 2\tau \\ 1 & t \geq 2\tau \end{cases}$$

$$B(x) = \begin{cases} 0 & x < 0 \\ \frac{1}{2} & 0 \leq x < 3\tau \\ 1 & 3\tau \leq x \end{cases}$$

Since we deal only with discrete random variables, let us define $a(k) = P[t_n = k\tau]$ and $b(k) = P[x_n = k\tau]$; we may then display these discrete functions as in Figure 2.2. Of course these also could have been represented as pdf's with impulses at these same points.

If we are to apply Eq. (2.76) we must find the probability distribution for u_n. We define $c(k) = P[u_n = k\tau]$; since $u_n = x_n - t_{n+1}$, we see that $c(k)$ must in general be given by the following discrete convolution:

$$c(k) = a(-k) \circledast b(k)$$

$$= \sum_{i=-\infty}^{\infty} a(-k+i)b(i)$$

So long as the representation $a(k)$ and $b(k)$ contain a small number of terms, then this convolution is easily carried out by hand; for our example it is trivial and leads to

$$c(k) = \begin{cases} \frac{1}{2} & k = -2 \\ \frac{1}{2} & k = 1 \\ 0 & \text{otherwise} \end{cases}$$

which is shown in Figure 2.3.

In order to apply the recursion in Eq. (2.76) we need a starting value, so let us assume for this example that $w_0 = 0$. Furthermore, we define

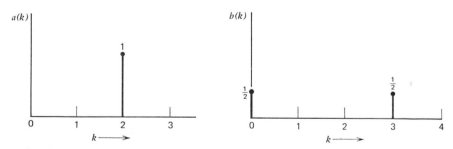

Figure 2.2   The discrete probabilities.

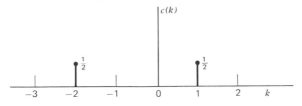

Figure 2.3   The probability $P[u_n = k\tau]$.

the probability (mass) function as

$$p_n(k) = P[w_n = k\tau]$$

We are now ready to apply the recursion, which means that we must carry out the operations described in Eq. (1.121). Assuming that we have applied this recursion up to the calculation of $p_n(k)$, we may proceed as follows. First we find the probability distribution for the random variable $w_n + u_n$, which means that we must convolve $p_n(k)$ with $c(k)$; then the effect of the operator π means that we must sweep the probability in the negative half-line up to the origin (which in our discrete problems is simply a matter of addition) giving the next stage of the recursion, namely, $p_{n+1}(k)$. Carrying this operation out for our problem we generate the sequence shown in Figure 2.4. Starting in the upper left-hand corner of this figure we see the initial waiting time distribution; forming its convolution with $c(k)$ we get the distribution shown in the upper right-hand corner. Sweeping the probability in the negative half-line up to the origin we then easily find $p_1(k)$; this convolved with $c(k)$ gives the figure to its right, which when its negative half-line is swept up to the origin gives us $p_2(k)$, and so on, as we follow the arrows through the sequence of probability (mass) functions. Our object is to find the limiting probability function defined as

$$p(k) \triangleq \lim_{n \to \infty} p_n(k)$$

In order for this ergodic distribution to exist we require that $\rho < 1$, which is equivalent to requiring that $E[u_n] < 0$; for this example we have $\bar{x} = \frac{3}{2}$, $\bar{t} = 2$ and so $\rho = \bar{x}/\bar{t} = \frac{3}{4}$ and $E[u_n] = \bar{x} - \bar{t} = -\frac{1}{2}$. The procedure for writing down the equations that describe the probabilities $p(k)$ is easily seen once we understand what is happening in Figure 2.4. We require that the stable distribution, after being convolved with $c(k)$ and then swept up to the origin, is exactly as it was before these two operations; that is, Eq. (1.122) must hold, which in our notation becomes $p(k) = \pi(p(k) \circledast c(k))$. Furthermore we note that the π operator only affects the term $p(0)$, and so the form for $c(k)$ tells us exactly how a given term $p(k)$ is related to its

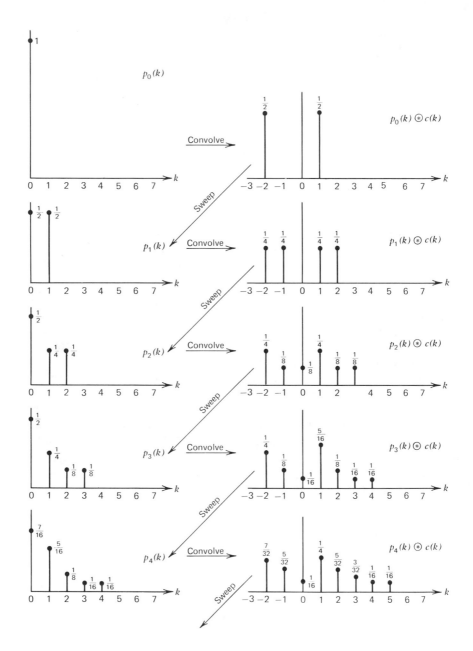

Figure 2.4 The recursion $p_{n+1}(k) = \pi(p_n(k) \circledast c(k))$.

neighbors. In particular, for our example we see that

$$p(k) = \tfrac{1}{2}p(k-1) + \tfrac{1}{2}p(k+2) \qquad k = 1, 2, 3, \ldots \qquad (2.77)$$

and that the boundary equation for $p(0)$ is

$$p(0) = \tfrac{1}{2}p(0) + \tfrac{1}{2}p(1) + \tfrac{1}{2}p(2) \qquad k = 0 \qquad (2.78)$$

We are now faced with the familar problem of solving a set of linear difference equations. The rest is straightforward (see Exercise 2.12). Certainly, the spectrum factorization method summarized in Section 1.10 leads to the same solution. The virtue of the method given here is that it explicitly takes advantage of the discrete nature of our random variables. In both cases the difficult part of the solution is in finding the roots of a polynomial [in the case here it is the roots of the denominator of $P(z) = \sum_k p(k)z^k$, whereas with the spectrum factorization method it is finding the roots of $A^*(-s)B^*(s) - 1$]. The point is, however, that we are suggesting an approximation scheme to convert continuous problems to discrete ones for which rather simple methods apply; the important question regarding how one generates an adequate approximation has not been discussed here. This issue of approximation has recently been studied by Wong [WONG 74]; he investigated how well matched were the distributions and moments of the input and waiting time variables. His results comparing the exact mean wait W with that of W_A as obtained from the iteration shown in Figure 2.4 and with Kingman's upper bound W_U are given in the table below; we note the excellent match between W and W_A.

System	ρ	W	W_A	W_U
M/M/1	2/3	0.13	0.13	0.22
M/E2/1	2/3	0.10	0.10	0.18
M/E3/1	2/3	0.09	0.09	0.17
M/E5/1	2/3	0.08	0.08	0.16
M/E10/1	2/3	0.07	0.07	0.16
E2/M/1	2/3	0.09	0.09	0.14
E3/M/1	2/3	0.08	0.08	0.12
E5/M/1	2/3	0.06	0.07	0.10
E10/M/1	2/3	0.06	0.06	0.08
M/D/1	2/3	0.07	0.07	0.15
D/M/1	2/3	0.05	0.05	0.07
E2/E2/1	2/3	0.06	0.06	0.11
M/H2/1	5/6	0.43	0.40	0.53
H2/M/1	4/5	0.28	0.29	0.36
H2/H2/1	5/9	0.12	0.14	0.26

We also comment that Cohen [COHE 69] gives a procedure for handling the case G/G/1 where he truncates the service time distribution and considers a new distribution $B_c(x)$ such that $B_c(x) = B(x)$ for $x \leq x_c$ and $B_c(x) = 1$ for $x > x_c$. Taking advantage of the simplifications derived from this truncated distribution, he then presents a method of solution and considers the implications as $x_c \to \infty$ to remove the effect of truncation. Neuts and Klimko also consider a G/G/1 system with truncated service times and further restrict their attention to discrete time (as we have done here) [NEUT 73]; they study the ease of numerical analysis for this case.

2.7. THE FLUID APPROXIMATION FOR QUEUES

When an engineer is faced with a systems analysis problem, the first thing he attempts to do is to estimate the gross behavior of the system, however crude that estimate may be. That is, he attempts to "size" the system behavior so that he may make some first-order engineering calculations. Once this is done his task is then to refine his estimates and his approximate analysis; this refinement need be carried only so far as is necessary to insure satisfactory operation within some bounds. The point is that he must come up with answers (estimates) on *all* aspects of the system behavior including transient response, overload conditions, and so on, and not only "nice" equilibrium results. Much of queueing theory is obsessed with "nice" results; even in the early parts of this chapter our bounds and inequalities have applied only to the equilibrium conditions. In this and the three following sections we adopt a different point of view, treating queueing systems as *continuous fluid flow* rather than as discrete customer flow. This enables us to study transients and overloads.

We know for any queueing system that both the number of customers and the unfinished work as functions of time are each stochastic processes with *discontinuous* jumps (for example, at instants of customer arrivals to the system). The approximation we wish to study takes advantage of the following observation: when the system is in the heavy-traffic condition (namely when the queue sizes are large compared to unity and when the waiting times are large compared to average service times) then it appears reasonable to replace these discontinuities by smooth continuous functions of time. The usefulness of this approximation lies in the fact that the magnitude of the original discontinuities is small relative to the average value of these functions. We are dealing with a case of small relative increments. Thus we are led to a continuous stochastic fluid flow approximation for the original discrete queueing system.* Much of the

---

* We comment that usually it is in the case of large queues and long delays that the analysis of queueing systems is important; the case of small queues and delays is usually less interesting since typically they pose no serious problem to system performance.

work reported upon here was developed by Newell [NEWE 65, 68, 71] and Gaver [GAVE 68]; others who have studied these approximation methods include Borovkov [BORO 64, 65], Iglehart [IGLE 69], Iglehart and Whitt [IGLE 70], Kingman [KING 64], Prohorov [PROH 63], and others.

Among the fundamental stochastic processes for queues are the arrival process and the departure process defined as

$$\alpha(t) \triangleq \text{Number of arrivals in } (0, t) \qquad (2.79)$$

$$\delta(t) \triangleq \text{Number of departures in } (0, t) \qquad (2.80)$$

A typical realization for these step-wise increasing processes is shown in Figure 2.5. Clearly at any instant of time their difference must represent $N(t)$, the number of customers present in the system (for $N(0) = 0$), that is,

$$N(t) = \alpha(t) - \delta(t) \qquad (2.81)$$

When $\alpha(t)$ gets large compared to unity then we expect only small percentage deviations from its average value $E[\alpha(t)] \triangleq \overline{\alpha(t)}$; that is by the law of large numbers we have

$$\lim_{t \to \infty} \frac{\alpha(t) - \overline{\alpha(t)}}{\overline{\alpha(t)}} = 0 \qquad (2.82)$$

with probability one. This suggests that a first-order approximation to the stochastic process is to *replace it by its average value as a function of time*. This amounts to what is known as the *fluid approximation* for queues in which the discontinuous *stochastic* process $\alpha(t)$ is replaced by the

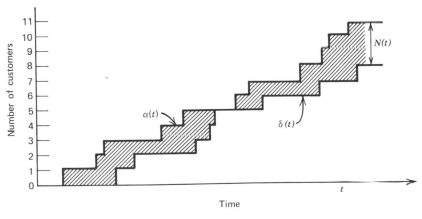

Figure 2.5   The discontinuous arrival and departure processes.

continuous *deterministic* process $\overline{\alpha(t)}$. As we show below, there is considerable merit in this approach. Similarly we let the discontinuous stochastic departure process $\delta(t)$ be replaced by its mean value $\overline{\delta(t)}$. Consequently, if we assume that $N(0)=0$, the fluid approximation predicts that the number in system at time t must be given by

$$N(t)=\overline{\alpha(t)}-\overline{\delta(t)} \tag{2.83}$$

which also is a deterministic continuous function of time. The analogy with fluid flow is complete and may be thought of in terms of the following example [see Figure 2.6(a)]. Consider a funnel with an adjustable valve controlling the rate at which fluid may pass out of this funnel. We pour fluid in at the top at a rate $d\overline{\alpha(t)}/dt \triangleq \lambda(t)$ (the arrival rate of "customers") and we permit fluid to discharge at a rate $d\overline{\delta(t)}/dt \triangleq \mu(t)$ (the service rate for queues); of course the total amount discharged must never exceed the total amount fed in. In this case, then, we see that the total fluid accumulated in the funnel by time t will be given through Eq. (2.83). Thus we have

$$\overline{\alpha(t)}=\overline{\alpha(0)}+\int_0^t \lambda(y)\,dy$$

$$\overline{\delta(t)}=\overline{\delta(0)}+\int_0^t \mu(y)\,dy$$

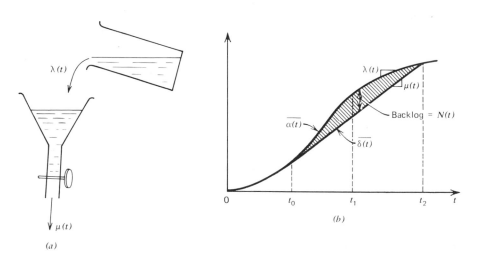

(a)

(b)

Figure 2.6   Fluid approximation to queues.

Consider, for the moment, the case where $\lambda(t)$ varies with time but where $\mu(t)$ is fixed* at μ; an example of this is shown in Figure 2.6(b). Here we see the case where $\lambda(t)$ increases from a small value at time 0 until it equals μ for the first time at t_0. At this instant the backlog begins to grow, reaching its maximum value at time t_1 when once again $\lambda(t) = \mu$; thereafter it decreases to 0 at time t_2. This simple approximation has a number of serious drawbacks. For example, we see that it claims that no queues form as the system approaches saturation from the left at time t_0; certainly we are aware that the size of the backlog at this time is strongly dependent on the manner in which $\lambda(t)$ approaches μ in the interval prior to t_0. As a result of this approximation, queue lengths may be badly underestimated.

In spite of the crudeness of this fluid approximation, it does lead to some worthwhile qualitative aspects of queueing behavior, which we now explore. Much of this material follows that of Newell [NEWE 71]. One of the important questions in the study of queueing theory is the way in which queues and delays grow during and after a "rush hour." The exact queueing analysis in these cases is abominably difficult even for the simplest assumptions for our stochastic processes. However, we can give a very gross picture through our fluid approximation (which will be refined in the following three sections). For example, consider Figure 2.7; here we show an idealized model of a rush-hour situation. In part (a) we show the arrival rate constant at one customer per second up until $t = 2$; then the arrival rate rises linearly at the onset of the rush hour, levels off at a constant value for a short while, drops linearly at the end of the rush hour to its former value at which time it levels off again, and maintains that value. We show the service rate constant (see footnote below) at a value $\mu = 2$. During the time interval $(2\frac{1}{2}, 7\frac{1}{2})$ we see that the system is overloaded in a serious way. In part (b) we show $\overline{\alpha(t)}$, the continuous arrival process, and $\overline{\delta(t)}$, the continuous departure process. The growth in the number of arrivals in the vicinity of the rush hour is evident [we assume $\alpha(0) = \delta(0) = 0$]. Up until $t = 2\frac{1}{2}$ we see that $\overline{\delta(t)} = \overline{\alpha(t)}$; however, for the next 5 units of time, the arrival rate exceeds the maximum departure rate and so the two curves separate forming a backlog $N(t)$. This backlog is shown in part (c) [on a scale twice as large as that in part (b)] and we observe that it grows quickly reaching its peak value at the instant when the arrival rate once again falls below the service rate at $t = 7\frac{1}{2}$. We emphasize that at the "end" of the rush hour [when once again

---

* That is,

$$\mu(t) = \begin{cases} \mu & \text{for } N(t) > 0 \\ \lambda(t) & \text{for } N(t) = 0 \end{cases}$$

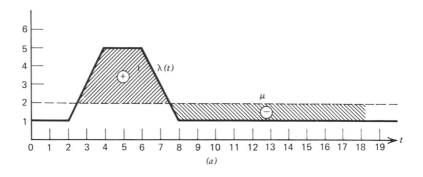

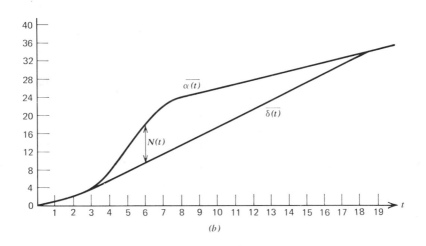

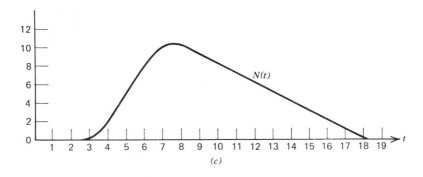

Figure 2.7 Fluid approximation to rush hour. (a) Rates of flow; (b) arrivals and departures; (c) number in system.

$\lambda(t) \leq \mu$] we have merely reached the *peak* of the backlog and the effect of the rush hour will continue for a (possibly long) while. From the figure we see that it takes until $t = 18\frac{1}{4}$ before the backlog disappears! It is easy to see what is happening by referring back to part (a) where the cross-hatched area labeled $\oplus$ is equal to the "deficit" between service rate and arrival rate and therefore represents the total number of customers backlogging in our system; on the other hand, once the arrival rate drops below the departure rate we can make up this deficit with the excess capacity shown as the cross-hatched area labeled $\ominus$. Only when the total negative area equals the total positive area will our backlog drop to zero. If the nonrush-hour value for $\lambda(t)$ is only slightly less than the departure rate μ, we see it will take quite a while for us to make up the deficit; this produces the "long tail" on the backlog $N(t)$. Conversely, if the rate of accumulation of negative area is large compared to that for the positive area, then the backlog will fall off rather quickly. Of course, these comments apply to any arrival and departure process such as, for example, shown in Figure 2.8; here we assume that the backlog is 0 just prior to time t_1 but that at this time it begins to grow since the arrival rate exceeds the departure rate. $N(t)$ will grow as fast as positive area is accumulated in the figure, finally reaching its peak at t_2; it then begins to decline by an amount equal to the accumulated negative area reaching zero when the two areas are equal.

Perhaps now we understand why the freeways remain saturated so long after the close of business. We express the strong caution that although the fluid approximation correctly predicts the long tail of the rush-hour effect, the backlog we have shown is, if anything, optimistically low since we have not included queues that arise due to the *variability* in the arrival and departure processes; these may be large compared to the fluid effects. For example, in Figure 2.6 we have shown $\lambda(t)$ slowly approaching the departure rate μ and finally equaling it at time t_0; in such a case we recognize from our earlier queueing results that as we approach t_0 we are also approaching $\rho \rightarrow 1$, and we expect our queues to grow large in this vicinity. These queues arise due to the random nature of our input

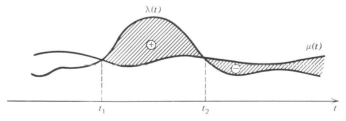

Figure 2.8   Making up the deficit.

processes. The fluid approximation (which we see is really a "continuous" D/D/1 approximation) assures us that there will be no backlog until after time t_0 and it is in this important sense that the approximation is deficient. In the next sections we improve our approximation to take account of such random effects.

2.8. DIFFUSION PROCESSES

In the previous section we used a first-order approximation for queues in which we replaced the arrival and departure processes by their mean values, thereby creating a deterministic continuous process—that is, the fluid flow approximation to queues. We realize, of course, that these processes are random in nature and in this section (and the following two sections) we improve that approximation by permitting $\alpha(t)$ and $\delta(t)$ to have *variations* about the mean. We do this by introducing the variances $\sigma^2_{\alpha(t)}$ and $\sigma^2_{\delta(t)}$ for the arrival and departure processes, respectively. A natural way for introducing these fluctuations about the mean value of our process is to represent these fluctuations by normal (Gaussian) distributions. We may justify this as follows. Observe that $\alpha(t)$ represents the total number of arrivals up to time t. If we ask for the probability that $\alpha(t) \geq n$, then that is the same as asking that customer C_n arrive at a time τ_n that occurs at or before t; that is,

$$P[\alpha(t) \geq n] = P[\tau_n \leq t] \qquad (2.84)$$

This is an important equivalence and is used extensively when one considers ladder indices and combinatorial methods in queueing theory [PRAB 65]. Here, however, we take advantage of this equivalence in the following way: The arrival time of C_n is merely the sum of n interarrival times, that is $\tau_n = t_1 + t_2 + \cdots + t_n$ where we assume $\tau_0 = 0$. For G/G/1 we assume that the set $\{t_i\}$ is a set of independent identically distributed random variables [each with a distribution function $A(t)$]. When the time t, and therefore the number n, get large, then τ_n is the sum of a large number of independent (and identically distributed) random variables. Thus we expect that the central limit theorem should apply and permits us to describe the random variable τ_n and therefore also the random process $\alpha(t)$ as Gaussian functions. This assumption of normality for $\alpha(t)$ [and for $\delta(t)$] is the cornerstone of our diffusion approximation and its details are derived in this section.

For the diffusion approximation, we propose that the arrival process $\alpha(t)$ and the departure process $\delta(t)$ are both to be approximated by continuous random processes (with independent increments) which at time t are normally distributed with means $\alpha(t)$ and $\delta(t)$ and variances $\sigma^2_{\alpha(t)}$ and

$\sigma^2_{\delta(t)}$, respectively. Once we determine these four parameters (the two means and two variances), then we will have completely described these two random processes since the Gaussian process with independent increments (yielding a trivial covariance function) is a two-parameter process (indeed, we have a Brownian motion process, i.e., integrated "white noise" with a nonzero mean [ITO 65]). As mentioned above, the variance terms are introduced in order to represent the random fluctuations of these processes about their means. We intend to use this approximation to make statements about the number of customers in the system $N(t)$ and the unfinished work in the system $U(t)$. As is well known [PAPO 65], if we have two independent normally distributed random processes, say $\alpha(t)$ and $\delta(t)$, then any linear combination of these two is also a normally distributed process (with some appropriate mean and variance). Of course one linear combination we are interested in is $\alpha(t) - \delta(t)$, which represents $N(t)$, the backlog expressed in number of customers. [We are also very much interested in the unfinished work, $U(t)$, which represents the backlog in units of time.] Indeed, we shall demonstrate in Eq. (2.132), below, that the backlog has a transient distribution that is the weighted difference of two Gaussian distributions. For $\rho < 1$, an equilibrium distribution exists which turns out to be the exponential distribution we obtained in Eq. (2.9) for the heavy-traffic approximation! For $\rho > 1$, of course, no equilibrium distribution exists; in this case, however, it turns out, for example, that the properly shifted and scaled waiting time w_n, namely,

$$\frac{w_n - nu}{\sigma_{\tilde{u}} \sqrt{n}}$$

satisfies the central limit theorem [KING 62b], permitting us to talk about this special kind of convergence.

When we attempt to take advantage of the linear combination [$N(t) = \alpha(t) - \delta(t)$] of two independent Gaussian processes, a difficulty arises immediately: $\delta(t)$, the departure process, clearly is dependent upon the arrival process [that is, $\delta(t) \le \alpha(t)$]. Fortunately, however, when $N(t) > 0$, then the departure process increases by one each time a service is completed and so the interdeparture times are distributed as service times, namely $B(x)$, independent of the arrival process [so long as $N(t) > 0$]. Thus when $N(t)$ is large, we have a departure process that is approximately independent of the arrival process, and it is this case which interests us. As a result we might expect that the approximation we are making is poor when the system is lightly loaded.

Thus we have the framework for our second-order approximation (the diffusion approximation) to our queueing system. Replacing $\alpha(t)$ by its mean $\overline{\alpha(t)}$ and its variance $\sigma^2_{\alpha(t)}$ is equivalent to making a Taylor expansion of this process about its mean value and throwing away all but the first two terms in this expansion [see Eq. (2.102)]. Let us now establish the relationship between the means and variances of our arrival and departure processes and the parameters of $A(t)$ and $B(x)$. We have already shown that $P[\alpha(t) \geq n] = P[\tau_n \leq t]$ for any t and n. Similarly if we let $X_n = x_1 + x_2 + \cdots + x_n$ represent the total time to service the first n customers then it is clear that $P[\delta(t) \geq n] = P[X_n \leq t]$, where we are assuming that the system never empties. Both the arrival and departure processes are similar in this sense and so the study of one gives us results for the other. Let us use the departure process in the following calculations. Given n, we have that

$$\bar{X}_n = n\bar{x}$$
$$\sigma_{X_n}{}^2 = n\sigma_b{}^2$$

using our former notation. Applying the central limit theorem we have that the normalized sum $(X_n - n\bar{x})/\sigma_b\sqrt{n}$ must be normally distributed with zero mean and unit variance as $n \to \infty$; that is, for $n \gg 1$,

$$P\left[\frac{X_n - n\bar{x}}{\sigma_b\sqrt{n}} \leq x\right] \cong \frac{1}{\sqrt{2\pi}} \int_{-\infty}^{x} e^{-y^2/2}\,dy \tag{2.85}$$

where the right-hand side is defined as $\Phi(x)$. If we say X is $N(m, \sigma^2)$, that is shorthand for saying that X is a random variable with a normal distribution of mean m and variance σ^2. Thus we say that X_n is $N(n\bar{x}, n\sigma_b{}^2)$ and that $(X_n - n\bar{x})/\sigma_b\sqrt{n}$ is $N(0, 1)$; we see that this second random variable is a shifted and scaled version of X_n so as to give us a new normalized variable with zero mean and unit variance. We observe that X_n itself has a mean that grows linearly with n and a standard deviation that grows like $\sqrt{n}$; therefore the ratio of the standard deviation to the mean is a decreasing function of n. This implies that the fluctuations about the mean become insignificant with respect to the mean as n approaches infinity.

Now the event $(X_n - n\bar{x})/\sigma_b\sqrt{n} \leq x$ is the same as the event $X_n \leq x\sigma_b\sqrt{n} + n\bar{x}$. Let us define

$$t \triangleq x\sigma_b\sqrt{n} + n\bar{x} \tag{2.86}$$

By our former arguments we have

$$P[X_n \leq t] = P[\delta(t) \geq n] \to \text{Gaussian} \tag{2.87}$$

for large n. We are interested in showing that $\delta(t)$ is indeed a normally distributed random process; this last equation almost does that but expresses it in terms of a given quantity n, and we must now use the relationship between t and n in Eq. (2.86) to express this properly. We note for large n that the dominant term is $t \cong n\bar{x}$ and so as an approximation [with the correction term from Eq. (2.86)], we write

$$n \cong \frac{t}{\bar{x}} - x\left(\frac{\sigma_b}{\bar{x}}\right)\sqrt{\frac{t}{\bar{x}}} \qquad (2.88)$$

Thus the event $\delta(t) \geq n$ may be written using this last approximation as

$$\frac{\delta(t) - (t/\bar{x})}{(\sigma_b/\bar{x})\sqrt{t/\bar{x}}} \gtrsim -x$$

If we apply this to Eq. (2.87) and use the symmetry given by $\Phi(x) = 1 - \Phi(-x)$ we find that

$$P\left[\frac{\delta(t) - (t/\bar{x})}{(\sigma_b/\bar{x})\sqrt{t/\bar{x}}} \leq x\right] \cong \Phi(x) \qquad (2.89)$$

for large t. Thus we conclude that the departure process is a normal random variable with mean $t/\bar{x}$ and standard deviation $(\sigma_b\sqrt{t/\bar{x}})/\bar{x}$; that is, $\delta(t)$ is $N(t/\bar{x}, \sigma_b^2 t/(\bar{x})^3)$ for large t. This same result applies for our arrival process $\alpha(t)$ if the mean and variance of service time are replaced by the mean and variance of interarrival time; that is, $\alpha(t)$ is $N(t/\bar{t}, \sigma_a^2 t/(\bar{t})^3)$. Note that these are both approximations that are good only for large t and for moderate-to-heavily loaded queueing systems.

Thus we conclude that the number of customers in the system at time t, given by $N(t) = \alpha(t) - \delta(t)$, is also given as a normal random process whose mean is $\overline{N(t)} = \overline{\alpha(t)} - \overline{\delta(t)}$ and with variance $\sigma_{N(t)}^2 = \sigma_{\alpha(t)}^2 + \sigma_{\delta(t)}^2$ since variances for independent processes must add. We note that the mean number in system for this second-order approximation is exactly the result for the first-order approximation (the fluid approximation) in the previous section. Thus, with this approximation, we have shown for the case G/G/1 that the mean and variance of $N(t)$ are

$$\overline{N(t)} = \frac{t}{\bar{t}} - \frac{t}{\bar{x}} = \left(\frac{\rho - 1}{\bar{x}}\right)t \qquad (2.90)$$

$$\sigma_{N(t)}^2 = \left[\frac{\sigma_a^2}{(\bar{t})^3} + \frac{\sigma_b^2}{(\bar{x})^3}\right]t \qquad (2.91)$$

We note that both the mean and variance grow linearly with time. However, for $\rho < 1$ our approximation shows that the mean number in system becomes negative; clearly we cannot tolerate this, and below we repair that defect by placing a "reflecting boundary" at the origin for $N(t)$. However for $\rho > 1$ we see that a reasonable approximation has developed where the mean number in system grows linearly with t and where the standard deviation of this number grows with $\sqrt{t}$ to represent the fluctuations about that growing mean.

Of course this approximation as a normal random process conceals the discrete nature of the arrival and departure processes. Nevertheless, in the case when $N(t)$ is large compared to unity, this approximation is useful. In order to gain more information about this diffusion approximation for the process $N(t)$, we will now study the partial differential equations for its probability distribution function. This will permit us to properly include the reflecting boundary at the origin as well as to make more explicit statements regarding transient and equilibrium behavior under this approximation. In addition to studying $N(t)$ we will also study $U(t)$, the unfinished work at time t, which will also be approximated as a normal random process. In particular we are interested in the way in which these random processes change during a small time interval; this time interval must be small enough so that the random process changes by a small fraction of its value but it must be large enough to permit enough discrete jumps to take place so that these two processes may be approximated by a continuum. We are thus led to the study of *continuous-time continuous-state Markov processes*. That is, we assume the processes to be Markovian in this time frame. We now launch into a derivation of the underlying partial differential equations for these continuous-time continuous-state Markov processes, and if the reader chooses to pass over this derivation he may do so and immediately proceed to Eq. (2.113), which is the key result of the following development.

In Section 1.3 we considered Markov processes that were continuous in time but with discrete state spaces and observed that the state-transition probabilities obeyed the Chapman–Kolmogorov equation given in Eq. (1.44). Since we intend to replace discrete and mixed random processes [that is, $N(t)$ and $U(t)$] with continuous ones we are naturally led to the consideration of a continuous-time continuous-state Markov process, which we denote by $X(t)$. In analogy with Eq. (1.43) for the conditional discrete-state transition probability, we consider the following conditional continuous-state transition probability:

$$F(x, t; y, \tau) \triangleq P[X(\tau) \leq y \mid X(t) = x] \qquad \text{for} \quad t < \tau \qquad (2.92)$$

Thus $F = F(x, t; y, \tau)$ merely gives the probability that the process takes on a value $\leq y$ at time τ given that it took on the value x at time t. This (possibly time-dependent) transition probability obviously satisfies the following Chapman–Kolmogorov equation:

$$F(x, t; y, \tau) = \int_{-\infty}^{\infty} F(w, u; y, \tau) \, d_w F(x, t; w, u)$$

where $t < u < \tau$.

We now introduce certain assumptions on the stochastic process $X(t)$ that are related to its continuity (as suggested above); in particular, we assume

$$\lim_{\Delta t \to 0} \frac{1}{\Delta t} \int_{|y-x| \geq \varepsilon} d_y F(x, t - \Delta t; y, t) = 0 \qquad (2.93)$$

for any $\varepsilon > 0$. Furthermore, we assume that $\partial F/\partial x$ and $\partial^2 F/\partial x^2$ exist and are continuous. We now introduce the conditional mean $M(x, t; \tau)$ and the conditional variance $V(x, t; \tau)$, where the condition is on the position (x) of the process at some previous time t $(t < \tau)$; these we define as follows:

$$M(x, t; \tau) \triangleq E[X(\tau) \mid X(t) = x] \qquad (2.94)$$

$$V(x, t; \tau) \triangleq E[\{X(\tau) - M(x, t; \tau)\}^2 \mid X(t) = x] \qquad (2.95)$$

We note that $M(x, t; t) = x$ and $V(x, t; t) = 0$. More interesting than this mean and variance are the *infinitesimal mean* $m(x, t)$ and *infinitesimal variance* $\sigma^2(x, t)$, which give the *rate of change* of these quantities with respect to τ at the point $\tau = t$, namely

$$m(x, t) \triangleq \frac{\partial M(x, t; \tau)}{\partial \tau} \bigg|_{\tau = t} \qquad (2.96)$$

$$\sigma^2(x, t) \triangleq \frac{\partial V(x, t; \tau)}{\partial \tau} \bigg|_{\tau = t} \qquad (2.97)$$

In these last two equations derivatives are taken for $\tau \geq t$; see Fig. 2.9. These infinitesimal quantities may be expressed in terms of the transition

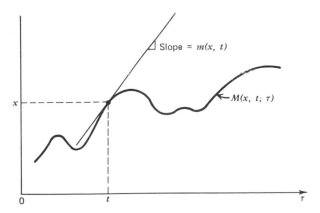

Figure 2.9   Relationship between the conditional mean and the infinitesimal mean.

probabilities as follows:

$$m(x, t) = \lim_{\Delta t \to 0} \frac{1}{\Delta t} \int_{-\infty}^{\infty} (y - x) \, d_y F(x, t - \Delta t; y, t) \tag{2.98}$$

$$\sigma^2(x, t) = \lim_{\Delta t \to 0} \frac{1}{\Delta t} \int_{-\infty}^{\infty} (y - x)^2 \, d_y F(x, t - \Delta t; y, t) \geq 0 \tag{2.99}$$

Thus for small positive Δt we have the following approximation:

$$M(x, t; t + \Delta t) \cong x + m(x, t) \, \Delta t$$

$$V(x, t; t + \Delta t) \cong \sigma^2(x, t) \, \Delta t$$

With this description of a continuous Markov process we now wish to derive the backward and forward equations for F. We begin with the backward equation. From the Chapman–Kolmogorov equation, we have

$$F(x, t - \Delta t; y, \tau) = \int_{-\infty}^{\infty} F(w, t; y, \tau) \, d_w F(x, t - \Delta t; w, t)$$

In a trivial way we may write F in the form

$$F(x, t; y, \tau) = \int_{-\infty}^{\infty} F(x, t; y, \tau) \, d_w F(x, t - \Delta t; w, t)$$

since upon factoring F out of the integral we are left with the integral of a pdf that goes to unity. Subtracting these last two equations and dividing by Δt we have

$$\frac{F(x, t - \Delta t; y, \tau) - F(x, t; y, \tau)}{\Delta t}$$

$$= \frac{1}{\Delta t} \int_{-\infty}^{\infty} [F(w, t; y, \tau) - F(x, t; y, \tau)] \, d_w F(x, t - \Delta t; w, t) \tag{2.100}$$

Now the integral on the right-hand side may be broken into two integrals, the first over the region $|w - x| \geq \varepsilon$ for $\varepsilon > 0$, and the second over the region $|w - x| < \varepsilon$. By Eq. (2.93) the first of these two integrals vanishes as $\Delta t \to 0$ and we may replace the integrand in the second of these two integrals by the following Taylor expansion:

$$F(w, t; y, \tau) - F(x, t; y, \tau)$$

$$= (w - x) \frac{\partial F}{\partial x} + \frac{1}{2} (w - x)^2 \frac{\partial^2 F}{\partial x^2} + o((w - x)^2) \tag{2.101}$$

If we now substitute Eq. (2.101) into Eq. (2.100) and take the limit as $\Delta t \to 0$ then from Eqs. (2.98) and (2.99) we arrive at the following partial differential equation for F:

$$-\frac{\partial F}{\partial t} = m(x, t) \frac{\partial F}{\partial x} + \frac{1}{2} \sigma^2(x, t) \frac{\partial^2 F}{\partial x^2} \tag{2.102}$$

This is the backward Kolmogorov equation for our continuous-time continuous-state Markov process; F will satisfy this equation except at points of accumulation (such as at the origin, $y = 0$).

We now derive the diffusion equation (also known as the *Fokker–Planck* equation) that is the forward equation for our process. The function satisfying this equation is the pdf associated with the final state y, namely,

$$f(x, t; y, \tau) \triangleq \frac{\partial F(x, t; y, \tau)}{\partial y} \qquad (2.103)$$

This density f also satisfies the Chapman–Kolmogorov equation

$$f(x, t; y, \tau) = \int_{-\infty}^{\infty} f(w, u; y, \tau) f(x, t; w, u)\, dw \qquad (2.104)$$

where $t < u < \tau$. We now consider an arbitrary function $Q(y)$ that (along with its derivatives) vanishes rapidly enough at $\pm\infty$ for the integral I to converge, where

$$I \triangleq \int_{-\infty}^{\infty} Q(y) \frac{\partial f(x, t; y, \tau)}{\partial \tau}\, dy \qquad (2.105)$$

Now from the definition of a derivative and from Eq. (2.104) we may rewrite I as

$$I = \lim_{\Delta\tau \to 0} \frac{1}{\Delta\tau} \int_{-\infty}^{\infty} Q(y)[f(x, t; y, \tau + \Delta\tau) - f(x, t; y, \tau)]\, dy$$

$$= \lim_{\Delta\tau \to 0} \frac{1}{\Delta\tau} \left[\int_{-\infty}^{\infty} Q(y) \int_{-\infty}^{\infty} f(x, t; w, \tau) f(w, \tau; y, \tau + \Delta\tau)\, dw\, dy \right.$$

$$\left. - \int_{-\infty}^{\infty} Q(w) f(x, t; w, \tau)\, dw \right] \qquad (2.106)$$

We examine the double integral in this last equation; interchanging orders of integration and using a Taylor expansion for Q about w we may write this double integral (denoted by I_2) as

$$I_2 = \int_{-\infty}^{\infty} f(x, t; w, \tau) \left[\sum_{n=0}^{\infty} \frac{d^n Q(w)}{dw^n} \frac{1}{n!} \right.$$

$$\left. \times \int_{-\infty}^{\infty} f(w, \tau; y, \tau + \Delta\tau)(y - w)^n\, dy \right] dw \qquad (2.107)$$

Now just as we defined the infinitesimal mean and variance in Eqs. (2.98) and (2.99) we now define the infinitesimal nth moments as

$$A_n(w, \tau) \triangleq \lim_{\Delta\tau \to 0} \frac{1}{\Delta\tau} \int_{-\infty}^{\infty} (y - w)^n f(w, \tau; y, \tau + \Delta\tau)\, dy \qquad (2.108)$$

which are assumed to exist as finite quantities. Clearly $A_1(w, \tau) = m(w, \tau)$ and $A_2(w, \tau) = \sigma^2(w, \tau)$. We note that the term for $n = 0$ in Eq. (2.107) involving $A_0(w, \tau)$ cancels the second (single) integral in Eq. (2.106). Now using the definition $A_n(w, \tau)$ and the expansion of I_2 we arrive at the following expression:

$$I = \sum_{n=1}^{\infty} \frac{1}{n!} \int_{-\infty}^{\infty} f(x, t; w, \tau) A_n(w, \tau) \frac{d^n Q(w)}{dw^n} \, dw$$

Let us integrate this last equation by parts (n times for the nth term) where $u = f(x, t; w, \tau) A_n(w, \tau)$ and $dv = (d^n Q(w)/dw^n) \, dw$. Since we have assumed that Q and all of its derivatives vanish rapidly enough at $\pm\infty$, the terms $uv|_{w=-\infty}^{w=+\infty}$ drop out in our integration by parts. Thus only the terms $-\int v \, du$ (and so on) remain, giving us

$$I = \int_{-\infty}^{\infty} Q(w) \sum_{n=1}^{\infty} \frac{(-1)^n}{n!} \frac{\partial^n}{\partial w^n} [A_n(w, \tau) f(x, t; w, \tau)] \, dw \qquad (2.109)$$

Subtracting Eq. (2.109) from (2.105) we have

$$I - I = 0 = \int_{-\infty}^{\infty} Q(w) \left\{ \frac{\partial f(x, t; w, \tau)}{\partial \tau} \right.$$

$$\left. - \sum_{n=1}^{\infty} \frac{(-1)^n}{n!} \frac{\partial^n}{\partial w^n} [A_n(w, \tau) f(x, t; w, \tau)] \right\} dw \qquad (2.110)$$

Now recall that $Q(w)$ was an arbitrary function, and so if Eq. (2.110) is to be satisfied, it must be that $f = f(x, t; w, \tau)$ satisfies

$$\frac{\partial f}{\partial \tau} = \sum_{n=1}^{\infty} \frac{(-1)^n}{n!} \frac{\partial^n}{\partial w^n} [A_n(w, \tau) f] \qquad (2.111)$$

Equation (2.111) holds for our continuous-time continuous-state Markov process, and is the basic equation for our process. It is from this equation that we may derive various of our approximations for queues. For example as we show below, we easily develop the fluid approximation by taking only the first term in this equation; that is, we assume $A_n(w, t) = 0$ for $n = 2, 3, 4, \ldots$, and this gives us

$$\frac{\partial f}{\partial t} = -\frac{\partial}{\partial w} [m(w, t) f] \qquad (2.112)$$

[recall that $A_1(w, t) = m(w, t)$], where now we have replaced τ by t for convenience. (Following the next paragraph, we study the solution of this equation for the case of constant arrival and departure rates.)

The approximation of more interest to this section is the second-order approximation in which we take the first two terms in our series to be nonzero and assume[*] $A_n(w, t) = 0$ for $n = 3, 4, 5, \ldots$, giving

$$\frac{\partial f}{\partial t} = -\frac{\partial}{\partial w}[m(w, t)f] + \frac{1}{2}\frac{\partial^2}{\partial w^2}[\sigma^2(w, t)f] \quad \blacksquare \quad (2.113)$$

[recall that $A_2(w, t) = \sigma^2(w, t)$]. This is known as a one-dimensional Fokker–Planck equation. Both equations (2.113) and (2.102) are referred to as *diffusion equations*, and the reader may observe that they are in fact the forward and backward Kolmogorov equations, respectively. Equation (2.113) is the one we use[†] in this section; it is satisfied except where f contains impulse functions. We comment that even were we to consider all terms in Eq. (2.111), this would still be only an approximation for queues since we have assumed that the underlying processes are continuous (which we know they are not)!

Let us begin simply by considering our deterministic fluid flow approximation to queues, namely Eq. (2.112). Here we see that only the infinitesimal mean of our process enters the picture, and this of course is equivalent to that in the previous section where we replaced our stochastic processes by their mean values. Since we studied the number in system $N(t)$ in that section, let us now take the unfinished work $U(t)$ as the related stochastic process of interest. Thus we consider

$$\Gamma(w_0, 0; w, t) = P[U(t) \leq w \mid U(0) = w_0]$$

which describes the time-dependent distribution of the unfinished work and where we have assumed an initial unfinished work of size w_0 at time $t = 0$. For simplicity we will assume that the mean arrival rate $\lambda(t) = \lambda$ and that the average departure rate $\mu(t) = \mu$ (both constant in time); as a result, $m(w, t)$ is also constant and independent of both w and t. Now, $m(w, t)$ may be calculated (in terms of the system parameters) as the average net rate of work accumulating in the system (for $w > 0$) as follows. Since we have on the average λ arrivals per second, and since each arrival carries with it an average unfinished work of magnitude $\bar{x}$ (the average service time), and since we assume that the service facility operates continuously and therefore clears work at a rate of 1 sec/second (this, too, is part of our approximation, namely, that the service facility

---

[*] The justification for neglecting these higher-order terms is that we expect the conditional pdf to be tightly concentrated around the value w. See Exercise 2.23 for a third-order approximation.

[†] Note that if $m(w, t) = m(t)$ and $\sigma(w, t) = \sigma(t)$, then these parameters may be moved outside the differential operators. This is the case where the arrival and service processes are independent of the backlog in the system, although they are permitted to vary with time.

never goes idle), then we find that $m(w, t)$ is a constant m given by

$$m(w, t) \, \Delta t \triangleq m \, \Delta t = E[U(t + \Delta t) - U(t) \mid U(t)]$$
$$= (\lambda \bar{x} - 1) \, \Delta t$$

and so

$$m = \rho - 1 \qquad\qquad (2.114)$$

Now, not only f, but also F (with different boundary conditions) must satisfy Eq. (2.112), and so we are asked to solve

$$\frac{\partial F(w, t)}{\partial t} = (1 - \rho) \frac{\partial F(w, t)}{\partial w} \qquad \blacksquare \quad (2.115)$$

where we have simplified our notation by suppressing the initial condition; that is, we write $F(w_0, 0; w, t) = F(w, t)$. In addition, we have the two natural boundary conditions, which are good for all t:

$$F(w, t) = 0 \qquad \text{for} \quad w < 0 \qquad\qquad (2.116)$$

$$F(\infty, t) = 1 \qquad\qquad (2.117)$$

We have already assumed the initial condition, which states that the waiting time at $t = 0$ is w_0 with probability 1, namely,

$$F(w, 0) = F_0(w) \triangleq \begin{cases} 0 & w < w_0 \\ 1 & w \geq w_0 \end{cases} \qquad (2.118)$$

The solution to Eq. (2.115) is an arbitrary function of $(w - mt)$, and once we apply the additional conditions stated above, we find the unique solution

$$F(w, t) = \begin{cases} F_0(w + (1 - \rho)t) & w \geq 0 \\ 0 & w < 0 \end{cases} \qquad (2.119)$$

If we examine this solution, we see for $\rho < 1$ that the unfinished work (namely, the virtual waiting time) begins with probability 1 at a value w_0 and decreases toward 0 at a rate $1 - \rho$, and finally at time $t = w_0/(1 - \rho)$ yields a zero backlog that persists forever (i.e., we force this boundary condition). On the other hand, for $\rho > 1$ we find that the backlog begins at a value w_0 and increases without bound at a rate of $\rho - 1$ sec/sec. Clearly, we have correctly described the transient behavior of deterministic queues as described in Section 2.7 above.

Let us temporarily leave behind the fluid approximation and proceed with an investigation of the diffusion approximation, which as we have seen is a second-order approximation including the mean and variance of the original process. (This corresponds to replacing the stochastic process with Brownian motion [ITO 65].) Once again the process we choose to look at is the time-dependent distribution of the unfinished work, that is, $F = F(w, t) = P[U(t) \leq w]$, where we have suppressed the initial condition and will introduce it only as needed. Equation (2.113) is the basic partial differential equation of motion that we must solve; we note that it is also satisfied by F with the appropriate boundary conditions. Again we are able to find the solution to this equation in the case when both the infinitesimal mean and the infinitesimal variance are independent of both w and t; therefore we assume

$$m(w, t) = m$$

$$\sigma^2(w, t) = \sigma^2$$

and we will assume that these are constants both in the original stochastic process $U(t)$ and in our diffusion approximation to it. We now have that F must satisfy the Fokker–Planck equation as follows:

$$\frac{\partial F}{\partial t} = -m \frac{\partial F}{\partial w} + \frac{1}{2} \sigma^2 \frac{\partial^2 F}{\partial w^2} \qquad \blacksquare \quad (2.120)$$

We have already calculated the value for m as given in Eq. (2.114). For σ^2 we carry out the following computation:

$$\sigma^2 \Delta t \triangleq \text{Var}\left[U(t + \Delta t) - U(t) \mid U(t)\right]$$
$$= E\{[(U(t + \Delta t) - U(t)) - m \Delta t]^2 \mid U(t)\}$$
$$= E\{[U(t + \Delta t) - U(t)]^2 \mid U(t)\} - m^2 (\Delta t)^2$$
$$= \lambda \, \Delta t \overline{x^2} + o(\Delta t) \qquad (2.121)$$

This last line comes from the fact that with probability $\lambda \Delta t$ (that is, the probability of an arrival) the second moment of the change in the unfinished work during $(t, t + \Delta t)$ will be $\overline{x^2}$ (namely, the second moment of service time). Thus we have

$$m = \rho - 1$$
$$\sigma^2 = \lambda \overline{x^2} \qquad (2.122)$$

A general solution to Eq. (2.120) is

$$F(w, t) = \frac{1}{\sqrt{2\pi}} \int_{-\infty}^{(w - w_0 - mt)/\sigma\sqrt{t}} e^{-x^2/2} \, dx = \Phi\left(\frac{w - w_0 - mt}{\sigma\sqrt{t}}\right) \qquad (2.123)$$

where $\Phi(x)$ is again the PDF for a standardized normal random variable as given in Eq. (2.85). However, this solution is unsatisfactory since it violates our boundary condition given in Eq. (2.116).

For the moment, let us simplify our task somewhat and ask merely for the *equilibrium* solution to Eq. (2.120) in the case when $\rho < 1$. That is, we seek

$$F(w) = \lim_{t \to \infty} F(w, t)$$

where F must satisfy Eqs. (2.116), (2.117), and (2.120). Clearly, the left-hand side of Eq. (2.120) goes to zero in this limiting case, and we see that the solution to the resulting first-order linear differential equation that also satisfies the boundary conditions is given by

$$F(w) = 1 - e^{2mw/\sigma^2} \qquad w \geq 0 \qquad \blacksquare \quad (2.124)$$

How good is this solution? We expect it should be a fairly good approximation in the heavy-traffic case, for then the backlog is typically large, and so the assumption that the service facility never goes idle is a fair approximation; the continuity assumption on the backlog is then reasonable since the truly discontinuous jumps are small in magnitude compared to the backlog itself. We have, in fact, already made a calculation for the system G/G/1 in the heavy-traffic case, and the solution is given in Eq. (2.9); it is identical in form to the result given above in Eq. (2.124). In that earlier solution, the exponent had a value s_0 as given in Eq. (2.7), namely,

$$s_0 \cong \frac{-2\bar{t}(1-\rho)}{\sigma_a{}^2 + \sigma_b{}^2}$$

However, this result due to Kingman was for the waiting time, whereas Eq. (2.124) is for the unfinished work; these two quantities are the same in the case of a first-come–first-serve M/G/1 queue. In that case we have $\sigma_a{}^2 = 1/\lambda^2$ and $\bar{t} = 1/\lambda$; so

$$s_0 \cong \frac{-2(1-\rho)}{(1/\lambda) + \lambda\sigma_b{}^2} = \frac{-2(1-\rho)}{(1/\lambda) + \lambda\overline{x^2} - \lambda\bar{x}^2}$$

However, in the heavy-traffic case $\rho = \lambda\bar{x} \cong 1$, and so $1/\lambda - \lambda\bar{x}^2 \cong 1/\lambda - \bar{x} \cong 0$, which yields

$$s_0 \cong \frac{-2(1-\rho)}{\lambda\overline{x^2}} \qquad (2.125)$$

But, from Eqs. (2.122) and (2.124), we see that s_0 is approximately equal to the exponent $2m/\sigma^2 = 2(\rho - 1)/\lambda\overline{x^2}$. *Thus the diffusion approximation agrees with Kingman's heavy-traffic approximation to queues!*

Equation (2.124) gives us the diffusion approximation to the equilibrium distribution for the waiting time. An analogous result of course holds for the limiting distribution of $N(t)$, where now m and σ^2 must be calculated for the number in system rather than for the unfinished work in system; in particular, we see from Eq. (2.90) that $m = \bar{N}(t)/t = (\rho - 1)/\bar{x}$, and from Eq. (2.91) that $\sigma^2 = \sigma_{N(t)}^2/t - (C_a^2/\bar{t}) \mid (C_b^2/\bar{x})$. Then, we find that the continuous diffusion approximation $F(w)$ for the equilibrium distribution for number in system is given again in Eq. (2.124). As suggested by Kobayashi [KOBA 74a], we may discretize this to obtain an approximation $\hat{p}_k$ to the distribution for the number in system:

$$\hat{p}_k = F(k+1) - F(k)$$
$$= (1-\hat{\rho})(\hat{\rho})^k$$

where

$$\hat{\rho} = e^{-2(1-\rho)/(\rho C_a{}^2 + C_b{}^2)}$$

Note that the solution for $\hat{p}_k$ reminds us of the M/M/1 solution given in Eq. (1.56) and, in fact, when $C_a = C_b = 1$ (M/M/1) then $\hat{\rho}$ is very close to ρ [KOBA 74a]. The predicted server utilization factor is $1 - \hat{p}_0 = \hat{\rho}$; however, we know from Eq. (1.26) that the exact value for the server utilization is ρ. From this observation, Kobayashi recommends an adjustment to $\hat{p}_k$ for $k = 0$, namely,

$$\hat{p}_k = \begin{cases} 1-\rho & k=0 \\ \rho(1-\hat{\rho})(\hat{\rho})^{k-1} & k \geq 1 \end{cases}$$

In [REIS 74], it is shown that the error in the equilibrium mean number in system, $\bar{N}$, due to this approximation is small for M/G/1 systems with $C_b \cong 1$ and that it grows as C_b deviates from one; however, the relative error in $\bar{N}$ goes to zero as $\rho \to 1$. The modification at $k = 0$ represents one method for reducing the errors that are caused by the compromise we have made in placing a reflecting barrier at the origin. Another approach to this problem is given by Gelenbe [GELE 74]; he places an absorbing boundary at the origin which collects mass (i.e., probability), allows the mass to remain absorbed for an exponentially distributed amount of time (with parameter λ, the arrival rate), and after this time, the mass "jumps" to unity on the real line. The mass collected at the origin corresponds to the probability of an empty system, and the jump to unity (at rate λ) corresponds to an arrival (of one customer) to the system. The solution to this diffusion approximation for the distribution of number in an M/G/1 system is

$$p_k{}^* = \begin{cases} 1-\rho & k=0 \\ K_1\hat{\rho} & k=1 \\ K_2(\hat{\rho})^k & k \geq 2 \end{cases}$$

where K_1 and K_2 are appropriate constants and $\hat{\rho}$ is as given earlier. It is interesting to note that $1 - p_0{}^* = \rho$ is the correct exact value for the server utilization. Also the mean number in the system, $\bar{N}$, predicted by this approximation differs from the known value [as given by the P-K mean value formula in Eq. (1.83)] by $\rho/2C_b^2$.

Of more interest to us is the time-dependent behavior of the mean wait. We have seen that Eq. (2.123) begins to yield the *transient* solution for the waiting time distribution, but, as we observed, it violates the boundary condition $F(w, t) = 0$ for $w < 0$; this boundary condition will be accounted for by the use of our reflecting boundary at the origin. Before proceeding, however, let us consider a convenient *scaling* transformation.

We consider once again the basic diffusion equation for the case of constant arrival and departure rates (that is, $\lambda(t) = \lambda$, $\mu(t) = \mu$), namely,

$$\frac{\partial F}{\partial t} = -m \frac{\partial F}{\partial w} + \tfrac{1}{2}\sigma^2 \frac{\partial^2 F}{\partial w^2} \qquad (2.126)$$

subject to the boundary equations (2.116) and (2.117) and the initial condition (2.118). Our objective is to render this equation dimensionless, solve for the pertinent system behavior, and then recover the appropriate performance measures in their original unscaled form. The basic transformation is to consider the following change of time variable and space variable:

$$t' = \frac{m^2}{\sigma^2} t \qquad (2.127)$$

$$w' = \frac{-m}{\sigma^2} w \qquad (2.128)$$

Thus we measure time in units of σ^2/m^2 and work in units of $-\sigma^2/m$ (recall that $m < 0$ for $\rho < 1$). The scaling operation produces the following dimensionless equation:

$$\frac{\partial F}{\partial t'} = \frac{\partial F}{\partial w'} + \frac{1}{2} \frac{\partial^2 F}{(\partial w')^2} \qquad \blacksquare \qquad (2.129)$$

where now $F = F(w', t')$ is the distribution of $U'(t') \triangleq -(m/\sigma^2)U(m^2 t/\sigma^2)$. Once we solve this last equation we will have solved all equations of the form (2.126).

We will soon give the solution to this dimensionless diffusion equation. At this point, however, we may draw an important conclusion from the fact that our transformation given in Eqs. (2.127) and (2.128) did indeed yield a dimensionless equation independent of m and σ^2. Observe that the natural unit in which we should measure "significant" values for the unfinished work is $-\sigma^2/m$ and that the natural unit for measuring time is σ^2/m^2. That is, significant changes ($\approx -\sigma^2/m$) in $U(t)$ occur during natural time units ($\approx \sigma^2/m^2$). This leads us to the conclusion that the basic "relaxation time" of the system is approximately

$$\text{Relaxation time} \cong \frac{\sigma^2}{m^2} = \frac{\lambda \overline{x^2}}{(1-\rho)^2} \qquad \blacksquare \quad (2.130)$$

A result of this form was noted early by Morse [MORS 55]. It is clear that when ρ is near unity, then the relaxation time may easily exceed the duration of a "rush hour" in practical problems. Also, note that this relaxation time is related to the average wait, W, in an M/G/1 queue as follows: $\sigma^2/m^2 = 2W/(1-\rho)$.

If we consider the equilibrium solution, $F(w') = \lim_{t' \to \infty} F(w', t')$ (assuming $m < 0$, that is, $\rho < 1$), then the dimensionless diffusion equation (2.129) yields

$$F(w') = 1 - e^{-2w'} \qquad w' \ge 0 \qquad\qquad (2.131)$$

with a mean value equal to one-half; this, of course, is the same as our previous result in Eq. (2.124).

Let us now return to the transient solution of Eq. (2.120). We saw earlier that $\Phi([w - w_0 - mt]/\sigma\sqrt{t})$ given in Eq. (2.123) satisfied this equation but unfortunately violated our boundary conditions. Let us denote this solution by $\alpha(w, t)$. Furthermore, we saw that the equilibrium solution to Eq. (2.120) was as given in Eq. (2.124). This leads us to consider the function $e^{2mw/\sigma^2}\alpha(w, t)$. If we let F take on this value, then Eq. (2.120) gives

$$\frac{\partial \alpha(w, t)}{\partial t} = m\frac{\partial \alpha(w, t)}{\partial w} + \frac{1}{2}\sigma^2\frac{\partial^2\alpha(w, t)}{\partial w^2}$$

But this is the same as Eq. (2.120) except for the sign of the first term on the right-hand side! This sign variation can be corrected by making the change of variable from w to $-w$. Thus we see that for any function $F(w, t)$ satisfying Eq. (2.120) there must correspond another solution of the form $e^{2mw/\sigma^2}F(-w, t)$. Now since the diffusion equation (2.120) is linear, it must be that any linear combination of these two solutions must also be a solution. In particular we wish to consider the combination $F(w, t) - e^{2mw/\sigma^2}F(-w, t)$. Now we are in a position to take advantage of our earlier solution $\alpha(w, t)$ in Eq. (2.123), which, when in this combination, quite fortunately also satisfies the previously violated boundary condition (2.116). Thus our time-dependent solution to the diffusion Eq. (2.120) is [NEWE 71]

$$F(w, t) = \Phi\left(\frac{w - w_0 - mt}{\sigma\sqrt{t}}\right) - e^{2mw/\sigma^2}\Phi\left(\frac{-w - w_0 - mt}{\sigma\sqrt{t}}\right) \qquad \blacksquare \quad (2.132)$$

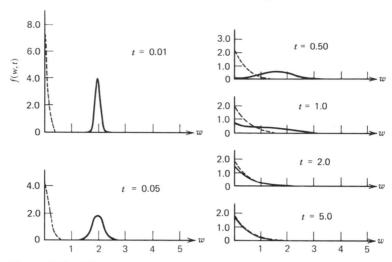

Figure 2.10   Time-dependent behavior of $f(w, t)$ for $\rho < 1$.

This solution corresponds to the case of constant-arrival-rate and constant-departure-rate processes and is good both for $\rho \leq 1$ and $\rho \geq 1$. (If w_0 is given in terms of a distribution function then our solution is obtained by integrating over this distribution.) Kobayashi [KOBA 74b] has evaluated the pdf $f(w, t) = \partial F(w, t)/\partial w$ for some examples of $\rho < 1$ (Figure 2.10) and the PDF $F(w, t)$ for $\rho > 1$ (Figure 2.11). In Figure 2.10 we see the approach of $f(w, t)$ to its exponential limit for $0 < t \leq 5$ with

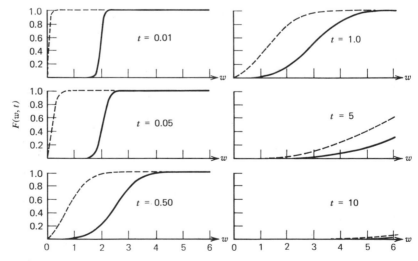

Figure 2.11   Time-dependent behavior of $F(w, t)$ for $\rho > 1$.

the two cases $w_0 = 0$ (dashed curves) and $w_0 = 2$ (solid curves). For $\rho > 1$, we see the unstable transient behavior of $F(w, t)$ in Figure 2.11; the same two cases are shown ($w_0 = 0$ as dashed curves and $w_0 = 2$ as solid curves) in the range $0 < t \leq 10$.

It is truly amazing that such a simple solution to our diffusion equation exists in this case since it must contain the elements of our fluid approximation, as well as our limiting equilibrium distribution, and must give the time-dependent solution for all values of ρ. We note that the first term in the solution involves a normal distribution with variance $\sigma^2 t$ and with mean $w_0 + mt$. For $\rho < 1$ ($m < 0$) we see that this mean drifts to the left and corresponds to our earlier solution using the fluid approximation as given in Eq. (2.119); for $\rho > 1$ ($m > 0$) we see that the mean drifts to the right again as in Eq. (2.119). The second term in our solution corresponds to a normal distribution drifting in a direction opposite to the first one but with an exponentially decreasing (increasing) weight for the case $\rho < (>) 1$. It is the second term that provides the reflection off the boundary at the origin. Moreover we note that as $t \to \infty$, and for $\rho < 1$ ($m < 0$), both of the Φ functions go to unity leaving us with the equilibrium distribution as calculated in Eq. (2.124)! On the other hand, when $\rho > 1$ ($m > 0$) we get behavior that does not settle down; in the following section we discuss this behavior in the context of M/G/1. The reader is referred to the excellent monograph by Newell [NEWE 71] for considerably more discussion of these matters. For now we wish to apply our diffusion approximation to M/G/1.

2.9. DIFFUSION APPROXIMATION FOR M/G/1 [GAVE 68]

In this section we study the time-dependent behavior of the unfinished work $U(t)$ for the first-come-first-serve M/G/1 system. In this system, $U(t)$ has a distribution that is the same as the waiting time distribution $W(y)$. We will use the continuous diffusion process as an approximation to our random sawtooth process $U(t)$, and in order to distinguish the approximation from the true process, we denote the former by $U_d(t)$. In the case when ρ is close to unity this is a good approximation. Much of this material is based upon the work of Gaver.

The general solution we obtained in Eq. (2.132) in the preceding section certainly provides the solution for the queue M/G/1. Nevertheless in this section we choose to study the behavior of M/G/1 using *transform* techniques for two reasons: first, because it gives an alternative approach to the solution and moreover provides additional insight into that solution; secondly, so that we may compare it with the exact time-dependent solution for M/G/1, which is given only in terms of transforms. Much of the algebra and "routine" development of these results is relegated to Exercise 2.21.

Using the notation and results developed in the previous section we therefore proceed to examine the behavior of the time-dependent distribution of waiting time $F(w, t)$ where F and its associated initial and boundary conditions are repeated here:

$$F(w, t) = P[U_d(t) \le w \mid U_d(0) = w_0]$$

$$F(w, 0) = \begin{cases} 0 & w < w_0 \\ 1 & w \ge w_0 \end{cases}$$

$$F(w, t) = 0 \quad \text{for} \quad w < 0$$

$$F(\infty, t) = 1$$

Also, as above, the infinitesimal mean m and variance σ^2 are independent of both w and t, and have values

$$m = \rho - 1$$

$$\sigma^2 = \lambda \overline{x^2}$$

where

$$\rho = \lambda \bar{x}$$

Furthermore, except where F takes jumps, this diffusion process must satisfy the Fokker-Planck equation, namely,

$$\frac{\partial F}{\partial t} = -m \frac{\partial F}{\partial w} + \tfrac{1}{2}\sigma^2 \frac{\partial^2 F}{\partial w^2} \tag{2.133}$$

In order to extract the behavior of F we find it convenient to use transform methods much as we did in Section 1.7. Therefore, as earlier, we define the double Laplace transform (on both w and t) as

$$F^{**}(r, s) \triangleq \int_0^\infty e^{-st} \int_{0^-}^\infty e^{-rw} \, d_w F(w, t) \, dt \tag{2.134}$$

which we assume exists certainly for Re $(s) > 0$ and Re $(r) > 0$. We must now apply this double transform to our partial differential equation (2.133). As developed in Exercise 2.21 we arrive at the following expression for this transform:

$$F^{**}(r, s) = \frac{2}{\sigma^2} \left[\frac{(r/\eta)e^{-\eta w_0} - e^{-rw_0}}{(r - r_1)(r - r_2)} \right] \tag{2.135}$$

where

$$r_1, r_2 = \frac{m}{\sigma^2} \left[1 \pm \left(1 + \frac{2s\sigma^2}{m^2} \right)^{1/2} \right] \tag{2.136}$$

and where r_1 takes the positive square root and r_2, the negative; also

$$\eta = \begin{cases} r_2 & \rho < 1 \\ r_1 & \rho > 1 \end{cases} \tag{2.137}$$

It is worthwhile to compare $F^{**}(r, s)$, which is the solution for our diffusion process, with the analogous result for the exact M/G/1 queueing system. We denote the latter by $F^{**}_{M/G/1}(r, s)$, which is defined as

$$F^{**}_{M/G/1}(r, s) \triangleq \int_0^\infty e^{-st} E[e^{-rU(t)} \mid U(0) = w_0] \, dt \tag{2.138}$$

This result was given in Section 1.7 as Eq. (1.97) and takes the form

$$F^{**}_{M/G/1}(r, s) = \frac{(r/\eta)e^{-\eta w_0} - e^{-rw_0}}{\lambda B^*(r) - \lambda + r - s} \tag{2.139}$$

where now η is the positive real root of the denominator. We notice the remarkable similarity between the solution for the original discrete stochastic process in Eq. (2.139) and the solution to the diffusion approximation given in Eq. (2.135).

We are now interested in inverting $F^{**}(r, s)$ on the transform variable r in order to obtain the time-transformed density for $U(t)$; this we define as

$$F^{\bullet *}(w, s) \triangleq \int_0^\infty e^{-st} \frac{\partial F(w, t)}{\partial w} \, dt \tag{2.140}$$

In Exercise 2.21 we show that this leads to

$$sF^{\bullet *}(w, s) = \begin{cases} -r_1 e^{r_1 w} & r_1 < 0 \quad \rho < 1 \\ -r_2 e^{r_2 w} & r_2 < 0 \quad \rho > 1 \end{cases} \tag{2.141}$$

where we have assumed temporarily that $w_0 = 0$. The reader should note that the dependence of this last equation upon s is through the value of the root r_i ($i = 1$ or 2) as given in Eq. (2.136).

Let us now comment on these last results. We begin by recalling from the method of collective marks (see Section 1.7) that the Laplace transform of a pdf evaluated at some (real) value s is equal to the probability that no Poisson-generated catastrophe will occur (where catastrophes occur at a rate of s per second) during a time interval whose duration is a random variable chosen from the pdf. In a similar fashion it is easily seen that the quantity $sF^{\bullet *}(w, s)$ in Eq. (2.141) may be interpreted as the pdf for the state of our diffusion process if it is observed at the instant of a catastrophe where catastrophes occur at the rate of s per second; in particular, if $w_0 = 0$ then the state (value) of our diffusion process $U_d(t)$ at the instant of a catastrophe has an exponential density as given by the right-hand side of Eq. (2.141). In the case when $w_0 > 0$ then a similar statement may be made which provides an explicit expression for the density of our diffusion process where it can be shown that this pdf will be a linear combination of exponentials. [We see from Eq. (2.139) that the solution to our exact process, namely the distribution of $U(t)$, enjoys no such simple interpretation.] Now we permit the rate of catastrophes to approach zero; therefore our random observation time (i.e., the time of occurence of a catastrophe) approaches infinity. This implies that $s \to 0$ and so we are interested in

$$\lim_{s \to 0} sF^{\bullet *}(w, s) = \lim_{t \to \infty} \frac{\partial}{\partial w} P[U_d(t) \leq w \mid U_d(0) = w_0]$$

which is nothing more than the final value theorem for Laplace transforms [KLEI 75]. In Exercise 2.21 we find for $\rho < 1$ that this leads to

$$\lim_{s \to 0} sF^{\bullet *}(w, s) = -\frac{2m}{\sigma^2} e^{2mw/\sigma^2} \qquad (\rho < 1) \tag{2.142}$$

which is the pdf corresponding to the equilibrium solution we found earlier in Eq. (2.124) as of course it must. (For $\rho > 1$ we find that the limit is zero indicating that no equilibrium solution exists.) Substituting in the values for m and σ^2 we see that the equilibrium distribution for the diffusion approximation to the unfinished work is given by

$$\lim_{t \to \infty} P[U_d(t) \leq w \mid U_d(0) = w_0] = 1 - e^{-2(1-\rho)w/\lambda \overline{x^2}} \tag{2.143}$$

so long as $\rho < 1$ [Eq. (2.124) again]. This distribution of course is independent of w_0 and corresponds to the diffusion approximation to the equilibrium distribution of waiting time for the stable M/G/1 system under a first-come–first-serve queueing discipline.

From Eq. (2.143) we immediately recognize that the mean unfinished work $E[U_d(t)]$ for our diffusion process, which also represents our approximation to the mean waiting time, has a limit for $\rho < 1$,

$$\lim_{t \to \infty} E[U_d(t)] = \frac{\lambda \overline{x^2}}{2(1 - \rho)} \tag{2.144}$$

and this is exactly the P-K formula for the mean wait W in M/G/1 as given in Eq. (1.82). Thus the limiting mean wait for our diffusion process is identical to the limiting mean wait for the exact process (for all $\rho < 1$)! The limiting value of the variance for the wait in our diffusion process, which we denote by $\sigma_{U_d}^2$ is easily calculated from the equilibrium distribution in Eq. (2.143) as

$$\sigma_{U_d}^2 = \left[\frac{\lambda \overline{x^2}}{2(1 - \rho)}\right]^2 \tag{2.145}$$

However, from Eq. (1.87) we have that the variance of the equilibrium waiting time, which we denote by σ_U^2, for our exact process is given by

$$\sigma_U^2 = \left[\frac{\lambda \overline{x^2}}{2(1-\rho)}\right]^2 + \frac{\lambda \overline{x^3}}{3(1-\rho)} \tag{2.146}$$

Now as $\rho \to 1$ we see that $\sigma_U^2 \to \sigma_{U_d}^2$. Thus our diffusion approximation gives an exact answer for the limiting *mean* wait and an answer for the *variance* of this wait that improves (to perfection) as ρ approaches 1 from below.

Let us now make use of our results to examine the *time-dependent* behavior of the mean waiting time, namely $E[U_d(t) \mid U_d(0) = w_0]$. To obtain this expression we first look at its Laplace transform,

$$\int_0^\infty e^{-st} E[U_d(t) \mid U_d(0) = w_0] \, dt$$

From the definition given in Eq. (2.134) and from the moment-generating properties of Laplace transforms we see that the expression we are looking for is obtainable from $-\partial F^{**}(r, s)/\partial r \mid_{r=0}$. Thus taking the partial differential with respect to r in Eq. (2.135) and then setting $r = 0$ we obtain (see Exercise 2.21)

$$\int_0^\infty e^{-st} E[U_d(t) \mid U_d(0) = w_0] \, dt = \frac{m}{s^2} + \frac{w_0}{s} + \frac{e^{-\eta w_0}}{s\eta} \tag{2.147}$$

Let us study Eq. (2.147) by first considering the case $\rho < 1$. We already know that as $t \to \infty$ the *equilibrium* mean waiting time is given through Eq. (2.144). The time-dependent behavior in this case is obtained through Eq. (2.147). For the case $\rho < 1$ and $w_0 = 0$, we have from Exercise 2.21 that

$$\int_0^\infty e^{-st} E[U_d(t) \mid U_d(0) = 0] \, dt = -\frac{1}{r_1 s}$$

$$= -\frac{\sigma^2}{sm\{1 + [1 + (2\sigma^2/m^2)s]^{1/2}\}} \tag{2.148}$$

If we invert Eq. (2.148) we will in fact obtain the time-dependent behavior of the mean waiting time in our diffusion approximation to the system M/G/1. It is appropriate to compare this approximation to exact results calculable for M/G/1 by inverting Eq. (2.139). Gaver [GAVE 68] has made this comparison and we reproduce his numerical results in the examples below. In these examples the Poisson arrival rate is taken at $\lambda = 0.95$ customers per minute and the expected service time is $\bar{x} = 1.0$

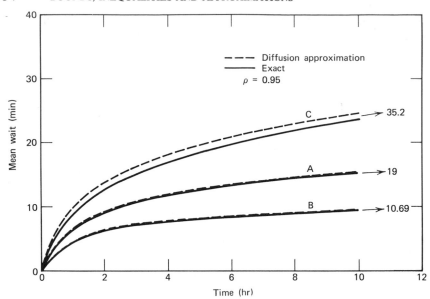

Figure 2.12   Comparison of mean wait versus elapsed time for exact and diffusion analyses: $\rho < 1$.

minute; therefore $\rho = 0.95$. Specifically, the three cases are

$$\text{A.} \quad b(x) = e^{-x} \qquad\qquad x \geq 0 \qquad\qquad (2.149)$$

$$\text{B.} \quad b(x) = \frac{8(8x)^7}{7!}\, e^{-8x} \qquad x \geq 0 \qquad\qquad (2.150)$$

$$\text{C.} \quad b(x) = 0.1\left[\frac{1}{4}\, e^{-x/4}\right]$$
$$+0.9[(6x)^3 e^{-6x}] \qquad x \geq 0 \qquad\qquad (2.151)$$

That is, A is exponential, B is Erlang distributed (with eight stages), and C is a mixture of an exponential (mean = 4) and a four-stage Erlang (mean = $\frac{2}{3}$). In Figure 2.12 we plot the mean wait in minutes as a function of time in hours for the diffusion approximation and for the exact result for the three cases. We also show the asymptotic value of the mean wait, which, as we have demonstrated above, coincides for the diffusion approximation and the exact analysis. In this figure we have taken $w_0 = 0$. We note how excellent the diffusion approximation is for the case $\rho < 1$. We observe once again for these M/G/1 systems that the approach to the equilibrium waiting time is rather slow [that is, from Eq. (2.130) we see that the relaxation time is 12.67, 7.13 and 23.43 hours for cases A, B, C, respectively].

Let us now consider the time-dependent behavior of the mean waiting time for the case $\rho > 1$. Temporarily we will consider once again the more general case of arbitrary w_0. For $\rho > 1$ we have from Eq. (2.137) that $\eta = r_1$. Thus from Eq. (2.147) we seek to invert

$$\frac{m}{s^2} + \frac{w_0}{s} + \frac{e^{-r_1 w_0}}{s r_1}$$

In order to simplify our task we will study the asymptotic behavior as $t \to \infty$ by permitting $s \to 0$; this allows us to make the replacement $r_1 = 2m/\sigma^2$ from Exercise 2.21. Thus inverting our expression we get the asymptotic $(t \to \infty)$ time-dependent behavior for the mean wait conditioned on an initial wait of w_0, namely,

$$E[U_d(t) \mid U_d(0) = w_0] \to (\rho - 1)t + w_0 + \frac{\lambda \overline{x^2} e^{-2(\rho - 1)w_0 / \lambda \overline{x^2}}}{2(\rho - 1)}$$

$$\blacksquare \quad (2.152)$$

This result demonstrates the linear growth of the mean wait predicted by the fluid approximation for the case $\rho > 1$ and large t, and provides an interpretation for the effect of the second term in Eq. (2.129) on the mean wait when $\rho > 1$. As we did for the case $\rho < 1$ we wish to provide some examples for this case to compare exact time-dependent behavior with our diffusion approximation to that behavior. Again we choose $w_0 = 0$, and therefore Eq. (2.152) becomes

$$E[U_d(t) \mid U_d(0) = 0] \to (\rho - 1)t + \frac{\lambda \overline{x^2}}{2(\rho - 1)} \quad (2.153)$$

Observe that this approximation includes the effect of the variance of the service time, an effect often omitted in such approximations (as for example in [COX 61, p. 66]). The examples here once again come from Gaver [GAVE 68], and we consider the case $\lambda = 1.1$, $\bar{x} = 1.0$, and therefore $\rho = 1.1$. In Figure 2.13 we show the time-dependent mean waiting time (in minutes) versus time (in hours) for cases A and B from our previous examples [see Eqs. (2.149) and (2.150)] and compare the exact results from Eq. (2.139) with the diffusion approximation given in Eq. (2.153). Once again we note the excellent approximation provided by our diffusion process. Note that cases A and B give slightly different results due to the variance term in Eq. (2.153).

Let us return once again to the time-dependent mean waiting time conditioned on an initial wait of zero for the case $\rho < 1$; that is, we are interested in the expression $E[U_d(t) \mid U_d(0) = 0]$ whose transform is given

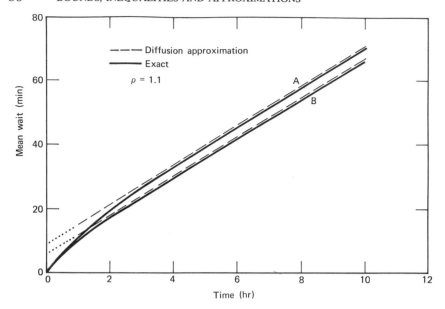

Figure 2.13   Comparison of mean wait versus elapsed time for exact and diffusion analyses: $\rho > 1$.

in Eq. (2.148). In particular we now wish to give a dimensionless form for this equation. Thus we will scale our time and unfinished work functions according to Eqs. (2.127) and (2.128), respectively, and in addition will scale the transform variable s (whose dimensions are 1/sec) as follows

$$s' \triangleq \frac{\sigma^2}{m^2} s \tag{2.154}$$

Thus we may rewrite Eq. (2.148) involving only properly scaled quantities as

$$\int_0^\infty e^{-s't'} E[U_d'(t') \mid U_d'(0) = 0] \, dt' = \frac{1}{s'[1 + \sqrt{1 + 2s'}]} \tag{2.155}$$

Using the final value theorem as earlier, we obtain the limiting expression for the mean wait in system,

$$\lim_{t \to \infty} E[U_d'(t') \mid U_d'(0) = 0] = \lim_{s' \to 0} s' \frac{1}{s'[1 + \sqrt{1 + 2s'}]} = \frac{1}{2} \tag{2.156}$$

Thus, as in Eq. (2.131), we see again that the equilibrium mean wait is equal to one-half (in scaled time units). If we invert expression (2.155) we

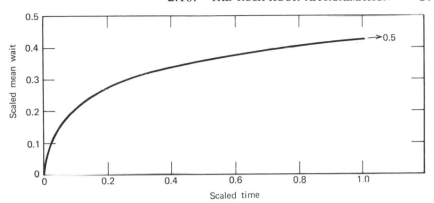

Figure 2.14   Mean wait in the scaled diffusion approximation to M/G/1 ($\rho < 1$).

obtain the time-dependent mean wait for our scaled diffusion approximation to the M/G/1 system as given by (see Exercise 2.22)

$$E[U'_d(t') \mid U'_d(0) = 0] = \left(1 + \frac{t'}{2}\right)[2\Phi(\sqrt{t'}) - 1] - \frac{t'}{2} - \frac{1}{2} P\left(\frac{3}{2}, \frac{t'}{2}\right) \quad \blacksquare \quad (2.157)$$

where $\Phi(x)$ is given in Eq. (2.85) and $P(a, x)$ is the incomplete gamma function defined as

$$P(a, x) \triangleq \frac{1}{\Gamma(a)} \int_0^x e^{-y} y^{a-1} \, dy \tag{2.158}$$

and $\Gamma(x)$ is the usual gamma function (see, for example, [ABRA 64]). This scaled mean wait is plotted in Figure 2.14. This figure gives a *universal* curve for the diffusion approximation to the scaled mean wait in the queue M/G/1; from it, we could have obtained the diffusion approximations in Figure 2.12.

A compact and interesting discussion of these and other asymptotic relations is given by Cohen [COHE 73]; for example, he studies further details regarding the approach to equilibrium.

This ends our specific investigation of the diffusion approximation for the stationary M/G/1 queue. In the following section, we return to the case where the arrival process and departure process are permitted to vary with time.

2.10.   THE RUSH-HOUR APPROXIMATION [NEWE 68, 71]

In applying the diffusion equation (2.113) we have so far emphasized the case where the infinitesimal mean and infinitesimal variance are both constant, that is, $m(w, t) = m$ and $\sigma^2(w, t) = \sigma^2$. Of course, this implies that the arrival and departure processes have average rates that are not time dependent, that is, $\lambda(t) = \lambda$, $\mu(t) = \mu$, and similarly for their variances. Moreover, from our previous studies, we know that in the case $\rho(=\lambda/\mu)<1$, an equilibrium distribution will exist; much of Volume I [KLEI 75] was devoted to a discussion of this equilibrium behavior. On the other hand, most interesting queueing systems are not stationary in time, and it is these we wish to discuss more fully in this section.

In particular, if we assume that the infinitesimal mean and variance are functions only of t, but not of w, then we may rewrite the Fokker-Planck equation as

$$\frac{\partial f}{\partial t} = -m(t)\frac{\partial f}{\partial w} + \frac{\sigma^2(t)}{2}\frac{\partial^2 f}{\partial w^2} \tag{2.159}$$

where we have set $m(w,t) = m(t)$ and $\sigma^2(w,t) = \sigma^2(t)$. We have already seen an "analysis" for time-varying arrival and departure rates if we are willing to accept the fluid approximation of Section 2.7; however, the diffusion equation (2.159) permits us to include the effect of fluctuations about the mean with time-varying rates. Let us begin by estimating the queueing behavior for some extreme cases that fall into the categories we have so far analyzed. First, if $m(t)<0$ (that is, the departure rate can keep up with the arrival rate) and if the variation in this infinitesimal mean is slow compared to $[\sigma(t)/m(t)]^2$, the relaxation time of our system [see Eq. (2.130)], then we expect that the queues and waiting times will "follow" this slow variation such that over a small number of relaxation times the system appears "stationary"; we shall refer to such a situation as being "quasistationary." The quasistationary distribution for the unfinished work is given as

$$F(w,t) = 1 - e^{2m(t)w/\sigma^2(t)} \qquad w \geq 0 \tag{2.160}$$

which is the natural extension of our former equilibrium solution in Eq. (2.124). Recall the definitions $m(t) = [\lambda(t)/\mu(t)] - 1$ and $\sigma^2(t) = \lambda(t)\overline{x^2}(t)$ [where $\overline{x^2}(t)$ is the second moment of the service time duration at the instant t]. Note that the relaxation time is a sensitive function of

$$\rho(t) = \frac{\lambda(t)}{\mu(t)}$$

To be a bit more precise, the quasistationary result in Eq. (2.160) will hold if no significant changes in the waiting time [as measured on the scale in Eq. (2.128), that is, changes in unfinished work of approximately $-\sigma^2(t)/m(t)$ sec] occur in intervals on the order of a relaxation time; thus our condition for quasistationarity becomes

$$\frac{|d\bar{U}(t)/dt|\ \text{(relaxation time)}}{\bar{U}(t)} \ll 1 \tag{2.161}$$

where $\bar{U}(t)$ is the average unfinished work at time t. From Eq. (2.160) we see that $\bar{U}(t) = -\sigma^2(t)/2m(t)$, which confirms our earlier choice of scale in Eq. (2.128). Thus our condition becomes

$$\left| 2\sigma(t)\frac{d\sigma(t)}{dt} - \frac{\sigma^2(t)}{m(t)}\frac{dm(t)}{dt} \right| [m^2(t)]^{-1} \ll 1 \qquad (2.162)$$

The interesting behavior of course is when $\rho(t)$ is close to 1 [that is, $m(t)$ close to zero], in which case the first term is usually insignificant compared to the other; using our expression for $m(t)$ the condition therefore becomes

$$\frac{\sigma^2(t)}{[1-\rho(t)]^3}\left| \frac{d\rho(t)}{dt} \right| \ll 1 \qquad (2.163)$$

We note for $\rho(t)$ close to unity that the left-hand side grows arbitrarily large, and so we can readily imagine situations in which this condition will not hold. Then of course our quasistationary solution will not describe the true picture; in fact the solution for $\bar{U}(t)$ given above for this case predicts that the average waiting time will grow to infinity at an enormous rate as $\rho(t) \to 1$. We know this cannot be the case since *the waiting time can grow no faster than the rate at which work enters the system* and this rate is finite (in spite of the fact that the system is overloaded).

Thus we see that as long as the time variations are slow and small, our quasistationary solution Eq. (2.160) will approximately describe the waiting time distribution in the case $\rho(t) < 1$. However, as ρ approaches and then perhaps exceeds 1, we find that the actual waiting time cannot grow as fast as the quasistationary solution would predict. Second, we observe that for stable queueing systems most of the delays arise because of the stochastic effects of the variability in the arrival and service processes. Recall from Chapter 1 that as long as the input rate is less than the capacity, no backlog should form when the flow is steady; thus, the backlog we see with unsteady flow is caused by the "random effects." On the other hand, from Chapter 1 we see that if the input rate exceeds the capacity ($\rho > 1$) then a huge backlog will develop in time and in such a situation *the stochastic effects become unimportant*! We might therefore expect that the fluid approximation given in Section 2.7 should describe the major part of the growth of delays and of queues for $\rho > 1$. In some sense we have already anticipated this result for the constant-parameter solution in Eq. (2.132) when $\rho > 1$. There we saw that the waiting time behaves as a normal distribution with a linearly growing mean and with a standard deviation that grows only as $\sqrt{t}$; thus the dominant effect is given by the mean with the fluctuations about that mean reducing in relative size. In the case of nonconstant parameters such as we are considering in this section, a similar statement can be made for queues that have been saturated for some time. Thus we can estimate the system behavior at both extremes [$\rho(t) \cong \rho < 1$ and $\rho(t) > 1$], and it is our intention in this section to describe what happens between these extremes.

In Figure 2.7 we illustrated a rush-hour condition and in parts b and c of that figure we gave the fluid approximation to its queueing behavior. We wish now to study the diffusion approximation to the onset of a rush hour; that is, we wish to consider the case where $\rho(t)$ grows with t from the stable case [$\rho(t) < 1$] through the critical value [$\rho(t) = 1$] and on into the overloaded case [$\rho(t) > 1$]. This "transition through saturation" as described by Newell [NEWE 68] (whose development we follow here) corresponds to the onset of a rush hour. Let us define our time axis such that $\rho(t)$ passes through the critical value (unity) at $t = 0$; therefore for $t < 0$ we have a stable case, whereas for $t > 0$ we clearly have an unstable situation. As described in the previous paragraph as long as $\rho(t)$ grows slowly toward its critical value then the quasistationary distribution given in Eq. (2.160) will approximately describe this system behavior. To grow slowly enough we mean that condition (2.163) is satisfied. However, as mentioned above, as $\rho(t)$ approaches unity this condition must certainly be violated and then the system behavior will depart from that of the quasistationary solution and the waiting time will not be able to grow as quickly as that result would imply. Well beyond $t = 0$ we expect the waiting time to grow much as the fluid approximation describes.

As Newell suggests [NEWE 68], let us make a Taylor expansion for $\rho(t)$ about the time origin, that is

$$\rho(t) = 1 + \alpha t + \frac{\alpha^2 t^2}{2!} + \cdots$$

In the vicinity of the critical value ($t = 0$) ρ behaves approximately as $1 + \alpha t$, and we will assume that this is a good approximation in the transition region between that point where the waiting time behavior begins to depart from the quasistationary solution and before it begins to behave as the fluid approximation would indicate. If we use this linear expression for $\rho(t)$ then condition (2.163) becomes

$$\left| \frac{\sigma^2(t)}{\alpha^2 t^3} \right| \ll 1$$

that is,

$$|t| \gg \left(\frac{\sigma^2}{\alpha^2} \right)^{1/3} \triangleq t_0 \qquad (2.164)$$

where we have assumed that $\sigma^2(t) \cong \sigma^2$ in this vicinity. When both sides of this inequality are approximately of the same order of magnitude, then we find that the quasistationary prediction for the waiting time will begin to diverge from its true value. At the time $t = -t_0$ we expect the average unfinished work to be approximately the mean of Eq. (2.160), namely, $-\sigma^2/2m(-t_0)$; however, since we found it convenient earlier to use $-\sigma^2/m$ as the unit of unfinished work in our scaled equations [see Eq. (2.128)], and since Newell also makes this choice, we too choose the following approximation for the mean work at $t = -t_0$,

$$\bar{U}(-t_0) \cong \frac{-\sigma^2}{m(-t_0)} = \left(\frac{\sigma^4}{\alpha}\right)^{1/3} \triangleq \bar{U}_0 \qquad (2.165)$$

During the time from $-t_0$ to 0 the average backlog should grow by an amount approximately equal to the work that arrives during $(-t_0, 0)$ less the work discharged (1 sec/sec) during this interval. Thus the expected increase in $\bar{U}$ during $(-t_0, 0)$ is

$$[\lambda(t)/\mu(t)]t_0 - t_0 = \alpha t_0^2 = \bar{U}_0$$

Therefore, we have defined a natural time unit t_0 and a natural backlog (virtual waiting time) unit $\bar{U}_0$. It is significant to note that the average backlog changes in proportion to $\alpha^{-1/3}$.

Now let us get a feeling for what the diffusion predicts in this transition through saturation. The equation of motion is given in Eq. (2.159); since $m(t) - \rho(t) - 1$ our approximation in the vicinity $t = 0$ is $m(t) \cong \alpha t$ (and moreover we have already assumed that the infinitesimal variance is essentially independent of t in this region, that is, $\sigma^2(t) \cong \sigma^2$), which then gives

$$\frac{\partial F}{\partial t} = -\alpha t \frac{\partial F}{\partial w} + \frac{\sigma^2}{2} \frac{\partial^2 F}{\partial w^2}$$

We have written this in terms of the distribution function rather than the pdf. We have prepared the way for transforming this last equation into a dimensionless equation and so we define the new scaled variables $t' = t/t_0$ and $w' = w/\bar{U}_0$; thus we now have $F(w', t')$ as giving the distribution of $U'(t') = U(t/t_0)/\bar{U}_0$. The scaled equation then becomes

$$\frac{\partial F}{\partial t'} = -t' \frac{\partial F}{\partial w'} + \frac{1}{2} \frac{\partial^2 F}{(\partial w')^2} \qquad (2.166)$$

We see that we have successfully scaled this equation to eliminate its dependence on the specific parameters of the problem; these parameters are now contained in the scaling factors $t_0 = \sigma^{2/3}/\alpha^{2/3}$ and $\bar{U}_0 = \sigma^{4/3}/\alpha^{1/3}$. This equation of course is subject to the boundary conditions that $F(\infty, t') = 1$, $F(0^-, t') = 0$; also for $t' \ll -1$ the solution must be the quasistationary solution $F(w', t') = 1 - e^{2w't'}$. As Newell points out, there is no

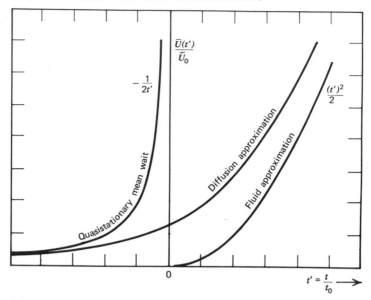

$\dfrac{\bar{U}(t')}{\bar{U}_0}$

$-\dfrac{1}{2t'}$

$\dfrac{(t')^2}{2}$

Diffusion approximation

Quasistationary mean wait

Fluid approximation

0

$t' = \dfrac{t}{t_0}\longrightarrow$

Figure 2.15   The diffusion approximation to the mean wait for the transition through saturation.

known simple analytic solution for Eq. (2.166) subject to these conditions. Fortunately this is a "universal" equation (that is, it is scaled to represent any set of parameter values), and thus we need solve it only once. Furthermore, if we carry out this computation numerically (as in Newell [NEWE 71]), we obtain the required system behavior in the transition through saturation. One way to demonstrate this behavior is to plot the average (virtual) waiting time (in units of $\bar{U}_0$ seconds) versus the normalized time scale t_0. A diagram showing the results of Newell's numerical computation is given in Figure 2.15. Here we see three curves, of which one, the quasistationary mean wait in the region $t' < 0$, forms the left asymptote for our diffusion approximation (which is the second curve); we also see the fluid approximation to the mean wait in the region $t' > 0$, which forms an approximate right asymptote to the diffusion approximation. From Eq. (2.160) we see that the quasistationary mean wait is merely

$$\bar{U}(t) = \frac{-\sigma^2(t)}{2m(t)} \qquad (2.167)$$

which when normalized with respect to $\bar{U}_0$ and expressed in terms of the scaled time t' becomes merely [for $\rho(t) = 1 + \alpha t$]

$$\frac{\bar{U}(t')}{\bar{U}_0} = -\frac{1}{2t'} \qquad (2.168)$$

Equation (2.168) could also have easily been obtained by inspecting the form for the quasistationary solution $F(w', t')$ given earlier as an exponential. For the fluid approximation we see for $t' > 0$ that we are aggravating the work deficit at a rate αt sec/sec, which gives an accumulated work backlog of size $\alpha t^2/2$ by time t; normalizing this with respect to $\bar{U}_0$ and scaling the time axis to t' we easily see that the scaled work backlog has a value $(t')^2/2$ at time t'. This then gives us the shape for the fluid approximation to the mean wait. Once we are deep into saturation (when the probability of the system emptying is insignificant) then our remarks from previous sections assure us that the change in the work backlog will be normally distributed over any interval of time. In fact we see that the average change in this backlog (under the diffusion approximation) will be the same as the average change predicted by the fluid approximation, and this calculation is just the integral of the overload during this time interval; this last statement does not depend upon the fact that the overload grows linearly in this region. As Newell has shown, the distribution F for $t' < -1$ is essentially of the exponential form given earlier, whereas for $t' > 1$, F begins to approach a normal distribution whose mean grows as predicted by the fluid approximation! The reader is urged to consult the fine monograph by Newell [NEWE 71] as well as his earlier article [NEWE 68] for more details.

And so we have a rather good understanding of the behavior of the waiting time as the system enters and continues in the rush hour. We see the important role played by the fluid approximation in this case. However, we again caution the reader that when the approach to saturation is slow, then the zero waiting time predicted by the fluid approximation prior to saturation is badly in error since the system has time to follow the quasistationary mean wait, which grows to large values in such a case. Nevertheless, the average change in queue size will follow the fluid approximation once we have been in saturation for a time on the order of one normalized time unit; the effect of a slow approach to saturation will be an offset between the diffusion and fluid asymptotes for $t' > 1$.

We now inquire into the waiting time behavior for the entire rush-hour cycle. This in some sense requires that we investigate the inverse to the problem we have just studied. We expect that the fluid approximation will be an accurate prediction while $\rho(t) > 1$ and as the system makes the transition back down from saturation into the range $\rho(t) < 1$ then it will settle down into a quasistationary mode. However, at the time when $\rho(t)$ first falls below 1 we recognize from the fluid approximation as shown in Figure 2.7 that the backlog at that time has a maximum value and therefore there will be a "long-tail" effect until the system has a possibility of going idle. During this long tail the behavior will be dominated by the fluid approximation, that is, there will be a normally distributed waiting time with a decreasing mean given by that approximation. After the tail expires then the quasistationary solution takes over and once again the stochastic effects are responsible for the occurrence of queues and delays. In his monograph Newell postulates a parabolic form for $\rho(t)$ during the rush hour merely as an example and shows in fact that the behavior just described does obtain; he derives another dimensionless diffusion equation to examine the transition region for this case as well and presents curves to display this behavior, which is very much what one would expect.

McNeil [McNE 73] nicely summarizes some diffusion models, including the rush-hour congestion process as well as some "almost stationary" situations.

Newell [NEWE 73] has also studied the approximate behavior of the G/G/m queue for $m \gg 1$. He classifies types of behavior, describes the qualitative properties of these types and discusses graphical and analytic (e.g., diffusion approximation) methods that might be used to obtain more quantitative behavior. He observes when the typical queue size is large and all servers are kept busy most of the time, that the system behaves like an effective G/G/1 system with service times $\{x_n/m\}$. Further, when $\rho(t)$ remains less than (approximately) $1 - 1/m$, then the system behaves like a G/G/∞ system. Between these two extremes, many different types of behavior may be observed, depending upon how $\rho(t)$ passes through the transition region, and these are discussed in [NEWE 73].

From a philosophical point of view the fluid approximation we have considered is extremely appealing in its simplicity both for calculation as well as for physical intuition. We find that it is not a bad approximation in the overloaded case, but we wonder if it has anything of interest to suggest about the stable case. We have seen that for the stationary case with $\rho < 1$ then the fluid approximation predicts a zero waiting time. This comes about because we have averaged the rate at which work arrives over the infinite time axis to find that on the average $\lambda \bar{x} < 1$, which means that work arrives at a rate less than the capacity of the system (which is 1 sec of work per second of elapsed time) and therefore no "fluid" will accumulate in our funnel. Of course one need not have averaged over the entire time axis, and it is this point of view we wish to take now. For example one may argue that as soon as a customer arrives to an empty system he "overloads" it in the sense that more work has arrived in the differential time interval surrounding his arrival than can be discharged in that differential time interval; thus the time derivative of work arriving is unbounded in this differential time interval, giving rise to an impulse whose area is equal to the service time of this customer and this

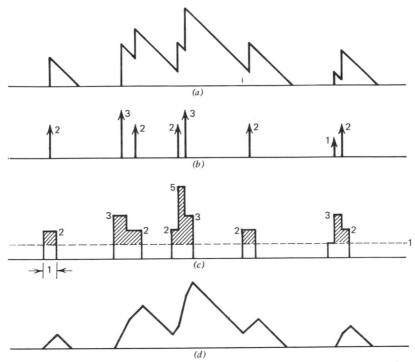

Figure 2.16   An intermediate fluid approximation   (*a*) $U(t)$; (*b*) true burst arrivals; (*c*) "smoothed" input; (*d*) intermediate fluid approximation.

represents the work backlog at that instant. If no other work arrives then this work is discharged at a unit rate until the customer departs; his time in system corresponds to the "long tail" of our fluid approximation. This is much like Figure 2.7 where the total positive area under the curve is considered to have arrived in zero time, giving rise to a step in Figure 2.7(*c*) rather than the smooth rise to its peak. Of course if more customers arrive before the backlog is discharged then the overload continues and takes vertical jumps equal to the service time of each arriving customer. What we are in fact describing in terms of this "instantaneous" fluid approximation is the unfinished work $U(t)$ itself! In Figure 2.16 we give an example of $U(t)$ and directly below it we show the impulses describing the arrival of work at the customer arrival instants; the number next to the impulse gives its area and is equal to the number of seconds of service brought in by each arrival.* Now, let $\omega(t)$ be the

---

* Thus $U(t)$ may be thought of as the output of a linear system whose impulse response is a linearly decaying ramp (slope, −1) with unit height (that is, a small triangle) and whose input is the sequence of work arrivals as shown in part (*b*) of Figure 2.16.

stochastic process representing the arrival of work to the system, that is, a sequence of impulses such as in Figure 2.16(b); to be precise we have

$$\omega(t) = \sum_{n=0}^{\infty} x_n u_0(t - \tau_n)$$

where, as usual, x_n and τ_n represent the service time and arrival time for C_n, and $u_0(y)$ is the unit impulse function occurring at the instant $y = 0$. Let us now consider a continuum of "intermediate" fluid approximations that lie between the original extreme fluid approximation in which the burst arrivals are averaged over an infinite interval and the *exact* situation in which the burst arrivals are averaged over an infinitesimal interval (that is, no average at all). Thus let us consider an averaging interval of length A. The continuum of smoothed input functions is defined as follows:

$$\omega_A(t) \triangleq \frac{1}{A} \int_{-\infty}^{\infty} \left[\omega\left(y + \frac{A}{2}\right) - \omega\left(y - \frac{A}{2}\right) \right] dy \qquad (2.169)$$

Thus we are taking the impulses and uniformly spreading their area over an interval of length A centered about the instant of their occurrence. Figure 2.16(c) shows an example with $A = 1$ for the arrival patterns shown in part (b). If this smoothed input is considered as the instantaneous rate of fluid flow into a queueing system whose departure capacity is 1 sec of work per elapsed second then we see that the cross-hatched region represents the short-term overloads to this system much as the positive area in Figure 2.7 did. Integrating this overload (which goes negative, with value -1 when the input drops to zero) gives us Figure 2.16(d), which corresponds to this intermediate fluid approximation for the original unfinished work $U(t)$ in Figure 2.16(a). Not a bad approximation at all! This relatively good fit results because A is on the order of an average service time; were A considerably larger than an average service time then we would begin to approach the original fluid approximation, which would have predicted zero backlog over most of the time axis. Of course one need not choose a uniform averaging as in Eq. (2.169). The usefulness of these intermediate fluid approximations is that they provide another point of view for understanding queueing systems and the formation of queues and delays.

In the remaining chapters of this book, we find use for many of these approximation techniques. It is perhaps fair to say that this field of generating clever approximations to the complex stochastic processes involved in queueing systems will provide the greatest impetus to the advancement of queueing theory and its applications in the next few years. There is great challenge and reward lying in that direction and the reader is urged to meet that challenge (and thus reap the reward).

REFERENCES

ABRA 64   Abramowitz, M., and I. A. Stegun, *Handbook of Mathematical Functions*, National Bureau of Standards (Wash., D.C.), 1964.

BARL 65   Barlow, R. E., and F. Proschan, *Mathematical Theory of Reliability*, Wiley (New York), 1965.

BORO 64   Borovkov, A., "Some Limit Theorems in the Theory of Mass Service, I," *Theory of Probability and Its Applications*, **9**, 550–565 (1964).

BORO 65   Borovkov, A., "Some Limit Theorems in the Theory of Mass Service, II," *Theory of Probability and Its Applications*, **10**, 375–400 (1965).

BROC 48   Brockmeyer, E., H. L. Halstrøm, and A. Jensen, "The Life and Works of A. K. Erlang," *Transactions of the Danish Academy of Technology and Science*, **2** (1948).

BRUM 71   Brumelle, S. L., "Some Inequalities for Parallel Server Queues," *Operations Research*, **19**, 402–413 (1971).

BRUM 73   Brumelle, S. L., "Bounds on the Wait in a GI/M/k Queue," *Management Science*, **19**, No. 7, 773–777 (1973).

BUZE 74   Buzen, J. P., "Structural Considerations for Computer System Models," *Proceedings of the Eighth Annual Princeton Conference on Information Sciences and Systems*, March 1974.

COHE 69   Cohen, J. W., *The Single Server Queue*, Wiley-Interscience (New York), 1969.

COHE 73   Cohen, J. W., "Asymptotic Relations in Queueing Theory," *Stochastic Processes and Their Applications*, **1**, No. 2, 107–124 (1973).

COX 61   Cox, D. R., and W. L. Smith, *Queues*, Methuen (London) and John Wiley and Sons (New York), 1961.

COX 65   Cox, D. R., and H. D. Miller, *The Theory of Stochastic Processes*, John Wiley (New York) 1965.

GAVE 68   Gaver, D. P., Jr., "Diffusion Approximations and Models for Certain Congestion Problems," *Journal of Applied Probability*, **5**, 607–623 (1968).

GELE 74   Gelenbe, E., "On Approximate Computer Systems Models," in E. Gelenbe and R. Mahl, eds., *Computer Architectures and Networks*, North-Holland Publishing Company (Amsterdam), 187–206 (1974).

GROS 73   Gross, D., "Sensitivity of Queueing Models to the Assumption of Exponentiality: I—Single-channel Queues," Technical Memorandum Serial 64121, Institute for Management Science and Engineering, The George Washington University (1973).

HARR 73   Harrison, J. M., "The Heavy Traffic Approximation for Single Server Queues in Series," *Journal of Applied Probability*, **10**, No. 3, 613–629 (1973).

HEYM 74   Heyman, D. P., "An Approximation for the Busy Period of the M/G/1 Queue Using a Diffusion Model," *Journal of Applied Probability*, **11**, 159–169 (1974).

IGLE 69   Iglehart, D., "Multiple Channel Queues in Heavy Traffic, IV: Law of the Iterated Logarithm," Technical Report No. 8, Dept. Operations Research, Stanford University, 1969.

IGLE 70   Iglehart, D., and W. Whitt, "Multiple Channel Queues in Heavy Traffic, I," *Advances in Applied Probability*, **2**, 150–177 (1970); "Multiple Channel Queues in Heavy Traffic, II: Sequences, Networks, and Batches," *Advances in Applied Probability*, **2**, 355–369 (1970).

ITO 65   Itô, K., and H. P. McKean, Jr., *Diffusion Processes and Their Sample Paths*, Academic Press (New York) 1965.

KIEF 55   Kiefer, J., and J. Wolfowitz, "On the Theory of Queues with Many Servers," *Transactions of the American Mathematics Society*, **78**, 1–18 (1955).

KING 61   Kingman, J. F. C., "The Single Server Queue in Heavy Traffic," *Proceedings of the Cambridge Philosophical Society*, **57**, 902–904 (1961).

KING 62a   Kingman, J. F. C., "On Queues in Heavy Traffic," *Journal of the Royal Statistical Society, Series B*, **24**, 383–392 (1962).

KING 62b   Kingman, J. F. C., "Some Inequalities for the Queue GI/G/1," *Biometrika*, **49**, 315–324 (1962).

KING 64   Kingman, J. F. C., "The Heavy Traffic Approximation in the Theory of Queues," in W. L. Smith and R. I. Wilkinson, eds., *Proceedings of the Symposium on Congestion Theory*, Univ. of North Carolina (Chapel Hill), 137–169 (1964).

KING 70   Kingman, J. F. C., "Inequalities in the Theory of Queues," *Journal of the Royal Statistical Society, Series B*, **32**, 102–110 (1970).

KLEI 75   Kleinrock, L., *Queueing Systems, Volume I: Theory*, Wiley-Interscience (New York) 1975.

KOBA 74a   Kobayashi, H., "Application of the Diffusion Approximation to Queueing Networks, I. Equilibrium Queue Distributions," *Journal of the Association for Computing Machinery*, **21**, No. 2, 316–328 (1974).

KOBA 74b   Kobayashi, H., "Application of the Diffusion Approximation to Queueing Networks, II. Nonequilibrium Distributions and Computer Modeling," *Journal of the Association for Computing Machinery*, **21**, No. 3, 459–469 (1974).

KOBA 74c   Kobayashi, H., "Bounds for the Waiting Time in Queueing Systems," in E. Gelenbe and R. Mahl, eds., *Computer Architectures and Networks*, North-Holland Publishing Company (Amsterdam), 263–274 (1974).

KOLL 74   Köllerström, J., "Heavy Traffic Theory for Queues with Several Servers. I," *Journal of Applied Probability*, **11**, 544–552 (1974).

MARC 74   Marchal, W. G., "Some Simple Bounds and Approximations in Queueing," Technical Memorandum Serial T-294, Institute for Management Science and Engineering, The George Washington University, January 1974.

MARS 68a   Marshall, K. T., "Some Inequalities in Queueing," *Operations Research*, **16**, 651–665 (1968).

MARS 68b   Marshall, K. T., "Bounds for Some Generalizations for the GI/G/1 Queue," *Operations Research*, **16**, 841–848 (1968).

MARS 68c   Marshall, K. T., "Some Relationships between the Distributions of Waiting Time, Idle Time and Interoutput Time in the GI/G/1 Queue," *SIAM Journal of Applied Mathematics*, **16**, 324–327 (1968).

McNE 73   McNeil, D. R., "Diffusion Limits for Congestion Models," *Journal of Applied Probability*, **10**, 368–376 (1973).

MORS 55   Morse, P. M., "Stochastic Properties of Waiting Lines," *Operations Research*, **3**, 255–261 (1955).

NEUT 73   Neuts, M. F., "The Single Server Queue in Discrete Time-Numerical Analysis," *Naval Research Logistics Quarterly*, Part I: **20**, No. 2, 297–304 (1973); Part II (with E. M. Klimko): **20**, No. 2, 305–320 (1973); Part III (with E. M. Klimko): **20**, No. 3, 557–568 (1973).

NEWE 65   Newell, G. F., "Approximate Methods for Queues with Application to the Fixed-Cycle Traffic light," *SIAM Review*, **7**, 223–240 (1965).

NEWE 68   Newell, G. F., "Queues with Time-Dependent Arrival Rates I–III," *Journal of Applied Probability*, **5**, 436–451, 579–606 (1968).

NEWE 71   Newell, G. F., *Applications of Queueing Theory*, Chapman and Hall, Ltd. (London), 1971.

NEWE 73   Newell, G. F., *Approximate Stochastic Behavior of n-Server Service Systems with Large n*, Springer-Verlag (Berlin), 1973.

PAPO 65   Papoulis, A., *Probability, Random Variables, and Stochastic Processes*, McGraw-Hill (New York), 1965.

PROH 63   Prohorov, Y., "Transient Phenomena in Processes of Mass Service," (in Russian), *Litovsk. Mat. Sb.*, **3**, 199–205 (1963).

REIS 74   Reiser, M., and H. Kobayashi, "Accuracy of the Diffusion Approximation for Some Queuing Systems," *IBM Journal of Research and Development*, **18**, 110–124 (1974).

ROSS 74   Ross, S. M., "Bounds on the Delay Distribution in GI/G/1 Queues," *Journal of Applied Probability*, **11**, 417–421 (1974).

SUZU 70   Suzuki, T., and Y. Yoshida, "Inequalities for Many-Server Queue and Other Queues," *Journal of the Operations Research Society of Japan*, **13**, 59–77 (1970).

SYSK 62   Syski, R., *Introduction to Congestion Theory in Telephone Systems*, Oliver and Boyd (London), 1962.

WONG 74   Wong, D. K., "A Discrete Approximation for G/G/1 Queue," M.S. Thesis, Computer Science Department, School of Engineering and Applied Science, University of California at Los Angeles, December 31, 1974.

EXERCISES

2.1. Show that the mean waiting time obtained from the heavy traffic approximation is too large by no more than a mean interarrival time for M/G/1.

2.2. Consider a G/D/1 system in which the constant service time has value $a+c$ and the interarrival time is uniformly distributed between c and $a+b+c$, where a, b, and c are all non-negative constants.
   (a) What relationship must exist among the constants if the system is to be stable?
   (b) Find W_U.
   (c) Find W_M.
   (d) In solving part (c), prove that the solution to Eq. (2.35) is unique in this case.

2.3. Joe is hired to measure the average length of the queue in front of the factory emergency room. Emergencies occur at an average of 6 emergencies per hour (with second moment $100\,\text{min}^2$) and they take 5 min on the average to treat (with variance $50\,\text{min}^2$). Assume this queue behaves as a stationary G/G/1 system.

   After many weeks, Joe puts out a report in which he claims that the average queue length he measured is 1.05. Do you believe him? Why?

2.4. Consider a G/G/1 system such that, for $0 \le \alpha \le 1$,

$$A(t) = \begin{cases} 0 & t < 0 \\ \alpha & 0 \le t < T_0 \\ 1 & T_0 \le t \end{cases} \qquad B(x) = \begin{cases} 0 & x < -(1-\alpha)\log\alpha \\ 1 & -(1-\alpha)\log\alpha \le x \end{cases}$$

   (a) Find $\bar{t}$, σ_a^2, $\bar{x}$, σ_b^2, and ρ in terms of α and T_0. What relation between α and T_0 must be true for stability?
   (b) Express the upper bound W_U in terms of α and T_0.
   (c) For a given value of T_0, find that value of α which minimizes W_U.
   (d) For this value of α, find W_U in terms of T_0.
   (e) For this value of α, find that value of T_0 which maximizes W_U. What value do we now get for W_U?

2.5. Consider a G/G/1 system with bulk arrivals, where the average bulk size is $\bar{g}$ and the variance is σ_g^2. Assume that $A(t)$ is $\bar{t}$-MRLA, where $\lambda = 1/\bar{t}$ is the mean arrival rate of groups. Find upper and lower bounds on W_g, the mean time a group spends in the queue until the first of the group's members begins service.

2.6. Let us derive W_U in an alternate fashion.

    (a) It is clear for G/G/1 that a_0 ($= P$ [arrival finds system empty] $= P[\tilde{y} > 0]$) is such that $0 < a_0 < 1$. Show that $E[\tilde{y}^k]$ is such that $\overline{y^k} = a_0 \overline{I^k}$. From these, establish a simple lower bound on $\bar{I}$. Also give a simple lower bound on $\overline{I^2}$ in terms of $\bar{I}$.

    (b) From (a), establish a lower bound on the mean residual life of the idle time in terms of λ and ρ only.

    (c) Using (b) in Eq. (1.132) prove the basic upper bound in Eq. (2.22).

2.7. Consider the waiting time variance, σ_w^2.

    (a) By first cubing Eq. (1.125) and then forming expectations in the limit as $n \rightarrow \infty$, express σ_w^2 in terms of the first three moments of $\bar{t}$, $\bar{x}$, and I.

    (b) From (a), proceed as in Sections 2.2 and 2.3 to show Eq. (2.42).

2.8. We wish to prove Eq. (2.47) for the queue γ-MRLA/G/1.

    (a) Let $\tilde{w} + \tilde{x}$ have the PDF $S(x) - P[\tilde{w} + \tilde{x} \le x]$. Prove that

$$P[\tilde{y} > y] = \int_0^\infty [1 - A(y + x)] \, dS(x)$$

    (b) Show that the idle time I must obey

$$P[I > y] = \frac{1}{a_0} P[\tilde{y} > y]$$

    (c) Using the γ-MRLA properties of $A(t)$ show the following, using (a) and (b) above:

$$\int_t^\infty P[I > x] \, dx \le \gamma P[I > t]$$

    (d) Form the mean residual life for I from (c) and show

$$\frac{\overline{I^2}}{2\bar{I}} \le \gamma$$

2.9. Using an approach similar to that in Exercise 2.8, prove for G/G/1 where $A(t)$ has DMRL, that when $t \geq 0$,

$$\int_t^\infty \frac{P[I>x]}{P[I>t]}\,dx \leq \int_t^\infty \frac{1-A(x)}{1-A(t)}\,dx$$

2.10. Now we consider a G/G/1 system where $A(t)$ has IFR.

(a) Beginning with the expression for $P[I>t]$ from part (b) of Exercise 2.8, find the failure rate for I.

(b) Using the IFR property for $A(t)$, show for $\varepsilon > 0$ that

$$\frac{P[I>t+\varepsilon]}{1-A(t+\varepsilon)} \leq \frac{P[I>t]}{1-A(t)}$$

(c) From (b), show that the following determinant must be nonpositive:

$$\det \begin{bmatrix} \int_t^\infty P[I>x]\,dx & \int_0^t P[I>x]\,dx \\ \int_t^\infty [1-A(x)]\,dx & \int_0^t [1-A(x)]\,dx \end{bmatrix} \leq 0$$

(d) From (c) show the final result for IFR/G/1,

$$\int_t^\infty \frac{P[I>x]}{\bar{I}}\,dx \leq \int_t^\infty \frac{1-A(x)}{\bar{t}}\,dx$$

2.11. Once again, we consider an IFR/G/1 system.

(a) Prove for any random variable X with second moment $\overline{X^2}$ and PDF $F(x)$, that

$$\int_0^\infty \int_t^\infty [1-F(x)]\,dx\,dt = \frac{\overline{X^2}}{2}$$

(b) Using the final result from Exercise 2.10, and (a) above, prove Eq. (2.50).

2.12. We wish to solve the system given in Eqs. (2.77)–(2.78).

(a) Find $P(z) = \sum_{k=0}^\infty p(k)z^k$ in terms of $p(0)$.

(b) Evaluate $p(0)$ and find $\{p(k)\}$ explicitly.

2.13. Repeat the solution of Exercise 2.12 (for the example of Section 2.6) using the method of spectrum factorization.

2.14. As in Section 2.6, consider a discrete queue for which

$$A(t) = \begin{cases} 0 & t<0 \\ \alpha & 0 \leq t < 2 \\ 1 & 2 \leq t \end{cases} \qquad B(x) = \begin{cases} 0 & x<1 \\ 1 & 1 \leq x \end{cases}$$

(a) Find ρ.

(b) Find $c(k)$.

(c) Assume $w_0 = 0$ with probability one. Find and draw $p_n(k)$ for $n = 0, 1, 2, 3$.

(d) Write the equilibrium equations for $p(k)$.

(e) Using z-transforms, solve for $p(k)$ explicitly.

2.15. Repeat the previous exercise for

$$A(t) = \begin{cases} 0 & t < 1 \\ \frac{1}{2} & 1 < t < 2 \\ 1 & 2 \le t \end{cases} \qquad B(x) = \begin{cases} 0 & x < 0 \\ \frac{1}{3} & 0 \le x < 1 \\ \frac{2}{3} & 1 \le x < 2 \\ 1 & 2 \le x \end{cases}$$

2.16. Repeat Exercise 2.14 for

$$a(k) = \begin{cases} \frac{1}{3} & k = 0 \\ \frac{2}{3} & k = 1 \end{cases} \qquad b(k) = \begin{cases} \frac{1}{2} & k = 0 \\ \frac{1}{2} & k = 1 \end{cases}$$

2.17. For the example of Section 2.6,

(a) Give an upper bound on W.

(b) Give the strongest lower bound you can for W, using the techniques from Section 2.3.

(c) Find W exactly (use the results from Exercise 2.12).

2.18. Consider the G/G/1 system of Exercise 2.4 with the values for α and T_0 as found in parts (c) and (e). If we try to solve this system by the method of Section 2.6, what problems do we encounter?

2.19. Express Eq. (2.132) in dimensionless form.

2.20. (a) Find an upper bound in terms of W_U for the root σ associated with G/M/1.

(b) Repeat for G/M/m.

2.21. In this exercise, we develop some of the material for the diffusion approximation to M/G/1.

(a) From the Fokker–Planck equation (2.133) form $F^{**}(r, s)$ given in Eq. (2.134) and show that

$$F^{**}(r, s) = \frac{2}{\sigma^2} \left[\frac{C_1 + rC_2 - e^{rw_0}}{r^2 - (2m/\sigma^2)r - (2/\sigma^2)s} \right]$$

where w_0 is the initial backlog, and C_1 and C_2 are constants with respect to r.

(b) Clearly, r_1 and r_2 as given in Eq. (2.136) are the denominator roots. Also define η as in Eq. (2.137). Establish that $\eta > 0$ for $0 \le \rho$.

(c) By setting $r = \eta$, find a relation between C_1 and C_2.

(d) What value must $sF^{**}(0, s)$ take on?

(e) Using (d), solve for C_1. From C_1 and (c), solve for C_2. We have now proven Eq. (2.135).

(f) Let $w_0 = 0$. Expand $F^{**}(r, s)$ in partial fractions. Observing that $r_1 r_2 = -2s/\sigma^2$, prove Eq. (2.141).

(g) Show that $\lim_{s \to 0} r_2 = -s/m$ and $\lim_{s \to 0} r_1 = 2m/\sigma^2$.

(h) From (g) show that Eq. (2.142) must hold for $\rho < 1$.

(i) By direct calculation, prove Eq. (2.147).

(j) For $w_0 = 0$ and $\rho < 1$, prove Eq. (2.148), noting that $\eta = r_2$ and $r_1 r_2 = -2s/\sigma^2$ again.

(k) Show that the scaled version of Eq. (2.148) is as given in Eq. (2.155).

(l) For $\rho > 1$, prove Eq. (2.152).

2.22. Show that Eq. (2.157) is the inverse of the transform given in Eq. (2.155). For this, use the common properties of transforms (see [KLEI 75] Table I.3) and the helpful transform pair

$$\frac{1}{\sqrt{\pi t}} - 2ae^{a^2 t}[1 - \Phi(a\sqrt{2t})] \Leftrightarrow \frac{1}{a + \sqrt{s}}$$

2.23. Consider the *third*-order approximation to Eq. (2.111) in which we permit the first three terms $A_n(w, t)$, for $n = 1, 2, 3$ to be nonzero, and assume all the rest to be zero ($n > 3$). Let us study this solution for the unfinished work in an M/G/1 system in equilibrium ($\rho = 1 - \varepsilon$ where $1 \gg \varepsilon > 0$) [COHE 73].

(a) Show that the general dimensionless [see Eqs. (2.127)–(2.128)] equilibrium solution must be

$$F(w') = C_1 + C_2 e^{s_1 w'} + C_3 e^{s_2 w'}$$

where s_1 and s_2 are the roots of

$$\frac{\gamma}{4} s^2 - \frac{1}{2} s - 1 = 0$$

and

$$\gamma = \frac{2}{3} \frac{\overline{x^3}}{\lambda (\overline{x^2})^2} (1 - \rho)$$

(b) Clearly, the two roots have opposite sign. Let $s_1 < 0$, $s_2 > 0$. Show that

$$F(w') = 1 - e^{s_1 w'} \qquad w' \geq 0$$

(c) Consider the approximation

$$s_1 \cong a + b\gamma + c\gamma^2$$

and find the best values a, b, c (note that $|\gamma| \ll 1$).

(d) From parts (b) and (c), find an explicit form for $F(w')$ and compare to Eq. (2.131).

2.24. Consider a diffusion approximation to $U(t)$ for M/G/1 [HEYM 74]. Consider a busy period initiated by a customer whose service time is x sec. Let $g(y; x)$, $G^*(s; x)$, and $g_k(x)$ be the pdf, its Laplace transform, and the kth moment of the duration of such a busy period when it is approximated by a diffusion process with mean $m = \rho - 1$ and variance $\sigma^2 = \lambda x^2$. It has been shown [COX 65] that

$$G^*(s; x) = \exp\left\{ -\frac{mx}{\sigma^2}\left[1 - \sqrt{1 + \frac{2\sigma^2 s}{m^2}} \right] \right\}$$

(a) Find $g_k(x)$ for $k = 1, 2, 3$.

(b) Let $g(y)$, $G^*(s)$, and g_k be the pdf, its Laplace transform, and the kth moment of the unconditional busy period duration under the diffusion approximation. Express $g(y)$ in terms of $g(y; x)$.

(c) From (a) find g_k for $k = 1, 2, 3$ and compare to the known values of the moments of the exact M/G/1 busy period.

(d) Express $G^*(s)$ in terms of $B^*(s)$.

(e) In what way is the expression in (d) superior to the corresponding expression for the exact M/G/1 system?

2.25. For the input work stream shown in Figure 2.16(b), redraw parts (c) and (d) of that figure in the case when the smoothing "filter" is such that it spreads a unit impulse uniformly over the two unit time slots surrounding that impulse (the height of this rectangular pulse will therefore be $\frac{1}{2}$. Repeat in the case where the unit impulse gets spread as a small triangle rising linearly from zero at $\frac{1}{2}$ sec prior to the impulse, to a value of 2 at the time of occurrence of the impulse, and then dropping linearly to zero at $\frac{1}{2}$ sec following the impulse.

3

Priority Queueing

Nobody likes to wait in line; however, some of us dislike it more than others. In fact, some of us dislike it so much that we are willing to do something about it. In order to improve one's position in line, one may cheat, bribe, push, or quit. A more cunning action might be to join a class structure that is afforded preferential treatment at the expense of others. Such schemes are referred to as priority queueing systems, and they form the subject of the present chapter.

Immediately when one considers priority queueing systems one naturally thinks in terms of minimizing some cost function with regard to delay for various customers. Surprisingly there exist relatively few works in the literature dealing with optimal queueing disciplines for priority groups such that some well-stated and realistic cost functions are minimized. Rather, what abounds is a vast literature on the construction of mathematical models for ingenious priority systems, followed by an analysis of the performance of such systems.*

Our purpose in this chapter is to discuss a few priority systems of interest (particularly to the author) and we in no way attempt or profess to cover the material in this field in any degree of completion. Rather, we raise questions and illustrate methods of approach that we feel are meaningful and general. These considerations lay the groundwork for some of our computer applications in Chapters 4, 5 and 6. For a much more complete work on the subject, the reader is referred to Jaiswal's book on priority queues [JAIS 68].

3.1. THE MODEL

A queueing discipline is nothing more than a means for choosing which customer in the queue is to be serviced next. This decision may be based

---

* On the other hand, a literature on the optimal *control* of queueing systems does exist. Here one is concerned with adjusting the service and arrival rates of the system under various cost structures. This material is summarized in [PRAB 73] and [CRAB 73], both of which contain useful bibliographies. Applications to closed queueing networks are given in [TORB 73]. A recent approach is also reported in [REED 74].

on any or all of the following:

1. some measure related to the relative arrival times for those customers in the queue;
2. some measure (exact value, estimate, pdf) of the service time required or the service so far received;
3. or some function of group membership.

The third case is usually referred to as a priority queueing discipline, although in this chapter we use the broader definition to include any of those three discriminators. Examples of queueing disciplines that depend only upon arrival time are first-come–first-serve (FCFS), last-come–first-serve (LCFS), and random order of service. Discrimination on the basis of service time only may take the following forms: shortest-job-first (SJF), longest-job-first (LJF), similar rules based on averages, and so on. Order of service based on an externally imposed priority class structure may take many forms as, for example, the head-of-the-line (HOL) system described below. Mixtures of these disciplines are also common, and we discuss one such mixture in Section 3.7.

We assume that arriving customers belong to one of a set of P different priority classes, indexed by the subscript p $(p = 1, 2, \ldots, P)$. We adopt the convention that the *larger* the value of the index associated with the priority group, the *higher* is the so-called priority associated with that group; that is, customers from priority group p are given preferential treatment in one form or another on the average over customers from priority group $p - 1$. We consider only equilibrium results here; however, we do encounter systems below in which some groups have no stable behavior while other groups do reach a limiting stable behavior, and it is the stable groups that we consider in such cases.

In general, then, we assume that an arriving customer is assigned a set of parameters (either at random or based on some property of the customer) that determine his relative position in the queue through the decision rule known as the queueing discipline. This position may vary as a function of time owing to the appearance of customers of higher or lower priority in the queue. At time t a customer's priority is calculated as a function of his assigned parameters, his service time and his time in the system. In fact, we associate with a customer from priority group p a numerically valued priority function $q_p(t)$ at time t. The higher the value obtained by this function, the higher is said to be the customer's priority; whenever the decision rule is called upon to select a customer for service, the choice is made in favor of that customer with the largest $q_p(t)$. All ties are broken on an FCFS basis.

We consider a fairly general model based on the system M/G/1 (in some cases, however, we constrain the system to be of the form M/M/1; at other times, we generalize to G/G/1). Thus we assume that customers from priority group p arrive in a Poisson stream at rate λ_p customers per second; each customer from this group has his service time selected independently from the distribution $B_p(x)$ with mean $\bar{x}_p$ sec. We define the following:

$$\lambda = \sum_{p=1}^{P} \lambda_p \tag{3.1}$$

$$\bar{x} = \sum_{p=1}^{P} \frac{\lambda_p}{\lambda} \bar{x}_p \tag{3.2}$$

$$\rho_p = \lambda_p \bar{x}_p \tag{3.3}$$

$$\rho = \lambda \bar{x} = \sum_{p=1}^{P} \rho_p \tag{3.4}$$

The interpretation of ρ here is, as usual, the fraction of time the server is busy (so long as $\rho < 1$). Moreover, ρ_p is the fraction of time the server is busy with customers from group p (again for $\rho < 1$). If a customer in the process of being served is liable to be ejected from service and returned to the queue whenever a customer with a higher value of priority appears in the queue, then we say that the system is a *preemptive* priority queueing system; if such is not allowed, then the system is said to be *nonpreemptive*. If only one customer is allowed in the (single) service facility at a time, then when there exists a tie between customers, the tie is broken on a first-come–first-serve basis.

3.2. AN APPROACH FOR CALCULATING AVERAGE WAITING TIMES

According to our earlier notation we have reserved the symbols W and T to denote a customer's average waiting time (in queue) and average total time in system (queue plus service), respectively; of course, the two are related through $T = W + \bar{x}$. We make the corresponding definitions for the case of priority classes, namely

$W_p \triangleq E[\text{waiting time for customers from group } p]$ (3.5)

$T_p \triangleq E[\text{total time in system for customers from group } p] = W_p + \bar{x}_p$ (3.6)

A customer's waiting time is easily decomposed into three parts: any delay he encounters due to the customer found in service upon his arrival; any delay he experiences due to customers he finds in the queue upon his arrival; and lastly, any delay due to customers who arrive after he does.

This is the basic observation from which we may establish a set of equations that define the quantities W_p or T_p.

We begin by considering the case of *nonpreemptive* systems and establish the equations for the average waiting times W_p. We study the system from the point of view of a newly arriving customer from priority group p (say); we shall refer to this customer as the "tagged" customer. We observe that the first part of the tagged customer's delay is due to the customer he finds in service; this delay will be equal to this other customer's residual life, the distribution of which will depend upon the priority group to which this other customer belongs. Let us denote by W_0 the *average delay to our tagged customer due to the man found in service.* Since ρ_i is the fraction of time that the server is occupied by customers from group i and since we have a Poisson process, then ρ_i is the probability that our tagged customer finds a type-i customer in service. In Section 1.7 we stated that with Poisson arrivals, the mean residual life of a service time as observed by an arrival is equal to the second moment of service divided by twice the first moment; these statements permit us to calculate W_0 as

$$W_0 = \sum_{i=1}^{P} \rho_i \frac{\overline{x_i^2}}{2\bar{x}_i}$$

$$= \sum_{i=1}^{P} \frac{\lambda_i \overline{x_i^2}}{2} \qquad \blacksquare \quad (3.7)$$

where $\overline{x_i^2}$ is the second moment of service time for a customer from group i.

Now we consider the second component of delay, namely, the delay due to customers found in the queue by our tagged customer who receive service before he does. We define

$$N_{ip} \triangleq \text{the number of customers from group } i \text{ found} \\ \text{in the queue by our tagged customer (from} \\ \text{group } p) \text{ and who receive service before our} \\ \text{tagged customer does} \qquad (3.8)$$

where the average of this quantity is $E[N_{ip}] \triangleq \bar{N}_{ip}$. Since the service time for any member from group i is drawn independently from $B_i(x)$, the second component of average delay to our tagged customer is given by

$$\sum_{i=1}^{P} \bar{x}_i \bar{N}_{ip} \qquad (3.9)$$

We may make similar statements regarding the third component (the delay to our tagged customer by later arrivals than he). Thus we define

$$M_{ip} \triangleq \text{ the number of customers from group } i \text{ who arrive}$$

to the system while our tagged customer (from group p) is in the queue and who receive service before he does   (3.10)

with average $\bar{M}_{ip}$. Thus we see that the third component of average delay is similar to that given in Eq. (3.9). Consequently, the total average delay in queue for our tagged customer may finally be written as

$$W_p = W_0 + \sum_{i=1}^{P} \bar{x}_i (\bar{N}_{ip} + \bar{M}_{ip}) \qquad p = 1, 2, \dots, P \qquad (3.11)$$

For any given priority queueing discipline the solution procedure then contains two steps: first, an evaluation of the averages $\bar{N}_{ip}$ and $\bar{M}_{ip}$; and second, a solution of the resulting set of equations (3.11).

In the general case both $\bar{N}_{ip}$ and $\bar{M}_{ip}$ may be expressed in terms of the average waiting times W_i, and therefore (3.11) leads to a set of simultaneous linear equations in the W_i. The simple approach herein described for calculating the average waiting times will be used in later sections of this chapter and is possible since the average of a sum is always equal to the sum of the averages. Higher moments are not so easily obtained and so in the next section we consider an approach for finding the *distribution* of waiting time for various priority groups.

The computation for *preemptive* queueing disciplines is similar to the above, but involves the additional complexity regarding how a customer recovers when he reenters service after having been preempted. Three cases are usually identified here. The first, where a customer picks up from where he left off (with perhaps a cost in time to either the customer or the system), is known as *preemptive resume*. The second and third cases assume that the customer loses credit for all service he has so far received: the second case assumes that a returning customer starts from scratch but with the same total service time requirement as he had upon his earlier visit, and this is known as *preemptive repeat without resampling*; the third case assumes that a *new* service time is chosen for our reentering customer and is referred to as *preemptive repeat with resampling*. (We study some examples of preemptive resume systems below and in Chapter 4.)

3.3. THE DELAY CYCLE, GENERALIZED BUSY PERIODS, AND WAITING TIME DISTRIBUTIONS

In this section we consider the analysis of "delay cycles," which permit us to calculate the Laplace transform for the pdf of "generalized" busy

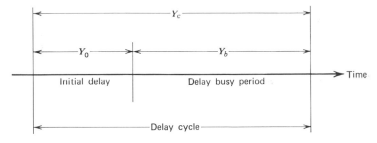

Figure 3.1 The delay cycle.

periods. From this we may obtain the Laplace transform for the waiting time density (as in Section 5.10 of Volume I [KLEI 75]) for a number of queueing disciplines. The concept of delay cycle analysis seems to have originated with Gaver [GAVE 62] (he used a notion known as "completion time"). Similar ideas appeared in Keilson [KEIL 62], who used the "basic server sojourn time," and in Avi-Itzhak and Naor [AVI 63], who used the "residence time." This work was extended by Miller [CONW 67], and it is his nomenclature (delay cycle) that we adopt here.

A delay cycle is similar to a busy period and is shown in Figure 3.1. The delay cycle Y_c consists of two portions: an initial delay of duration Y_0 and a delay busy period of duration Y_b. The initial delay is usually some special interval that may correspond to the completion of some partly completed customer or may correspond to some other "special" task. The delay busy period corresponds to the servicing of "ordinary" customers and may be viewed as a sequence of sub-busy periods*; the delay busy period ends when there are no more ordinary customers to be serviced. The generalization here over that of an ordinary busy period is that in the former we permit an arbitrary distribution for the initial delay, whereas in the latter we require that the initial delay be a service time distributed the same as the individual service times making up the elements of the delay busy period. In all cases, however, we have

$$Y_c = Y_0 + Y_b$$

The ordinary customers arrive according to a Poisson process; for purposes of this section we will assume that such customers arrive at a rate λ. We note that when the initial delay terminates, there may have accumulated during this initial delay a number of ordinary customers awaiting

---

* A sub-busy period is that interval of time which is required to service an arbitrary customer and all those (his "descendents") who enter the system during his service time or during the service time of any of his descendents. In the system M/G/1, the pdf for the duration of a sub-busy period is the same as that for a busy period. A sub-busy period is said to be "generated" by the customer who initiates the period. See Chapter 5 in [KLEI 75].

service; *each of these will generate his own sub-busy period, which, taken together, form the delay busy period.*

Now for some notation. Previously we had defined $B^*(s)$ and $G^*(s)$ as the transform for the pdf of service time and busy period durations, respectively, and the basic equation relating these two is given as Eq. (1.89). For the random variables Y_0, Y_b, and Y_c, we now define the PDF's $G_0(y)$, $G_b(y)$, and $G_c(y)$ along with the transforms (of the corresponding pdf's) denoted by $G_0^*(s)$, $G_b^*(s)$, and $G_c^*(s)$, respectively.

We assume that we are given $G_0^*(s)$, or if not we usually can calculate it. We are interested in solving for $G_b^*(s)$ and $G_c^*(s)$ in terms of the known functions $B^*(s)$, $G^*(s)$, and $G_0^*(s)$ where the relation given in Eq. (1.89) will be required to evaluate $G^*(s)$ from $B^*(s)$. The derivations we give here are rather abbreviated since they closely parallel the development of Section 5.8 of Volume I [KLEI 75]. We begin with the calculation of $G_b^*(s) = E[e^{-sY_b}]$. Let N_0 be the number of ordinary customer arrivals during the interval Y_0. We condition the transform we are seeking on Y_0 and N_0 to arrive at the simple expression

$$E[e^{-sY_b} \mid Y_0 = y, N_0 = n] = [G^*(s)]^n$$

This last follows from the fact that all n sub-busy periods (one is generated by each of the arrivals during Y_0) are independent and each is distributed exactly the same as a busy period. We proceed to uncondition first on N_0,

$$E[e^{-sY_b} \mid Y_0 = y] = \sum_{n=0}^{\infty} [G^*(s)]^n \frac{(\lambda y)^n}{n!} e^{-\lambda y}$$

$$= e^{-[\lambda - \lambda G^*(s)]y}$$

and finally on Y_0 to give

$$E[e^{-sY_b}] \triangleq G_b^*(s) = \int_{y=0}^{\infty} e^{-[\lambda - \lambda G^*(s)]y} \, dG_0(y)$$

This integral we recognize as the transform of the pdf for Y_0, and we therefore have the final result

$$G_b^*(s) = G_0^*(\lambda - \lambda G^*(s)) \qquad\blacksquare \qquad (3.12)$$

Now for $G_c^*(s)$. Proceeding as above we have

$$E[e^{-sY_c} \mid Y_0 = y, N_0 = n] = e^{-sy} [G^*(s)]^n$$

and removing the conditions on N_0 and Y_0 we have

$$G_c^*(s) \triangleq E[e^{-sY_c}] = \int_{y=0}^{\infty} e^{-sy} \sum_{n=0}^{\infty} [G^*(s)]^n \frac{(\lambda y)^n}{n!} e^{-\lambda y} \, dG_0(y)$$

which yields the result

$$G_c^*(s) = G_0^*(s + \lambda - \lambda G^*(s)) \qquad \blacksquare \quad (3.13)$$

Thus Eqs. (3.12) and (3.13) provide the defining equations for our unknowns where, of course, $G^*(s)$ is given in Eq. (1.89), that is,

$$G^*(s) = B^*(s + \lambda - \lambda G^*(s)) \qquad (3.14)$$

We note in the special case when Y_0 is distributed as an ordinary customer's service time that Y_c will merely be a regular busy period and Eq. (3.13) will then reduce to Eq. (3.14).

As we shall soon see, the delay cycle analysis is an extremely powerful method for obtaining results in many queueing systems, especially those with priorities.

3.4. CONSERVATION LAWS

In most physical systems, "you don't get something for nothing." So too in priority queueing systems—preferential treatment given to one class of customers is afforded at the expense of other customers. In a real sense then we "borrow from Peter to pay Paul." In this section we investigate such invariances or conservations within priority queueing systems.

Our conservation relations are based upon the fact that the unfinished work $U(t)$ during any busy period is independent of the order of service so long as the system is "conservative." By conservative we mean that no work (service requirement) is created or destroyed within the system; for example, destruction of work would occur if a customer were to leave the system before completing his service and the creation of work might correspond to a server standing idle in the face of a nonempty queue. Thus we consider only work-conserving systems in this section. The simplest case to consider is the FCFS system about which we know so much already. Most priority queueing systems are compared to the FCFS system and we see below that its performance enters our conservation relationships in a very natural way.

We begin by observing that the *distribution of waiting time* will indeed depend upon the order in which service is given. However, we now show that *so long as the queueing discipline selects customers in a way that is independent of their service time (or any measure of their service time) then the distribution of the number in the system will be invariant to the order of service; the same will also be shown to be true for the average waiting time of customers.* Let us consider the M/G/1 queue. For this system we have the basic relation given in Eq. (1.81). The definition for q_n was given as the number of customers left behind by the departure of C_n. Let us change our point of view now and redefine this quantity to refer to the number of customers left behind by the departure of the nth *departing* customer (thereby allowing arbitrary order of service). Similarly v_n is to be interpreted as the number of customers arriving during the service of the nth customer to be served. It is clear that the relationship (1.81) now holds even for these more general queueing disciplines (and reduces to our former interpretation for the FCFS system). The identical steps (see [KLEI 75]) that take us from this relationship to an expression for the z-transform $[Q(z)]$ of the number of customers in an FCFS system will now take us from that defining equation to $Q(z)$ for a system with arbitrary order of service. Thus we can state for any queueing discipline whose decision rules are independent of a customer's service time that we must have the following as the z-transform for the number of customers in system:

$$Q(z) = B^*(\lambda - \lambda z) \frac{(1-\rho)(1-z)}{B^*(\lambda - \lambda z) - z} \tag{3.15}$$

where the notation here is the same as in Section 1.7. Therefore we immediately have complete information about the number in system. Bear in mind that this independence of order of service for number in system has only been shown to hold when the decision rule is itself independent of any aspect of service time of the customers.

Let us now conserve the unfinished work $U(t)$. From its definition, $U(t)$ is a function which (a) decreases at a rate of 1 sec/sec whenever $U(t) > 0$, (b) remains saturated at zero when it hits the horizontal axis, and (c) takes vertical jumps at the arrival instants in amounts equal to the service requirements brought in by the arrivals. Thus *it is clear that regardless of the order of service (service-dependent or not) $U(t)$ will not change; this is true for $G/G/1$.* For M/G/1 the following conservation law was first stated and proven in [KLEI 64a, 65]:

The M/G/1 Conservation Law.   For any M/G/1 system and any non-preemptive work-conserving queueing discipline it must be that

$$\sum_{p=1}^{P} \rho_p W_p = \begin{cases} \dfrac{\rho W_0}{1-\rho} & \rho < 1 \\ \infty & \rho \geq 1 \end{cases} \quad \blacksquare \tag{3.16}$$

[Recall from Section 1.7 and Eq. (3.7) that W_0 represents the residual life of the customer found in service upon an arrival's entry.] Thus this weighted sum of the waiting times W_p *can never change* no matter how sophisticated or elaborate the queueing discipline may be. Let us prove the validity of this conservation law. If at time t there are $N_p(t)$ customers from group p in the queue and if the ith of these $[i-1, 2, \ldots, N_p(t)]$ is to have a service time x_{ip} and if x_0 represents the work yet to be done on the man in service at time t (that is, his residual service time) then we may say

$$U(t) = x_0 + \sum_{p=1}^{P} \sum_{i=1}^{N_p(t)} x_{ip}$$

regardless of the order of service. Then taking expectations on both sides we have*

$$E[U(t)] = W_0 + \sum_{p=1}^{P} \sum_{n_p=0}^{\infty} P[N_p(t) = n_p] \sum_{i=1}^{n_p} E[x_{ip}]$$

We observe that $E[x_{ip}] = \bar{x}_p$ independent of the index i. With t taken at random (and large) we may write $\bar{U} \triangleq \lim_{t \to \infty} E[U(t)]$, which will be the limiting average of the unfinished work. Thus we may write

$$\bar{U} = W_0 + \lim_{t \to \infty} \sum_{p=1}^{P} \sum_{n_p=0}^{\infty} n_p P[N_p(t) = n_p] \bar{x}_p$$

$$= W_0 + \sum_{p=1}^{P} \bar{x}_p E[N_p]$$

However, Little's result [Eq. (1.31)] tells us that $E[N_p] = \lambda_p W_p$ since this result is valid for individual priorities as well. Thus we conclude that

$$\bar{U} = W_0 + \sum_{p=1}^{P} \rho_p W_p \qquad (3.17)$$

Now since $\bar{U}$ is independent of the order of service we may as well use our FCFS result, which states that for Poisson arrivals the average unfinished work (the average virtual waiting time) must equal the average waiting time for customers, which we denote by W. This quantity is given in Eq. (1.82); here the second moment of service time is easily expressed in terms of the second moment associated with each group's service time, namely

$$\overline{x^2} = \sum_{p=1}^{P} \frac{\lambda_p}{\lambda} \overline{x_p^2} = \frac{2W_0}{\lambda}$$

and so we may write

$$\bar{U} = W = \frac{W_0}{1 - \rho} \qquad (3.18)$$

---

* Here we take $E[x_0] = W_0$, where W_0 is defined in Eq. (3.7). This value for $E[x_0]$, which is the average unfinished work for the customer in service, is correct even for G/G/1 since we are *not* averaging over customer arrival instants, but are averaging uniformly over all time; as we know, Poisson arrivals also observe the system uniformly over all time and for this reason our result is the same as the mean residual service time seen by Poisson arrivals.

If we use this value for $\bar{U}$ in Eq. (3.17) we have the conservation law given in Eq. (3.16) (where for $\rho \geq 1$ the value of ∞ is obvious). Q.E.D.

Thus the conservation law puts a linear equality constraint on the set of average waiting times W_p. We see that any attempt to modify the queueing discipline so as to reduce one of the W_p will force an increase in some of the other W_p; however, this need not be an "even trade" since the weighting factors for the W_p are generally distinct. Now in the special case where $\bar{x}_p = \bar{x}$ for all p then the conservation law gives (for $\rho < 1$)

$$\sum_{p=1}^{P} \lambda_p W_p = \frac{\lambda W_0}{1-\rho} \qquad \bar{x}_p = \bar{x} \qquad (3.19)$$

However, Little's result gives us $\lambda_p W_p = E[N_p]$ again and so

$$\sum_{p=1}^{P} E[N_p] = \frac{\lambda W_0}{1-\rho}$$

But this sum is merely the average total number in queue for which we have the notation $\bar{N}_q = E[\text{number in queue}]$ giving

$$\bar{N}_q = \frac{\lambda W_0}{1-\rho} = \text{constant} \bigg|_{\substack{\text{queue} \\ \text{discipline}}} \qquad \bar{x}_p = \bar{x} \qquad (3.20)$$

and of course from Little's result we further have

$$W = \frac{W_0}{1-\rho} = \text{constant} \bigg|_{\substack{\text{queue} \\ \text{discipline}}} \qquad \bar{x}_p = \bar{x} \qquad (3.21)$$

Thus in the special case where $\bar{x}_p = \bar{x}$, the average number in queue and the average waiting time in queue are independent of the queue discipline. Note the correspondence between this statement and our statement above regarding the invariance of the distribution of number of customers in the system (when the order of service was independent of service time). When $\bar{x}_p = \bar{x}$, then, at least regarding first moments, all customers behave the same with regard to service time and therefore order of service does not depend on average service time; this apparently leads to the invariance properties mentioned in Eqs. (3.20) and (3.21). If the average service times are not equal then it is not true in general that the average queue size and average waiting time are independent of queue discipline [which clearly depends upon (average) service time]. These same statements, of course, apply to the average number in system (since we know that $\bar{N}_q + \rho = \bar{N}$) and for the average time in system (since $W + \bar{x} = T$). Note further that we are not claiming $W_p = W$ but merely that the sum in Eq. (3.19) is constant.

This conservation law has been extended [SCHR 70] to the case G/G/1 where not only are we dropping the Poisson arrival assumption but also no assumption regarding independence is required; what is required is that equilibrium distributions exist. In the above proof for M/G/1, we first used the Poisson arrival assumption following Eq. (3.17); however, that equation itself is good for G/G/1, and this gives us the generalized version of the conservation, namely

The G/G/1 Conservation Law

$$\sum_{p=1}^{P} \rho_p W_p = \bar{U} - W_0 \qquad \blacksquare \quad (3.22)$$

Of course for each problem $\bar{U}$ must be evaluated since this quantity is in general unknown for the system G/G/1! However, the notion of conservation stands out: Given a specific work-conserving G/G/1 queueing system with a nonpreemptive priority queueing discipline then the linear equality constraint given in Eq. (3.22) must be satisfied regardless of that queueing discipline. We note of course that Eq. (3.22) reduces to Eq. (3.16) in the case M/G/1.

It is not hard to conceive of priority queueing disciplines in which some groups experience finite waiting times while other groups find themselves in the abominable situation of experiencing infinite average waiting times; in fact the system considered below in Section 3.6 is an example of this. One inquires as to whether there exists a form of conservation law for those groups that do experience finite waits even in this unstable situation. We find in fact there do exist appropriate conservation laws that tell us more than does Eq. (3.16) in the case $\rho \geq 1$. This material may be found in [KLEI 65] and is elaborated on in Exercises 3.5 and 3.6 at the end of this chapter.

Equation (3.22) certainly applies for nonpriority G/G/1 systems as well. In this case from Eq. (3.7) we see that $W_0 = \lambda \overline{x^2}/2$ and so the conservation law for the nonpriority G/G/1 system becomes

$$\bar{U} = \rho W + \frac{\overline{x^2}}{2\bar{t}} \qquad \blacksquare \quad (3.23)$$

where we have written $\bar{t} = 1/\lambda$ to correspond with our more usual notation for G/G/1. Brumelle [BRUM 69] also derives this as a special case of his general class of formulas of the form, "time average = $\lambda \cdot$ customer average." The most famous example of this form of equation is of course Little's formula; so too is Eq. (3.23), which is also rather important and for some reason seems to be somewhat obscure in the queueing literature. Brumelle permits the case where there is dependence among the basic processes, and in this case the term ρW is replaced by $E[\tilde{x}\,\tilde{w}]/\bar{t}$.

3.5. THE LAST-COME–FIRST-SERVE QUEUEING DISCIPLINE

Let us return to the M/G/1 queue again and consider the case in which service is given to the most recent arrival on a nonpreemptive basis. Here we have $P = 1$ (no externally assigned priorities). This order of service is not uncommon, strange as it may appear; for example, any push-down stack operates in this fashion. Since the decision rule is independent of service time, we see immediately that the average queue size and the average waiting time must be the same as for FCFS [see Eqs. (3.20) and (3.21)]. Moreover, we know that Eq. (3.15) gives the distribution of number in the system. However, we suspect that the waiting time distribution differs from the FCFS case, and it is this which we solve for below. Our intuiton correctly suggests that this rule will give a large variance of waiting time even though the average is the same as FCFS.

This queueing discipline lends itself especially well to analysis. We observe that a new arrival is in no way affected by the queue size he finds upon his entry to the system; only the customer found in service can make him wait and the balance of his delay is due to arrivals that enter the system after he does but prior to his initiation of service. This is a perfect set-up for the delay cycle analysis of Section 3.3 where the initial delay is the residual life of the customer found in service and the delay busy period is the interval required to empty the system of all those arrivals who follow him prior to his entry into service, at which point his service commences. The Laplace transform for the residual life pdf is given in the footnote on page 16; we rewrite this transform using our notation for delay cycle analysis as

$$G_0{}^*(s) = \frac{1 - B^*(s)}{s\bar{x}}$$

Moreover, $G_c{}^*(s)$, the Laplace transform for the delay cycle pdf, is given in terms of $G_0{}^*(s)$ and $G^*(s)$ in Eqs. (3.13) and (3.14). The delay cycle here corresponds to the waiting time for our customer in this LCFS system, and so using these transform relations we may write down the conditional transform for waiting time as

$$E[e^{-s\tilde{w}} \mid \text{system busy upon arrival}] = G_c{}^*(s)$$
$$= G_0{}^*(s + \lambda - \lambda G^*(s))$$
$$= \frac{1 - B^*(s + \lambda - \lambda G^*(s))}{[s + \lambda - \lambda G^*(s)]\bar{x}}$$

Then from Eq. (3.14) we may simplify the numerator to give

$$E[e^{-s\tilde{w}} \mid \text{system busy upon arrival}] = \frac{1 - G^*(s)}{[s + \lambda - \lambda G^*(s)]\bar{x}}$$

If we now uncondition this expression, we find that with probability $1 - \rho$ our customer has a waiting time of zero and with probability ρ he has a waiting time whose transform is given in this last equation. Thus

$$W^*(s) = E[e^{-s\tilde{w}}] = 1 - \rho + \frac{\lambda[1 - G^*(s)]}{s + \lambda - \lambda G^*(s)} \qquad \blacksquare \qquad (3.24)$$

and this is the result we were seeking. We note that it differs significantly from the Pollaczek–Khinchin transform equation for FCFS given in Eq. (1.85). In Exercise 3.4 we compare the mean and variance of waiting time for these two systems. We find that the first moments are of course the same (as we stated earlier) but that the variance for LCFS is larger than for FCFS.

Let us now consider the case where an *external* priority structure is imposed.

3.6. HEAD-OF-THE-LINE PRIORITIES

Among the queuing disciplines that impose an *external* priority structure on the arriving customers, the head-of-the-line (HOL) priority queueing system is perhaps the most common and most natural. This system, first studied in 1954 by Cobham [COBH 54] is known also by the name of strict priority queueing or fixed priority queueing. The system structure is given in Figure 3.2. In this system customers queue according to priority groups and are strictly separated on the basis of the group to which they belong. Thus an arrival from group p joins the "torso" of the queue behind all customers from group p (and higher) and in front of all customers from group $p - 1$ (and lower). The value of one's priority in this case remains constant in time and so we may take the priority

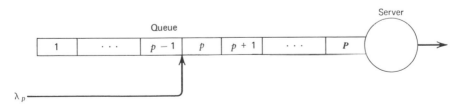

Figure 3.2   Head-of-the-line priority queue.

function to be

$$q_p(t) = p \tag{3.25}$$

Let us now use the method of Section 3.2 to derive the average waiting time W_p for members of the pth priority group in the case of a non-preemptive HOL system. Equation (3.11) is our point of departure and we must evaluate the two functions $\bar{N}_{ip}$ and $\bar{M}_{ip}$ that represent the average number of customers from priority group i who get served before our tagged customer (from group p) and who are present in the queue upon his arrival ($\bar{N}_{ip}$) or who arrive while he is in the queue ($\bar{M}_{ip}$). Because of the strict order of queueing and under the assumption that customers within the same priority group get served according to an FCFS rule, it is clear that

$$\bar{N}_{ip} = 0 \qquad i = 1, 2, \ldots, p-1$$
$$\bar{M}_{ip} = 0 \qquad i = 1, 2, \ldots, p$$

All customers from group p and higher who are present in the queue upon our tagged customer's arrival must certainly get served before he does; from Little's result we know that on the average there will be $\lambda_i W_i$ customers from the ith group present in the queue when our tagged customer arrives and therefore,

$$\bar{N}_{ip} = \lambda_i W_i \qquad i = p, p+1, \ldots, P \tag{3.26}$$

Similarly all customers from groups $p+1, p+2, \ldots, P$ who join the system while our tagged customer is in the queue will also be served before he is; since he spends on the average W_p sec in the queue and since each group's arrival process is independent of queue size we know that there will on the average be $\lambda_i W_p$ customer arrivals from the ith group while our tagged customer waits on queue. Therefore,

$$\bar{M}_{ip} = \lambda_i W_p \qquad i = p+1, p+2, \ldots, P \tag{3.27}$$

Thus for the nonpreemptive HOL system Eq. (3.11) becomes

$$W_p = W_0 + \sum_{i=p}^{P} \bar{x}_i \lambda_i W_i + \sum_{i=p+1}^{P} \bar{x}_i \lambda_i W_p \qquad p = 1, 2, \ldots, P \tag{3.28}$$

By straightforward arguments we have arrived at the set of defining equations for our unknowns W_p. Solving Eq. (3.28) for W_p we have

$$W_p = \frac{W_0 + \sum_{i=p+1}^{P} \rho_i W_i}{1 - \sum_{i=p}^{P} \rho_i} \qquad p = 1, 2, \ldots, P \tag{3.29}$$

This set of equations may be solved recursively with no difficulty since we have a triangular set; that is, we first find W_P and from this find W_{P-1}, and so on. We find it convenient to define

$$\sigma_p = \sum_{i=p}^{P} \rho_i \qquad (3.30)$$

Solving recursively, we obtain the solution

$$W_p = \frac{W_0}{(1-\sigma_p)(1-\sigma_{p+1})} \qquad p = 1, 2, \ldots, P \qquad \blacksquare \quad (3.31)$$

This last equation was one of Cobham's principal contributions to this problem, and its form is rather suggestive. In particular we see the effect of those customers of equal or higher priority *present* in the queue when our customer arrives as given by the denominator term $1-\sigma_p$ and also the effect of customers of higher priority *arriving* during our customer's queueing time as given by the denominator term $1-\sigma_{p+1}$. Furthermore we notice that W_p does not depend on customers from lower priority groups (that is, for $i = 1, 2, \ldots, p-1$) except for their contribution to the numerator W_0. Thus the solution given in Eq. (3.31) demonstrates that some W_p may be finite (for $p \geq$ some critical value) while other lower priority groups may be experiencing unstable (unbounded) queueing times.† Figure 3.3(a) demonstrates this stable unstable behavior for a system with $P = 5$ groups. In this figure we have plotted the normalized waiting time $W_p/\bar{x}$ since this is a useful dimensionless form. Also in this figure we gain our first experience with the conservation law [Eq. (3.16)] by observing the dashed line that represents the average waiting time for the FCFS system; this dashed curve is a plot of Eq. (3.18). In Figure 3.3(b) the same curves are shown on an expanded scale; there one may observe the way in which the conservation law is functioning. In particular if one measures these curves, it will be seen that $\sum_p (\rho_p/\rho)W_p = W_0/(1-\rho)$; for $\rho > 1$ it is clear that this average blows up to infinity.

From the method described in Section 3.3 it is possible to find the distribution of waiting time for each priority group. Let us denote the Laplace transform for the pth group's waiting time in queue by $W_p^*(s)$. The solution is [CONW 67, KEST 57]:

$$W_p^*(s) = \frac{(1-\rho)[s + \lambda_{II} - \lambda_{II}G_{II}^*(s)] + \lambda_L[1 - B_L^*(s + \lambda_H - \lambda_H G_H^*(s))]}{s - \lambda_p + \lambda_p B_p^*(s + \lambda_H - \lambda_H G_H^*(s))}$$

$$(3.32)$$

† We note that the pth group experiences finite waiting time so long as $\rho < 1 + \rho_1 + \rho_2 + \cdots + \rho_{p-1}$.

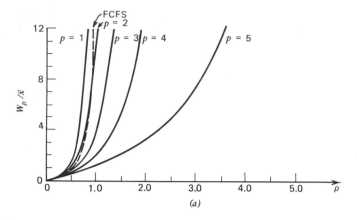

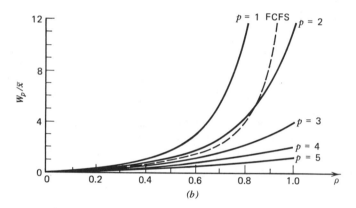

Figure 3.3   HOL with no preemption: $P = 5$, $\lambda_p = \lambda/5$, $\bar{x}_p = \bar{x}$.

where

$$\lambda_H = \sum_{i=p+1}^{P} \lambda_i \tag{3.33}$$

$$\lambda_L = \sum_{i=1}^{p-1} \lambda_i \tag{3.34}$$

$$B_H{}^*(s) = \sum_{i=p+1}^{P} \frac{\lambda_i}{\lambda_H} B_i{}^*(s) \tag{3.35}$$

$$B_L{}^*(s) = \sum_{i=1}^{p-1} \frac{\lambda_i}{\lambda_L} B_i{}^*(s) \tag{3.36}$$

$$G_H{}^*(s) = B_H{}^*(s + \lambda_H - \lambda_H G_H{}^*(s)) \tag{3.37}$$

and of course by our usual notation $B_i^*(s)$ corresponds to the Laplace transform of the pdf for the ith group's service time. In these definitions we have used the subscript H to denote the set of priority classes higher than our tagged unit and the subscript L to denote those lower. We observe that the definition in Eq. (3.37) is the same as the functional equation given in (3.14); also the form for $W_p^*(s)$ is surprisingly similar to that for the Pollaczek–Khinchin transform equation for waiting time in the FCFS system. Specifically we note that in the case $P = 1$ then $\lambda_H = \lambda_L = B_H^*(s) = B_L^*(s) = G_H^*(s) = 0$ and $\lambda_p = \lambda$, and so $W_1^*(s)$ reduces to the P-K transform equation for the FCFS system, as of course it must. From Eq. (3.32) one may calculate the mean waiting time given in Eq. (3.31) by differentiation.

Since the various priority groups receive different grades of service, and since they may each have different distributions of service time, then in a real sense we are discriminating on the basis of (distribution of) service time; therefore, whereas the conservation law certainly holds, the distribution for number in system will differ from that of the FCFS system. Let us in fact attempt to discriminate completely on the basis of *exactly known* service times. We may accomplish this as follows with the HOL model defined above. In particular let us create a priority queueing discipline in which highest priority is given to that job with the shortest service time (that is, an SJF system). This was considered in [PHIP 56] in which a continuum of priority classes was defined such that the class index p was defined to be any strictly decreasing function of the service time $\bar{x}$. Thus we have a model in which an entering customer whose service time is known to be exactly x sec joins the queue behind all other customers with service times less than (or equal to) x and in front of all customers in the queue with service times greater than x (note that in the case where the pdf of overall service time has impulses, then any ties are broken by the FCFS rule). This is an M/G/1 queueing system in which we assume that customer service times are chosen from $B(x)$ prior to their arrival and therefore they may be ordered in the queue immediately upon entry as described above. Customers whose service times fall in the interval $x < \bar{x} \le x + dx$ are all grouped into the same priority group, and the service time density associated with this group is merely a unit impulse at x sec of service; of course the fraction of customers who fall in this group is merely $b(x)\, dx$, and this will be an infinitesimal quantity unless $B(x)$ has a discontinuity at x. Such is the nature of our continuum of priority groups. Let us now calculate the average waiting time $W(x)$ for a customer whose service time lies in the interval $(x, x + dx)$. Recall that the priority index p is a strictly decreasing function of service time $\bar{x}$.

Therefore in the limit Eq. (3.30) becomes

$$\sum_{i=p}^{P} \rho_i \rightarrow \int_{y=0}^{x^+} \rho(y) \, dy$$

where $\rho(x) = \lambda(x)x$ and $\lambda(x) = \lambda b(x)$; this is the correct expression for ρ since the average service time for such customers is exactly x sec and the average arrival rate of such customers is $\lambda \, dB(x)/dx$. Therefore Eq. (3.31) takes on the following limiting value for $W(x)$:

$$W(x) = \frac{W_0}{\left[1 - \lambda \int_0^{x^-} yb(y) \, dy\right]\left[1 - \lambda \int_0^{x^+} yb(y) \, dy\right]} \quad \blacksquare \quad (3.38)$$

and we note that the denominator reduces to $[1 - \lambda \int_0^x yb(y) \, dy]^2$ when $B(x)$ is continuous at x. Here, as in the discrete case, our solution applies only for those "priority" groups that enjoy finite average waiting times. Thus Eq. (3.38) gives the average wait in an SJF queueing discipline for a customer whose service time is x. Note for a customer with a very long service time that $\lim_{x \to \infty} W(x) = W_0/(1-\rho)^2$, whereas for extremely short customers, we have $\lim_{x \to 0} W(x) = W_0$.

Let us now consider an HOL system with a *preemptive* queueing discipline. For this case we assume that the preemption is of the preemptive resume type. The approach here is much like that described in Section 3.2 and proceeds as follows. Recalling that T_p is the average of the total time spent in system by our tagged customer from group p, we recognize that his average delay consists of three components. The first is his average service time $\bar{x}_p$. Second, there is the delay due to the service (work) required by those customers of equal or higher priority whom he finds in the system; by our conservation results, we see that our tagged customer finds an average amount of work in the system equal to $(\sum_{i=p}^{P} \lambda_i \overline{x_i^2}/2)/(1 - \sigma_p)$, which must be done before he gets served [the mean work backlog is equal to the mean wait in M/G/1 and so we recognize this term as equal to the expression $W_0/(1-\rho)$ for a system fed only by groups $p, p+1, \ldots, P$; the other groups are completely "invisible" to our tagged customer!]. Third, he will be delayed by any customers who enter the system before he leaves and who are members of strictly higher priority groups; the average number of such arrivals from group i must be $\lambda_i T_p$, each of which delays our tagged customer by an average of $\bar{x}_i$ sec. Thus we may say

$$T_p = \bar{x}_p + \frac{\sum_{i=p}^{P} \lambda_i \overline{x_i^2}/2}{1 - \sigma_p} + \sum_{i=p+1}^{P} \rho_i T_p$$

The solution for T_p is therefore

$$T_p = \frac{\bar{x}_p(1-\sigma_p) + \sum\limits_{i=p}^{P} \lambda_i \overline{x_i^2}/2}{(1-\sigma_p)(1-\sigma_{p+1})}$$ ■   (3.39)

Let us now pose an interesting *optimization* problem whose solution is within our grasp. It seems natural for us to ask for some guidance in assigning external priorities to customers. We consider the nonpreemptive case below. Let us assume that there is a system cost (rate) of C_p dollars for each second of delay suffered by each customer from priority group p; it is then clear that the average cost per second to the system, which we denote by C, must be

$$C = \sum_{p=1}^{P} C_p \bar{N}_p$$

where $\bar{N}_p$ is merely the average number of type p customers in the system. Of course from Little's result we know that regardless of the queueing discipline it must be that $\bar{N}_p = \lambda_p T_p = \lambda_p [W_p + \bar{x}_p]$ and so we have

$$C = \sum_{p=1}^{P} \rho_p C_p + \sum_{p=1}^{P} C_p \lambda_p W_p \qquad (3.40)$$

We desire to find that nonpreemptive (work-conserving) queueing discipline which minimizes C. We will solve this problem for a given M/G/1 system with P priority groups, an average arrival rate of λ_p type-p customers per second, and a service time distribution for type-p customers given by $B_p(x)$. Let us rewrite Eq. (3.40) to bring the *constant* sum to the left-hand side as follows:

$$C - \sum_{p=1}^{P} \rho_p C_p = \sum_{p=1}^{P} (C_p/\bar{x}_p)(\rho_p W_p)$$

and it is the sum on the right-hand side that we must minimize by an appropriate choice of queueing discipline. Let $f_p = C_p/\bar{x}_p$ (a given quantity) and $g_p = \rho_p W_p$ (a design variable through W_p), and so we are asking to minimize the following sum of products $\sum_{p=1}^{P} f_p g_p$. However, from the conservation law given in Eq. (3.16) we know that

$$\sum_{p=1}^{P} g_p = \text{constant with respect to queue discipline} \qquad (3.41)$$

That is, we wish to minimize the "area" under the product of two functions, one of which itself has a constant area. Now if we reorder the subscripts so that

$$f_1 \leq f_2 \leq \cdots \leq f_P \qquad (3.42)$$

we see that the optimum way in which we can match the terms g_p against the terms f_p [under the constraint in Eq. (3.41)] is to assign as little "mass" as possible in g_P to match the largest term f_P. Having done this we will then assign as little of the remaining mass in g_{P-1} against the next largest term f_{P-1}, and so on. Now from its definition ($g_P = \rho_P W_P$) we see that since ρ_P is a given constant then we will minimize g_P by minimizing W_P. The nonpreemptive work-conserving queueing discipline that minimizes W_P is clearly HOL with the highest priority group corresponding to P (as usual). Having accomplished this, W_{P-1} may next be minimized by making this the second highest priority group in an HOL system, and so on. Thus we see that *the solution to our optimization problem is that, of all the possible nonpreemptive work-conserving queueing disciplines, the HOL discipline with the ordering given in Eq. (3.42)* is that which minimizes the average cost given in Eq. (3.40)!* This proof depended upon the conservation law but also could have been established by the more usual interchange argument [COX 61] (in which only stationary disciplines are considered—a sufficient class as recently shown [LIPP 75]). It is truly amazing that such a result is obtainable and so simply.

3.7. TIME-DEPENDENT PRIORITIES

The reason for imposing a priority structure on the customer arrivals is to provide preferential treatment to the "higher priority" groups at the expense of the "lower priority" groups. We have shown above in the case of linear cost rates that the HOL priority system is optimum in that it minimizes the average (linear) cost. However, this linear cost function is not always suitable and in fact the appropriate cost function is often unknown. In spite of this, the practical world is abundant with examples of priority queueing systems for which decisions have been made regarding the relative desired performance among classes (and which therefore imply some form of cost function, perhaps unknown to the users or the system). For example, most military systems use an HOL discipline with preemption permissible by the highest priority groups and usually with four or five groups in all. Another example is in the servicing of automobiles at the repair station in which the mechanic selects from among those automobiles waiting for service that one with perhaps the shortest (or perhaps the most expensive) service requirement. Often, therefore, rather than specifying the cost function, one is willing to specify the relative waiting times among the various priority groups. For example, a system designer may be required to synthesize a priority queueing

* Taking $\bar{x}_p = 1/\mu_p$, we have $f_p = \mu_p C_p$, and so this optimum ordering is referred to as "the μC rule."

discipline in which the desired performance is given as the ratios W_{p+1}/W_p $(p = 1, 2, \cdots, P-1)$. The designer is then faced with achieving these ratios for a given specification of the customer behavior; that is, we assume the arrivals are of the M/G/1 type with given quantities λ_p, $\bar{x}_p$, and $B_p(x)$. Therefore from their definitions the quantities ρ_p, σ_p, and W_0 are also specified [in fact, $B_p(x)$ need not be specified but only $\bar{x}_p$ and $\overline{x_p^2}$ are necessary to determine these three system parameters]. From Eq. (3.31), therefore, the behavior of W_p is completely determined for HOL and as a consequence so is the behavior of the ratios W_{p+1}/W_p. Unfortunately this freezes the design and so the independently specified ratios cannot, in general, be achieved for the HOL system!

Therefore we must introduce some additional degrees of freedom into our priority queueing discipline if we are to meet the required specifications. The time-dependent priority system described below provides a set of variable parameters b_p where $0 \le b_1 \le b_2 \le \cdots \le b_P$; these parameters are at the disposal of the designer in adjusting these relative waiting times [KLEI 64a, b].

Let us assume that some tagged customer arrives at time τ and is assigned at time t a priority $q_p(t)$ calculated from

$$q_p(t) = (t - \tau)b_p$$

where t ranges from τ until the time at which this customer's service is completed. We consider the following nonpreemptive system. Whenever the service facility is ready for a new customer, that customer with the highest instantaneous priority $q_p(t)$ is then taken into service [that is, at time t, a customer with priority $q'(t)$ is given preferential treatment over a customer with priority $q(t)$ where $q't > q(t)$]. Whenever a tie for the highest priority occurs, the tie is broken by an FCFS rule. We note that higher priority customers gain priority at a faster rate (b_p) than lower priority customers.

Figure 3.4 shows an example of the manner in which this priority

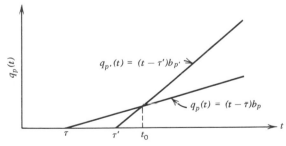

Figure 3.4   Interaction between priority functions for the delay-dependent priority system.

structure allows interaction between the priority functions for two customers. Specifically, at time τ, a customer from priority group p arrives, and attains priority at a rate equal to b_p. At time τ' a different customer enters from a higher priority group p'; that is, $p' > p$. When the service facility becomes free, it next chooses that customer in the queue with the highest instantaneous priority. Thus, in our example, the first customer will be chosen in preference to the second customer if the service facility becomes free at any time between τ and t_0 (in spite of the fact that the first customer is from a "lower" priority class); but, for any time after t_0, the second customer will be chosen in preference to the first.

We study this system for the case of exponential service times. We use the method in Section 3.2 and are faced immediately with calculating the quantities $\bar{N}_{ip}$ and $\bar{M}_{ip}$. We begin with the calculation of $\bar{M}_{ip}$ and refer the reader to Figure 3.5. Clearly, $\bar{M}_{ip} = 0$ for $i \leq p$ since no later arrivals with smaller (or equal) slope can ever "catch up" to the tagged customer. Now consider the arrival of a p-type customer, the tagged customer, at time 0. Since W_p is its expected waiting time, the expected value of its attained priority at the expected time it is accepted for service is $b_p W_p$, as shown in Figure 3.5. In looking for $\bar{M}_{ip}$, we must calculate how many i-type customers (for $i > p$) arrive, on the average, after time 0 and reach a priority of at least $b_p W_p$ before time W_p. It is obvious from the figure that type-i customers that arrive in the time interval $(0, V_i)$ will satisfy these conditions. Let us calculate the value of V_i. Clearly,

$$b_p W_p = b_i (W_p - V_i)$$

and so

$$V_i = W_p \left[1 - \frac{b_p}{b_i} \right]$$

Therefore, with an input rate λ_i for the type-i customers, we find that

$$\bar{M}_{ip} = \lambda_i V_i$$

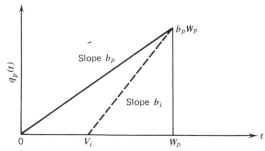

Figure 3.5   Diagram of priority, $q_p(t)$, for obtaining $\bar{M}_{ip}$.

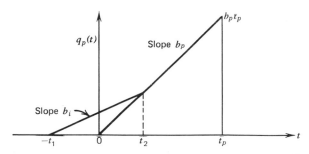

Figure 3.6   Diagram of priority, $q_p(t)$, for obtaining $\bar{N}_{ip}$.

and so

$$\bar{M}_{ip} = \lambda_i W_p\left[1 - \frac{b_p}{b_i}\right] \qquad \text{for all } i > p \qquad (3.43)$$

We now prove that $\bar{N}_{ip} = \lambda_i W_p(b_i/b_p)$ for $i \le p$. Consider again that a type-p customer, the tagged customer, arrives at time $\tau = 0$ and spends a total time t_p in the queue. His attained priority at the time of his acceptance into the service facility will be $b_p t_p$, as shown in Figure 3.6. Upon his arrival, the tagged customer finds n_i type-i customers already in the queue. Let us consider one such type-i customer, as shown in the figure, which arrived at $t = -t_1$. In looking for $\bar{N}_{ip}$, we must calculate how many type-i customers arrive before $t = 0$, are still in the queue at $t = 0$, and obtain service before the tagged customer does. It is obvious from the figure that a type-i customer that arrives at time $-t_1$ ($t_1 > 0$) and that waits in the queue a time $w_i = w_i(t_1)$ such that $t_1 < w_i(t_1) \le t_1 + t_2$ will satisfy these conditions. Obviously, $w_i(t_1)$ must not exceed $t_1 + t_2$ since otherwise the i-type customer will be of lower priority than the tagged customer, and will therefore fail to meet the conditions stipulated above. Let us solve for t_2. Clearly,

$$b_p t_2 = b_i(t_1 + t_2)$$

and so

$$t_2 = \left[\frac{b_i}{b_p - b_i}\right]t_1$$

or

$$t_1 + t_2 = \left[\frac{b_p}{b_p - b_i}\right]t_1 \qquad (3.44)$$

It is therefore clear that the expected number, $\bar{N}_{ip}$, of i-type customers that are in the queue at $t = 0$ and that also obtain service before the tagged customer does, can be expressed as

$$\bar{N}_{ip} = \int_0^{\infty} \lambda_i P\left\{t < w_i(t) \le \left[\frac{b_p}{b_p - b_i}\right]t\right\}\,dt \qquad (3.45)$$

where $\lambda_i \, dt$ is the expected number of i-type customers that arrived during the time interval $(-t - dt, -t)$ and where $P\{t < w_i(t) \leq [b_p/(b_p - b_i)]t\}$ is the probability that a customer who arrived in that interval spends at least t and at most $[b_p/(b_p - b_i)]t$ sec in the queue. Equation (3.45) can be written as

$$\bar{N}_{ip} = \lambda_i \int_0^\infty [1 - P(w_i \leq t)] dt - \lambda_i \int_0^\infty \left[1 - P\left\{ w_i \leq \left[\frac{b_p}{b_p - b_i} \right] t \right\} \right] dt$$

$$= \lambda_i \int_0^\infty [1 - P(w_i \leq t)] \, dt - \lambda_i \left[1 - \left(\frac{b_i}{b_p} \right) \right] \int_0^\infty [1 - P(w_i \leq y)] \, dy$$

where we have made the change of variable $y = [b_p/(b_p - b_i)]t$. Now, since

$$E[w_i] = \int_0^\infty [1 - P(w_i \leq x)] \, dx$$

(for w_i a non-negative random variable), and since in our notation $W_i = E[w_i]$, we obtain

$$\bar{N}_{ip} = \lambda_i W_i - \lambda_i \left[1 - \frac{b_i}{b_p} \right] W_i$$

and therefore

$$\bar{N}_{ip} = \lambda_i W_i \frac{b_i}{b_p} \qquad \text{for all } i \leq p \tag{3.46}$$

Furthermore, it is clear from Eq. (3.26) that $\bar{N}_{ip} = \lambda_i W_i$ for $i \geq p$ since our tagged customer can never catch up with these higher priority customers (all of whom are present upon his arrival).

Having derived expressions for $\bar{N}_{ip}$ and $\bar{M}_{ip}$ we may now substitute for these quantities into Eq. (3.11) and obtain

$$W_p = \frac{W_0 + \sum_{i=p}^{P} \rho_i W_i + \sum_{i=1}^{p-1} \rho_i W_i (b_i/b_p)}{1 - \sum_{i=p+1}^{P} \rho_i [1 - (b_p/b_i)]} \qquad p = 1, 2, \dots, P \tag{3.47}$$

This set of P linear equations in the W_p's is sufficient to solve our problem. However, it is possible to create a much simpler triangular set of equations from these by making use of the conservation law given in Eq. (3.16). Specifically we may rewrite the first sum in the numerator from Eq. (3.47) as follows:

$$\sum_{i=p}^{P} \rho_i W_i = \frac{\rho W_0}{1 - \rho} - \sum_{i=1}^{p-1} \rho_i W_i$$

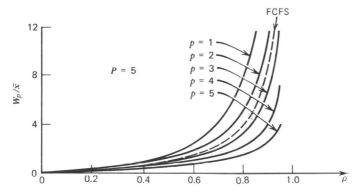

Figure 3.7 $W_p/\bar{x}$ for the time-dependent priority system with no preemption. $P = 5$, $\lambda_p = \lambda/5$ $\bar{x}_p = \bar{x}$.

and when this substitution is made we have

$$W_p = \frac{[W_0/(1-\rho)] - \sum\limits_{i=1}^{p-1} \rho_i W_i[1-(b_i/b_p)]}{1 - \sum\limits_{i=p+1}^{P} \rho_i[1-(b_p/b_i)]} \qquad p = 1, 2, \ldots, P$$

■ (3.48)

which is the main result for this nonpreemptive time-dependent discipline. It is interesting to note the extremely simple dependence that W_p has on the parameters b_i, namely these parameters only appear as ratios [KLEI 64a, b].

The typical behavior for this time-dependent queueing discipline is shown in Figure 3.7. The dashed curve shown is that for the FCFS system and once again shows the effect of the conservation law on priority queueing disciplines.

Thus we have analyzed a queueing discipline that provides a free set of parameters b_p/b_{p+1} that may be used to meet the specified system performance requirements given as W_p/W_{p+1} for $p = 1, 2, \ldots, P-1$. We see that only $P-1$ performance ratios may be specified and that we have exactly this many degrees of freedom for meeting that specification; the Pth condition is forced upon us as a scaling factor (that is, the conservation law) for all the waiting times and of course is a function of the utilization of the system (it is also clear that W_P for this class of systems can never lie below the corresponding curve for the HOL system, because of the ultimate preference given to the highest priority group in HOL).

A natural extension to this discipline is one in which a customer's priority increases in proportion to some *arbitrary power* of his elapsed

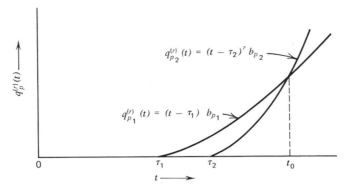

Figure 3.8   Coupling of different priority units.

time, rather than the first power as above [KLEI 67a]. We now address ourselves to this "generalization." Thus we define, for any non-negative number r, an rth-order time-dependent priority discipline as one that calculates the priority $q_p^{(r)}(t)$ at time t associated with a customer arriving at time τ as follows:

$$q_p^{(r)}(t) = (t - \tau)^r b_p$$

Further define $W_p^{(r)}$ as the expected value of the time spent in the queue of an rth-order system for a unit from group p.

The coupling between customers of different priority classes is illustrated in Figure 3.8. As for $r = 1$, we see that it is possible for customers to change their relative positions in the queue. It should be noted that there can be at most only *one* interchange between any two customers and it is this property which simplifies the analysis.

Consider two generalized time-dependent priority systems, one of order r with a set of parameters $\{b_p\}$ and the other of order r' with parameter set $\{b_p'\}$. In Exercise 3.14, we prove that if we choose

$$\left(\frac{b_p}{b_{p+1}}\right)^{1/r} = \left(\frac{b_p'}{b_{p+1}'}\right)^{1/r'} \qquad p = 1, 2, \ldots, P-1 \qquad (3.49)$$

then

$$W_p^{(r)} = W_p^{(r')} \qquad (3.50)$$

Thus all rth-order systems may be characterized (with respect to average waiting times) by an r_0th-order system (for any $r_0 > 0$) through a suitable change of parameters as given by Eq. (3.49). The case for $r_0 = 1$ has already been treated above and so, in order to obtain $W_p^{(r)}$, we appeal to these results and obtain (see Exercise 3.14) the main result: for an rth-order delay-dependent priority system without preemption, and $0 \leq \rho < 1$,

$$W_p^{(r)} = \frac{[W_0/(1-\rho)] - \sum_{i=1}^{p-1} \rho_i W_i [1 - (b_i/b_p)^{1/r}]}{1 - \sum_{i=p+1}^{P} \rho_i [1 - (b_p/b_i)^{1/r}]} \qquad p = 1, 2, \ldots, P$$

(3.51)

This result we see is in basically the same form as $r = 1$. Furthermore from the result in Eq. (3.50) we see that no greater generality for W_p is afforded with arbitrary r since they are all equivalent to our earlier result for $r = 1$.

However, there is insight to be gained from this generalization. In Figure 3.9 we show $q_p^{(r)}(t)$ for a customer arriving at time τ, with r as a parameter. We see that

$$\lim_{r \to 0} q_p^{(r)}(t) = b_p u_{-1}(t - \tau)$$

(3.52)

where $u_{-1}(t - \tau)$ is the unit step function occurring at time τ. Thus for $r = 0$, an entering customer from group p is assigned a fixed value of priority equal to b_p. This is the HOL system studied above. Moreover, as $r \to \infty$, $q_p^{(r)}(t)$ becomes a step function of infinite height at time $\tau + 1$. Thus, customers that have been in the system for more than 1 sec have infinite priority and those that have been in the system for less than 1 sec have zero priority. Remembering that an FCFS criterion is used to break a tie, the limit as r approaches infinity is seen to be a strict FCFS system. These

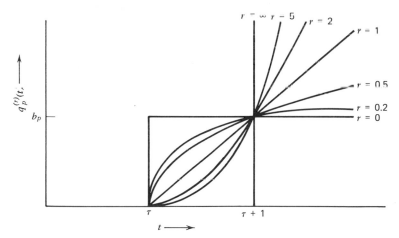

Figure 3.9 $q_p^{(r)}(t)$ for several r.

two limiting cases can also be obtained by taking the limit of $W_p^{(r)}$ as follows. From Eq. (3.51), for $b_p < b_{p+1}$ $(p = 1, 2, \ldots, P-1)$,

$$\lim_{r \to 0} W_p^{(r)} = \lim_{(b_i/b_{i+1})^{1/r} \to 0} W_p^{(r)} = \frac{W_0/(1-\rho) - \sum_{i=1}^{p-1} \rho_i W_i}{1 - \sum_{i=p+1}^{P} \rho_i}$$

Solving this last set of recursive equations yields

$$\lim_{r \to 0} W_p^{(r)} = \frac{W_0}{\left(1 - \sum_{i=p}^{P} \rho_i\right)\left(1 - \sum_{i=p+1}^{P} \rho_i\right)} \tag{3.53}$$

which is the same result obtained for HOL. We also note that $\lim_{r \to \infty} (b_p/b_{p+1})^{1/r} = 1$ and so

$$\lim_{r \to \infty} W_p^{(r)} = \frac{W_0}{1 - \rho} \tag{3.54}$$

which is the result for FCFS.

We now consider $\{b_p\}$ to be fixed and display the dependence of $W_p^{(r)}$ on r. As discussed above, as $r \to 0$ we obtain the HOL system and as $r \to \infty$ we obtain the FCFS system. For $r = 1$ we have the first-order time-dependent system. In Figure 3.10 we illustrate the general behavior of the expected wait on queue as we vary our priority discipline over the class of rth-order systems for $0 \le r$. We show the case with $P = 5$, $b_p/b_{p+1} = \frac{1}{2}$, $\rho_p = \rho/5$, $\bar{x}_p = \bar{x}$, for $p = 1, 2, \ldots, 5$, $\rho = 0.95$, and $W_0 = 1$. The dashed line in this figure demonstrates the conservation law for this particular case. The wide dispersion of $W_p^{(r)}$ among the priority groups shown in Figure 3.10 is due to the large value of $\rho = 0.95$ which causes considerable interaction among conflicting arrivals. For smaller values of ρ, the dispersion is not nearly as great. However, as we have shown, the relative waiting times can be adjusted by varying r for a given $\{b_p\}$; moreover for a given r, variation of the $\{b_p\}$ accomplishes the same adjustment of relative waiting times.

It is interesting to note that the class of rth-order delay-dependent priority systems covers the spectrum from that queueing discipline which separates priority groups in the greatest possible extent (that is, HOL) to the discipline that does not separate them at all (that is, FCFS).

Comparing the performance of the HOL system in Figure 3.3 with the performance of the time-dependent system in Figure 3.7 we observe that a certain generality seems to be lacking in the latter curves; namely all priority groups appear to saturate at the same point $\rho = 1$. This lack of generality is only an illusion. Indeed stable performance for higher

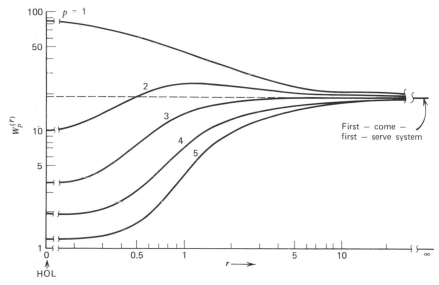

Figure 3.10 $W_p^{(r)}$ versus $r(b_p/b_{p+1} = 0.5, \ \rho = 0.95)$.

priority groups while lower groups are experiencing infinite average waits may be realized with the time-dependent discipline by permitting certain of the ratios b_p/b_{p+1} to approach zero; this effectively separates the $(p+1)$st group (and all higher) from the pth group (and all lower) in an HOL fashion [KLEI 66]. So for example in Figure 3.11 we show the performance of this more generalized time-dependent priority queueing discipline for which we have chosen the parameters $P = 25$, $\lambda_p = \lambda/25$, $\bar{x}_p = \bar{x}$ and have forced the creation of five HOL groups; within each HOL group are five priority groups that interact in a fashion similar to that shown in Figure 3.7.

Other "dynamic" priority queueing disciplines have been considered and the reader is referred to references [JACK 60, 61, 62].

3.8.   OPTIMUM BRIBING FOR QUEUE POSITION

For those queueing disciplines so far studied (and for most of those described in the literature) the relative priority given to any customer is completely out of his individual control. The customer in effect has no choice as to which priority group he must join.

In this section we shift the emphasis somewhat, and allow each entering customer to "buy" his relative priority by means of a bribe [KLEI 67b]. The size of the bribe will be determined, in general, from certain economic factors inherent in the population of customers; in particular, the greater

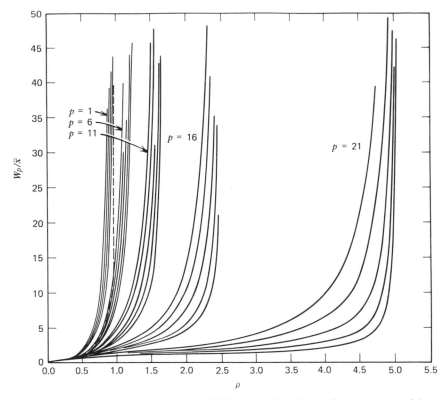

Figure 3.11 $W_p/\bar{x}$ for the mixed HOL and time-dependent system with no preemption.

the wealth of a customer and the greater his dislike of waiting on queue, the greater will be his bribe.

We consider an M/G/1 system with Poisson arrivals at a mean rate of λ customers per second, and an arbitrary PDF for service time $B(x)$ with a mean service time of $\bar{x}$ sec. Let a customer's bribe, given by Y, be a random variable with an arbitrary distribution function $\beta(y) = P[Y \le y]$. We assume that the arrival time, the service time, and the bribe are all independent random variables for each customer and are independent of the values chosen for all other customers.

The system operates as follows: A new arrival to the system offers* a non-negative bribe Y to the "queue organizer." This customer is then

_____

* This bribe may be thought of as given before the customer sees the length of the queue [in which case the distribution $\beta(x)$ reflects his measure of wealth and impatience] and is therefore independent of queue length.

placed in position on the queue so that all those customers whose bribes $Y' \geq Y$ are in front of him and all these with bribes $Y'' < Y$ are behind him. Newly entering customers may therefore be placed in front of, or behind this customer, depending upon their bribe. Each time the service facility completes work on some customer (who then leaves the system), it then accepts into service the customer at the front of the queue. Once in service, a customer cannot be ejected until he is completely serviced (that is, nonpreemptive*). Whenever customers give identical bribes, they are serviced in a first-come–first-serve order.

We define, for $\varepsilon > 0$, the left and right limits of $\beta(y)$ as

$$\beta(y^-) = \lim_{\varepsilon \to 0} \beta(y - \varepsilon)$$

$$\beta(y^+) = \lim_{\varepsilon \to 0} \beta(y + \varepsilon)$$

Let $W(y) \triangleq$ average waiting time (in queue) for a customer whose bribe $Y = y$. For such a customer, (say, the tagged customer), we now calculate $W(y)$ using the method as described in Section 3.2. The tagged customer must, on the average, wait a time W_0 before the customer who is in service upon his arrival is finished. In addition, he must wait until service is given to all those customers still in queue who arrived before he did and whose bribes equaled or exceeded his. The expected number of such customers whose bribes lie in the region $(u, u + du)$ is, by Little's result,

$$\lambda(u)W(u)\, du$$

where

$$\lambda(u) = \frac{\lambda\, d\beta(u)}{du}$$

Each such customer causes the tagged customer to wait an average of $\bar{x}$ sec. Furthermore, the tagged customer must wait until service is given to all those who enter the system while he is on the queue and whose bribes exceed his. The expected number whose bribes lie in the interval $(u, u + du)$ and who arrive during his average wait $W(y)$ is

$$\lambda(u)W(y)\, du$$

Each such customer also adds $\bar{x}$ sec to the tagged customer's average wait. Combining these three contributions to the tagged customer's

* A preemptive case is studied in [KLEI 67b].

average wait, we get*

$$W(y) = W_0 + \int_{y^-}^{\infty} \bar{x}\lambda(u)W(u)\,du + \int_{y^+}^{\infty} \bar{x}\lambda(u)W(y)\,du$$

or

$$W(y) = \frac{W_0 + \int_{y^-}^{\infty} \rho W(u)\,d\beta(u)}{1 - \int_{y^+}^{\infty} \rho\,d\beta(u)}$$

where $\rho = \lambda\bar{x}$. Since $\beta(\infty) = 1$, we have

$$W(y) = \frac{W_0 + \rho \int_{y^-}^{\infty} W(u)\,d\beta(u)}{1 - \rho + \rho\beta(y^+)} \tag{3.55}$$

In Exercise 3.15 we ask the reader to show that the solution to this integral is simply

$$W(y) = \frac{W_0}{[1 - \rho + \rho\beta(y^+)][1 - \rho + \rho\beta(y^-)]} \quad \blacksquare \quad \tag{3.56}$$

We also note that at those bribes of value y where $\beta(y)$ is continuous the solution becomes

$$W(y) = \frac{W_0}{[1 - \rho + \rho\beta(y)]^2} \tag{3.57}$$

It is interesting to note once again that the only way in which the service time distribution $B(x)$ enters the solution is through its first and second moments; this is no surprise to us for M/G/1 systems.

Note, in general, that we obtain finite average waits for all those customers offering bribes greater than y_{crit} where

$$y_{crit} = \begin{cases} 0^- & \rho < 1 \\ \beta^{-1}\left(\dfrac{\rho - 1}{\rho}\right) & \rho \geq 1 \end{cases} \tag{3.58}$$

and where $\beta^{-1}(u)$ is that value of y for which $\beta(y) = u$. This behavior is not unlike that of the HOL system for $\rho \geq 1$. We note in the limit as the bribe approaches infinity that the average waiting time approaches W_0.

Let us consider some special cases now. First in the case of a constant bribe y_0 (the same for all customers) as given by

$$\beta(y) = \begin{cases} 0 & y < y_0 \\ 1 & y \geq y_0 \end{cases}$$

* The lower limits of y^- and y^+ come about since all ties are broken on a first-come–first-serve basis. (Also, W_0 is merely $\lambda\overline{x^2}/2$ as usual.)

we have

$$W(y_0) = \frac{W_0}{1-\rho} \qquad (3.59)$$

Since all bribes are the same (resulting in no effective bribe at all), Eq. (3.59) should correspond to the well-known result for an FCFS M/G/1 queue, which it does. Second, for the case where $\beta(y)$ is continuous at the origin, giving $\beta(0) = 0$, we see that

$$W(0) = \frac{W_0}{(1-\rho)^2} \qquad (3.60)$$

This behavior at zero bribe should describe the waiting time for the lower priority group of a $P = 2$ HOL system where the arrival rate of this lower priority group is negligible compared to the total arrival rate. Indeed, as can be seen from Eq. (3.31) with $P = 2$ and $\rho_1 \ll \rho_2$, the equations above are consistent. We also observed this behavior following Eq. (3.38) for $\lim_{x \to \infty} W(x)$. Last, when only a finite (countable) set of bribes are allowed (say at the values y_p), then we have a discrete distribution which yields

$$W(y_p) = \frac{W_0}{\left[1 - \rho \sum\limits_{i=p+1}^{P} \Delta\beta(y_i)\right]\left[1 - \rho \sum\limits_{i=p}^{P} \Delta\beta(y_i)\right]} \qquad (3.61)$$

where $\Delta\beta(y_i) \triangleq \beta(y_i^+) - \beta(y_i^-)$. This equation corresponds exactly to the result for HOL.

As soon as we introduce the notion of a bribe, we must then consider other cost factors as well. In particular, let us define a random impatience factor $\tilde{\alpha}$ (≥ 0) that measures how many dollars* it costs a customer for each second that he spends in the queue. We now introduce a cost function $C(\alpha)$ defined as

$$C(\alpha) = y_\alpha + \alpha W(y_\alpha) \qquad (3.62)$$

where, again,

α = value taken on by a customer's impatience factor $\tilde{\alpha}$ (dollar/sec),

y_α = bribe offered by a customer whose impatience factor $\tilde{\alpha} = \alpha$,

$W(y_\alpha)$ = average waiting time (in queue) for a customer whose bribe is y_α.

Thus $C(\alpha)$ is the sum of the customer's bribe (in dollars) and his cost of waiting (in dollars). We assume that customers have (self-) assigned values of $\tilde{\alpha}$ before they enter the system, and that the population of customers, as a whole, produces a probability distribution $P(\alpha)$ on the

---

* This cost may be measured in terms of customer inconvenience or impatience, if you will.

random variable $\tilde{\alpha}$, that is, $P(\alpha) = P[\tilde{\alpha} \leq \alpha]$. The queueing models here are the same as those considered earlier where now the bribe is some (deterministic) function of the random variable $\tilde{\alpha}$. We have thus shifted emphasis from the situation in which a customer offers a random bribe to a situation where the customer's bribe is functionally related to his (random) impatience factor $\tilde{\alpha}$.

We pose the following optimization problem: Find that function y_α which minimizes the expected cost C, that is,

$$\underset{y_\alpha}{\text{minimize}} \left[C \triangleq \int_0^\infty C(\alpha) \, dP(\alpha) \right] \tag{3.63}$$

subject to an average bribe constraint equal to B, that is,

$$B = \int_0^\infty y_\alpha \, dP(\alpha) \tag{3.64}$$

We must therefore choose y_α to minimize

$$C = \int_0^\infty [y_\alpha + \alpha W(y_\alpha)] \, dP(\alpha) \tag{3.65}$$

Because of the average bribe constraint, this is equivalent to minimizing

$$C - B = \int_0^\infty \alpha W(y_\alpha) \, dP(\alpha)$$

Define

$$\rho(\alpha) = \rho \frac{dP(\alpha)}{d\alpha} \tag{3.66}$$

We may interpret the quantity $\rho(\alpha) \, d\alpha$ as the fraction of time that the server is busy serving customers whose impatience factor lies in the interval $(\alpha, \alpha + d\alpha)$. Using Eq. (3.66) we then find that for $0 < \rho$,

$$C - B = \frac{1}{\rho} \int_0^\infty \alpha \rho(\alpha) W(y_\alpha) \, d\alpha \tag{3.67}$$

Now, the continuous form of the conservation law may be written as

$$\int_0^\infty \rho(\alpha) W(y_\alpha) \, d\alpha = \frac{\rho}{1-\rho} W_0 \tag{3.68}$$

We note that minimizing Eq. (3.67) involves finding that function y_α such that the product of $\rho(\alpha) W(y_\alpha)$ and α has minimum area. However, Eq. (3.68) states that the first of these functions must itself have a *constant* area. Since $\rho(\alpha)$ is independent of y_α, we must look for conditions on $W(y_\alpha)$. Now, using the same argument as that for proving HOL optimum for linear costs (in which we matched an increasing sequence against a decreasing sequence) we see that a necessary and achievable (see below) condition on $W(y_\alpha)$ is that it decreases with α, that is,

$$\frac{dW(y_\alpha)}{d\alpha} < 0 \tag{3.69}$$

for all $\alpha \notin S$ [where the set S has the property $\int_S dP(\alpha) = 0$]. Here the increasing function is α itself. Condition (3.69) may be rewritten as

$$\frac{dW(y_\alpha)}{dy_\alpha}\frac{dy_\alpha}{d\alpha} < 0 \tag{3.70}$$

From Eq. (3.56) we have (letting $y_\alpha = y$)

$$\frac{dW(y)}{dy} = -\rho W_0 \frac{A(y^+)[d\beta(y^-)/dy] + A(y^-)[d\beta(y^+)/dy]}{[A(y^+)A(y^-)]^2}$$

where $A(u) \triangleq 1 - \rho + \rho\beta(u)$. Now since $\beta(u) \le 1$ and $\rho < 1$ then $A(y^+)A(y^-) > 0$. Also since $\beta(u)$ is a distribution function then $d\beta(u)/du \ge 0$. This implies that for all values of its argument $W(y_\alpha)$ has a nonpositive derivative, that is

$$\frac{dW(y_\alpha)}{dy_\alpha}\begin{cases} < 0 \text{ at those } y \text{ for which } \dfrac{d\beta(y)}{dy} > 0 \\ = 0 \text{ at those } y \text{ for which } \dfrac{d\beta(y)}{dy} = 0 \end{cases} \tag{3.71}$$

From Eqs. (3.70) and (3.71), then, we have that our necessary condition on y_α becomes

$$\frac{dy_\alpha}{d\alpha} > 0 \qquad \text{for } \alpha \notin S \tag{3.72}$$

That this last is achievable is obvious for a large family of functions (for example, $y_\alpha = \alpha$). From this family, however, we may use only those functions satisfying Eq. (3.64); such functions clearly exist [for example, see Eq. (3.73) below].

Consider an interval $\alpha_1 < \alpha < \alpha_2$ in which $P(\alpha)$ is constant. Clearly y_α can be arbitrary in any such interval without affecting C; the same is true at any *point* α for which $P(\alpha)$ is continuous. But such regions are in the set S. However, for the sets* S_1 (defined by $\alpha_1 - \varepsilon \le \alpha \le \alpha_1$) and S_2 (defined by $\alpha_2 \le \alpha \le \alpha_2 + \varepsilon$), in which $P(\alpha)$ is assumed to be increasing, we require that Eq. (3.69) holds and also that

$$W(y_a) > W(y_b)$$

* Here $\varepsilon > 0$.

where $a \in S_1$ and $b \in S_2$. This last is true for the same reasons leading up to Eq. (3.69), namely, that in order to minimize $C - B$, we must reduce $W(y_\alpha)$ as α increases.

To demonstrate that Eq. (3.72) is also sufficient we consider Eq. (3.65). The first term merely gives B, which is independent of y_α, and the second term depends only upon the *relative* size of the bribes and not upon the absolute bribe itself. However, Eq. (3.72) gives a complete description of the rank ordering of the bribes. Consequently the necessary and sufficient condition for y_α to be an optimum bribing function is merely that it satisfy Eqs. (3.64) and (3.72).

Thus the solution to the minimization problem set forth in Eqs. (3.63) and (3.64) restricts y_α to be a strictly increasing function of α for $\alpha \notin S$. Having constrained only the mean bribe, we get only a *condition* on y_α rather than an explicit functional form; indeed, the solution is independent of the exact form of y_α, as long as it is strictly increasing with α. Thus for the purposes of calculation and example we may choose some (simple) relation, such as the following linear one:

$$y_\alpha = K\alpha$$

Applying the mean bribe constraint, we get

$$B = K \int_0^\infty \alpha \, dP(\alpha)$$

Letting A be the average impatience factor, we get from the last two equations

$$y_\alpha = \frac{B}{A} \alpha \qquad (3.73)$$

This then is an optimal bribing function.

In order to obtain some insight into the behavior of the optimum bribing procedure and the cost function, we offer the following example. Consider a system with an exponentially distributed bribe, namely,

$$\beta(y) = 1 - e^{-\sigma y} \qquad \sigma \geq 0, \, y \geq 0 \qquad (3.74)$$

We may immediately calculate the waiting time $W(y)$ from Eq. (3.57) as

$$W(y) = \frac{W_0}{(1 - \rho e^{-\sigma y})^2} \qquad (3.75)$$

Using our optimum bribing rule given in Eq. (3.73) we find that the distribution of impatience factor $P(\alpha)$ that gives rise to the bribing distribution in Eq. (3.74) must be

$$P(\alpha) = 1 - e^{-(B/A)\sigma\alpha}$$

and then the average cost gives

$$C = \frac{1}{\sigma} + \frac{AW_0}{\rho} \log_e \frac{1}{1-\rho} \tag{3.76}$$

where, of course, A is the average impatience factor and $B = 1/\sigma$ is the average bribe.

The optimization described above is a global optimization and seeks bribing functions that minimize the total average cost. Recently, in [BALA 72], a similar (nonpreemptive) bribing system was studied in which conditions were found for which a customer would offer a bribe so as to minimize his own expected cost (disregarding the global minimum). Most of the considerations in [BALA 72] center around the discussion of *stable* bribing policies; roughly speaking, a bribing policy is stable if when all customers follow this policy then it does not pay for any individual customer to deviate from it. Upon their arrival and prior to making their bribe, customers are informed of the bribes given by all other customers in the system (and therefore, the queue size is also given). The first result obtained is that in the system G/G/1 for $\rho < 1$, with finite second moment of service time and with $C(\alpha)$ as given in Eq. (3.62) then a global optimal policy is one in which all customers should give zero bribe; however, this policy is clearly unstable since an infinitesimal bribe puts a customer at the head of the queue. For the M/M/1 queue the bribing policy b_k (bribe size when the queue size is k) is stable if and only if

$$(1-\rho) \max_k \Delta b_k \le \alpha \bar{x} \le \frac{(1+\rho)(1-\rho)}{\rho} \min_k \Delta b_k \tag{3.77}$$

where $\Delta b_k = b_k - b_{k-1}$. We are here implying that each customer has the same impatience factor $\bar{\alpha} = \alpha$. Furthermore, it can be shown for the system G/M/1 that the bribing policy in which $b_{2k} = b_{2k+1}$ but which is strictly increasing on the even integers $(2k)$ is stable if and only if

$$\frac{1}{2} \max_k \Delta b_k \le \frac{\alpha \bar{x}}{1-\rho} \le \frac{1}{1+p_1+p_2} \min_k \Delta b_k \tag{3.78}$$

and p_i is the probability that at least i new customers arrive between an arrival instant and the completion of the service in progress. Considerations for the M/G/1 system are considerably more complex. In [BALA 72] the preemptive resume case is also considered. See also [BALA 73]. In [ADIR 72] locally optimal bribing policies are described for M/M/1 (both with and without preemption) as well as optimal pricing policies for the server.

3.9. SERVICE-TIME-DEPENDENT DISCIPLINES

We have seen earlier in Section 3.4 that queueing disciplines that do not discriminate in any way on the basis of service time must all have the same average waiting time. Beyond that we have seen some examples (e.g., HOL) in which priority depends in some incidental way on service time. In this section we mention some results in which more explicit use is made of a customer's service time. This discussion is rather abbreviated in this chapter but forms a point of departure for Chapter 4 when we consider models for computer time-sharing in which great effort is devoted to creating strong discrimination on the basis of required service.

One feels intuitively that giving preferential treatment to shorter jobs tends to reduce the overall average waiting time as well as the average number in a priority queueing system. In fact we have seen one example of this result in Section 3.6 where Eq. (3.42) determined the correct ordering of priorities in the optimum HOL system under linear costs; we note that if all costs are identical (that is, $C_p = C_0$ for $p = 1, 2, \ldots, P$) then this ordering is strictly on the basis of shortest average job. In fact in the continuous case [see Eq. (3.38)] we found that the SJF discipline gave the smallest possible average waiting time for nonpreemptive disciplines (i.e., apply the μC rule to this case).

There are a number of interesting disciplines based on customer service time and we list some of these below along with those we have studied (we use notation common in the applications and in the scheduling theory literature [CONW 67]):

1. FCFS: first-come–first-serve
2. LCFS: last-come–first-serve
3. SPT: shortest-processing-time-first (same as SJF)
4. SRPT: shortest-remaining-processing-time-first
5. SEPT: shortest-expected-processing-time-first
6. SERPT: shortest-expected-remaining-processing-time-first

We now state, without proof, results among these various disciplines. Since these studies arise not only from the study of queueing systems but also from the study of scheduling systems, we also consider the case where no arrivals are permitted to enter (the case for scheduling) but rather all jobs that require processing are available at the start of the "busy period." (We choose to refer to service now as "processing time" as is common in the literature.) For the case of no arrivals, there is an additional discipline that we must consider which has recently been introduced by Sevcik [SEVC 74], namely,

7. SIPT: shortest-imminent-processing-time-first

The SIPT discipline operates as follows: if the ith customer has a distribution of processing time $B_i(x)$ and has already received x sec of service, and if the permitted points in time when this ith customer may be preempted are $t_{i1}, t_{i2}, \ldots$, then the priority $q_i(x)$ of this job is calculated under this discipline as follows:

$$\frac{1}{q_i(x)} = \min_{\{j : t_{ij} > x\}} \left[\frac{\int_x^{t_{ij}} [1 - B_i(y)] \, dy}{B_i(t_{ij}) - B_i(x)} \right]$$

The customer whose function $q_i(x)$ is maximized is defined to have the largest priority; the above ratio is the expected time spent on customer i if he is allowed at most an amount of service equal to $t_{ij} - x$.

For some of these seven disciplines, we are interested in defining related disciplines in which *cost* enters the picture. Often, we are concerned with linear cost functions (that is, costs which are linear with average waiting time). The new class of disciplines we wish to introduce involves forming one of the measures mentioned above (as for example, SEPT) and dividing this measure for each job by the cost rate associated with that job (thus, for example, forming the new discipline SEPT/C). We will need this additional definition for the case of SPT/C, SEPT/C (note that this is the μC rule), SRPT/C, and SIPT/C.

Let us now comment on some of the known results. In all cases, we assume a conservative system (no creation or destruction of work—specifically, no cost for preemption and no idle server when jobs are present).

First, we consider the case of no arrivals and costs that are linear with average waiting time. In the case of exactly known service times, it is known (see, for example, p. 26ff [CONW 67] for a proof and discussion of this result) that SPT/C is optimum (i.e., it minimizes the average cost). If only the distribution of service time for each job is known, then, for the nonpreemptive case, SEPT/C minimizes the average cost [SMIT 56]; in the preemptive case (where the set of permissible preemption points may be specified), then SIPT/C scheduling is optimum [SEVC 74].

Let us now consider the more interesting case of arrivals (i.e., queueing systems). We seek the optimum scheduling rule (but restrict ourselves to the case of priority disciplines only—that is, rules that evaluate a job's priority based only on that job's parameters) for given types of cost functions [linear, convex, concave, minimum variance of time in system (σ_s^2), maximum σ_s^2]. In a recent paper by Schrage [SCHR 74], a very nice summary of several of these optimal scheduling disciplines is given. He considers three possible information states, namely, (1) exact service time information given; (2) distribution of service time only given, or (3) no

Table 3.1.
Optimal Service-Time-Dependent Scheduling Algorithms

	(1) PROCESSING TIMES KNOWN	(2) ONLY DISTRIBUTION OF PROCESSING TIMES KNOWN	(3) NO INFORMATION REGARDING PROCESSING TIMES
Nonpreemptive	SPT/C for M/G/1 linear costs [PHIP 56, ACZE 60, FIFE 65, HARR 72] [a] SPT/C for G/G/1 linear costs two classes only [WOLF 70, SCHR 74]	SEPT/C for M/G/1 linear costs [PHIP 56, ACZE 60, FIFE 65, HARR 72] [a] SEPT/C for G/G/1 linear costs two classes only [WOLF 70, SCHR 74]	FCFS for G/G/1 convex costs [HAJI 71] FCFS for G/G/1 minimum σ_s^2 [KING 62] LCFS for G/G/1 maximum σ_s^2 [TAMB 68]
Preemptive	SRPT for G/G/1 identical linear costs ($C_p = C_o$) [SCHR 68] SRPT/C for M/G/1 linear costs ([AVI 64, ETSC 66, JAIS 68] for two classes) [SCHR 74] [b]	SEPT/C for G/M/1 linear costs [SCHR 74] SIPT/C for M/G/1 linear costs [SEVC 74] [b]	FCFS for G/IFR/1 convex costs [JACK 61, SCHR 74] [b] LCFS for G/M/1 concave costs [SCHR 74] [b] FB for G/DFR/1 [c] linear costs [KALR 71]—for M/DFR/1 [SCHR 74] [b]

[a] Also see heuristic proof that the μC rule is optimal in Section 3.6.
[b] Conjecture only (with heuristic proof in some cases).
[c] See Chapter 4 below for a definition of FB.

information regarding servivicc times given. Observe that case (2) is the most general; case (1) is clearly a special (degenerate distribution) example of (2).* Also, case (3) is the situation with only one class of customer. The results and appropriate references to these results are given in the table on p. 146; the reader is urged to see these references for further details and restrictions on these optimality results.

Another comparison among some of these queueing disciplines is given in [SUZU 70]. Let $D_1 \to D_2$ denote the fact that the average wait using discipline D_1 is greater than or equal to that for D_2. Then, the following relationships hold:

$$LRPT \to LPT \to FCFS \to SPT \to SRPT$$

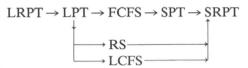

where LPT(LRPT) means longest-(remaining)-processing-time-first, and RS means random order of service.

There are cases in which exact processing times are unavailable but more than just the distribution of service times is known. For example it may be possible to separate customers' required processing times into "large" and "small." In particular, many examples indicate [CONW 67] that this separation into two groups provides a considerable reduction in mean waiting times as opposed to the FCFS system (see Exercises 3.8 and 3.9).

As mentioned earlier, other processing-time-dependent queueing disciplines will be considered next in Chapter 4.

REFERENCES

ACZE 60   Aczel, M. A., "The Effect of Introducing Priorities," *Operations Research*, **8,** 730–733 (1960).

ADIR 72   Adiri, I., and V. Yechiali, "Optimal Pricing and Priority Purchasing Policies," IBM Research Report RC-3581, September 2, 1972.

AVI 63   Avi-Itzhak, B., and P. Naor, "Some Queueing Problems with the Service Station Subject to Breakdown," *Operations Research*, **11,** 303–320 (1963).

AVI 64   Avi-Itzhak, B., I. Brosh, and P. Naor, "On Discretionary Priority Queueing," *Zeitschrigt für angewandte Mathematik und Mechanik*, **6,** 235–242 (1964).

BALA 72   Balachandran, K. R., "Purchasing Priorities in Queues," *Management Science*, **18,** No. 5, 319–326 (1972).

* Note that SEPT/C becomes SPT/C and SERPT/C becomes SRPT/C for known service times.

BALA 73   Balachandran, K. R., and J. C. Lukens, "Stable Pricing Policies in Service Systems," Report MS 1, College of Industrial Management, Georgia Institute of Technology, October 1973.

BRUM 69   Brumelle, S. L., "Some Inequalities for Multi-Server Queues," ORC 69-17, Operations Research Center, University of California, Berkeley, 1969.

COBH 54   Cobham, A., "Priority Assignment in Waiting Line Problems," *Operations Research*, **2**, 70–76 (1954).

COLE 71   Cole, G. C., *Computer Network Measurements: Techniques and Experiments*, School of Engineering and Applied Science, University of California, Los Angeles, UCLA-ENG-7165, 1971.

CONW 67   Conway, R. W., W. L. Maxwell, and L. W. Miller, *Theory of Scheduling*, Addison-Wesley (Reading, Mass.), 1967.

COX 61   Cox, D. R., and W. L. Smith, *Queues*, Methuen (London) and Wiley (New York), 1961.

CRAB 73   Crabill, T. B., D. Gross, and M. J. Magazine, "A Survey of Research on Optimal Design and Control of Queues," Serial T-280, School of Engineering and Applied Science, The George Washington University, June 1, 1973.

ETSC 66   Etschmaier, "Discretionary Priority Processes," M.S. Thesis, Case Institute of Technology, Cleveland, Ohio, 1966.

FIFE 65   Fife, D. W., "Scheduling With Random Arrivals and Linear Loss Functions," *Management Science*, **11**, No. 3, 429–437 (1965).

GAVE 62   Gaver, D. P., Jr., "A Waiting Line with Interrupted Service, Including Priorities," *Journal of the Royal Statistical Society, Series B*, **24**, 73–90 (1962).

HAJI 71   Haji, R., and G. F. Newell, "Optimal Strategies for Priority Queues with Nonlinear Costs of Delay," *SIAM Journal of Applied Mathematics*, **20**, 224–240 (1971).

HARR 72   Harrison, J. M., "Dynamic Scheduling of a Multi-Class Queue, I: Problem Formulation and Descriptive Results," Technical Report #36, and "Dynamic Scheduling of a Multi-Class Queue, II: Discount Optimal Dynamic Policies," Technical Report #37, Department of Operations Research, Stanford University, Stanford, California, June 1972.

JACK 60   Jackson, J. R., "Some Problems in Queueing with Dynamic Priorities," *Naval Research Logistics Quarterly*, **7**, 235–249 (1960).

JACK 61   Jackson, J. R., "Queues with Dynamic Priority Discipline," *Management Science*, **8**, No. 1, 18–34 (1961).

JACK 62   Jackson, J. R., "Waiting-Time Distributions for Queues with Dynamic Priorities," *Naval Research Logistics Quarterly*, **9**, 31–36 (1962).

JAIS 68   Jaiswal, N. K., *Priority Queues*, Academic Press (New York), 1968.

KALR 71   Kalro, A. L., "Optimal Processor Scheduling in a Computer Time Shared System," ORC 71-25, University of California, Berkeley, California, September 1971.

KEIL 62   Keilson, J., "Queues Subject to Service Interruption," *Annals of Mathematical Statistics*, **33**, 1314–1322 (1962).

KEST 57   Kesten, H. and J. Th. Runnenberg, "Priority in Waiting-Line Problems I and II," *Nederlandse Akademie van Wetenschappen, Amsterdam, Proceedings, Series A*, **60**, 312–324, 325–336, (1957).

KING 62   Kingman, J. F. C., "The Effect of the Queue Discipline on Waiting Time Variance," *Proceedings of the Cambridge Philosophical Society*, **58**, 163–164 (1962).

KLEI 64a   Kleinrock, L., *Communication Nets: Stochastic Message Flow and Delay*, McGraw-Hill (New York), 1964, reprinted by Dover Publications, Inc., (New York), 1972.

KLEI 64b   Kleinrock, L., "A Delay Dependent Queue Discipline," *Naval Research Logistics Quarterly*, **11**, 329–341 (1964).

KLEI 65   Kleinrock, L., "A Conservation Law for a Wide Class of Queueing Disciplines," *Naval Research Logistics Quarterly*, **12**, 181–192 (1965).

KLEI 66   Kleinrock, L., "Queueing with Strict and Lag Priority Mixtures," *Proceedings of the 4th International Conference on Operational Research*, Boston, Mass. K-I-46 to K-I-67, 1966.

KLEI 67a   Kleinrock, L., and R. P. Finkelstein, "Time Dependent Priority Queues," *Operations Research*, **15**, 104–116 (1967).

KLEI 67b   Kleinrock, L., "Optimum Bribing for Queue Position," *Operations Research*, **15**, 304–318 (1967).

KLEI 75   Kleinrock, L., *Queueing Systems, Vol. I: Theory*, Wiley Interscience, (New York), 1975.

LIPP 75   Lippman, S. A., "On Dynamic Programming with Unbounded Rewards," *Management Science*, 21, No. 11, 1225–1233 (1975).

PHIP 56   Phipps, T. E., Jr., "Machine Repair as a Priority Waiting-Line Problem," *Operations Research*, **4**, 76–85 (1956).

PRAB 73   Prabhu, N. U., and S. Stidham, Jr., "Optimal Control of Queueing Systems," Technical Report No. 186, Dept. of Operations Research, Cornell University, June 1973.

REED 74   Reed, F. C., "Difference Equations and the Optimal Control of Single Server Queueing Systems," Technical Report No. 23, Stanford University, Dept. of Operations Research, March 22, 1974.

SCHR 68   Schrage, L., "A Proof of the Optimality of the Shortest Remaining Processing Time Discipline," *Operations Research*, **16**, 687–690 (1968).

SCHR 70   Schrage, L., "An Alternative Proof of a Conservation Law for the Queue G/G/1," *Operations Research*, **18**, 185–187 (1970).

SCHR 74   Schrage, L., "Optimal Scheduling Disciplines for a Single Machine Under Various Degrees of Information," Working Paper, Graduate School of Business, University of Chicago, 1974.

SEVC 74   Sevcik, K., "A Proof of the Optimality of 'Smallest Rank' Scheduling," *Journal of the Association for Computing Machinery*, **21**, 66–75 (1974).

SMIT 56   Smith, W. E., "Various Optimizers for Single-Stage Production," *Naval Research Logistics Quarterly*, **3**, 59–66 (1956).

SUZU 70   Suzuki, T., and K. Hayashi, "On Queue Disciplines," *Journal of the Operations Research Society of Japan*, **13**, 43–58 (1970).

TAMB 68   Tambouratzis, D. G., "On the Property of the Variance of the Waiting Time of a Queue," *Journal of Applied Probability*, **5**, 702–703 (1968).

TORB 73   Torbett, E. A., "Models for the Optimal Control of Markovian Closed Queueing Systems with Adjustable Service Rates," Technical Report No. 20, Dept. of Operations Research, Stanford University, January 15, 1973.

WOLF 70   Wolf, R. W., "Work Conserving Priorities," *Journal of Applied Probability*, **7**, 327–337 (1970).

EXERCISES

3.1. Using the notion of residual life, show that $W_0 = \lambda \overline{x^2}/2$.

3.2. Consider an M/G/1 system with 2 priority groups and some unspecified queueing discipline which is work-conserving. We are given that

$$W_2 = \frac{W_0}{1 - \alpha\rho_1 - \beta\rho_2}$$

where $\rho_p = \lambda_p \bar{x}_p$ is the utilization factor for the pth group ($p = 1, 2$) and $o < \alpha < 1, 0 < \beta < 1$. Find W_1 in terms of ρ_1, ρ_2, α, β, and W_0.

3.3. Find the mean and variance for Y_b and Y_c in Figure 3.1.

3.4. For M/G/1 we wish to compare FCFS with LCFS.
   (a) Show that $W_{FCFS} = W_{LCFS}$ ($= W$) by using the moment-generating properties of $W^*(s)$.
   (b) Similarly show that $\sigma^2_{FCFS} = (1 - \rho)\sigma^2_{LCFS} - \rho W^2$ ■

3.5. Consider an M/M/1 nonpreemptive HOL system. Let j be the smallest integer such that $\sum_{i=j}^{P} \rho_i < 1$. Solve for W_p ($p = j, j+1, \cdots, P$). Note that $W_p = \infty$ for $p < j$.

3.6. For the system described in the previous problem, establish a conservation for the sum

$$\sum_{p=j}^{P} \rho_p W_p$$

3.7. Calculate W_p for the nonpreemptive HOL system from Eq. (3.32).

3.8. Consider a nonpreemptive HOL system with $P=2$ constructed as follows. We assume that service times for all customers are drawn from $B(x)$, *but are known when a customer arrives.* Let x_0 be a number defining the boundary between the two groups, that is, if $x < x_0$, then a job falls in group $p=2$ and if $x \geq x_0$ it is placed in group $p=1$.

(a) Show that

$$W = \sum_{p=1}^{2} \frac{\lambda_p}{\lambda} W_p = \frac{W_0}{1-\rho}\left[\frac{1-\rho B(x_0)}{1-\rho_2}\right]$$

(b) Prove that this simple discrimination is an improvement for W over FCFS.

3.9. Consider an M/G/1 system with a $P=2$ nonpreemptive HOL priority discipline. Let

$$W = \frac{\lambda_1}{\lambda} W_1 + \frac{\lambda_2}{\lambda} W_2$$

Assume $\rho < 1$ [COLE 71].

(a) Prove that

$$W = W_{\text{FCFS}} \frac{1-\lambda_2\bar{x}}{1-\lambda_2\bar{x}_2}$$

where

$$\bar{x} = \frac{\lambda_1}{\lambda}\bar{x}_1 + \frac{\lambda_2}{\lambda}\bar{x}_2$$

and W_{FCFS} is the mean wait in an FCFS system.

(b) Suppose now that group 2 consists of all jobs with service time $\leq \tau$, and group 1 has $\tilde{x} > \tau$, where $\tilde{x}$ has the general distribution $B(x)$. Let τ_0 be the optimum value of this threshold such that W is minimized, and let $W_{\min}$ be this minimum average wait.

(i) Show that

$$\frac{W_{\min}}{W_{\text{FCFS}}} = \frac{\bar{x}}{\tau_0}$$

(ii) Show that τ_0 is defined through

$$\frac{1}{\rho} = 1 + \frac{\frac{\lambda_1}{\lambda}(\bar{x}_1 - \tau_0)}{\tau_0 - \bar{x}}$$

3.10. Consider a two priority M/G/1 system for which $W_0 = 2$, $\rho_1 = \rho_2 = \frac{1}{4}$.

  (a) Suppose $W_1 = 5$. Find W_2.

  (b) If the system is HOL (nonpreemptive), find W_1 and W_2.

  (c) If the system is FCFS, find W_1 and W_2.

  (d) If the system is LCFS, find W_1 and W_2.

3.11. Consider a $P = 2$ nonpreemptive priority queueing system with $\lambda_1 = \lambda_2 = 1$ and $\bar{x}_1 = \frac{1}{2}$ and $\bar{x}_2 = \frac{1}{4}$.

  (a) Design a system which achieves a performance ratio $W_2/W_1 = \alpha < 1$.

  (b) Suppose a customer enters at some random time and must wait for service until the system empties. Give an expression for the ratio of this customer's average waiting time to his average wait in an FCFS system with the same input.

3.12. Consider a delay-dependent discipline for which $0 \geq b_1 \geq b_2 \geq \cdots \geq b_P$. Find the set of simultaneous equations similar to Eq. (3.48) that define W_p $(p = 1, 2, \ldots, P)$.

3.13. Consider a $P = 2$, rth-order time-dependent priority discipline with the following cost function (for some constant, m):

$$C = \sum_{p=1}^{2} C_p \lambda_p [W_p^{(r)}]^{m+1}$$

  (a) For a given pair (b_1, b_2), express $W_p^{(r)}$ in terms of W_0, ρ_p, b_p, and r $(p = 1, 2)$.

  (b) For $0 \leq (b_1/b_2)^{1/r} < 1$, find the optimum value for b_1/b_2 so as to minimize C.

3.14. Consider the rth-order time-dependent priority discipline. We wish to prove that Eq. (3.50) follows from Eq. (3.49).

  (a) Let $r' = 1$ with no loss of generality. Consider any three customers whose priority functions (may) intersect each other. Show that the intersection times for the general r case will be exactly the same for the $r' = 1$ case if Eq. (3.49) holds.

  (b) Use induction to establish the equivalence of all intersection times for the general case of M customers to establish Eq. (3.50) (i.e., equivalence of intersection times implies equivalence of order of service).

  (c) Now, in addition prove Eq. (3.51).

3.15. **(a)** Show that Eq. (3.56) is indeed the solution (for the average wait conditioned on a bribe of value x) to Eq. (3.55).

(b) Show that the average cost C given in Eq. (3.76) is correct for M/M/1.

(c) For bribes uniformly distributed over the interval $[0, M]$, find the average cost C.

3.16. Consider a nonpreemptive M/M/m system in which the average service time for the ith server is $1/\mu_i$ where $1/\mu_1 < 1/\mu_2 < \cdots < 1/\mu_m$. That is, the ith server is faster than the $(i+1)$st on the average. Customers join the queue in order of arrival. When the ith server becomes free, he offers his services to the first queued customer; if this customer refuses service from him (i.e., the customer is holding out for a faster server), he then offers his services to the second queued customer, and so on. If no queued customer accepts, he remains idle until possibly some newly arriving customer accepts him. Each customer uses a (local) strategy that minimizes his average time in system. No service times are known to any customers (not even their own service time). Let k_i be that position in the queue at which a customer should first accept the ith server [KLEI 64a].

Show that the critical positions k_i must satisfy

$$k_i < \frac{S_{i-1}}{\mu_i} \le k_i + 1$$

where $S_i = \mu_1 + \mu_2 + \cdots + \mu_i$ and $k_1 - 1$.

3.17. Consider an M/G/1 system with N queues that are labeled $Q1$, $Q2, \ldots, QN$. Arriving customers join the tail of $Q1$ and after receiving service [from distribution $B(x)$], they join the tail of $Q2$, and so on, finally they join QN, receive service, and then depart at last. Each customer therefore receives N independent services from $B(x)$. We also assume a priority ordering among the queues, which we denote by $P = \{q_1, q_2, \cdots, q_N\}$, which implies that Qi has priority over Qj if $q_i < q_j$; for example, $P = \{1, 3, 2\}$ implies that $Q1$ has highest priority, $Q3$ is next, and $Q2$ is last. This is the order in which a queue is selected for service whenever a service completion occurs. Within a given queue, service is FCFS.

Let $p_k = P[k$ "services" in system at departure instants], where each newly arriving customer counts for N "services." Let $Q(z) = \sum_{k=0}^{\infty} p_k z^k$. Let $\rho = \lambda \bar{x} N$

(a) Find $Q(z)$.

Now let $N = 2$ and $P = \{2, 1\}$. Let

$$R(z_1, z_2) = \sum_{k_1=0}^{\infty} \sum_{k_2=0}^{\infty} p(k_1, k_2) z_1^{k_1} z_2^{k_2}$$

where $p(k_1, k_2) = P[k_1$ customers in $Q1$ and k_2 customers in $Q2]$ (this too is calculated at departure instants).

(b) For the given priority ordering, what possible values can k_2 take?

(c) Show that the answer to (b) allows us to write

$$p(k_1, k_2) = p_{2k_1 + k_2}$$

(d) Show that

$$R(z_1, z_2) = \frac{(1-\rho)B^*(\lambda - \lambda z_1)[z_2 - z_1 + (1-z_2)B^*(\lambda - \lambda z_1)]}{[B^*(\lambda - \lambda z_1)]^2 - z_1}$$

[*Hint:* Consider the series obtained for the sum $Q(z) + Q(-z)$ and for the difference $Q(z) - Q(-z)$.]

(e) From (d), find $E[k_2]$.

Let us now consider $P = \{1, 2\}$.

(f) Find T, the average time a customer spends in the system.

3.18. As a variation on the cost function given in Eq. (3.65), consider an FCFS M/M/1 system for which $C = \mu C_s + \lambda T\alpha$ where C_s is the cost per unit of service rate for a server. C is thus the average cost of running the facility. Find the optimum value of μ that minimizes C.

3.19. Show that W_{SPT}, the mean wait for a shortest-processing-time-first (same as SJF) discipline is related to T_{SRPT}, the mean time in system for a shortest-remaining-processing-time-first discipline for M/M/1 as follows:

$$W_{SPT} = \rho T_{SRPT}$$

For an M/M/1 system with $\lambda = 0.9$ and $\mu = 1.0$, calculate and compare W_{SPT}, W_{SRPT}, and W_{FCFS}.

3.20. Consider an M/G/1 system operating under the shortest-remaining-processing-time-first (SRPT) discipline in which all customer service times are known ahead of time. A new customer will preempt a customer in service only if his service time is less than that which remains for the customer in service. At a service completion, the shortest job is served next.

Show that the mean time in system T_{SRPT} is

$$T_{SRPT} = \int_{y=0}^{\infty} \int_{x=0}^{y} \frac{dx}{1 - \lambda f_1(x)} \, dB(y)$$
$$+ \frac{\lambda}{2} \int_{y=0}^{\infty} \frac{f_2(y) + y^2[1 - B(y)]}{[1 - \lambda f_1(y^-)][1 - \lambda f_1(y^+)]} \, dB(y)$$

where

$$f_i(y) = \int_0^y u^i \, dB(u)$$

3.21. Consider a sequence of arrival times $\{1, 3, 4, 5, 7, 8, 13, 15, 17, 23, 27\}$ and a corresponding sequence of service times $\{5, 6, 2, 3, 2, 1, 2, 2, 2, 3, 1\}$. Assume the system is empty prior to the arrival at time 1.

    (a) Calculate the average wait for these eleven customers when the discipline is FCFS.

    (b) Repeat for LCFS.

    (c) Repeat for SPT (SJF).

    (d) Repeat for LPT.

    (e) Repeat for SRPT.

    (f) Repeat for LRPT.

    (g) Confirm the $D_1 \rightarrow D_2$ relationship given in the text for these six cases.

4

Computer Time-Sharing and Multiaccess Systems

The information processing industry is one of the fastest growing, most dynamic and most glamorous industries on the scientific scene today. It is now inexpensive and convenient to gain ready access to computing power. Perhaps in the near future we will see this power provided to the public through a computer utility. As the computational capacity grows and as the data bases grow and as the number of users requesting access to these resources grows, so also must grow the rate of conflict among these users for simultaneous access to the system. These conflicting demands naturally give rise to queues and delays, and it is therefore appropriate that we include in this text a chapter such as this, which exemplifies the successful application of queueing theory to the analysis of multiaccess computer systems.

In the early 1950s when high-speed digital computers first appeared on the scene it was not uncommon to see a programmer sitting at the console of a computer and interacting with it on a one-for-one basis for hours at a time. The user would spend a period thinking and generating a request for computation, then wait while that computation was carried out by the machine, then upon receipt of the result would proceed into the thinking state again, etc. This mode of operation was extremely efficient from the user's point of view but highly wasteful of the machine's resources. As a result, in the late 1950s when larger, faster, and more expensive machines appeared on the market, we find that the user was thrown out of the machine room and was forced to submit his request (typically in the form of punched cards) to an operator; hours (and sometimes days) later the operator would then return reams of printed paper representing the results of computation. The operator, of course, had queued up a number of such tasks and processed them through the computer in a "batch" mode. In this mode, input (and output) was typically transferred to (from) magnetic tapes, which themselves could transfer data to (from) the main

156

system much more rapidly than card readers (line printers). This operational method kept the machine quite busy and was therefore efficient in terms of machine utilization but was woefully inefficient in terms of satisfying the response requirements of the user. Moreover, the excess system capacity available during a given job's processing could not be used for some other job. In the early 1960s a major breakthrough occurred with the advent of time-sharing whereby the user was once again permitted direct access to the computer through a remote console; in this mode many users at many consoles are permitted access to the computer simultaneously as each goes through his cycle of thinking and computing. The principle behind time-sharing is that significantly more time is spent in the thinking phase than in the computing phase (in most cases of interactive computing) and therefore, computation may take place on one (or a small number of) user requests while most of the other users are in their thinking phase. If properly handled, a time-shared computer system is capable of offering excellent response-time characteristics to many users while maintaining high utilization efficiency for the many resources of the facility itself through concurrent use of devices. The ability to service many jobs simultaneously came about with the introduction of autonomous data channels, larger and cheaper core memories that could store several programs simultaneously, interrupt capability, memory protection, and the supervisory system programs to keep the flow orderly and efficient through the application of sophisticated resource allocation and sharing. The next natural step in the evolution of computer systems appeared in the late 1960s when it was recognized that many of these geographically distributed time-shared computer facilities could profitably be interconnected in a computer-communication network so that the resources of each could be shared by all; we postpone discussion of these networks until the next chapter and focus here only on queueing models for individual multiaccess computer systems.

In this chapter, then, we are concerned with a collection of remote terminals from which users gain simultaneous access to a collection of finite-capacity resources commonly known as a computer system. Refering to Figure 4.1 we see that the resources include: the terminals themselves; the communication lines connecting these terminals to the computer system; the *communications-oriented-processor* controlling the flow of traffic between the users and the computer system (which we refer to as a traffic COP); the storage capacity in the main memory and in the secondary memory devices; the processing capacity of both the central processing unit (CPU) and of the input/output processor (I/O PROC) that controls the flow of data between main memory and the secondary devices; other I/O devices; and all of the other channels interconnecting

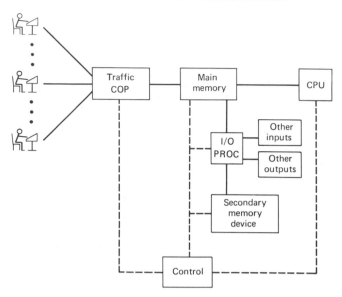

Figure 4.1   Computer system structure, (———— denotes data flow, – – – – de-notes control).

these resources. In addition, there is a control function (basically a supervisory program) that governs the activity and assignment of the various resources in the system. It is clear that numerous conflicts will arise when many user programs access these various resources, and it is our purpose in this chapter to describe the degree of success that has so far been achieved in modeling and analyzing such structures.

We have identified three kinds of resources: communication capacity, processing capacity, and storage capacity. It is clear, therefore, that we are dealing with a multiple-resource problem which requires that we consider the sequence of resource demands created by jobs as they flow through the system; this then becomes a problem involving networks of queues and represents a rather ambitious task for analysis. We defer consideration of the multiple-resource case until later in this chapter where we present some of the significant advances recently obtained. We begin with a simple case of competition among many users for a *single* resource. Indeed, in 1964 [KLEI 64a, 64b] the first model for a time-shared computer system was presented in just this fashion (as a single-resource model) and this analysis ushered in the field of computer systems modeling and analysis as we know it today. A number of references summarize various of these results, including [CHNG 70,

ESTR 67, KLEI 66b, KLEI 67b, KLEI 68a, KLEI 69, KLEI 72a, KLEI 72b, KLEI 73a, McKI 69, MUNT 74b, OMAH 73, SCHR 69].

4.1.   DEFINITIONS AND MODELS

We view a time-shared computer system as a collection of resources and a population of users who compete at various times for the use of these resources. Conflicts arise when simultaneous demands are placed upon these resources. In order to resolve these conflicts, a *scheduling algorithm* is required that allocates resources to users; this function is usually called a scheduler. In our first model we assume that *only one resource* (assumed to be the CPU) is under demand. Our second model generalizes this to the case of multiple resources, which we consider in Sections 4.12 and 4.13.

A very general model to describe the way in which computer systems resolve conflicting requests for attention of the CPU by means of time-sharing is shown in Figure 4.2. This model of a time-sharing system consists of the *single resource* (CPU) and a *system of queues* that holds those customers awaiting service. In addition, there exists a *scheduling algorithm*, which is a set of decision rules determining which user will next be serviced and for how long. Thus a newly entering request is placed in the system of queues and, when the scheduling algorithm finally permits, is given a turn in the processing facility. The interval of time during which the customer is' permitted to remain in service is referred to as his *quantum* (the quantum size may vary). The quantum offered may or may not be enough to satisfy the request. If sufficient, the customer departs from the system; if not, he reenters the system of queues as a partially completed task and waits within the system of queues until the scheduling algorithm decides to give him a second quantum, and so on. Eventually, after a sufficient number of visits to the service facility, the customer will have gained enough service and will depart. This is a (highly) preemptive resume priority queueing rule. For obvious reasons we refer to this general model as a *feedback queueing system*.

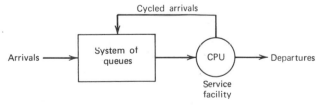

Figure 4.2   The feedback queueing model.

It is generally accepted that one of the major advantages of time-shared systems is that they permit interactive use of the computer by many users simultaneously (an excellent example of "resource sharing"). The goal is to provide to each of these interactive users with what he thinks is a computer all to himself, since interactive jobs require rapid response to their frequent requests for small computational demands. At the same time, those jobs that make large demands on the CPU need not be serviced as quickly since typically such requests do not expect this interactive type of response. Thus we wish to give preferential treatment to short jobs (at the expense of the longer ones!). However, we also take the point of view that we *do not know* how large a demand each arriving customer places on the CPU. How then can we service the jobs giving priority in relation to their demands? The solution is an implicit one, namely, we permit each job to demonstrate to us that indeed it is small! We see from the feedback queueing system described above that if a job is long, it will require many visits to the CPU. Thus, by continually "testing" the collection of jobs demanding service, we successfully *discover* those that are short. By varying the scheduling algorithm, it is possible to effect various degrees of preferential treatment for short jobs. How this is accomplished is discussed below.

Let us now describe the means for specifying the nature of the customer demands as well as the details of the system structure for the model shown in Figure 4.2. As usual, we define $A(t)$ to be the interarrival time distribution* and $B(x)$ to be the required service time distribution. The quantum of time offered to a customer may depend on an externally applied priority as well as on the number of visits that customer has already made to the service facility. Accordingly, we define

$$q_{pn} \triangleq \text{the quantum offered to a customer from priority class}$$
$$p \text{ upon his } n\text{th entry into service} \qquad (4.1)$$

When a customer's quantum expires, we must remove him from service and insert the next customer; typically, the overhead in time required to perform this operation is a quantity referred to as the *swap time*. Consequently, not all of the assigned quantum may be available to a customer for useful processing. Having specified $A(t)$, $B(x)$, and q_{pn}, it remains to describe the internal structure of the system of queues (which is equivalent to specifying the scheduling algorithm). The nature of some specific scheduling algorithms is discussed later.

Once we specify the above quantities, we are then in a position to

---

* Until we reach Section 4.11, we shall assume that the customers arrive from an infinite population so that $A(t)$ is independent of the state of the system.

analyze the feedback queueing system. Typically, one would like to solve for the distribution of the total time a customer spends in the system (this is his *response time*). We certainly should condition this distribution on the amount of service $\tilde{x}$ required by this job since we wish to expose the preferential tratment given to short jobs at the expense of longer ones. We define this conditional distribution as

$$S(y \mid x) \triangleq P[\text{response time} \le y \mid \tilde{x} = x] \qquad (4.2)$$

In some cases we are able to solve for this distribution; however, often we ask only for the *average* of the response time, and this we define as

$$T(x) \triangleq \text{average response time for a customer} \\ \text{requiring } x \text{ sec of processing} \qquad (4.3)$$

$T(x)$ is generally accepted as the *single most important performance measure* for these systems. As usual, we define the waiting time as the difference between the response time and the service time, so that the average conditional waiting time (conditioned on the required service of x sec) is given by

$$W(x) = T(x) - x \qquad \qquad ▬ \quad (4.4)$$

We also refer to $W(x)$ as the average "wasted" time (which in fact it is!); in the processor-sharing systems introduced below, this is the more appropriate interpretation.

A variety of analytical models for time-shared computer systems have appeared in the published literature, along with results of many kinds. We select here certain of those models and results that we feel characterize the behavior of many time-shared systems and at the same time demonstrate the underlying methods from queuing theory that are useful in solving these systems.

Among our assumptions are the following. First, we assume no knowledge of a job's exact service time, but only knowledge of the distribution of service. Second, we will usually assume that the swap time is some fixed percentage of the quantum offered. This is then approximated by assuming a zero swap time, but where the average service rate is decreased by this same percentage[*] [KLEI 70b]. On the other hand, if the swap time on the nth visit to the CPU is θ_{pn} sec, then we can model this

---

[*] Thus, if from each quantum of q sec, we find that θ sec are devoted to swapping, then a fraction $\phi = \theta/q$ is lost. We may model this by replacing each required service time of x sec by a service time of $x/(1-\phi)$ sec.

exactly by replacing $B(x)$ with $B_\theta(x)$ as follows:

$$B_\theta(x) = \begin{cases} B\left(\sum_{i=1}^{n-1} (q_{pi} - \theta_{pi})\right) & \sum_{i=1}^{n-1} q_{pi} \le x \le \theta_{pn} + \sum_{i=1}^{n-1} q_{pi} \\ B\left(x - \sum_{i=1}^{n} \theta_{pi}\right) & \theta_{pn} + \sum_{i=1}^{n-1} q_{pi} \le x \le \sum_{i=1}^{n} q_{pi} \end{cases}$$

If we use $B_\theta(x)$ in place of $B(x)$ in the models below, then we completely account for this form of swap time. Thus, for our purposes, we may omit any further consideration of swap time (the exercises consider swap time in more detail). Third, we adopt the point of view that all quanta shrink to zero, resulting in what is commonly known as the *processor-sharing* model for time-shared systems; this concept was first introduced in [KLEI 67a]. The justification for this zero quantum limit is simply one of great analytic convenience. The finite quantum studies in the literature suffer from the annoying situation that a customer may depart before his current quantum has fully expired; this results in considerable mathematical complexity to account for the unused portion. When one reduces all quanta to zero, this difficulty disappears and the resulting performance measures are extremely good approximations to the finite quantum results. When we study the round-robin systems below, it will be clear why we have adopted the name processor-sharing. This chapter consists mainly of the processor-sharing results.*

Unless otherwise stated, the models below are of the form M/G/1 with parameters λ, $\bar{x}$, and $\rho = \lambda\bar{x} < 1$ as usual.

4.2. DISTRIBUTION OF ATTAINED SERVICE [KLEI 67c]

In all of the feedback queueing systems to be studied below, it is clear that at any instant of time we will have jobs scattered through the system, each of which has received various amounts of useful (i.e., attained) service. It is of interest to us to solve for the distribution of this attained service for those customers still in the system for various of the scheduling algorithms considered below. Fortunately, it is possible to solve for this distribution in all of the feedback queueing models to be described, and the solution is especially simple; in fact, the results of this section are

---

* Not only is the analysis simple, but also the solutions themselves are simply expressed. This is vital if one intends the results to be of use to system designers. We must always be wary of solutions that are so complex that more computational effort is required for their numerical evaluation than would be required to conduct a (more realistic) simulation of the system itself!

good for G/G/1. Accordingly, let us define*

$n(x) \triangleq$ average *density* of customers still in the
system who have so far received x sec
of service ..... (4.5)

The units of this density are customers per second and the interpretation is in the usual sense of a density function, namely,

$$\int_{x_1}^{x_2} n(x)\, dx = \text{average number of customers still in the}$$
system who have so far received between
x_1 and x_2 sec of service ..... (4.6)

This density gives one a description of the composition of the various queues and a measure of the relative state of partial service received by those customers still in the system. We consider two cases: first the case in which the quanta are finite, for which we define

$N(x) \triangleq$ average *number* of customers in the system of queues
whose attained service is exactly x sec ..... (4.7)

and second, the processor-sharing case where all quanta shrink to zero and the density $n(x)$ is of interest. Furthermore, for convenience we suppress the external priority structure that may be imposed upon the system and therefore consider only the case $q_{pn} = q_n$ so that all customers on their nth visit to the CPU are granted the same quantum q_n (see Exercise 4.2 for the priority case). Let us now derive $N(x)$ and $n(x)$.

In the case of a feedback queueing system for which $q_n > 0$, it is clear that all customers who have visited the service facility exactly n times must so far have received an amount of service equal to

$$Q_n \triangleq \sum_{i=1}^{n} q_i \qquad (4.8)$$

As a result, we see that for any service requirement lying in the interval $Q_{n-1} < x \le Q_n$ we must have that $W(x) = W(Q_n)$ since a customer requiring x sec of service in this range must join the system of queues exactly n times; the only distinction among customers with such service times is that they spend different amounts of time in their final visit to the CPU. Moreover, $W(Q_n)$ is the average time spent in the system of queues prior to the nth visit to the CPU for *any* customer whose service requirement exceeds Q_{n-1} sec. It is clear, then, that the average time spent in the system of queues after the $(n-1)$st but prior to the nth visit to the CPU

---

* We introduce this density in anticipation of our limiting results for processor-sharing.

will on the average be $W(Q_n) - W(Q_{n-1})$. Since the arrival rate is λ it is clear that $\lambda[1 - B(Q_n)]$ is merely the average arrival rate of jobs back to the system of queues after they have made n visits to the CPU and therefore each of these customers will so far have received exactly Q_n sec of service. Little's result then allows us to write down immediately that

$$N(Q_n) = \lambda[1 - B(Q_n)][W(Q_{n+1}) - W(Q_n)] \qquad (4.9)$$

This, then, is the principle result for finite quanta, giving the expected number of customers in the system of queues with an attained service of Q_n sec.

For our second case, we consider the natural limit where all quanta approach zero; we find the following analogous result expressed in terms of the average density of customers;

$$n_q(x) = \lambda[1 - B(x)]\frac{dW(x)}{dx} \qquad \blacksquare \qquad (4.10)$$

where $n_q(x)$ refers to the system of queues alone; on the other hand, $n(x)$ refers to the total system (including queues as well as CPU) and so we may relate it to the time in system rather than to the time in queues, which gives immediately

$$n(x) = \lambda[1 - B(x)]\frac{dT(x)}{dx} \qquad \blacksquare \qquad (4.11)$$

This principle result expresses the distribution of attained service in terms of known quantities [λ and $B(x)$] and in terms of the major performance measure for such systems, namely, the average conditional response time $T(x)$. It is indeed amazing that the derivative of this performance measure appears in the solution. We will need this general result shortly. For now, we observe that $dT(x)/dx$ is the inverse of the average rate of service given to a job with an attained service of x sec.

4.3. THE BATCH PROCESSING ALGORITHM

The simplest scheduling algorithm to consider is the case of batch processing. In particular, we assume a first-come–first-serve discipline whereby newly entering jobs join the tail of a single queue. The CPU always selects for service that job at the head of the queue and offers a quantum of service to this job. The simplest way to shape this into the classical FCFS system is to assume that all quanta are infinite in duration, thereby giving us an M/G/1 FCFS system; an alternative way is to assume infinitesimal quanta (i.e., processor-sharing) but with the added restriction that customers ejected from service due to the termination of their quanta

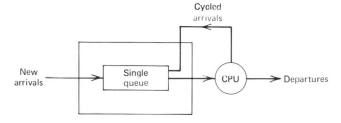

Figure 4.3   The FCFS system with an infinitesimal quantum.

are fed back directly to the head of the single queue and are thereby immediately taken back into service (see Figure 4.3). In any case, we have a system whose solution was given in Section 1.7; moreover, we pointed out in Section 3.4 that all queueing disciplines which take no account of service time in determining order of service will give rise to the same distribution for *number* in system; the generating function for this distribution is of course given in Eq. (3.15). For FCFS we have in addition that the transform for the *waiting* time density is given by Eq. (1.85) (the P-K transform equation); we note that this expression is independent of the job's service time in the CPU. Now defining the Laplace transform of the conditional response time density as

$$S^*(s \mid x) \triangleq \int_0^\infty e^{-sy} \, dS(y \mid x)$$

and recognizing that the response time is the sum of the waiting time plus service time [where the transform of the density of this *given* service time (x) is simply e^{-sx}] we have immediately that

$$S^*(s \mid x) = \frac{s(1-\rho)e^{-sx}}{s - \lambda + \lambda B^*(s)} \qquad \blacksquare \quad (4.12)$$

Also, one sees that the conditional average response time must be the sum of the given service time, x, and the average waiting time in an M/G/1 system, that is

$$T(x) = \frac{W_0}{1-\rho} + x \qquad \blacksquare \quad (4.13)$$

where we recall that $W_0 = \lambda \overline{x^2}/2$.

The significant observation to be made in this case is that batch processing offers *no* discrimination among jobs on the basis of their required service time x. Indeed, the average wait is independent of service time, and so *this is the least discriminatory system possible!* Clearly, it is a poor candidate for a time-shared system in which we seek to improve the response time for short jobs.

4.4. THE ROUND-ROBIN SCHEDULING ALGORITHM

Perhaps the most well-known and widely used scheduling algorithm for time-shared computer systems is the round-robin (RR) algorithm. The structure of this extremely interesting system is given in Figure 4.4. We consider the case $q_{pn} = q \to 0$; that is, all quanta are the same size and shrink to zero. Newly arriving customers join the single queue, work their way up to the head of this queue in a first-come–first-serve fashion, and then finally receive a quantum of service. When that quantum expires and if they need more service, they then return to the tail of that *same* queue and repeat the cycle. It is clear in this processor-shared system that a customer is required to make an infinite number of cycles each infinitely quickly and each time receiving infinitesimal service, until finally his attained service equals his required service, at which time he departs.

Consider for a moment the case where a customer enters an empty system. In this situation he receives service at the rate of 1 sec/sec. When a second customer arrives and joins him in this frantic cycling, we see that each customer will be receiving service at the rate of $\frac{1}{2}$ sec/sec since they are leapfroging with each other in and out of service. Thus when k customers are in the system, each is receiving service at the rate of $1/k$ sec/sec, and hence the name *processor-sharing* where indeed all customers are sharing the capacity of the processor equally. Thus we may take two points of view, the one being that customers are given the *full capacity* of the processor on a *part-time* basis, and the second being that customers are given a *fractional-capacity* processor on a *full-time* basis; the former is referred to as time-sharing and the latter as processor-sharing.

This system was studied first by the author in 1964 [KLEI 64a] as a discrete-time Markovian model* and later studied by him as a processor-sharing case [KLEI 67a]. Chang [CHNG 70], Shemer [SHEM 67], and others have also pursued this model, in all cases assuming M/M/1. Sakata

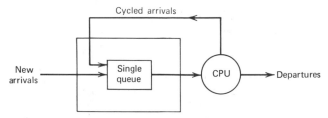

Figure 4.4   The round-robin system.

---

* The approach used in [KLEI 64a], in which one calculates the delay to a job for each pass through the cycle, has been followed by many others in evaluating more complex models.

et al. [SAKA 69, 71] were perhaps the first to study the M/G/1 case, and they found that the solution there for the average response time was *independent* of the form of the service time distribution (depending only on the mean service $\bar{x}$). Moreover, Coffman et al. [COFF 70] solved for the waiting time distribution for M/M/1. The results of some of these analyses are given below. We begin by solving for the conditional response time $T(x)$ for the M/G/1 case and proceed in a fashion much simpler than that in [SAKA 69].

From Section 1.7 we recall that the (instantaneous) failure rate of a component given its attained age is related to its lifetime density $f(x)$ in a rather simple way. Of interest here is the rate of completing service given an attained service (age) of x sec, which we denote by $\mu(x)$; we have that

$$\mu(x) = \frac{b(x)}{1 - B(x)} \tag{4.14}$$

We now consider $n(x)$, the average density of customers in the system with an attained service of x sec. For the moment, we "back off" to a finite (but tiny) quantum of size Δx. We observe that customers who have an attained service of x sec will, after one "pass" through the queue, then have attained $x + \Delta x$ sec; of these customers, those who remain in the system (they need more than $x + \Delta x$) will make up the complete set of customers who have so far attained $x + \Delta x$. Thus the density $n(x + \Delta x)$ must be equal [to within $o(\Delta x)$] to the value of the density at x times the probability $[1 - \mu(x) \Delta x]$ that a customer requires more service than $x + \Delta x$, given that he has already received x. We may therefore write

$$n(x + \Delta x) = n(x)[1 - \mu(x) \Delta x] + o(\Delta x)$$

If we now let $\Delta x \to 0$ again, we obtain

$$\frac{dn(x)}{dx} = -\mu(x)n(x)$$

This first order differential equation has the solution

$$n(x) = n(0) \exp \left[- \int_0^x \mu(y) \, dy \right]$$

The integral in the exponent is easily evaluated by using Eq. (4.14) and the simple substitution $v(x) = 1 - B(x)$ to give

$$- \int_0^x \mu(y) \, dy = \log_e[1 - B(x)]$$

Finally, we have

$$n(x) = n(0)[1 - B(x)] \tag{4.15}$$

This is a general relationship for the distribution of attained service in a processor-sharing RR system.

We may equate the expression for $n(x)$ in Eq. (4.15) to that in Eq. (4.11) to yield

$$\frac{dT(x)}{dx} = \frac{n(0)}{\lambda}$$

Recognizing that $T(0) = 0$ in the RR system we find the solution to this last equation as

$$T(x) = \frac{n(0)}{\lambda} x \qquad x \geq 0 \tag{4.16}$$

In order to evaluate the unknown constant $n(0)$, we examine the response time for the case of very large service times. In this case we know that all those jobs that arrive while our long "test" job is in the system must be served to completion before the test job departs; this test job therefore appears to be the lowest priority customer in a preemptive HOL priority system (with $\bar{x}_1 = x \to \infty$). Thus, from Eq. (3.39), we have

$$T(x) = \frac{x(1-\rho) + W_0}{(1-\rho)^2}$$

and in the limit,

$$\lim_{x \to \infty} T(x) = \frac{x}{1-\rho}$$

This permits us to evaluate the constant $n(0)$ from Eq. (4.16) as $\lambda/(1-\rho)$ to finally yield the exact solution for the RR processor-sharing system

$$T(x) = \frac{x}{1-\rho} \qquad\qquad \blacksquare \tag{4.17}$$

This is the conditional response time we were looking for, and we note that it is of extremely simple form. Clearly, we also have

$$W(x) = \frac{\rho x}{1-\rho} \tag{4.18}$$

We now list the delightful properties of this solution:

1. The discrimination is *linear*. That is, the response time depends on service time in a strictly linear fashion, which implies that a job twice as long as some other will spend on the average twice as long in the system. This is, perhaps, the *simplest discrimination* one could hope for.

2. The mean response is *independent* of the service time distribution $B(x)$ and depends only upon the mean value of service time through $\rho = \lambda \bar{x}$. Thus the RR processor-sharing system is one that removes any dependence of the average response time upon the variance (or higher moments) of the service time. This shows that the average wait $W = \int_0^\infty W(x)b(x)\ dx = \rho\bar{x}/(1-\rho)$ is finite so long as $\rho < 1$ and $\bar{x} < \infty$; we note that the FCFS system can have unbounded average wait under these conditions if $\overline{x^2} \to \infty$.

3. If we form the ratio of wasted time to service time, we obtain

$$\frac{W(x)}{x} = \frac{\rho}{1-\rho}$$

Since this ratio measures how much time must be sacrificed in waiting per unit of service time received, we may consider the ratio (in general) as some form of *penalty function* imposed upon a customer for receiving x sec of service. In the RR case we have the very nice property that the penalty function is independent of a customer's service time and is in that sense eminently "fair." On the other hand, imagine a penalty function which is increasing with service time. In this case customers with longer service requirements are penalized more heavily *per unit of service time;* in such a case the obvious countermeasure [COFF 68a] to the scheduling algorithm producing such a function would be for a user to partition his job into a number of smaller jobs, thereby enjoying the preferred treatment (per unit of required service) offered to short jobs! Alternatively, consider the case of some scheduling algorithm that produces a penalty function which decreases with increasing service time (e.g., FCFS). Here we see that the longer jobs enjoy the preferred treatment per unit of required service, and so the obvious countermeasure is for many jobs to pool their programs and offer one monstrously long job to the computer. Such tactics on the part of users* typically tend to increase the overhead to the system and are generally undesirable; in view of this it appears that the RR system removes the motivation for any such unusual action on the part of the user.

4. Another very nice property of the RR scheduling algorithm for the case of exponential service time is that a job whose service requirement is of average size will spend on the average as much time in the RR system as he would in the batch processing system. Thus, for exponential service, jobs less than average in length

---

* Of course, these same tactics severely modify the M/G/1 assumptions, in which case the analysis presented is no longer suitable.

receive better treatment in the RR as opposed to the batch processing system; similarly, jobs greater than average receive poorer treatment. This breakpoint between RR and batch processing for the M/G/1 case is easily seen to occur at $x = \overline{x^2}/2\bar{x}$ (the mean residual service time!).

For the case of exponential service times it is possible to solve the RR processor-sharing model for the distribution of response time [COFF 70]. The solution is

$$S^*(s \mid x) = \frac{(1-\rho)(1-\rho r^2) \exp\{-[\lambda(1-r)+s]x\}}{(1-\rho r)^2 - \rho(1-r)^2 \exp[-\mu x(1-\rho r^2)/r]} \qquad (4.19)$$

where r is taken as the smaller root of

$$\lambda r^2 - r(\lambda + \mu + s) + \mu = 0 \qquad (4.20)$$

From this we find that the average conditional response time is that given in Eq. (4.17) and also that the variance of the conditional response time is given as

$$\text{Var (response time} \mid \bar{x} = x) = \frac{2\rho x}{\mu(1-\rho)^3} - \frac{2\rho}{\mu(1-\rho)^4}[1 - e^{-\mu(1-\rho)x}] \qquad (4.21)$$

We note that this variance is linear with service time for large values of service time. We also note that the response time variance in the RR system is potentially much greater than that in the batch processing system.

Thus the RR system provides for us a very natural, rather common, and analytically simple system for the implementation of scheduling algorithms on time-shared computer systems. Let us now consider a seemingly unrelated scheduling discipline, the last-come–first-serve.

4.5. THE LAST-COME–FIRST-SERVE SCHEDULING ALOGRITHM

An interesting scheduling algorithm is the last-come–first-serve (LCFS) system in which the most recently arrived customer *preemptively* captures the use of the complete processor until *he* is either preempted by a newly arriving job or until he receives his total required service. We had earlier (in Section 3.5) considered the nonpreemptive LCFS system; the preemptive resume case considered here has an interesting property, as we shall see below. The internal structure of the system of queues for this system is shown in Figure 4.5. This, too, may be structured as a processor-sharing system in which the quanta have shrunk to zero; whenever the

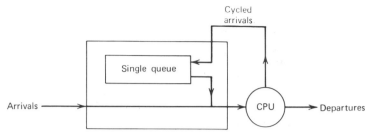

Figure 4.5   The preemptive LCFS system.

CPU empties, that customer at the head of the single queue is then permitted into service, thereby implementing the LCFS rule.

For this case let us first solve for the average conditional response time. For a customer requiring x sec of service, his response time will equal his service time x plus his waiting time. On the average, this last is merely the average number (λx) of customers who preempt him from service, times the average busy period duration generated by each such. The length of each busy period is on the average equal to $\bar{x}/(1-\rho)$ as we know from Eq. (1.90). Therefore, the average conditional response time is given by

$$T(x) = x + \frac{\lambda x \bar{x}}{1-\rho}$$

or

$$T(x) = \frac{x}{1-\rho} \qquad \blacksquare \quad (4.22)$$

It is amazing that this average response time is the same as that for the RR system! (Both results have been established for the M/G/1 system.) Thus we see that all the properties of the average response time shared by the RR system are also enjoyed by the LCFS system. One may therefore conclude that the average response time by itself is not a very good indicator of system performance.

To find the response time distribution here is easy. First, we recall from Section 3.3 that each new arrival generates his own sub-busy period; for LCFS with preemption, the duration of this sub-busy period is simply the response time for the arrival! For a job requiring x sec of service, we may use the delay cycle analysis of Section 3.3 where $Y_0 = x$ and Y is the sub-busy period (response time) just mentioned. Thus $S^*(s \mid x) = G_c^*(s)$ with $G_0^*(s) = e^{-sx}$, which, by Eq. (3.13), gives

$$S^*(s \mid x) = e^{-[s+\lambda-\lambda G^*(s)]x} \qquad (4.23)$$

where $G^*(s)$ is given through Eq. (3.14).

Once again we have a method for giving favorable treatment to the short jobs at the expense of the longer ones.

4.6. THE FB SCHEDULING ALGORITHM

The systems discussed heretofore give varying degrees of preference to short jobs. As mentioned earlier, this preference comes about by implicit discovery of which are the short jobs. One may inquire as to which scheduling algorithm is capable of discriminating *most* in favor of the very short jobs. The answer is immediately apparent: it is that scheduling algorithm that *next gives service to that job (or jobs) which has so far received least service* of all. This is commonly referred to as a generalized foreground–background (FB) scheduling algorithm*. The generalized case we consider here is a processor-sharing model since the quantum size is effectively zero; more than one job may share the processor as follows. Consider a new arrival who finds an empty system. The processor gives him its full attention and he receives service at the rate of 1 sec/sec. At some time, say t_1, later, a new job arrives; the server then ceases to serve the first customer and turns its attention to the second customer providing 1 sec/sec of service to this new customer. If no customers arrive within the next t_1 sec, then the second customer and the first customer will each have received exactly the same amount of service, at which point they share the processor and receive service at the rate of $\frac{1}{2}$ sec/sec each. This will continue until some new customer enters or until one of the two customers has had enough service and departs. And so it goes with the processor always devoting its complete attention to *all* those customers who have so far received the least amount of service.

Let us calculate the average conditional response time for this system. In doing so we make the following observation. When a "tagged" job requiring x sec of service enters the system, he finds other jobs in the system, some of which have received less and some greater then x sec

---

* The name originally comes from a finite quantum algorithm in which there exist two queues. A newly arriving customer joins the first queue and waits in a first-come–first-serve fashion to receive his first quantum of service; when that (finite) quantum expires, he then joins the tail of the second queue, and each time through the system thereafter he will join the tail of the same second queue. The server always gives attention to the first queue as long as anyone there needs service and only when it is empty does he give attention to the second queue. Thus the first queue may be thought of as the foreground jobs (hopefully those requiring one quantum or less) and the second queue may be referred to as the background queue (containing jobs that require more service and therefore may be considered background operations). If we permit N queues (rather than two) and let $N \to \infty$ and the quantum $q \to 0$, then we arrive at the (generalized) FB algorithm studied in this section.

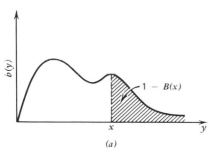
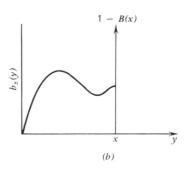

Figure 4.6   The truncated service time pdf.

already. It is clear that all those jobs who have an attained service equal to or greater than x sec will in no way interfere with our tagged job. In a similar fashion any job with attained service less than x sec will be permitted to receive up to, but not exceeding, x sec before our tagged job receives his x sec of service; thus from the point of view of our tagged job, any job requiring in excess of x sec of service may as well leave the system when it has received that much service. We are therefore led to consider the following truncated distribution of service:

$$B_x(y) \triangleq \begin{cases} B(y) & y < x \\ 1 & y \geq x \end{cases}$$

The pdf $[b_x(y)]$ associated with this is shown in Figure 4.6(b), where we see for service times less than x the density is the same as the original density $b(y)$ of the original M/G/1 system; all service times equal to or in excess of x have been accumulated into the impulse of area $1 - B(x)$ at $y = x$. Of interest to us are the moments of this truncated distribution defined as

$$\overline{x_x^n} \triangleq \int_0^x y^n \, dB(y) + x^n[1 - B(x)] \tag{4.24}$$

We may therefore define the utilization factor associated with this truncated pdf as

$$\rho_x \triangleq \lambda \bar{x}_x \tag{4.25}$$

We note that $\overline{x_\infty^n} = \overline{x^n}$ and $\rho_\infty = \rho$. If we temporarily consider an M/G/1 system operating under the FCFS rule but with service time truncated as above, then we recognize that W_x, the average waiting time (or backlog) in that system, is given simply by the P-K mean value formula [Eq. (1.82)] and may in this (truncated) case be expressed as

$$W_x = \frac{\lambda \overline{x_x^2}}{2(1 - \rho_x)} \tag{4.26}$$

It is easy to see that this is the average amount of "work" found by a new arrival to the modified system (the unfinished work is a Markov process for M/G/1). Now in the original FB system we know that a new arrival finds an average amount of work equal to $W = \lambda \overline{x^2}/[2(1-\rho)]$ in the system upon his arrival; this is true by our conservation law from Chapter 3. Of this work, only an amount W_x must be completed before our new arrival (who requires x sec of service) leaves the system.

We are now in a position to calculate $T(x)$. Our tagged arrival is delayed on the average by W_x sec for the backlog he finds and is further delayed by his own service time of x sec. In addition, during his stay in the system, there will on the average be $\lambda T(x)$ new arrivals, each of which will delay him an average of $\bar{x}_x$ sec, as discussed above. We have

$$T(x) = W_x + x + \lambda T(x)\bar{x}_x$$

which immediately gives

$$T(x) = \frac{W_x + x}{1 - \rho_x} \qquad \blacksquare \qquad (4.27)$$

as the final solution for the average conditional response time in the FB system. Thus we see how simple it is to handle these processor-sharing systems!

In [COFF 66, COFF 68b, SCHR 67] the finite quantum FB system was studied. The transform for the response time distribution is derived in [SCHR 67, ADIR 69b, ADIR 69c]. The derivation there depends basically on the same observations we used to get the average response time, wherein one views the system as a head-of-the-line preemptive priority system with three priority groups. The high priority group corresponds to customers whose service time distribution is $B_x(y)$, whereas the intermediate priority group corresponds to the tagged customer requiring exactly x sec of service (the "low" priority group is ignored). The solution as given by Schrage [SCHR 67, also personal communication] is basically as follows. We define $B_x^*(s)$ as the transform of the truncated density shown in Figure 4.6. Next we define a busy period transform $G_x^*(s)$ recursively as in Eq. (1.89) as

$$G_x^*(s) = B_x^*(s + \lambda - \lambda G_x^*(s))$$

Last, we define the transform of the pdf of the work found in system by an arrival (whose service time is x sec), and which must be processed before he leaves as

$$W_x^*(s) = \frac{s(1 - \rho_x)}{s - \lambda + \lambda B_x^*(s)}$$

This is the usual M/G/1 waiting-time transform with a truncated service time pdf as given in Figure 4.6. Finally, we get the FB response time transform conditioned on x as

$$S^*(s \mid x) = W_x^*(s + \lambda - \lambda G_x^*(s)) \exp\left[-x(s + \lambda - \lambda G_x^*(s))\right] \quad \blacksquare \quad (4.28)$$

Let us now return to an examination of the average conditional response time. We note that the slope of this function is easily calculated as

$$\frac{dT(x)}{dx} = \frac{1}{1 - \rho_x} + 2\frac{\lambda x[1 - B(x)]}{(1 - \rho_x)^2} + \lambda^2 \overline{x_x^2}\frac{1 - B(x)}{(1 - \rho_x)^3}$$

We then have

$$\lim_{x \to 0} \frac{dT(x)}{dx} = 1 \qquad (4.29)$$

$$\lim_{x \to \infty} \frac{dT(x)}{dx} = \frac{1}{1 - \rho} \qquad (4.30)$$

The interpretation of these last two equations is the following. We recall that $dT(x)/dx$ is the inverse of the average rate of service given to a job with x sec of attained service. Since in this system, service is always given to that job with the least attained service, it is clear that a newly entering job waits not at all, and so a job with extremely short service requirements has the largest possible service rate of unity [Eq. (4.29)]. Similarly, an extremely long job waits in the system until all jobs present and all new jobs that arrive during his time in system have been fully processed; it is clear then that his performance looks similar to that in either a preemptive LCFS system (in which case he gravitates to the position of the oldest customer in the system) or in an RR system (since there too he will wait until all other customers have passed through their complete service cycle). In all cases the inverse of the rate at which he gains service is given by Eq. (4.30), which is the same as that for the RR and LCFS, as can be seen from Eq. (4.17). Another way to think about this result is to recognize that a very long job only gets served during the "idle time" of the system if he were not present (i.e., $1 - \rho$ of the time).

In Figure 4.7 we see an example of the average response time for the case of exponential service times. We note that the FB system is most discriminatory in *favor* of short jobs and therefore we might expect that it is most discriminatory *against* long jobs; this is indeed the case, as we shall see in Section 4.9.

In a recent simulation experiment of an IBM APL/360 system (a dedicated interactive system serving a population of experienced users), the performance of the FB algorithm was studied [ANDE 74]. A finite quantum ($q = 0.1$ sec) FB algorithm with an infinite number of queues

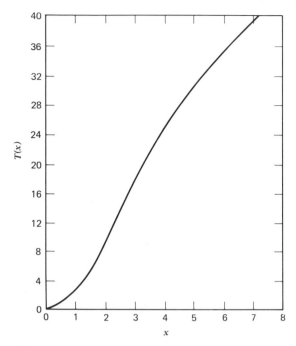

Figure 4.7   Response time for FB with $\lambda = 0.75$, $\bar{x} = 1$.

$(N = \infty$; see footnote on p. 172) was shown to yield the smallest response time averaged over all jobs requiring less than 2 sec of service (the average response time was 0.75 sec serving a population of 35 users) as compared to some other algorithms; it also yielded the highest through-put. The service times were drawn from a distribution that had been accumulated through measurement on the actual system. It is interesting to note that the service time distribution had a mean of 2.064 sec and a variance of 387.8 sec^2; this gives a coefficient of variation equal to 9.54. In fact, only 6% of the service times exceeded 2 sec (indeed, the upper 4% of the service times contributed 90% of the estimated mean); therefore the mean service time was greatly overestimated for a typical job. It was further found that the measured $b(x)$ had an increasing mean residual life, and so one is not surprised that the FB algorithm performed so well (i.e., that job with the smallest attained service is most likely to need the least additional service on the average); in fact, it has been shown [VAND 69] that when $b(x)$ has an increasing mean residual life, then the processor-sharing FB algorithm produces the minimum average response time as compared to all finite quantum FB algorithms. However, with a decreasing mean residual life, the FB algorithm can be disastrous as regards the average service time; for example, the reader might consider how jobs behave with FB when all service times are equal to a constant!

4.7. THE MULTILEVEL PROCESSOR-SHARING SCHEDULING ALGORITHM [KLEI 70c, 71b, 72c]

Here we describe a mixed scheduling algorithm that allows one to define a very large class of algorithms. Again we assume Poisson arrivals, arbitrary service time distribution, zero quantum size, and a system of queues as follows. Define a set of attained service times $\{a_i\}$ such that

$$0 = a_0 < a_1 < a_2 < \cdots < a_N < a_{N+1} = \infty \qquad (4.31)$$

We also define $N+1$ scheduling disciplines where D_i is the discipline followed for a job when it has attained service, τ, in the interval (level)

$$a_{i-1} \leq \tau < a_i \qquad i = 1, 2, \ldots, N+1 \qquad (4.32)$$

We permit D_i to be either FCFS (batch processing), FB, or RR. In addition, the FB rule is used to select the next interval to be served; that is, the processor will give its complete attention to those jobs in the lowest nonempty level and will schedule them according to the discipline appropriate for that interval (see Figure 4.8). For example, when $N = 0$,

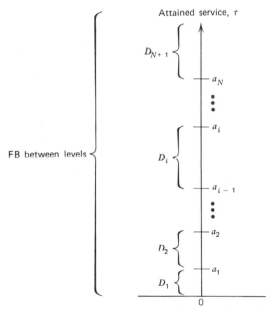

Figure 4.8   Intervals of attained service, with disciplines, D_i.

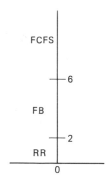

Figure 4.9   Example of $N = 2$.

we have the usual single-level case of either FCFS, RR, or FB. For $N = 1$, we could have any of nine paired disciplines (FCFS followed by FCFS, ..., RR followed by RR); note that FB followed by FB is just a single FB system (due to the overall FB policy between levels).

As an illustrative example, consider the $N = 2$ case shown in Figure 4.9. Any new arrivals begin to share the processor in an RR fashion with all other customers who so far have less than 2 sec of attained service. Customers in the range $2 \le \tau < 6$ may get served only if no customers present have had less than 2 sec of service; in such a case, that customer (or customers) with the least attained service will proceed to occupy the server in an FB fashion until they either leave, or reach $\tau = 6$, or some new customer arrives (in which case the overall FB rule provides that the RR policy at level 1 preempts). If all customers have $\tau \ge 6$, then the "oldest" customer will be served to completion unless interrupted by a new arrival. The history of some customers in this example system is shown in Figure 4.10. As usual, we denote customer n by C_n. Note that the slope of attained service varies inversely with the number of customers simultaneously being serviced. We see, for example, that C_2 requires 5 sec of service and spends 14 sec in system (i.e., a response time of 14 sec).

So much for the system specification. We may summarize by saying that we have an M/G/1 queueing system model with processor-sharing and with a generalized multilevel (ML) scheduling structure.

The quantity we wish to solve for is $T(x)$. We make the further definition:

$$T_i(x) = E\{\text{time spent in the interval } i \ [a_{i-1}, a_i) \text{ for customers}$$
$$\text{requiring a total service of } x \text{ sec}\} \qquad (4.33)$$

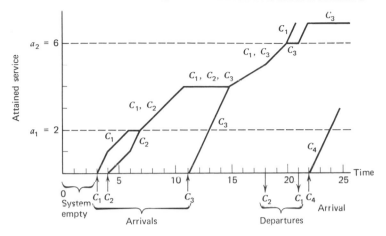

Figure 4.10   History of customers in example.

We note that

$$T_i(x) = T_i(x') \qquad \text{for } x, x' \geq a_i$$

Furthermore, we have, for $a_{k-1} \leq x < a_k$, that

$$T(x) = \sum_{i=1}^{k} T_i(x) \tag{4.34}$$

We wish to find an expression for $T(x)$, the mean response time for jobs with service time x such that $a_{i-1} \leq x < a_i$, that is, for jobs which leave the system in the ith level. To accomplish this it is convenient to isolate the ith level to some extent. We make use of the following two facts.

1. By the assumption of preemptive priority of lower level queues (i.e., FB discipline between levels) it is clear that jobs in levels higher than the ith level can be ignored. This follows since these jobs cannot interfere with the servicing of the lower levels.
2. We are interested in jobs that will reach the ith-level queue and then depart from the system before passing to the $(i+1)$st level. The system time of such a job can be thought of as occurring in two parts. The first portion is the time from the job's arrival to the queueing system until the group at the ith level is serviced for the first time after this job has reached the ith level. The second portion starts with the end of the first portion and ends when the job leaves the system. It is easy to see that both the first and second portions of the job's system time are unaffected by the

exact choice of service disciplines used in levels 1 through $i-1$. Therefore, we can assume any convenient disciplines. In fact, all these levels can be lumped into one equivalent level that services jobs with attained service between 0 and a_{i-1} sec using any service discipline.

From the above facts it follows that we can solve for $T(x)$ for jobs that leave the system from the ith level by considering a three-level system. The lowest level services jobs with attained service between 0 and a_{i-1} whereas the middle level services jobs with attained service between a_{i-1} and a_i. Jobs that would have passed to the $(i+1)$st level after receiving a_i sec of service in the original system are now assumed to leave the system at that point (this is actually the ignored third level). In other words, the service time distribution is truncated at a_i.

We have three cases to consider, depending upon which discipline is in effect at the customer's ith level (his departure level).

We begin with the case where the ith-level discipline is FB. Consider the three-level system with the middle level corresponding to the ith level of the original system. Since we are free to choose the discipline used in the lower level, we can assume that the FB discipline is used in this level as well. Clearly, the system now behaves like a single-level FB system with service time distribution truncated at x. Thus the solution for such a system is exactly that given in Eq. (4.27) above.

Now let us study the case in which the discipline at the ith level is FCFS. Consider a three-level system with breakpoints at a_{i-1} and a_i. Regardless of the discipline in the first (lowest) level, a tagged job entering the system will be delayed by the sum of the work currently in the first two levels plus a delay due to any new arrivals to the first level queue during the interval this job is in the system. Due to the FCFS discipline in the middle level, we know that all of the "current" work (whose average is W_{a_i}) which the arrival finds in the first two levels must be done before our job leaves the system and certainly, of course, he must be delayed by his own service time x. On the average $\lambda T(x)$ new customers will arrive during his time in system and each of these will delay him on the average by $\bar{x}_{a_{i-1}}$ sec (since none of these later arrivals will receive more than a_{i-1} sec of service due to the FCFS policy in the ith level).* Thus we may write as usual

$$T(x) = W_{a_i} + x + \lambda T(x) \bar{x}_{a_{i-1}}$$

---

* These new arrivals form a Poisson process with parameter λ and the contribution of each to the delay is a random variable whose first and second moments are $\bar{x}_{a_{i-1}}$ and $\overline{x^2_{a_{i-1}}}$, respectively.

which gives

$$T(x) = \frac{W_{a_i} + x}{1 - \rho_{a_{i-1}}} \qquad \blacksquare \quad (4.35)$$

as the result for the average response time when the ith level is FCFS. It is also possible to use methods such as these for solving the LCFS and random order of service at the last level for a given customer.

When the customer's ith level is RR the situation becomes more complex. If he departs during the first level, that is, $0 \le x < a_1$, then the result is simply

$$T(x) = \frac{x}{1 - \rho_{a_1}} \qquad\qquad (4.36)$$

which is just like a pure RR system where the service time distribution is truncated at $x = a_1$.

When the ith-level discipline is RR and $i \ge 2$ then the analysis increases in difficulty. Again the approach we take is that of a head-of-the-line analysis in which we consider an $N = 2$ system with break points at a_{i-1} and a_i. We consider the busy periods of the lower level. During each such busy period there may be a number of jobs that reach the doorway of the middle level. We choose to consider these arrivals to the middle level as occurring at the end of the lower-level busy period (llbp) so that there is a *bulk* arrival to the middle level at this time. We also choose to create a *virtual* time axis telescoped to delete the llbp's from the time axis. Since the time from the end of one llbp to the start of the next is exponentially distributed (Poisson arrivals!), the arrivals to the middle level appear in virtual time to be bulk arrivals at instants generated from a Poisson process with parameter λ. Now, consider a job that requires $x = a_{i-1} + \tau$ sec of service ($0 \le \tau < a_i - a_{i-1}$). Let α_1 be the mean *real* time the job spends in the system until its arrival (at the end of the llbp) to the middle level queue. Let $\alpha_2(\tau)$ be the mean *virtual* time the job spends in the middle level queue. α_1 will be calculated using a delay cycle analysis (see Section 3.3). The initial delay is equal to the mean work the job finds in the lower level on arrival plus the a_{i-1} sec of work that it contributes to the lower level. This initial delay is "expanded" by a factor of $1/(1 - \rho_{a_{i-1}})$ due to new jobs arriving at the lower level,* therefore

$$\alpha_1 = \frac{1}{1 - \rho_{a_{i-1}}} (W_{a_{i-1}} + a_{i-1}) \qquad (4.37)$$

---

* That is, the usual argument for α_1 includes the delay term $\lambda \alpha_1 \bar{x}_{a_{i-1}}$, which is the average number of arrivals ($\lambda \alpha_1$) times their average service ($\bar{x}_{a_{i-1}}$) for the truncated distribution. This gives $\alpha_1 = W_{a_{i-1}} + a_{i-1} + \lambda \alpha_1 \bar{x}_{a_{i-1}}$ and leads to Eq. (4.37).

If $\alpha_2(\tau)$ is the mean *virtual* time the job spends in the middle level, we may easily convert this to the mean *real* time spent in this level as follows. In the virtual time interval $\alpha_2(\tau)$ there are an average of $\lambda\alpha_2(\tau)$ llbp's that have been ignored. From Eq. (1.90), each of these has a mean length of $\bar{x}_{a_{i-1}}/(1-\rho_{a_{i-1}})$. Therefore, the mean real time the job spends in the middle level is given by

$$\alpha_2(\tau) + \lambda\alpha_2(\tau)\frac{\bar{x}_{a_{i-1}}}{1-\rho_{a_{i-1}}} = \frac{\alpha_2(\tau)}{1-\rho_{a_{i-1}}} \qquad (4.38)$$

Combining these results, we see that a job requiring $x = a_{i-1} + \tau$ sec of service has a mean response time given by

$$T(a_{i-1}+\tau) = \frac{1}{1-\rho_{a_{i-1}}}[W_{a_{i-1}} + a_{i-1} + \alpha_2(\tau)] \qquad (4.39)$$

The only unknown quantity in this equation is $\alpha_2(\tau)$. To solve for $\alpha_2(\tau)$ we must, in general, consider an M/G/1 system with *bulk* arrivals and RR processor-sharing. The only exception is the case of RR at the first level, which has only single arrivals and whose solution was given in Eq. (4.36).

Let us now consider the bulk arrival RR system in isolation in order to solve for the virtual time spent in the middle-level queue, $\alpha_2(\tau)$, which we temporarily write as $\alpha(\tau)$. This mean conditional virtual response time must then be substituted into Eq. (4.39) to yield our final solution for the mean conditional real response time in our multilevel system at the ith level that corresponds to an RR discipline; on the other hand the quantity $\alpha(\tau)$ that we seek will correspond to the solution for the mean conditional real response time if we had a bulk arrival RR processor-sharing system operating in isolation by itself, and so the reader should think of this quantity in the same way he has been thinking of $T(x)$. Regarding the input, we let $\tilde{a}$ be the size of a bulk whose average will be denoted by $\bar{a}$. Moreover we let b be the mean number of arrivals that enter with (and in addition to) our tagged job (the one requiring τ sec of service). In Exercise 4.14 we derive an integral equation for $\alpha(\tau)$ using a finite quantum argument and then take the processor-sharing limit; here we give an intuitive but direct derivation of that integral equation for the processor-sharing case. The difference $\alpha(\tau+\Delta\tau)-\alpha(\tau)$ is just the mean time it takes our tagged job to receive an additional $\Delta\tau$ sec of service after it has received its first τ sec. This difference is composed of four delay terms: first, the time required by those jobs that were initially found by our tagged job upon its arrival and that are still there after it has received τ sec; secondly, the time required by those jobs which have arrived since our job entered the system and that are still there; thirdly, a delay due to the tagged job itself; and lastly, the delay due to those jobs that arrived along with our tagged job in the original bulk and that require more than τ sec of service. The first of these terms may be calculated as follows. We know from Eq. (4.5) that the expected number of jobs found by our arrival who have an attained service which lies in the interval $(x, x + dx)$ is just $n(x)\, dx$. The probability that any one of these is still in the system after our tagged job has received τ sec of service, given that such a job had reached x sec of service by the time our tagged job arrived, is merely $[1 - B(x + \tau)]/[1 - B(x)]$ (recall that in the RR processor-sharing system all jobs receive service at the same rate); therefore the total number of these "old" jobs that interfere after our tagged job has received τ sec of service is merely

$$\int_0^\infty n(x)\frac{1 - B(x + \tau)}{1 - B(x)}dx$$

Now for the second term. The average time it takes our job to receive its next Δx sec of service given that it has received x sec $(0 \le x < \tau)$ is merely $\alpha(x + \Delta x) - \alpha(x)$; during this time an average of $\lambda \bar{a}[\alpha(x + \Delta x) - \alpha(x)]$ new jobs will arrive, any one of which will still be there as an interference after our job has received a total of τ sec of service with probability $1 - B(\tau - x)$. If we now multiply and divide by Δx and then let Δx shrink to zero we may then integrate over all possible values of x obtaining the average number of new arrivals that still interfere with our tagged job after it has received τ sec of service, namely,

$$\lambda \bar{a} \int_0^\tau \alpha^{(1)}(x)[1 - B(\tau - x)]\, dx$$

where

$$\alpha^{(1)}(x) \triangleq \frac{d\alpha(x)}{dx}$$

The third term corresponds to the job itself, and the average number of jobs that contribute to the fourth term is merely $b[1 - B(\tau)]$. Since each of these interfering jobs will receive $\Delta \tau$ sec of service while our tagged job receives this same amount, we must multiply each of the four quantities described above by $\Delta \tau$; dividing by this quantity we will then have an expression for $[\alpha(\tau + \Delta \tau) - \alpha(\tau)]/\Delta \tau$, and taking the limit we immediately find the following integral equation, which describes the mean response time we were seeking:

$$\alpha^{(1)}(\tau) = \int_0^\infty n(x)\frac{1 - B(\tau + x)}{1 - B(x)}\, dx$$

$$+ \lambda \bar{a} \int_0^\tau \alpha^{(1)}(x)[1 - B(\tau - x)]\, dx$$

$$+ 1 + b[1 - B(\tau)]$$

Lastly, we use Eq. (4.11) to replace $n(x)$ and obtain an integral equation involving only the derivative of the function we are looking for:

$$\alpha^{(1)}(\tau) = \lambda \bar{a} \int_0^\infty \alpha^{(1)}(x)[1 - B(\tau + x)] \, dx$$

$$+ \lambda \bar{a} \int_0^\infty \alpha^{(1)}(x)[1 - B(\tau - x)] \, dx$$

$$+ 1 + b[1 - B(\tau)] \tag{4.40}$$

Defining the generating function* $G(z)$ for the bulk size $\tilde{a}$,

$$G(z) = \sum_{n=0}^\infty P[\tilde{a} = n] z^n$$

we know immediately that $\bar{a} = G^{(1)}(1)$. The value of b may also be expressed in terms of $G(z)$ as follows. Assume that the tagged job is selected at random from the arrivals to the queueing system. Then the probability that the job was selected from a bulk of size n may be calculated as $nP[\tilde{a} = n]/\bar{a}$, which is the discrete form of the pdf for the sampled interval lifetime as given in the footnote on p. 16. Now since $1 + b$ is the average size of the bulk from which our tagged job is selected and since we have just given the probability distribution for this "sampled" bulk size, we must have

$$1 + b = \sum_{n=0}^\infty n \frac{nP[\tilde{a} = n]}{\bar{a}}$$

$$= \frac{\overline{a^2}}{\bar{a}}$$

$$= \frac{G^{(2)}(1) + G^{(1)}(1)}{G^{(1)}(1)}$$

and so finally we may evaluate b from

$$b = \frac{G^{(2)}(1)}{G^{(1)}(1)} \tag{4.41}$$

Our remaining task is to solve the integral equation (4.40). To date this equation has been solved only with the restriction that the service time

---

distribution $B(\tau)$ for the ith level be of the following form:

$$1 - B(\tau) = \begin{cases} q(\tau)e^{-\mu\tau} & 0 \le \tau < a_i - a_{i-1} \triangleq t_1 \\ 0 & t_1 \le \tau \end{cases} \tag{4.42}$$

where $q(\tau)$ is a polynomial of degree d [restricted only in that it must render $B(\tau)$ a distribution function]. In the original multilevel system this distribution transforms into a similar form over the interval $a_{i-1} \le x < a_i$. The solution to the integral equation in this case is developed in Exercise 4.15 as reported in [KLEI 71b]. That solution may be written as

$$\alpha^{(1)}(\tau) - \frac{1}{1 - \lambda \bar{a} \bar{\tau}}$$

$$- \frac{b}{\lambda \bar{a}} \sum_{i=1}^{d+1} \frac{(\mu^2 - \beta_i^2)^{d+1}}{Q_2^{(1)}(\beta_i)} \left[\frac{Q_0(\beta_i)e^{\beta_i\tau} - Q_1(\beta_i)e^{\beta_i(t_1-\tau)}}{Q_0(\beta_i) + Q_1(\beta_i)e^{-\beta_i t_1}} \right] \tag{4.43}$$

where

$$Q_0(y) = (y + \mu)^{d+1} - \lambda \bar{a} \sum_{i=0}^{d} q^{(i)}(0)(y + \mu)^{d-i} \tag{4.44}$$

$$Q_1(y) = \lambda \bar{a} \sum_{i=0}^{d} e^{-\mu t_1} q^{(i)}(t_1)(y + \lambda)^{d-i} \tag{4.45}$$

$$Q_2(y) = Q_0(y)Q_0(-y) - Q_1(y)Q_1(-y) \tag{4.46}$$

and where the roots of the equation

$$Q_2(y) = 0$$

occur in pairs which we denote by $(\beta_i, -\beta_i)$ for $i = 1, 2, \ldots, d + 1$. $\bar{\tau}$ is the expectation with respect to $B(\tau)$ in Eq. (4.42). This solution may now be inserted back into Eq. (4.39) [with $\alpha_2(\tau) = \alpha(\tau)$], which completes our solution for the multilevel system when the ith-level discipline is RR.

Since the solution with RR is so complex, let us consider some of the behavior possible by means of examples. We begin with four examples of exponential service time. For the case $N = 1$ we have the nine possible disciplines mentioned earlier. Since the behavior of the average conditional response time in any particular level is independent of the discipline in all other levels, we show the behavior of a two-level ($N = 1$) system displaying each of the three disciplines in each level in Figure 4.11; these may be combined in any way to create the nine disciplines. In Figure 4.12 we show the case for $N = 3$ where $D_1 = \text{RR}$, $D_2 = \text{FB}$, $D_3 = \text{FCFS}$ (batch processing), and $D_4 = \text{RR}$. Also shown (dashed) is the case of FB over the entire range, which serves as a reference curve for

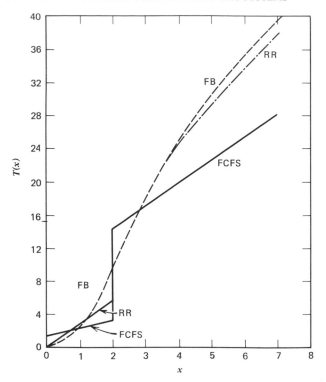

Figure 4.11 Response time possibilities for $N = 1$, M/M/1, $\bar{x} = 1$, $\lambda = 0.75$, $a_1 = 2$.

comparison with this more complicated discipline. Figure 4.13 is the case for the iterated structure $D_i = \text{FCFS}$. Again we show the FB case as a reference curve. Our fourth example is given in Figure 4.14 for the iterated structure $D_i = \text{RR}$. Again the FB curve is shown. Note in all these examples that the FB reference curve never lies above the response function at the beginning of an interval (that is, just beyond a_i) nor does it ever lie below the response curve just before the end of an interval (just before a_i).*

For nonexponential service time distributions we offer two examples. First we choose a two-stage Erlangian and the disciplines we consider are $D_1 = \text{RR}$ and $D_2 = \text{FCFS}$. In Figure 4.15 we show the response time

---

* That this must always be true can be shown by the conservation law given in Eq. (4.57) below.

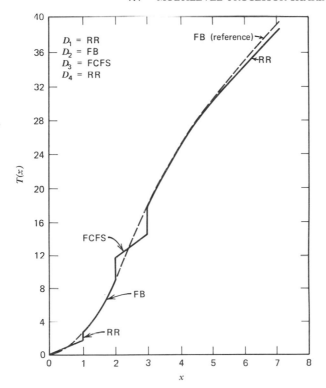

Figure 4.12 Response time for an example of $N = 3$, $\bar{x} = 1$, $\lambda = 0.75$, $a_1 = i$.

where the break point a_1 is chosen with five different values. For our second example we choose

$$
b(x) = \begin{cases} 1 & 0 \leq x \leq \dfrac{1}{2} \\[2ex] e^{-2[x-(1/2)]} & \dfrac{1}{2} \leq x \end{cases}
$$

The discipline considered is $D_1 = \text{FCFS}$, $D_2 = \text{RR}$, and $D_3 = \text{FCFS}$. The performance of this system is given in Figure 4.16, where now we show RR as a reference curve.

The versatility made available with this multilevel discipline is considerable. We do note, however, the tendency of the mean response time curves to "hover" about the FB performance (this is due to the overall FB policy among levels).

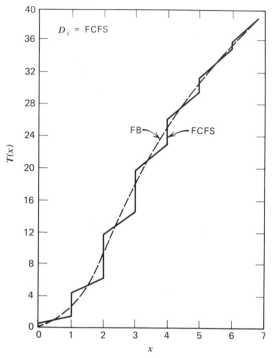

Figure 4.13   Response time for an example of $N = \infty$, $D_i = FCFS$, M/M/1, $\bar{x} = 1$, $\lambda = 0.75$, $a_i = i$.

4.8.   SELFISH SCHEDULING ALGORITHMS [KLEI 70a, HSU 71, KLEI 73b]

We have so far described scheduling algorithms which provide: no discrimination with respect to service time (batch processing); linear discrimination (RR and LCFS); the ultimate discrimination in favor of short jobs (FB); and in the last section, a large family of multilevel scheduling algorithms (ML). In this section we generalize all of these (and all others now known or yet to be invented) by creating a *continuum* of "selfish" scheduling algorithms (SSA). These systems were first studied in [KLEI 70a] and the principle behind the model is that all customers in the computer system are divided into two groups: those in a "queue box" waiting for service, and those in a "service box" sharing the facility in a fashion that we shall refer to as the "raw" scheduling algorithm (see Figure 4.17). This raw scheduling algorithm may be of any type, as for example any of those already discussed. An arrival always enters the queue box where his priority (a numerical value) increases from zero at a

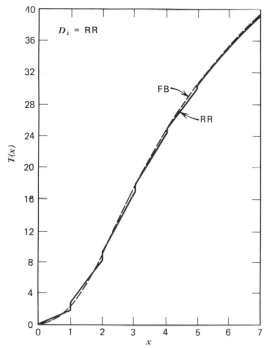

Figure 4.14   Response time for example of $N = \infty$, $D_i = $ RR, M/M/1, $\bar{x} = 1.0$, $\lambda = 0.75$, $a_i = 1$.

positive rate α; similarly, whenever a job is in the service box, his priority increases at a positive rate β (regardless whether he is sharing the CPU or merely waiting in the service box for access to the CPU). All customers possess the same parameters α, β and we restrict our attention to the case $0 \leq \beta \leq \alpha$. Furthermore, the members of the class of queueing systems we consider are of the form M/G/1.

A customer will pass from the queue box to the service box when his priority climbs to a value equal to the priority of the customers in the service box (note that all customers in the service box have the same priority, which grows at a rate β). Since $0 \leq \beta \leq \alpha$, a customer who was originally placed in the queue box must sooner or later catch up with those customers in the service box and join them to share the service facility in a fashion dictated by the raw scheduling algorithm*; note that

* In any case when the service box empties completely, then that customer (if any) with the highest value of priority in the queue box will immediately pass to the service box, at which time his priority growth rate drops from α to β. A customer arriving to an empty system passes directly into the service box (and his priority then grows from zero at a rate β).

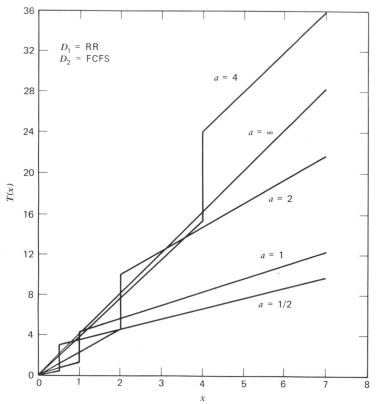

Figure 4.15   Response time for RR, followed by FCFS in $M/E_2/1$ with $\bar{x} = 1$, $\lambda = 0.75$, and $a_1 = \frac{1}{2}$, 1, 2, 4, ∞.

there is no feedback from the service box to the queue box. Since those customers in the service box are attempting to "run away" with the service facility, we choose to call these "selfish" scheduling algorithms. The parameters α and β will appear in our results as a ratio (β/α) whose value may be varied in a continuous way in order to achieve a variety of system behaviors as we now show.

Let us follow a tagged customer through the system and condition his service requirement to be x sec. In Figure 4.18 we have portrayed the case for which our tagged customer arrives at an instant t_1. At time t_3 his priority has increased to that of those in the service box, at which point he joins the service box. The arrival rate of customers to the service box conditioned on the presence of our tagged customer in that box will be denoted by λ' (typically different from λ, which is the arrival rate of jobs to the overall system) but will still be a Poisson arrival process. Thus we

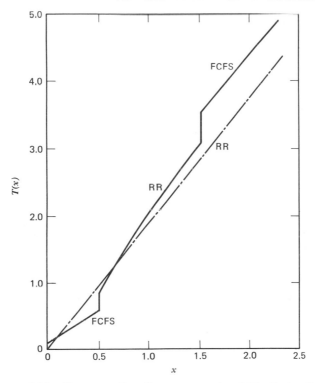

Figure 4.16   Response time for an example of $N = 2$, $\lambda = 0.75$.

see that the service box itself (conditioned on the presence of our tagged customer in that box) is also an M/G/1 system with average arrival rate λ' and with service distribution $B(x)$, which is the same as for the overall system. This turns out to be the key to analyzing the selfish scheduling algorithm systems.

Let us now calculate λ'. Referring once again to Figure 4.18 we have

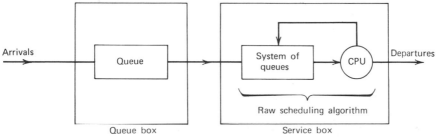

Figure 4.17   The SSA system.

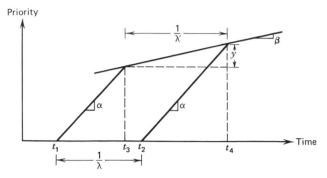

Figure 4.18   Calculation of the conditional arrival rate to the service box.

indicated a second arrival to the system at time t_2 where on the average the interarrival time is of course $1/\lambda$. These two adjacent arrivals will arrive to the service box at a slightly slower rate, namely λ' where their interarrival time is shown in the figure. We are sure that the service box does not empty during this interval since we are calculating the service box arrival rate conditioned on the presence of the tagged customer in that box. We may calculate the vertical offset y in two different ways as shown from the geometry of that diagram:

$$y = \frac{\beta}{\lambda'}$$

$$y = \left(\frac{1}{\lambda'} - \frac{1}{\lambda}\right)\alpha$$

which therefore gives

$$\lambda' = \lambda\left(1 - \frac{\beta}{\alpha}\right) \tag{4.47}$$

We now derive the system behavior.

We have already defined $S^*(s \mid x)$ as the transform for the response time density conditioned on x sec of service whose mean of course is given by $T(x)$. Let us further define: $W^*(s \mid x)$ as the transform of the conditional wasted time density whose mean is $W(x)$; $Q^*(s \mid x)$ as the transform of the density of the total conditional time our tagged customer spends in the queue box whose mean will be denoted by $W_q(x)$; $V^*(s \mid x)$ as the transform of the conditional density of time wasted in the service box with mean $V(x)$; and $Y^*(s \mid x)$ as the transform of the density of the conditional time our tagged customer spends in the service box with mean $Y(x) = V(x) + x$. It is intuitively clear (and can be shown rigorously [HSU

71]) that the time our tagged customer spends in the queue box is independent of the time he wastes in the service box. As a result $Q^*(s \mid x)$ will be independent of x and may therefore be written as $Q^*(s \mid x) = Q^*(s)$. Furthermore the total wasted time in the system for our tagged customer will be the sum of the time he wastes in the queue box and the time he wastes in the service box; since these are independent we have immediately

$$W^*(s \mid x) = Q^*(s)V^*(s \mid x) \qquad (4.48)$$

Moreover the time spent in the queue box will be independent of the raw scheduling algorithm (so long as it is a conservative algorithm, the case of interest here) and will depend only on α and β. Independent of the raw scheduling algorithm, then, the flow of customers from the queue box to the service box will occur at a rate λ' if the service box is not idle (and at an infinite rate if the service box goes idle when the queue box is not empty). The functions $W^*(s \mid x)$ and $V^*(s \mid x)$ will depend upon the scheduling algorithm but their ratio $Q^*(s)$ will be independent of that algorithm.

Let us assume that we have solved for the behavior of the raw scheduling algorithm in an isolated M/G/1 time-sharing system. In particular suppose that we have calculated the transform of the conditional wasted time density, which we denote by $\hat{W}_\lambda^*(s \mid x)$ where we have explicitly noted the dependence of this result upon the Poisson arrival parameter λ. In this case then it is clear that for the service box of the SSA system we must have, using this raw scheduling algorithm in that box,

$$V^*(s \mid x) = \hat{W}_\lambda^*(s \mid x) \qquad (4.49)$$

We must now solve for $Q^*(s)$ in order to complete our solution. This task is simple since the time spent in the queue box is independent of the raw scheduling algorithm and also is independent of the tagged customer's service time. Therefore let us choose a specific raw scheduling algorithm, the FCFS algorithm, in order to solve for $Q^*(s)$. In this case the overall SSA system really becomes one large FCFS system since the oldest customer in this system will receive the full attention of the server. Furthermore the service box itself behaves as an FCFS M/G/1 system with an arrival rate λ'. From these two observations and from Eq. (1.85) we may write down immediately

$$W^*(s \mid x) = \frac{s(1-\rho)}{s - \lambda + \lambda B^*(s)}$$

$$V^*(s \mid x) = \frac{s(1-\rho')}{s - \lambda' + \lambda' B^*(s)}$$

where $\rho = \lambda\bar{x}$, $\rho' = \lambda'\bar{x}$. These last two equations hold only when the raw scheduling algorithm is FCFS, but we have already established that the ratio $Q^*(s)$ will be independent of this algorithm. Thus we have immediately from Eq. (4.48) that

$$Q^*(s) = \left(\frac{1-\rho}{1-\rho'}\right)\left(\frac{s-\lambda'+\lambda'B^*(s)}{s-\lambda+\lambda B^*(s)}\right)$$

Using this last general result and Eq. (4.49) we may then substitute back into Eq. (4.48) to yield the final solution for the transform of the conditional waiting time density in an arbitrary SSA system, namely

$$W^*(s\mid x) = \left(\frac{1-\rho}{1-\rho'}\right)\left(\frac{s-\lambda'+\lambda'B^*(s)}{s-\lambda+\lambda B^*(s)}\right)\hat{W}_{\lambda'}^*(s\mid x) \quad \blacksquare \quad (4.50)$$

where we must recall that $\hat{W}_{\lambda'}^*(s\mid x)$ is the solution for the raw scheduling algorithm in isolation in an M/G/1 time-sharing system whose Poisson arrival rate has parameter λ'.

We may use the result in Eq. (4.50) to obtain the mean conditional response time for any SSA system as

$$T(x) = \frac{\lambda\overline{x^2}}{2(1-\rho)} - \frac{\lambda'\overline{x^2}}{2(1-\rho')} + \hat{T}_{\lambda'}(x) \quad \blacksquare \quad (4.51)$$

where again $\hat{T}_{\lambda'}(x)$ is the mean conditional response time for the raw scheduling algorithm with input rate λ'; a similar equation holds for the wasted time in the obvious way.

From the basic structure of the selfish scheduling algorithms we see when $0 < \beta = \alpha$ that the system behaves in a pure FCFS fashion since those in the queue box can never catch those in the service box and therefore the oldest arrival will be served to completion. At the other extreme we note that when $0 = \alpha = \beta$ then customers enter with and always maintain a priority value of zero; therefore all customers in the system will always be in the service box and our SSA algorithm reduces to the raw scheduling algorithm. These limits are easily seen from Eqs. (4.50)–(4.51).

Let us demonstrate the behavior of these SSA systems by means of some examples. First, we consider the selfish round robin (SRR) system. From Eq. (4.51) we have immediately that

$$T(x) = \frac{\lambda\overline{x^2}}{2(1-\rho)} - \frac{\lambda'\overline{x^2}}{2(1-\rho')} + \frac{x}{1-\rho'}$$

which is easily converted to the form

$$T(x) = \frac{\beta/\alpha}{1-\rho'}\,T_{\text{FCFS}}(x) + \left[1 - \frac{\beta/\alpha}{1-\rho'}\right]T_{\text{RR}}(x) \quad (4.52)$$

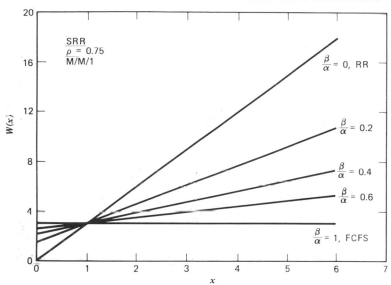

Figure 4.19 Average waiting time functions for the SRR systems with exponential service distribution. $\lambda = 0.75$, $\bar{x} = 1.0$.

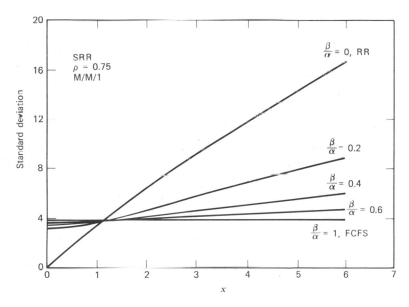

Figure 4.20 Standard deviations of the waiting time for the SRR systems with exponential service distribution. $\lambda = 0.75$, $\bar{x} = 1.0$.

where $T_{FCFS}(x)$ is the expression given in Eq. (4.13) and $T_{RR}(x)$ is the expression given in Eq. (4.17); this exposes the nature of the linear mixing of the SRR system. Recall that $T_{RR}(x_0) = T_{FCFS}(x_0)$ for $x_0 = \overline{x^2}/2\bar{x}$. In Figure 4.19 we give an example for the mean conditional wasted time in the SRR system for the case of exponential service time. In Figure 4.20 we give the standard deviation of the wasted time for the same system.

As a second example we consider the SSA system with an FB raw scheduling algorithm. This is the SFB system whose solution is

$$T(x) = \frac{\lambda \overline{x^2}}{2(1-\rho)} - \frac{\lambda' \overline{x^2}}{2(1-\rho')} + \frac{\lambda' \overline{x_x^2}}{2(1-\rho_x')^2} + \frac{x}{1-\rho_x'} \qquad (4.53)$$

where $\rho_x' = \rho_x(1 - \beta/\alpha)$. For the same system as in the two previous figures we find that the SFB system produces Figures 4.21 and 4.22.

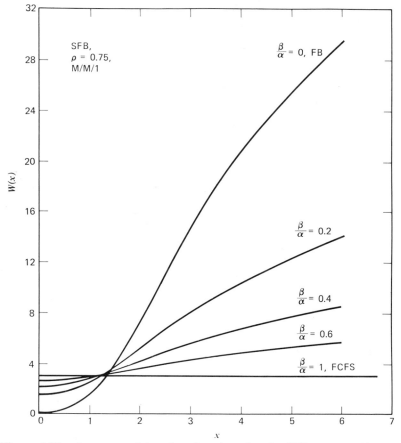

Figure 4.21   Average waiting time functions for the SFB systems with exponential service distribution. $\lambda = 0.75$, $\bar{x} = 1.0$.

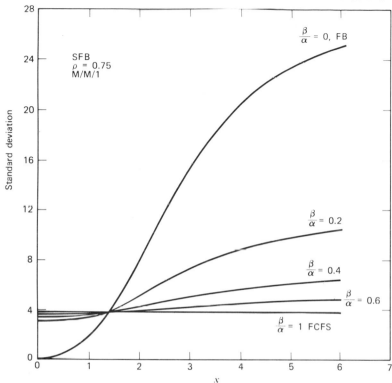

Figure 4.22   Standard deviations of the waiting time for the SFB systems with exponential service distribution. $\lambda = 0.75$, $\bar{x} = 1.0$.

Thus we see that the SSA algorithm permits us to take *any* solved system (or any newly solved ones that appear in the future) and immediately display the results for an SSA system in which this original system appears as the raw scheduling algorithm; this provides a continuum of systems that depend upon the parameter β/α and that will give performance ranging smoothly from the FCFS system all the way through to the raw scheduling algorithm itself. This generality may be obtained with almost no cost of implementation in a computer system. The implementation merely consists of counting at a rate α for units in the queue box and at a rate β for all those in the service box.

4.9.   A CONSERVATION LAW FOR TIME-SHARED SYSTEMS [KLEI 71a]

As with priority queueing systems we observe that if our scheduling algorithm improves the response time for the short jobs then there must intuitively be a conservation relationship that expresses how badly the long jobs get treated. It is this conservation law that we derive below.

Of one thing we are sure, namely, that the average unfinished work in the system must be invariant to the scheduling algorithm (so long as it is conservative). In the M/G/1 case we already know from Eq. (3.18) that the mean work, $\bar{U}$ is merely the mean waiting time in an FCFS system, expressed as

$$\bar{U} = \frac{\lambda \overline{x^2}}{2(1-\rho)} = W_{\text{FCFS}} \tag{4.54}$$

On the other hand we may also calculate this mean work as

$$\bar{U} = \int_{0^-}^{\infty} n(x)E \text{ [remaining service time for a job}$$
$$\text{with attained service time of } x] \, dx$$

where $n(x)$ is merely the attained service time density. Thus

$$\bar{U} = \int_{0^-}^{\infty} n(x) \int_x^{\infty} (\tau - x) \frac{dB(\tau)}{1 - B(x)} \, dx$$

Substituting from Eq. (4.11), we have

$$\bar{U} = \lambda \int_{0^-}^{\infty} [W^{(1)}(x) + 1] \int_x^{\infty} (\tau - x) \, dB(\tau) \, dx$$

By changing the order of integration we get

$$\bar{U} = \lambda \int_{0^-}^{\infty} \left[\int_{0^-}^{\tau} [W^{(1)}(x) + 1](\tau - x) \, dx \right] dB(\tau) \tag{4.55}$$

Integrating the inner integral by parts, yields

$$\int_{0^-}^{\tau} [W^{(1)}(x) + 1](\tau - x) \, dx = (\tau - x)[W(x) + x] \Big|_{0^-}^{\tau} + \int_{0^-}^{\tau} [W(x) + x] \, dx$$
$$= \int_{0^-}^{\tau} [W(x) + x] \, dx$$

Substituting into Eq. (4.55) we have

$$\bar{U} = \lambda \int_0^{\infty} \int_{0^-}^{\tau} [W(x) + x] \, dx \, dB(\tau)$$

Again changing the order of integration, we find

$$\bar{U} = \lambda \int_{0^-}^{\infty} [W(x) + x] \int_x^{\infty} dB(\tau) \, dx$$
$$= \lambda \int_{0^-}^{\infty} [W(x) + x][1 - B(x)] \, dx \tag{4.56}$$

But in general, we have that

$$\int_{0^-}^{\infty} x[1-B(x)]\,dx = \frac{\overline{x^2}}{2}$$

which is easily proved by integration by parts. Also, from Eq. (4.54) we have $\bar{U}$, which gives finally the following *conservation laws* for $T(x)$ and $W(x)$:

$$\int_{0^-}^{\infty} T(x)[1-B(x)]\,dx = \frac{\overline{x^2}}{2(1-\rho)} \qquad \blacksquare \quad (4.57)$$

and

$$\int_{0^-}^{\infty} W(x)[1-B(x)]\,dx = \frac{\rho\overline{x^2}}{2(1-\rho)} \qquad \blacksquare \quad (4.58)$$

Usually we write the lower limit merely as 0.

We refer to Eqs. (4.57) and (4.58) as Conservation Laws since they are based on the conservation of the average unfinished work in the system. This places an integral constraint on $W(x)$ [and $T(x)$] as a necessary condition, regardless of the scheduling discipline. The implications of the conservation law may be seen by recognizing that $[1-B(x)]$ is a nonincreasing function of x. Thus if one had a given $W(x)$ as a result of some scheduling algorithm, and then changed the algorithm so as to reduce $W(x)$ over some interval $(0, x_0)$, then the conservation law would require that the new $W(x)$ be considerably above the old value for some range above x_0. This follows since the weighting factor, $1-B(x)$, is smaller for larger x.

4.10.   TIGHT BOUNDS ON THE MEAN RESPONSE TIME [KLEI 71a]

In the previous sections of this chapter we have given results for a large number of scheduling algorithms for time-shared computer systems. We have in fact defined an infinite continuum of algorithms from any given algorithm! It is natural that we should seek some order in this embarrassment of riches. For example in the previous section we have established an invariant (our Conservation Laws) for *all* conservative algorithms. In this section we establish *tight bounds* on the possible range of performance for all such algorithms regardless of their structure.

In Figure 4.23 we display the wasted time $W(x)$ for a variety of algorithms. This figure happens to correspond to the case of exponential service with $\lambda = 0.75$, $\bar{x} = 1.0$, and therefore $\rho = 0.75$. We have deliberately superimposed the performance curves for numerous scheduling

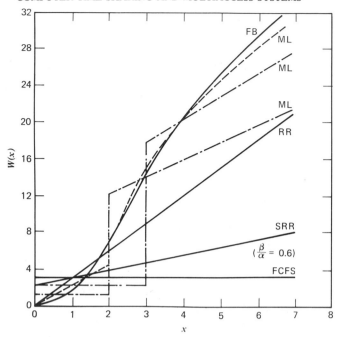

Figure 4.23   A set of response curves for M/M/1, $\bar{x} = 1.0$, $\lambda = 0.75$, $\rho = 0.75$.

disciplines and find that we are confronted with quite a selection of possible performance functions. We note, for example, the linear behavior of the RR system as contrasted with the nondiscriminatory behavior of batch processing (FCFS) and the ultimate discrimination in favor of short jobs displayed by the FB system. We note that most of these curves tend to fall in a given range of the $(W(x), x)$ plane. Were the reader to display many more of these curves he would then begin to see a pattern whereby he could sketch out the boundary within which these curves would fall. Below we establish the exact form for this boundary once and for all; specifically we will show that for any $W(x)$ we must have

$$W_L(x) \leq W(x) \leq W_U(x) \tag{4.59}$$

Let us first establish the lower bound. We claim that to *minimize* $W(x)$ the scheduling discipline must:

1. never service jobs with attained service time greater than or equal to x while there are jobs in the system with attained service time less than x, and
2. never preempt a job once it has been selected for service until it has at least x sec of attained service time.

Under these conditions the response function in the interval $(0, x)$ is just the response function for a nonpreemptive system with service times truncated at x. For convenience we will assume an FCFS scheduling discipline. We may view this as a two-level system of the sort described in Section 4.7, and so we know that the response function [denoted $W_{FCFS-x}(\tau)$] must take on a constant value as given in Eq. (4.26) over the interval $(0, x)$. The scheduling of jobs with attained service time greater than x is of no concern in this argument as long as condition (1) is maintained.

Let $\bar{U}_x$ be the mean work in the system* excluding work to be done on jobs beyond providing x sec of attained service to each. In other words, if a job requires $y > x$ sec of service and has received $\tau < x$ sec of service, its contribution to $\bar{U}_x$ is $x - \tau$. By the same method used to derive Eq. (4.57) it can be shown that

$$\bar{U}_x = \lambda \int_0^x [W(\tau) + \tau][1 - B(\tau)] \, d\tau \qquad (4.60)$$

Note that $\bar{U}_\infty = \bar{U} = W_0/(1 - \rho)$.

Since scheduling decisions are made only on the basis of a job's attained service time and its elapsed waiting time then we see that the history of a job requiring $x_1 \geq x$ sec of service from the time of its arrival to the system until it has received x sec of service is independent of the particular value x_1. Therefore we have the immediate result that $W(x)$ must be a nondecreasing function, that is,

$$\frac{dW(x)}{dx} \geq 0 \qquad \blacksquare \qquad (4.61)$$

Now since $W_{FCFS-x}(\tau)$ has minimum slope (that is, 0) over the interval $(0, x)$, if any other response curve $W(\tau)$ is such that $W(x) < W_{FCFS-x}(x)$ it must be such that $W(\tau) < W_{FCFS-x}(\tau)$ for $0 \leq \tau \leq x$. But under condition (1) above $\bar{U}_x$ has the minimum value at x when the discipline is FCFS (with truncated service) since work in this class is continuously decreased at maximum rate whenever there is such work in the system; therefore, for any $W(\tau)$,

$$\lambda \int_0^x [W(\tau) + \tau][1 - B(\tau)] \, d\tau \geq \lambda \int_0^x [W_{FCFS-x}(\tau) + \tau][1 - B(\tau)] \, d\tau$$

Thus we conclude that $W(\tau) < W_{FCFS-x}(\tau)$ in $(0, x)$ is impossible and therefore $W(x) \geq W_{FCFS-x}(x)$.

* Note that, in general, $\bar{U}_x$ differs from W_x since the former is the work (up to x) in a system with an arbitrary algorithm, whereas the latter applies only to systems (such as FCFS) whose scheduling algorithms are independent of any measure of service time.

The lower bound, $W_L(x)$, is therefore given by the waiting time for the FCFS discipline with service times truncated at x as in Eq. (4.26), namely

$$W_L(x) = \frac{\lambda \overline{x_x^2}}{2(1-\rho_x)} \qquad (4.62)$$

Note that $W_L(0) = 0$ and that $W_L(\infty) = \bar{U}$; also $W_L^{(1)}(0) = W_L^{(1)}(\infty) = 0$.

Now let us derive the upper bound. In this case we begin with a discrete-time system and then take the limit. Assume that the service time distribution is of the form

$$P[\text{service time} = kq] = p_k \qquad k = 1, 2, 3, \ldots$$

where q is the (infinitesimal) quantum as discussed earlier. Therefore, the only possible service time requirements are multiples of q. We shall also assume that arrivals may take place only during the instant before the end of a quantum and that the processor is assigned to a job for a quantum at a time. The probability that an arrival takes place at the end of a quantum is λq so that the mean arrival rate is λ. It should be clear that any continuous service time distribution can be approximated arbitrarily closely by a discrete-time distribution by letting q approach 0. Also, these restrictions on the service discipline and arrival mechanism are effectively eliminated when $q \to 0$ leading to an M/G/1 processor-sharing system. In this discrete-time model our goal is to *maximize* $W(kq)$.

We claim that the following scheduling rule is necessary and sufficient to maximize $W(kq)$: no allocation of a kth quantum is made to any job when there is some other job in the system waiting for its jth quantum where $j \neq k$. We note in passing that many scheduling disciplines will satisfy this rule.

We relabel the time axis so that $t = 0$ at an arbitrary point in some idle period. The times at which some job is allocated a kth quantum we call *critical times*. Let c_i be the epoch of the ith critical time. We wish to maximize $\bar{c}_l$ (the average of c_l) for some fixed l, and we will show that to accomplish this it is necessary and sufficient to satisfy the condition that at the lth critical time no job is waiting for a jth quantum where $j \neq k$. Certainly this condition is necessary since if a proposed scheduling discipline did not have this property then c_l can easily be increased as follows: use the proposed schedule until the point where the lth critical time would occur and then assign a quantum to a job waiting for its jth ($\neq k$) quantum.

Since we have already shown necessity, to prove the sufficiency of the condition for maximizing $\bar{c}_l$, we need only show that any schedule satisfying the condition yields the same value for $\bar{c}_l$. Let A be any scheduling algorithm that satisfies the rule that at the lth critical time no job is waiting for a jth quantum where $j \neq k$. Let a_l be the time at which the lth job arrives that will require at least kq sec of service. The state of the system at a_l will, in general, depend on the algorithm A. In particular the number of critical times that have occurred prior to a_l (let this be s) is a function of A. Let $E_A[c_l - a_l \mid \text{state of system at } a_l]$ be the expected value of $c_l - a_l$ under algorithm A conditioned on the state of the system at a_l. The state of the system is given by the number of jobs in the system, the attained service time of each job in the system, and s, the number of critical times that have occurred. Thus by our usual arguments we have

$$E_A[c_l - a_l \mid \text{state of system at } a_l]$$
$$= E_A[\text{work in system at } a_l \mid \text{state at } a_l] + (k-1)q$$
$$+ \lambda x_{(k-1)q} E_A[c_l - a_l \mid \text{state of system at } a_l] \quad (4.63)$$

Removing the condition on the state of the system at a_l we have

$$E_A[c_l - a_l] = E_A[\text{work in the system at } a_l]$$
$$+ (k-1)q + \lambda \bar{x}_{(k-1)q} E_A[c_l - a_l]$$

or

$$E_A[c_l - a_l] = \frac{E_A[\text{work in system at } a_l] + (k-1)q}{1 - \lambda \bar{x}_{(k-1)q}}$$

But $E_A[\text{work in system at } a_l]$ is not a function of the particular scheduling algorithm, and therefore $E_A[c_l - a_l]$ does not depend on A. Since $E[c_l] = E[c_l - a_l] + E[a_l]$ and the right-hand side is independent of A, $E[c_l]$ is independent of A which is what we set out to prove. Note that the form of Eq. (4.63) depended on A having the property that at c_l there are no jobs in the system waiting for a jth quantum where $j \neq k$. We have now shown that this condition is necessary and sufficient to maximize $E[c_l]$ ($\triangleq \bar{c}_l$). We now show that the general scheduling rule to maximize $W(kq)$ is the same rule that maximizes $\bar{c}_l$ applied for all l. Since $E[c_l - a_l] = \bar{c}_l - \bar{a}_l$ is the average time the lth critical customer waits for his kth quantum, we may take the limit

$$W(kq) = \lim_{n \to \infty} \frac{\sum_{l=1}^{n} \bar{c}_l - \sum_{l=1}^{n} \bar{a}_l}{n}$$

The $\bar{a}_l$ are independent of the scheduling discipline, and the proposed scheduling rule is necessary and sufficient to *individually* maximize the $\bar{c}_l$. Therefore, the same rule is necessary and sufficient to maximize $W(kq)$, which establishes our earlier claim for discrete time. However, we do not wish to constrain ourselves only to discrete quanta. It should be clear that in a continuous-time system we can approach the maximum of $W(x)$ by the following rule (the continuous version of the discrete rule above): no job with attained service time in the interval $(x - \varepsilon, x]$ (for $\varepsilon > 0$) is serviced while there is a job waiting for service that has attained service time outside this interval. By permitting ε to shrink to zero, we approach the maximum for $W(x)$. One scheduling discipline that maximizes $W(x)$ is a two-level system of the multilevel type (Section 4.7) with $a_1 = x$ (that is, the first level provides service up to but not including x sec of service) and where $D_1 = D_2 = \text{FCFS}$. From Eq. (4.35) we therefore have immediately that

$$W_U(x) = \frac{\lambda \overline{x^2}}{2(1 - \rho_x)(1 - \rho)} + \frac{x \rho_x}{1 - \rho_x}$$

We note that $W_U(0) = W_L(\infty) = \bar{U}$, $W_U^{(1)}(0) = \lambda \bar{U}$ and that $W_U^{(1)}(\infty) = \rho/(1 - \rho)$. Thus we have, finally, the tight* bounds on $W(x)$:

$$\frac{\lambda \overline{x_x^2}}{2(1 - \rho_x)} \leq W(x) \leq \frac{\lambda \overline{x^2}}{2(1 - \rho_x)(1 - \rho)} + \frac{x \rho_x}{1 - \rho_x} \qquad \blacksquare \quad (4.64)$$

In Figure 4.24 we repeat Figure 4.23 with the tight upper and lower bounds superimposed. At $x = 0$, the upper bound and FCFS start at the same point because, under the constraint of the conservation law, no other scheduling algorithm can give a longer average waiting time at $x = 0$ than FCFS. The upper bound approaches the FB response asymptotically as x approaches infinity; therefore, a customer with a very long requested service time (as compared to the mean) cannot be delayed much more than he is with FB. The lower bound starts at zero (as does the FB curve), increasing less rapidly with x than the upper bound. It approaches the FCFS curve asymptotically as x goes to infinity. We see that FCFS is the worst possible for extremely short jobs and is the best possible for extremely long jobs; conversely, FB is the best possible for extremely short jobs and the worst possible for extremely long jobs. Thus we note that the least discriminating scheduling algorithm (FCFS) touches the upper bound at $x = 0$ and forms the asymptote for the lower bound as x approaches infinity; conversely, the most discriminating scheduling algorithm (FB) touches the lower bound at $x = 0$, leaves the origin tangent to the lower bound (at zero slope), and forms the asymptote for the upper bound as x approaches infinity. The above-mentioned behavior of the upper and lower bounds applies not only for the M/M/1 system shown, but also holds true for any M/G/1 system in general, although the rate of convergence for the bounds to their respective asymptotes varies for different service-time distributions.

These upper and lower bounds on $W(x)$ immediately give upper and lower bounds on the penalty rate $W(x)/x$. An example of these bounds is

---

* These bounds are tight in the sense that every point on these bounds can actually be achieved by some feasible algorithm, and therefore no tighter general bounds can be found.

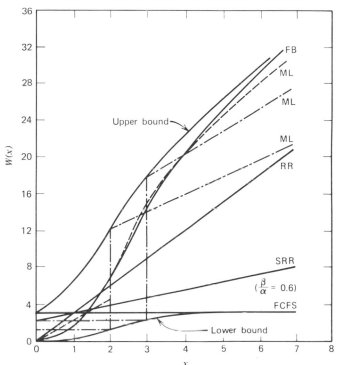

Figure 4.24   Bounds on response for M/M/1, $\bar{x} - 1.0$, $\lambda - 0.75$, $\rho - 0.75$.

given in Figure 4.25, which is the same example as for the two previous figures. We note the interesting variation of this function for several of the algorithms; as always, the RR system demonstrates its eminently fair property.

A very simple application of some of these bounding results is as follows. Imagine that you, as a system designer, are asked to invent a scheduling algorithm to provide a mean response time of no more than 4 sec for all jobs requiring 3 sec or less of service, and assume that the conditions for our three previous figures apply. You may immediately use the relationship $W(x) = T(x) - x$ and recognize that you are being asked to design a system whose mean wasted time must be no more than 1 sec for jobs requiring up to 3 sec of service; we easily see that this point falls *below* the lower bound of Figure 4.24, and you may therefore respond immediately that there is *no possible* scheduling algorithm that can provide this performance. On the other hand had the requirement fallen within the feasible range, then not only can you claim it is feasible but you can also suggest a two-level algorithm that will meet such a specification.

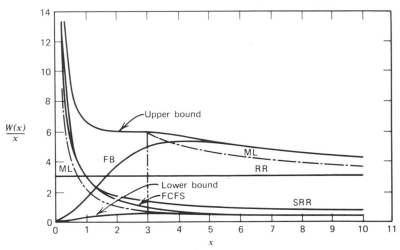

Figure 4.25   Bounds on penalty rate for M/M/1, $\bar{x} = 1.0$, $\lambda = 0.75$, $\rho = 0.75$.

We point out that the more general question (unanswered at the time of this writing) is, "What are the necessary and sufficient conditions for a given response function to be feasible?" We have presented some important necessary conditions.

So far in this chapter we have given a veritable "menu" of scheduling algorithms along with their behavior. The natural next step in this development would be to find the "optimum" algorithm from among these. Unfortunately, there has so far been precious little useful work published in this regard, mostly because no acceptable definition of "optimum" has yet been put forward. A step in this direction is reported in [KLEI 75b] and forms the substance of Exercise 4.28. Let us now leave the infinite population models and study the more realistic case of a finite user population.

4.11.   FINITE POPULATION MODELS

All of the computer models so far described have assumed that the input population was infinite and, in fact, that the arrival process was Poisson. Of course, the real world contains no infinite populations, and so we may inquire as to when we may approximate an input population by an infinite one. The answer is merely that this is a good approximation when the nature of the arrival process depends only in a negligible way upon the number of customers already in the system. Such is the case certainly for telephone and telegraph traffic and many other cases of interest. Moreover, when time-shared computing becomes available to the public at large, then perhaps that population of users will be able to be considered as infinite; thus what we have described so far in this chapter is the behavior for futuristic systems and certainly provides some insight into existing and planned systems.

Nevertheless, we must study the case of a finite input population. Consider Figure 4.26. Here we have M users, or computer consoles, that make demands upon the time-shared computer system. The dashed lines surround a feedback queueing model similar to that shown in Figure 4.2. What we are therefore describing is a view of the world external to that of the feedback queueing system, which is itself a larger feedback loop. The finite population model operates as follows: when a user at a console makes a request for service of the computer, the request "enters" the dashed box and proceeds to receive service according to the scheduling algorithm for this time-shared processor. During this time the user "goes to sleep" (i.e., he cannot generate any new requests). When finally that request is complete, the response is fed back to the console at which point the user at the console "wakes up" and then begins to generate a new request for the computer. The time spent by the user in generating this new request is referred to as his "thinking time." Thus alternating periods of thinking and processing take place as discussed in the introduction to this chapter.

We assume that the thinking time is exponentially distributed with a mean of $1/\lambda$ sec; if $M \to \infty$ and $\lambda \to 0$ such that $M\lambda = \lambda'$ (a constant) then we create a Poisson arrival process at average rate λ' from this finite population. The scheduling algorithm is assumed to be RR processor-sharing.

When the service time is exponentially distributed and when the scheduling algorithm is FCFS then this is exactly the same as the finite customer population system whose solution was given in Eq. (1.71). From the memory-less property of the exponential it is easy to see that the distribution for number in system must be independent of the scheduling algorithm. Scherr [SCHE 67] considered this case with exponential service time. He

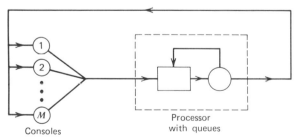

Consoles

Processor
with queues

Figure 4.26   Finite input population.

solved for the mean response time in the system (defined as time spent within the dotted box of Figure 4.26) without conditioning his result on the service time required; as usual we denote this by T. His result is (see Exercise 4.22 for a trivial derivation)

$$T = \frac{M/\mu}{1 - p_0} - \frac{1}{\lambda}$$ ▬ (4.65)

where $1/\mu$ is the average processing time required per request and where p_0 [given in Eq. (1.71) with $k = 0$] as usual is the probability and there are no customers in the dashed box. Scherr found this to be a surprisingly excellent model for the measured response time in the MIT Compatible Time Shared System (CTSS).* We note that the minimum possible average response time occurs when $M = 1$ and then $T = 1/\mu$. Normalizing T with respect to this minimum (that is, forming μT), we may plot the normalized response time as a function of the number of consoles M (see Figure 4.27). From this figure we note the very slow rise in the response time as the number of consoles increases from 1; however, after passing through a transition region, the normalized response function becomes linear, in fact, with a slope of unity. This behavior is readily understandable. In the region where the response time is growing very slowly, it is clear that the number of users is so small that the periods when a customer needs service are usually the periods when other customers are

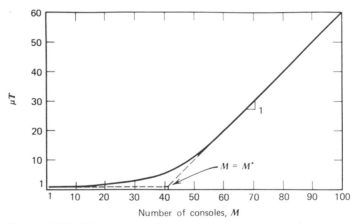

Figure 4.27   Finite population performance and saturation.

---

* CTSS violated most of his assumptions. It did not have exponential thinking times, nor exponential service, nor the structural details he assumed; neither was it conservative. In spite of this, the model predicted well. This "robustness" is more than just an accident as we shall see below.

thinking and therefore not interfering with him. However, when the number of customers increases we can see from Eq. (4.65) that the normalized response time (μT) must become linear (with a slope of unity) since p_0 approaches zero (the probability of an empty processor will go to zero when the number of active consoles increases to a "large enough" value).

Because of the finite value of M, one questions whether it is possible to "saturate" the system. Indeed if we define saturation as that point where the system becomes unstable in some sense, such as average response time growing to infinity, then we see immediately that our system is never saturated (for $\lambda/\mu < \infty$). Nevertheless there does exist an appropriate definition for saturation, which we denote by M^* and which is given by

$$M^* \triangleq \frac{1/\mu + 1/\lambda}{1/\mu} = \frac{\mu + \lambda}{\lambda} = 1 + \frac{\mu}{\lambda} \qquad \blacksquare \quad (4.66)$$

We are here defining the saturation number as the cycle time (i.e., the sum of the average thinking time plus the average required service time) divided by the average service time. Indeed, if nature were kind and all jobs required *exactly* $1/\mu$ sec of processing and *exactly* $1/\lambda$ sec for thinking (a deterministic system), then M^* is exactly the maximum number of these that could be scheduled to cause *no* mutual interference. Each user beyond M^* would cause all other users to be delayed by a time equal to his *entire* processing time ($1/\mu$ sec)—such users are certainly not absorbed "gracefully." In Figure 4.27, we may extrapolate the normalized linear asymptote for the response time ($\mu T = M - \mu/\lambda$) back to the position where this normalized response time equals unity (that is, $M - \mu/\lambda = 1$); this occurs at a number of consoles exactly equal to the saturation number, $M^* = 1 + \mu/\lambda$. The asymptotic response may then be expressed as $\mu T = M - M^* + 1$. The behavior of this asymptote shows, for $M \gg M^*$ (the "large enough" case from the previous paragraph), that with M active consoles, all users experience a normalized response time that is $M - M^* + 1$ times greater than it would be if each user had an entire processor to himself; thus M^* users appear to have been "absorbed" and converted into one user (see Exercise 4.29). Moreover, for $M \gg M^*$, each additional user causes all other users to be delayed by his entire average service time; thus we see that the saturated system behaves like a deterministic system. These important effects and numerous others are given in [KLEI 68b].

Greenberger [GREE 66] also considered a system of this sort in which he permitted nonzero swap time and finite quanta. He solved for the average response time conditioned on the service time and found a very good approximation to the behavior of this function. Adiri and Avi-Itzhak

[ADIR 69a] also considered such a case and solved for the (extremely complex) average conditional response time exactly. If we take the zero swap-time limit of Greenberger's approximation and drive the quantum to zero, we find that the conditional response time may be given approximately as

$$T(x) \cong \mu Tx \qquad \blacksquare \quad (4.67)$$

where T, the average unconditional response time, is given through Eq. (4.65). This behavior is also striking from the graphical results presented in [ADIR 69a]. Once again, we see that the round-robin scheduling algorithm gives basically linear behavior with respect to service time!

Let us now generalize some of these results for RR as was done in [BASK 71]. The generalization has to do with the exponential thinking time and service time assumptions in our basic finite population model. As discussed in Section 4.7 of Vol. I, we know that all rational distribution functions (those with rational Laplace transforms) can be synthesized by using exponential stage-type devices. The particular structure we are interested in is shown in Figure 4.28. Using this structure for the service facility gives rise to the following transform for the service time pdf in terms of the parameters β_i, γ_i, μ_i, and r (we let $\gamma_{r+1} = 1$):

$$B^*(s) = \gamma_1 + \sum_{i=1}^{r} \beta_1 \beta_2 \cdots \beta_i \gamma_{i+1} \prod_{j=1}^{i} \left(\frac{\mu_j}{s + \mu_j} \right) \qquad (4.68)$$

In a similar manner we will use this stage-type approach in describing the thinking time distribution where the corresponding parameters will be denoted by b_i, g_i, λ_i and with the total number of such (thinking) stages equal to n. We assume (as above) $g_{n+1} = 1$. Whereas in the exponential case an appropriate state description was to give the number of jobs in the dotted box of Figure 4.26, we now require a more complex state description, which we shall denote as $(k, \mathbf{u}, \mathbf{v})$; here k is the number of jobs being served in the dotted box (recall that the scheduling algorithm is RR), $\mathbf{u}$ is an n-component vector whose ith element u_i is the number of jobs in their ith stage of thinking, and where $\mathbf{v}$ is an r-component vector whose ith element v_i is the number of jobs currently receiving service at stage i

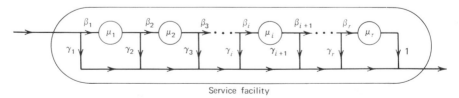

Service facility

Figure 4.28   Another stage-type server.

within the service facility in the dotted box. We let $P(k, \mathbf{u}, \mathbf{v})$ be the equilibrium probability of being in state $(k, \mathbf{u}, \mathbf{v})$. In order to write down the equilibrium balance equations we must introduce notation similar to that in Section 4.8, Vol. I and to that in [JACK 63]; to this end we define $\mathbf{u}(i^+) = \mathbf{u}$ except for its ith component, which takes on the value $u_i + 1$, $\mathbf{u}(i^-) = \mathbf{u}$ except for its ith component, which takes on the value $u_i - 1$, and $\mathbf{u}(i, j) = \mathbf{u}$ except for its ith component, which is $u_i - 1$, and its jth component, which is $u_j + 1$ [we define $\mathbf{u}(i, i) = \mathbf{u}$]. Similar notation will be used for the vector $\mathbf{v}$. By our usual methods (e.g., see Vol. I), then, we may write down the following equilibrium balance equations:

$$\left[\sum_{i=1}^{n} u_i \lambda_i + \sum_{i=1}^{r} \frac{v_i \mu_i}{k} \right] P(k, \mathbf{u}, \mathbf{v})$$

$$= \sum_{i=1}^{n-1} b_{i+1}(u_i + 1)\lambda_i P(k, \mathbf{u}(i+1, i), \mathbf{v})$$

$$+ \sum_{i=1}^{r-1} \beta_{i+1} \frac{(v_i + 1)}{k} \mu_i P(k, \mathbf{u}, \mathbf{v}(i+1, i))$$

$$+ \sum_{i=1}^{n} g_{i+1}(u_i + 1)\lambda_i \left(\frac{\beta_1}{1 - g_1 \gamma_1} \right) P(k-1, \mathbf{u}(i^+), \mathbf{v}(1^-))$$

$$+ \sum_{i=1}^{n} g_{i+1}(u_i + 1)\lambda_i \left(\frac{\gamma_1 b_1}{1 - g_1 \gamma_1} \right) P(k, \mathbf{u}(1, i), \mathbf{v})$$

$$+ \sum_{i=1}^{r} \frac{\gamma_{i+1}(v_i + 1)}{k + 1} \mu_i \left(\frac{b_1}{1 - g_1 \gamma_1} \right) P(k+1, \mathbf{u}(1^-), \mathbf{v}(i^+))$$

$$+ \sum_{i=1}^{r} \frac{\gamma_{i+1}(v_i + 1)}{k} \mu_i \left(\frac{g_1 \beta_1}{1 - g_1 \gamma_1} \right) P(k, \mathbf{u}, \mathbf{v}(1, i)) \qquad 0 \le k \le M \qquad (4.69)$$

where $P(k, \mathbf{u}, \mathbf{v})$ is considered to be zero whenever k is negative or greater than M or when u_i or v_i is negative. We note that the left-hand side of this equation is merely the rate of flow out of state $(k, \mathbf{u}, \mathbf{v})$ and the right-hand side corresponds to flow into the state; the appearance of k as a divisor is merely the effect of the RR processor-sharing when there are k customers in the service facility. One may verify that the solution to this set of equations factors into

$$P(k, \mathbf{u}, \mathbf{v}) = f(\mathbf{u}) P(k, \mathbf{v})$$

where

$$f(\mathbf{u}) = (M - k)! \lambda^{M-k} \prod_{i=1}^{n} \frac{x_i^{u_i}}{u_i!}$$

and

$$P(k, \mathbf{v}) = \frac{\lambda^k M! \, k!}{(M-k)!} \prod_{i=1}^{r} \frac{y_i^{v_i}}{v_i!} P(0, 0)$$

where $P(0, \mathbf{0}) = p_0$, the probability that there is no one in the dotted box, and

$$y_i = \left(\prod_{j=1}^{i} \beta_j\right) \frac{1}{\mu_i}$$

$$\frac{1}{\mu} = \sum_{i=1}^{r} y_i = \text{average service time}$$

$$x_i = \left(\prod_{j=1}^{i} b_j\right) \frac{1}{\lambda_i}$$

$$\frac{1}{\lambda} = \sum_{i=1}^{n} x_i = \text{average thinking time}$$

We may now form p_k by summing over $\mathbf{u}$ and $\mathbf{v}$ as follows:

$$p_k = \sum_{\mathbf{u}} \sum_{\mathbf{v}} P(k, \mathbf{u}, \mathbf{v})$$

The amazing thing is that this calculation yields

$$p_k = \frac{M!}{(M-k)!} \left(\frac{\lambda}{\mu}\right)^k p_0 \qquad 0 \le k \le M \tag{4.70}$$

and

$$p_0 = \left[\sum_{k=0}^{M} \frac{M!}{(M-k)!} \left(\frac{\lambda}{\mu}\right)^k \right]^{-1} \tag{4.71}$$

which is precisely the solution we obtained for the pure Markovian finite population case in Eq. (1.71)! Thus we find that the equilibrium distribution for the number of customers in the RR dotted box for the case of rational thinking and service time distributions is identical to that obtained for the strictly exponential case with parameters λ and μ, respectively; this result depends only upon the average thinking time $(1/\lambda)$ and the average service time $(1/\mu)$! This is reminiscent of the situation in Section 4.4 where we found that $T(x)$ was independent of the form of the service time distribution for the RR processor-sharing case.

4.12. MULTIPLE-RESOURCE MODELS

In the introductory comments to this chapter, we pointed out that the appropriate model for time-shared computer systems was one which accounted for *multiple* resources. Until now we have focused our attention on the single-resource model wherein only the CPU was under demand. In this section, we generalize to the multiple-resource case which we model as a *queueing network*.

Let us review the progress in queueing networks so far. (This material was introduced in Section 4.8 of our first volume [KLEI 75a] and was

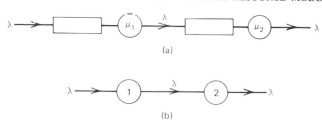

Figure 4.29   The two-tandem net. (*a*) Details of the two-node tandem net. (*b*) Network representation.

summarized in Section 1.6 in this volume.) Over twenty years ago, R. R. P. Jackson studied the two-node tandem network shown in Figure 4.29(*a*) [JACK 54] in which the first node is an M/M/1 queue (with parameters λ, μ_1, and $\rho_1 = \lambda/\mu_1 < 1$) whose output feeds a second queueing system (the second "node") with a single exponential server (of rate μ_2 and with $\rho_2 = \lambda/\mu_2 < 1$). In part (*b*) of this figure, we represent this network simply as two nodes (omitting the structure of the queues and of the servers). Jackson showed that the (equilibrium) probability, $p(k_1, k_2)$, of finding k_1 customers in the first node and k_2 customers in the second node was simply

$$p(k_1, k_2) = (1 - \rho_1)\rho_1^{k_1}(1 - \rho_2)\rho_2^{k_2}$$

This expression is simply the product of the state probabilities for two independent M/M/1 queues. In 1956, Burke showed [BURK 56] that the output of an M/M/1 queue is a Poisson process (see Section 4.8 of [KLEI 75a]); in fact, he showed that the only FCFS queueing system whose interdeparture times form an independent Poisson process is the M/M/m queue. This implies that *any* feedforward network of exponential queues (i.e., customers may not return to previously visited nodes) that is fed from independent Poisson sources, will generate an independent arrival process at each node which is Poisson; therefore, the joint probability distribution over all nodes will simply be the product of marginal distributions each of which is the solution to an M/M/m queue (given in Section 1.5). A simple example of a feedforward network is the two-node network discussed above. These feedforward networks are themselves special cases of the feedback networks first studied by J. R. Jackson in 1957 [JACK 57]. He considered arbitrary *open* Markovian networks of queues (open in the sense of permitting external arrivals and departures) an example of which is shown in Figure 4.30. This model was described in Section 1.6 and consists of an N-node network, the ith node of which contains m_i exponential servers each with a mean service time of $1/\mu_i$ sec. A Poisson source generates external arrivals to the ith node at a rate γ_i

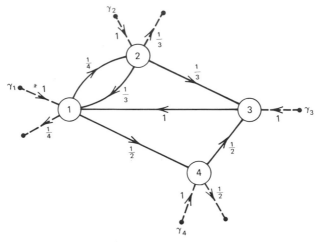

Figure 4.30   Example of an open network (branch label is r_{ij}).

customers/sec (shown as dashed inputs in the example). After receiving service at the ith node, a customer goes next to node j with probability r_{ij} (these are the branch labels in the figure) or leaves the network with probability $1 - \sum_{j=1}^{N} r_{ij}$ (shown as dashed outputs in the example); note that feedback to previously visited nodes is permitted in this model.

We let $\mathbf{R} = [r_{ij}]$ be the $N \times N$ matrix containing these transition probabilities. For the example of Figure 4.30, we have

$$\mathbf{R} = \begin{bmatrix} 0 & \frac{1}{4} & 0 & \frac{1}{2} \\ \frac{1}{3} & 0 & \frac{1}{3} & 0 \\ 1 & 0 & 0 & 0 \\ 0 & 0 & \frac{1}{2} & 0 \end{bmatrix}$$

The total (internal plus external) arrival rate of customers to the ith node is defined to be λ_i (customers per sec). To find the λ_i, we must solve the linear equations (1.75) repeated here $(i = 1, 2, \ldots, N)$

$$\lambda_i = \gamma_i + \sum_{j=1}^{N} \lambda_j r_{ji} \qquad \blacksquare \quad (4.72)$$

If we denote $\boldsymbol{\lambda} = [\lambda_1, \lambda_2, \ldots, \lambda_N]$ and $\boldsymbol{\gamma} = [\gamma_1, \gamma_2, \ldots, \gamma_N]$, then Eq. (4.72) becomes

$$\boldsymbol{\lambda} = \boldsymbol{\gamma} + \boldsymbol{\lambda}\mathbf{R} \qquad (4.73)$$

For example, if $\boldsymbol{\gamma} = [1, 0, 5, 2]$ in the network of Figure 4.30, then we find that $\boldsymbol{\lambda} = [12, 3, 10, 8]$. We note that these calculations are independent of the exponential assumptions, and of the parameters m_i and μ_i; that is, $\boldsymbol{\lambda}$

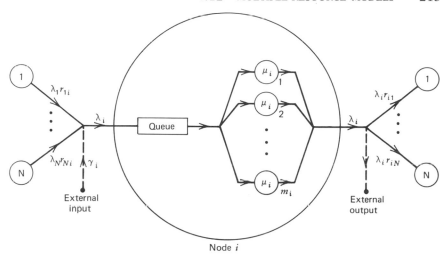

Figure 4.31   The total structure of node i (branch label is customer traffic).

depends *only* upon γ and **R**. In Figure 4.31, we show the detailed structure of the multiple-server queueing system at node i.

It can be shown for the networks with feedback that the arrival processes to the various nodes will *not* be Poisson in general. However, the amazing result [JACK 57] known as *Jackson's Theorem* states that each node behaves *as if its input were Poisson!* Specifically, if we let $p(k_1, k_2, \ldots, k_N)$ denote the equilibrium probability that there are k_i customers in the ith node ($i = 1, 2, \ldots, N$), then for $\lambda_i/m_i\mu_i < 1$ for all i, the theorem states that

$$p(k_1, k_2, \ldots, k_N) = p_1(k_1)p_2(k_2) \cdots p_N(k_N) \qquad \blacksquare \quad (4.74)$$

where $p_i(k_i)$ is simply the solution for the equilibrium probability of finding k_i customers in an M/M/m_i queue with input λ_i and with average service time $1/\mu_i$ for each of the m_i servers; this solution is given in Section 1.5. Thus we see the product solution form once again.

Next, the Gordon and Newell model [GORD 67] for arbitrary *closed* Markovian networks appeared in the literature in 1967; in fact, this was an elaboration of a special case considered by J. R. Jackson [JACK 63] in 1963. These networks are the same as the open Markovian networks except that K customers are contained (trapped) in the network and no external arrivals (or departures) are permitted; that is, $\sum_{j=1}^{N} r_{ij} = 1$ for all i.

For example, let us restructure the open network of Figure 4.30 by removing all external arrivals and departures (i.e., the dashed lines) and by changing the transition probabilities r_{ij} to yield the closed network

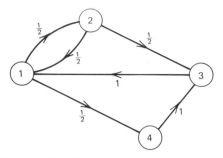

Figure 4.32   Example of a closed network.

shown in Figure 4.32. Since $\gamma_i = 0$ (for all i) in these closed networks, we see that Eq. (4.73) becomes

$$\boldsymbol{\lambda} = \boldsymbol{\lambda}\mathbf{R} \tag{4.75}$$

This equation does not uniquely define $\boldsymbol{\lambda}$, but only determines the relative value of its components λ_i (i.e., it solves for λ_i to within a multiplicative constant). For the example of Figure 4.32, we find that the ratios $\lambda_1:\lambda_2:\lambda_3:\lambda_4$ are given by 4:2:3:2. In fact, λ_i gives the relative number of visits to the ith node as customers rattle around the closed network. Gordon and Newell introduce the notation x_i where

$$\lambda_i = \mu_i x_i \tag{4.76}$$

We further note that $x_i/m_i = \lambda_i/m_i\mu_i$ is the relative utilization of the ith node. For the moment, let us assume that $m_i = 1$ for $i = 1, 2, \ldots, N$. Then the equilibrium probability $p(k_1, k_2, \ldots, k_N)$ is given by [GORD 67]

$$p(k_1, k_2, \ldots, k_N) = \frac{1}{G(K)} \prod_{i=1}^{N} x_i^{k_i} \qquad \blacksquare \tag{4.77}$$

where $G(K)$ is a constant that guarantees that the probabilities sum to unity and is given by

$$G(K) = \sum_{\mathbf{k} \in A} \prod_{i=1}^{N} x_i^{k_i} \tag{4.78}$$

where $\mathbf{k} = [k_1, k_2, \ldots, k_N]$ and A is that set of vectors $\mathbf{k}$ for which $k_i \geq 0$ and for which

$$\sum_{i=1}^{N} k_i = K \tag{4.79}$$

Again we see the product solution form in Eq. (4.77). (For the solution in the case when $m_i \geq 1$, see Section 1.6.) The evaluation of Eqs. (4.77) and (4.78) is important in calculating properties of closed networks; as shown by Buzen in [BUZE 73], if we calculate $G(1), G(2), \ldots, G(K)$, we may then derive many useful network properties. In particular (see Exercise 4.35) if we let $\tilde{k}_i$ be the number of customers in the ith node, then

$$P[\tilde{k}_i = k_i] = \frac{x_i^{k_i}}{G(K)}[G(K-k_i) - x_iG(K-k_i-1)] \qquad \blacksquare \quad (4.80)$$

$$E[\tilde{k}_i] = \sum_{k_i=1}^{K} x_i^{k_i}\frac{G(K-k_i)}{G(K)} \qquad \blacksquare \quad (4.81)$$

Following Buzen, we show an iterative algorithm in Exercise 4.36 which permits us to calculate $G(1)$, $G(2)$, ..., $G(K)$ using a total of NK multiplications and NK additions. Buzen also gives a more complex algorithm when the service rate at a node is allowed to be dependent on the number of customers present in that node [BUZE 73]; for example, this permits $m_i \geq 1$. Another approach for finding $G(K)$ is described by F. R. Moore [MOOR 72], who takes advantage of a combinatorial lemma to arrive at an explicit expression for $G(K)$ in the case when $m_i = 1$ ($i = 1, 2, \ldots, N-1$) and $m_N \geq 1$ and when $x_i \neq x_j$ ($i \neq j$).

All these results lay dormant in the literature until 1971 when C. G. Moore [MOOR 71] recognized that closed Markovian networks were suitable for modeling the behavior of multiple-resource computer systems in which each resource is modelled as a network node.* (In fact, the model in Section 4.11 [KLEI 68b] partially motivated Moore's application.) This application is a rather natural one since we may picture jobs flowing from one resource to another within the computer system as they place successive demands upon these resources; simultaneous conflicting demands on a resource, of course, are resolved by the formation of a queue in front of the resource. The surprising result at that time was that the exponential service times underlying the Markovian network models were capable of predicting system behavior that could be validated by direct measurement on the Michigan Terminal System (MTS), a system which did not enjoy exponential service times. MTS is a dual-processor system (two IBM 360/67 CPU's) with 1.5 megabytes of main memory, two swapping drums, three IBM 2314 disk storage units and other secondary memory devices, collectively serving over 100 terminals. The system was monitored for many 10–15-minute intervals in order to obtain the μ_i and the x_i. Moore found that the closed network model predicted the mean response time of the system to within 10% of the measured response time in a given time interval, when the measured parameters for that time interval were used in the model. A similar situation had been found by Scherr [SCHE 67] in his use of Markovian models for the MIT CTSS

---

* In 1964, Wallace [WALL 64] suggested the use of Jackson's open networks for the analysis of very large computer systems, but never elaborated with an application.

time-sharing system as discussed in the previous section; there we came to understand the robust behavior of the model through the surprising result in Eq, (4.70).

The Gordon and Newell model used by Moore even predicts the "bottleneck" phenomenon. The bottleneck node is that one having the largest ratio x_i/m_i which we recall is the relative utilization of that node. In the limit as $K \to \infty$, an infinite queue will form at the bottleneck node while only finite queues form at the other nodes. Furthermore, Moore found that balancing the system (that is, making all the ratios x_i/m_i approximately equal) did, in fact, remove the bottleneck and caused all nodes to act as limiting resources simultaneously.

Shortly after the applicability of closed Markovian networks to computer system modeling was demonstrated, a number of papers rapidly appeared in the literature extending this application. In this section, we describe some of the useful results that have been obtained. The observant reader will note that the classic finite population model discussed in the previous section is already an example of a closed Markovian network as long as the service times and thinking times are exponentially distributed; the system contains two nodes (the single CPU node and the M-server terminal node) with $K = M$, $m_1 = 1$, and $m_2 = M$.

When we consider adapting these open and closed network models for the multiple-resource case, we find that the previous investigations (by Jackson, Gordon and Newell, and Moore) are limited in at least the three following ways: all customers behave in an identical fashion (service times and transition probabilities are drawn from the same distribution for each); all service time distributions are exponential (Markovian queueing networks); and the scheduling algorithm in each node is basically FCFS (or at best ones that make no use of attained service for given customers). Our studies earlier in this chapter regarding RR processor-sharing suggest that perhaps one can remove these limitations. In particular, recall the two "anomalous" results in that study which we shall soon see appearing in our generalizations: first, that the RR processor-sharing system (and also the preemptive LCFS system) has a mean conditional response time that is independent of the service time distribution in the case of an open system (M/G/1); secondly, that a similar result holds for the finite population model (a closed network) in that when RR processor-sharing is used for the scheduling algorithm, the distribution for number of jobs in the system is independent of the (rational) thinking time and service time distributions. There has been recent activity with considerable progress in removing some of the above-mentioned limitations. Ferdinand [FERD 71] considered a system that allowed different classes of customers in a closed system with two nodes. Posner and Bernholtz [POSN 68]

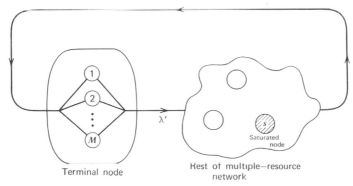

Figure 4.33   Decomposition of the multiple-resource model.

also considered a closed network where different customers were permitted different transition probabilities r_{ij}, as well as their own set of exponentially distributed service times;[*] in addition each customer was permitted an arbitrarily distributed "travel time" between nodes. Chandy [CHAN 72] applied local balance conditions to networks (see Section 4.8 of our first volume [KLEI 75a]). Some useful limit theorems for the rate at which various nodes perform work in closed networks are developed in [CHNG 72]. Before we describe the results that carefully remove the three aforementioned restrictions, we shall study some asymptotic results which are of great interest.

Let us begin by obtaining some asymptotic behavior of closed networks [MUNT 74c]. The model is identical to the Gordon and Newell model with the important exception that we do *not* require the Markovian assumptions, that is, we permit arbitrary distributions of service at each node (with a mean $1/\mu_i$ for each of the m_i servers at node i). For these asymptotic results, we assume that all customers behave in the same statistical fashion as above. Assume further that we have a time-shared system serving M terminals. Our approach is to study the system from the terminal's point of view as shown in Figure 4.33. This system is not unlike that shown in Figure 4.26; here we use the closed network model for multiple-resource time-shared systems, and obtain results similar to those in the previous section. We have a closed network with a total of $K = M$ customers, each of which generates a job from the terminal at a rate λ jobs/sec whenever the customer is in the "terminal node" (that is, each customer has a mean thinking time of $1/\lambda$ sec drawn from an arbitrary distribution). Each such generated job then enters the "rest of the

_____

* They required RR processor-sharing at those nodes where different exponential distributions for the customers were allowed.

multiple-resource network" (after all, the terminals are resources also), and moves around from resource to resource according to the transition probabilities r_{ij}, eventually returning to the terminal at which time the user generates (thinks of) a new job as in Section 4.11; note that the "rest of the network" here corresponds to the dashed box in Figure 4.26 where we now permit numerous resources (not only a CPU) and each of the service centers (resources) has an arbitrary service time distribution. We let T be the average (response) time to pass through the rest of the network, and as earlier $1/\lambda$ be the average time in the terminal node. The average cycle time is then $T + 1/\lambda$ and the system throughput is simply $\lambda' = M/(T + 1/\lambda)$ customers/sec. Similarly, we let $\bar{N} = E[$number of jobs in the rest of the network$]$ and $\bar{M} = E[$number of jobs in the terminal node$]$. By Little's result, we have

$$T = \frac{\bar{N}}{\lambda'}$$

However, since $M = \bar{N} + \bar{M}$, we get

$$T = \frac{M}{\lambda'} - \frac{\bar{M}}{\lambda'}$$

Moreover, if we apply Little's result to the terminal node, we have $1/\lambda = \bar{M}/\lambda'$ and so

$$T = \frac{M}{\lambda'} - \frac{1}{\lambda} \tag{4.82}$$

Let us now consider the saturated or bottleneck node (shown shaded and denoted by s in Figure 4.33) in the rest of the network. As mentioned above, this node is such that x_s/m_s is the largest ratio (x_i/m_i) in the system. To find x_i we must solve Eqs. (1.78), where we recall that $\mu_i x_i$ may be interpreted as the relative number of visits a job makes to the ith node in passing through the network. For the case of Figure 4.33, we see that $\mu_s x_s/\mu_N x_N$ is merely the average number of times the bottleneck is visited for each entry into the rest of the network (that is, for each visit to the terminal node); we have arbitrarily let the subscript N correspond to the terminal node and so we have the identity $\lambda = \mu_N$. Next we note that so long as $M \gg M^*$ (deeply saturated), then the output rate from the saturated node should approximately be $\mu_s m_s$. Thus the output rate of customers from the rest of the network will be $\mu_s m_s/(\mu_s x_s/\mu_N x_N)$ and this is merely λ', that is, $\lambda' = m_s \mu_N x_N/x_s$. Using this in Eq. (4.82), we finally have that T_a, the asymptotic behavior of T, is simply

$$T \cong T_a = \frac{M x_s}{m_s \mu_N x_N} - \frac{1}{\lambda} \qquad M \gg M^* \qquad \blacksquare \tag{4.83}$$

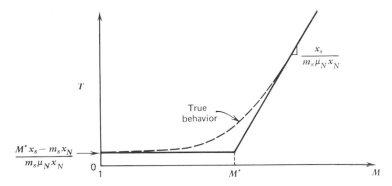

Figure 4.34   Asymptotic behavior for delay in closed networks.

This is the asymptotic behavior for T when $M \gg M^*$ and is of the same form as Eq. (4.65) (again for $M \gg M^*$). In Figure 4.34, we plot this asymptote (as in Figure 4.27). Note that it is linear with M at a slope $x_s/m_s\mu_N x_N$ for $M \gg M^*$. We also show the horizontal asymptote and a sketch of the true behavior.

To find the saturation number M^*, let us take the same approach here as we did in Eq. (4.66), and use the usual deterministic system argument. That is, we argue that M^* must be equal to the maximum number of perfectly scheduled jobs that cause no mutual interference. For each of the m_s servers in the saturated node, we can schedule a maximum number of jobs equal to the service required by a job in a cycle divided by the service time spent by a job in the saturated node per cycle (all service times are temporarily assumed to be deterministic). The total service per cycle is simply $\sum_{i=1}^{N} (\mu_i x_i/\mu_N x_N)(1/\mu_i)$, that is, the sum over each node of the number of visits to that node per cycle times the average service time per visit. The time spent in the saturated node per cycle is simply $(\mu_s x_s/\mu_N x_N)(1/\mu_s)$ for the same reason. Since we may schedule m_s such servers, then the ratio of the total service per cycle to the time spent in the saturated node per cycle may be multiplied by m_s to obtain

$$M^* = \frac{m_s}{x_s} \sum_{i=1}^{N} x_i \qquad \blacksquare \quad (4.84)$$

So far we have only a definition for M^*, and the asymptotic behavior of T. We now show the relationship between these. First, we note that the average cycle time is simply $T + (1/\lambda)$; when $M = 1$, the cycle time is simply the service time in a cycle (no queueing), which was calculated above. From this and Eq. (4.84) we see that at $M = 1$, we must have $T = (M^* x_s/m_s\mu_N x_N) - (1/\lambda)$ and this is the horizontal asymptote shown in

Figure 4.34 (we have again used $\lambda = \mu_N$). Comparing this to the saturation asymptote in Eq. (4.83), we see that the two asymptotes cross at exactly $M = M^*$, just as in Figure 4.27. Thus M^* is indeed the saturation number again, and enjoys all of its interesting properties!

It is natural to inquire as to the improvement in performance if we "remove" the bottleneck by increasing the service rate in the saturated node. If we do so, some other node, denoted by s', will become the new bottleneck, and the asymptotic behavior will again be similar to that in Eq. (4.83) with a new slope $x_{s'}/m_{s'}\mu_N x_N$ and new saturation number $M^{*(t)}$. In fact, if we continue this procedure of removing bottlenecks, we will always expose a new one with slope $x_i/m_i\mu_N x_N$ as sketched in Figure 4.35, where M^* must be recalculated with the new service rates. The behavior shown in this figure is very important in computer system design since we must be able to predict the improvement in system performance as we invest in more powerful resources. For example, if we see that the first and second bottleneck asymptotes are close to each other, then very little system improvement will be gained by removing only the first bottleneck (and it could be extremely expensive to remove the first bottleneck if it requires the addition, say, of another CPU); clearly, such a cost-ineffective change can be avoided with an analysis of this type.

Let us study the closed network model of Figure 4.33 a bit further to obtain more asymptotic results [MUNT 74c]. We consider a general closed network with N nodes; at the ith node there are m_i servers (resources) each with an arbitrary service time whose mean is $1/\mu_i$ sec independent of the system state. The Nth node is the "terminal" node

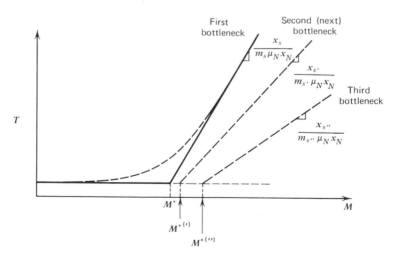

Figure 4.35   Sequence of asymptotic bottlenecks.

with $m_N = M$. The closed system contains K customers where $K = M$. As we know, $\mu_i x_i$ is the relative number of visits a job makes to the ith node in moving around the network, and so x_i/m_i is the relative utilization of each resource at node i; therefore, x_i is simply the relative utilization of the collection of m_i resources at node i. Recall that s denotes the most saturated node in the "rest of the network," that is, $x_s/m_s \geq x_i/m_i$ ($i = 1, 2, \ldots, N-1$). We let

$$f_i(M) = \text{fraction of time each resource (i.e.,}$$
$$\text{each server) in node } i \text{ is busy when}$$
$$\text{the closed net contains } M \text{ jobs}$$

We observe for $M = 1$ that there is no queueing for any system resource (i.e., no conflicts) and so

$$m_1 f_1(1) + m_2 f_2(1) + \cdots + m_N f_N(1) = 1$$

since exactly one resource is occupied at any time (note that $m_N = 1$ for $M = 1$). From this last equation and from our interpretation of x_i, we see that

$$f_i(1) = \frac{x_i/m_i}{x_1 + x_2 + \cdots + x_N} \qquad \blacksquare \qquad (4.85)$$

Let us now show that $f_i(M) \leq M f_i(1)$. The ratio x_i/x_j is independent of M since it depends only upon the mean service times and on the job flow patterns that are generated by the transition probabilities $\{r_{ij}\}$. This ratio of relative utilizations may be expressed as

$$\frac{x_i}{x_j} = \frac{m_i f_i(M)}{m_j f_j(M)}$$

Now, if $f_i(M) > M f_i(1)$ for $i = 1, 2, \ldots, N$, then $m_1 f_1(M) + \cdots + m_N f_N(M) > M$; however, this is an impossibility since the expected number of busy resources $[m_1 f_1(M) + \cdots + m_N f_N(M)]$ cannot exceed the number of customers (M). Therefore, it cannot be true for all i that $f_i(M) > M f_i(1)$. Now, if there is some i for which

$$f_i(M) > M f_i(1)$$

then

$$\frac{m_i f_i(M)}{m_j f_j(M)} > M \frac{m_i f_i(1)}{m_j f_j(M)}$$

But since $m_i f_i(M)/m_j f_j(M)$ is a constant independent of M, then also

$$\frac{m_i f_i(1)}{m_j f_j(1)} > M \frac{m_i f_i(1)}{m_j f_j(M)}$$

from which it follows that

$$f_i(M) > Mf_i(1)$$

Thus, if $f_i(M) > Mf_i(1)$ for any i, then it is true for all i; however, we showed above that this is impossible. Therefore we conclude that

$$f_i(M) \leq Mf_i(1) \qquad (4.86)$$

for all i. [We also know that $f_i(M) \leq 1$.] Note that since $m_N = M$, then $f_N(M) \leq x_N/(x_1 + \cdots + x_N)$; since this is independent of M, we see that the terminal node can never be fully utilized. Now, as we let M grow, there will be a critical (minimum) value $M = M_c$ such that $M_c f_s(1) = 1$; that is, we have selected M such that the upper bound on the utilization $f_s(M)$ for the saturated node is unity, and the value of M which achieves this is

$$M_c = \frac{1}{f_s(1)} = \frac{x_1 + x_2 + \cdots + x_N}{x_s/m_s}$$

We recognize this expression from Eq. (4.84), and therefore find that $M_c = M^*$, a most happy circumstance! For $M \geq M^*$, the upper bound on $f_s(M)$ must remain at unity since it can increase no further. Now since the ratios $m_i f_i(M)/m_j f_j(M)$ are constant with M, then the upper bound on $f_i(M)$ cannot increase beyond its value at M^* for all $M \geq M^*$ and all i for which m_i is constant (i.e., $i = 1, 2, \ldots, N-1$). In particular, for $i = 1, 2, \ldots, N-1$

$$f_i(M) \leq \begin{cases} Mf_i(1) & M \leq M^* \\ M^* f_i(1) = \dfrac{x_i/m_i}{x_s/m_s} & M \geq M^* \end{cases} \qquad \blacksquare \quad (4.87)$$

This upper bound behavior is shown in Figure 4.36; the true typical behavior is sketched in as dotted lines. Moreover, since x_N/x_i is constant, then so is $Mf_N(M)/m_i f_i(M)$; but since $\lim_{M \to \infty} f_i(M) = M^* f_i(1)$, then we see that $\lim_{M \to \infty} Mf_N(M) = M^* f_N(1)$ which gives the asymptotic behavior for the case $i = N$. Again, we may interpret M^* as before. That is, if all service times exactly equalled their mean ($1/\mu_i$ at node i) and transitions between nodes were *perfectly* coordinated in a fashion that guaranteed no mutual interference among jobs, then M^* would be the maximum number of jobs that could be so scheduled. Nature is not as kind as that, and so our true utilization falls below the asymptote; however, as $M \to \infty$, we approach the horizontal asymptote for each resource in the "rest of the network" and $Mf_N(M) \to M^* f_N(1) = m_N M^* f_N(1)$ (since $m_N = 1$ for $M = 1$), and so we may sum the utilization of all resources to obtain

$$\lim_{M \to \infty} \sum_{i=1}^{N} m_i f_i(M) = \sum_{i=1}^{N} m_i M^* f_i(1) = M^* \qquad (4.88)$$

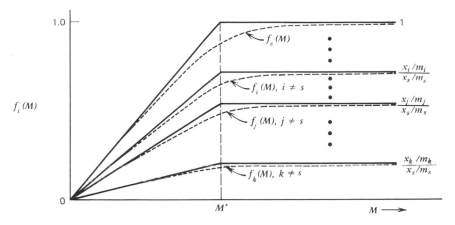

Figure 4.36   Utilization and saturation in a closed network.

Thus M^* has the additional interpretation of being equal to the limiting (maximum) utilization summed over all network resources! Note that $\lim_{M \to \infty} M f_N(M) = M^* f_N(1) = x_N/(x_s/m_s)$ and therefore, of the M^* active resources in the limit, $x_N/(x_s/m_s)$ are active terminals and the rest, say G^*, are active resources in the "rest of the network;" we see that $G^* = M^* - x_N/(x_s/m_s)$, and this is sometimes referred to as the "multiprogramming gain." Often, the degree of multiprogramming is limited to some maximum (say, D) due to the constraints of finite main memory. This may be modeled by restricting the number of customers (the "active set") in the "rest of the network" to be exactly D (a good assumption when M is large), and forcing any other ready jobs to wait at their terminals until some job leaves from the active set, at which time one of the ready jobs enters the "rest of the network" and is allowed to compete . for system resources. The asymptotic analysis of this "limited memory" model is not unlike that given above, and may be found in [MUNT 74c]; these and other asymptotic results may also be found in [WONG 75].

We shall see another application of closed networks for modeling a multiprogramming computer system in Section 4.13 below. For now, we return to the further generalization of queueing networks for computer systems modelling and discuss exact results rather than asymptotic results.

A more general queueing network model that overcomes many of the shortcomings of the original models mentioned earlier has been formulated and solved in [MUNT 72a] (this material also appears in [BASK 75]). The model is as follows. N nodes (referred to as service centers) are considered and a finite number, L, of different customer classes is permitted. As usual, customers travel through the network according to transition probabilities where, for $l = 1, 2, \ldots, L$

$$r_{ij}(l) = P[\text{next node is } j \mid \text{current node is } i$$
$$\text{and customer class is } l] \tag{4.89}$$

Since we have different customer classes, then both open and closed systems may be considered simultaneously. In particular, if the network is "closed" for customers of class l, then the number of such customers within the network is fixed at a constant value equal to $K(l)$. For those classes that see an "open" network, then such customers arrive from outside the network to node i in a Poisson fashion with an average rate $\gamma_i(l)$. Thus we see that customers from open classes leave the system upon their departure from node i with probability $1 - \sum_{j=1}^{N} r_{ij}(l)$ and for closed systems, this probability is zero. Thus far the generalization consists of introducing customer classes and thereby permits open and closed networks simultaneously. We now wish to introduce generalizations regarding order of service, number of servers, and distribution of service time at a given node. If we plan to use balance equations (as in Chapter 4, Vol. I) we must preserve the Markovian nature of our network in spite of the non-Markovian service time distributions; we achieve this by the use of stage-type servers (as shown in Figure 4.28). Finally, then, the model we consider premits service centers, some of which have arbitrary (but rational) service time distributions, with any of the following four types of scheduling algorithms:

Type 1 Node*: FCFS ·/M/1.   Nodes of this type contain a single server using an FCFS discipline; the service time distribution is exponentially distributed with parameter μ_{ik}, which may be a function of k, the number of customers in that node (say node i).

Type 2 Node*: RR ·/G/1.   Nodes of this type contain a single server using an RR processor-sharing algorithm; the service time distribution is arbitrary except that it must have a rational Laplace transform and is synthesized as in Figure 4.28.

Type 3 Node*: ·/G/∞.   Nodes of this type have an infinite number of servers (or at least equal to the maximum number of customers which can be demanding service simultaneously at this node); arbitrary service time distributions with rational Laplace transforms are permitted.

Type 4 Node*: LCFS ·/G/1.   Nodes of this type have a single server with a preemptive-resume LCFS scheduling algorithm; arbitrary service time distributions are permitted such that they have rational Laplace transforms.

---

* Nodes of types 2, 3, and 4 permit the service time distribution to vary with the customer class; this is not true of type 1 nodes.

This is the complete model. One should observe that type 1 nodes are those considered by Jackson and Gordon and Newell (with single servers). Type 2 nodes remind us of the fact that RR processor-sharing gives results that are independent of the service time distribution [see Eq. (4.17)]. Type 3 nodes take advantage of the fact that the queue $M/G/\infty$ has a solution independent of the service time distribution [see Eq. (1.98)]. Type 4 nodes again take advantage of the fact that preemptive-resume LCFS gives behavior similar to the RR processor-sharing algorithm [see Eq. (4.22)].

To solve these more general networks, it has been found useful to incorporate the concept of local balance (as discussed in Section 4.8, Vol. I). Chandy [CHAN 72] has considered networks of this type but the most general type to date is that in [MUNT 72a]. For a class l customer we let $e_i(l)$ be the solution to the following set of equations:

$$e_i(l) = \gamma_i(l) + \sum_{j=1}^{N} e_j(l)r_{ji}(l) \qquad \blacksquare \quad (4.90)$$

For those customer classes that see a *closed* system we must have that $\gamma_i(l) = 0$; in such a case the quantity $e_i(l)$ is merely the relative frequency of visits to node i by a class l customer and corresponds to Eq. (4.75) [where $e_i(l)$ corresponds to $\mu_i x_i$ broken down by customer class]. $e_i(l)/e_j(l)$ is the relative number of times a customer from class l visits node i compared to his number of visits to node j. The solution to this set of equations in the closed case leaves an arbitrary multiplicative constant in the solution. For those customer classes that see an *open* system, then some of the $\gamma_i(l)$ will be nonzero, and this will correspond to the set of equations (4.72) [again where $e_i(l)$ corresponds to λ_i broken down by customer class]; in this case $e_i(l)$ is the average arrival rate of class l customers to node i (from internal and external sources combined).

The states of the system involve a rather complete and complex description giving locations of customers, their class, and their stage of attained service. When one attempts to write down the global balance equations here, they quickly become unmanageable and it is at this point that local balance saves the day. In fact, even if one were to guess the solution, the local balance equations provide a much easier means for checking the validity of that solution than do the global balance equations. The main theorem proved in [MUNT 72a] states that the equilibrium probabilities for the system states may be written as

$$P(\alpha_1, \alpha_2, \ldots, \alpha_N) = Cg_1(\alpha_1)g_2(\alpha_2) \cdots g_N(\alpha_N) \qquad \blacksquare \quad (4.91)$$

where the state vector $(\alpha_1, \alpha_2, \ldots, \alpha_N)$ consists of components α_i that represent the conditions prevailing at node i; this representation depends upon the node type. C is a normalizing constant that conserves probability and each g_i is a function that depends only on the node type and the state of that node. Thus this main result once again exposes the *product-form* of solution for these very complex networks whereby there is a factoring of functions, one for each node in the network! To simplify the results, one solves for the marginal distribution for the number of class l customers in node i (eliminating the distinction as to the stage of service) and finds that it too obeys an equation of the form given in (4.91). Further, forming the marginal distribution merely on the total number of customers k_i in the ith node (finally eliminating customer class distinction) we see that this probability also factors in the expected way as follows:

$$p(k_1, k_2, \ldots, k_N) = Ch_1(k_1)h_2(k_2) \cdots h_N(k_N) \qquad \blacksquare \quad (4.92)$$

where if the node is of type 1 then

$$h_i(k_i) = \left(\sum_l \frac{e_i(l)}{\mu_{ik_i}} \right)^{k_i} \qquad \blacksquare \quad (4.93)$$

and if the node is of type 2 or 4 then

$$h_i(k_i) = \left(\sum_l \frac{e_i(l)}{\mu_{il}} \right)^{k_i} \qquad \blacksquare \quad (4.94)$$

and if the node is of type 3 then

$$h_i(k_i) = \frac{1}{k_i!} \left(\sum_l \frac{e_i(l)}{\mu_{il}} \right)^{k_i} \qquad \blacksquare \quad (4.95)$$

where the sum is over all customer classes which may enter node i and where $1/\mu_{il}$ is the average service time of class l customers at node i and $1/\mu_{ik_i}$ is the mean service time of the exponential distribution at node i when k_i customers are present. The constant C is once again a normalizing constant and must be calculated over all possible states. When the system is open this constant can be evaluated as shown in [MUNT 72a]. Furthermore in the case of an open system, if we let $p_i(k_i)$ denote the equilibrium probability of finding k_i customers in node i [as in Eq. (4.74)] then it can be shown that

$$p(k_1, k_2, \ldots, k_N) = p_1(k_1)p_2(k_2) \cdots p_N(k_N) \qquad \blacksquare \quad (4.96)$$

where

$$p_i(k_i) = \begin{cases} (1-\rho_i)\rho_i^{k_i} & \text{if node type is 1, 2, or 4} \\ \dfrac{\rho_i^{k_i}}{k_i!} e^{-\rho_i} & \text{if node type is 3} \end{cases} \qquad \blacksquare \quad (4.97)$$

and where

$$\rho_i = \begin{cases} \sum_l \dfrac{e_i(l)}{\mu_i} & \text{if node type is 1 and} \\[2mm] & \qquad \mu_{ik_i} = \mu_i \\[3mm] \sum_l \dfrac{e_i(l)}{\mu_{il}} & \text{if node type is 2, 3, or 4} \end{cases} \qquad \blacksquare \quad (4.98)$$

This is an amazing result! We see that if the node type is 1, 2, or 4 then the marginal distribution is the same as the distribution for the number of customers in an M/M/1 system (with an appropriately defined ρ_i). Of course we require that $\rho_i < 1$ as always. Moreover if the node type is 3 then we see that the solution corresponds with that of an M/G/∞ system. That is, the solution given in Eq. (4.96) states that $p(k_1, k_2, \ldots, k_N)$ factors into a product with one term for each node; this term is given as $p_i(k_i)$, which is the solution for that node *in isolation* with a Poisson input and with an exponential service time such that ρ_i is as given in Eq. (4.98). Note that only ρ_i (and *not* the form of the service time distribution) counts. A recent simulation study [LEE 73] successfully demonstrated the application of these concepts. Computationally efficient algorithms for closed networks with different customer classes have been developed by Muntz and Wong [MUNT 74a]; this is a generalization of the work reported above by Buzen [BUZE 73].

The solution described above may also be obtained by studying the condition required for the output of a queueing system to be a Poisson (departure) process [MUNT 72b]. A sufficient condition for this to be true is rather simple. Consider a queueing system with a countable state space whose state X is given by the number of customers in the system; we denote the probability of a given state by $\pi_k = P[X = k] = P[E_k]$ and by q_{jk}, we denote the departure rate out of E_j and into E_k. Then the sufficient condition for the system to have a Poisson departure process (at rate λ) is simply that for all states E_k,

$$\sum_{j \in J} \frac{\pi_j q_{jk}}{\pi_k} = \lambda \qquad (4.99)$$

where J is the set of all states in which the number of customers in the system is one greater than k. This condition is easily generalized to many classes of customers by considering state *vectors* and freezing all components not corresponding to the class of interest; the rate λ will then correspond to the departure rate for that class. If the input process is Poisson, and if then the output process is also Poisson, we say that the system has the $M \Rightarrow M$ (Markov implies Markov) property. Using this approach, it has been shown by Muntz that the four node types described above satisfy the Poisson departure condition. More important, if each node in a queueing network has the $M \Rightarrow M$ property in isolation, then the network will have a product form of solution as given in Eq. (4.91) where $g_i(\alpha_i)$ is the equilibrium state probability for the ith node placed in isolation with a Poisson input (but with the same mean arrival rate as seen when it is imbedded in the network).

We recall our reference in Chapter 2 to the work of Kobayashi [KOBA 74] and his approximation to both open and closed networks which is an additional method for studying these networks. (He requires nonpreemptive single-server nodes with one class of customer.) Also, the concurrent use of a series of processors was studied as a simple cyclic queueing network in [KLEI 66a].

Thus we see that here is the solution to a rather general* network of queues that permits different customer classes, rather interesting scheduling disciplines, and rather general service time distributions. This provides a fairly general model for multiple-resource computer systems and is currently being exploited heavily in the area of computer applications.

4.13. MODELS FOR MULTIPROGRAMMING

In this section we consider two different models for multiprogrammed computer systems. The first takes advantage of the original Gordon and Newell closed queueing network model and therefore requires the use of exponential service times throughout. The second model makes use of our diffusion approximation from Chapter 2, which permits the use of arbitrary distributions but yields only approximate answers. A nice survey of queueing models for multiprogramming is given by Adiri [ADIR 72].

With goals similar to that of time-sharing, a multiprogrammed computer system attempts to utilize its resources more efficiently than does a batch processing operation. In particular a number of jobs are permitted simultaneous access to the resources of the system in such a way that the CPU is busy processing one job while various input–output peripheral units are processing some of the others concurrently; of course, queues form in fron of each of these resources, and it is of interest to study the utilization and throughput for such a system. In our first model below we consider the *central-server model* that permits the inclusion of a number of peripheral devices. In our second model, we allow only one peripheral device (to be referred to as a data transmission unit).

---

* One major weakness in the model is that a single job (process) that requires more than one resource simultaneously (e.g., main memory space and CPU time) still cannot be represented except as mentioned earlier with the fixed number, D, of active jobs in a multiprogramming system with $D =$ degree of multiprogramming.

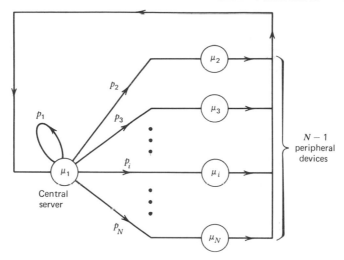

Figure 4.37   The central-server model.

An interesting model studied by Buzen [BUZE 71], which he refers to as a central-server model, is shown in Figure 4.37. The central-server model is constructed to describe the behavior of multiprogrammed computer systems (with a fixed number of partitions) where exactly K jobs (programs) are permitted into the computer system and circulate endlessly sharing the N resources of that system. The central server, node 1, is meant to represent the CPU and the other $N-1$ nodes represent peripheral devices, for example, a rotating disk memory, a swapping drum, a magnetic tape unit, a data cell, a card reader, and so on. In such a multiprogramming environment jobs do indeed circulate among these devices in such a way that they require the attention of the CPU followed by the need for some peripheral device, after which they again require the service of the CPU before again requiring the attention of perhaps some other peripheral device; thus we see the constant circulation back to the central-server, the CPU. The transition probabilities r_{ij} once again represent the probability of going next to node j upon leaving node i, and in this central-server model we have

$$r_{ij} = \begin{cases} p_j & i=1, \quad 1\leq j\leq N \\ 1 & 2\leq i\leq N, \quad j=1 \\ 0 & \text{otherwise} \end{cases}$$

and where of course $\sum_{j=1}^{N} p_j = 1$. In a true multiprogramming environment, most jobs eventually leave,* at which time a new job will be

_____
* One job that does not leave, for example, is the resident supervisor.

inserted into the system; this is represented in the model by permitting the job to return directly back to the CPU (with probability p_1), which represents the departure of an old job and the insertion of a new job to replace it, with the recognition that this new request for service in the CPU is the demand by the supervisory system for this job interchange. Thus the number of jobs in the system will remain constant at a value of K. In the central-server model each node has a single server ($m_i = 1$) and the ith node has an exponential service rate of value μ_i. Recalling that $\mathbf{R}$ is the matrix of transition probabilities (r_{ij}) among nodes in our network, then in the central server case we have the simple matrix

$$
\mathbf{R} =
\begin{bmatrix}
p_1 & p_2 & p_3 & \cdots & p_N \\
1 & 0 & 0 & \cdots & 0 \\
1 & 0 & 0 & \cdots & 0 \\
\cdot & \cdot & \cdot & & \cdot \\
\cdot & \cdot & \cdot & & \cdot \\
\cdot & \cdot & \cdot & & \cdot \\
1 & 0 & 0 & \cdots & 0
\end{bmatrix}
$$

The problem reduces to solving the eigenvalue problem posed in Eq. (4.75), which with our special matrix has the solution

$$
\mu_i x_i =
\begin{cases}
\mu_1 & i = 1 \\
\mu_1 p_i & i = 2, 3, \ldots, N
\end{cases}
\tag{4.100}
$$

The general solution to any closed Markovian network, with single-server nodes is given in Eq. (4.77) in terms of the x_i; for this central-server model we then have the special case

$$
p(k_1, k_2, \ldots, k_N) = \frac{1}{G(K)} \prod_{i=2}^{N} \left(\frac{\mu_1 p_i}{\mu_i} \right)^{k_i}
\qquad \blacksquare \tag{4.101}
$$

where as usual

$$
G(K) = \sum_{k \in A} \prod_{i=2}^{N} \left(\frac{\mu_1 p_i}{\mu_i} \right)^{k_i}
\tag{4.102}
$$

Thus we have a straightforward solution expressed simply in terms of the system parameters μ_i and p_i. Now let A_i be the equilibrium probability that the ith node is not empty. We may show that (see Exercise 4.35)

$$
A_i =
\begin{cases}
G(K-1)/G(K) & i = 1 \\
\dfrac{\mu_1 p_i}{\mu_i} A_1 & i = 2, 3, \ldots, N
\end{cases}
\tag{4.103}
$$

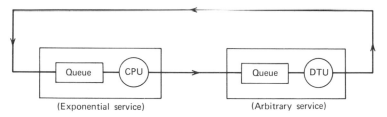

Figure 4.38 The cyclic queue model for multiprogramming.

From this last equation one notes that $A_1\mu_1 p_i = A_i\mu_i$ $(i \geq 2)$, which is easily seen to be equating the rate at which customers enter the ith node to the rate at which they depart from that node and is therefore a flow balance equation. Buzen defines the extent to which the ith node is creating a "bottleneck"* as being proportional to the rate of change of throughput with respect to an increase in the service rate of that node; throughput is defined as the average number of jobs processed per unit of time and is easily shown to be $A_1\mu_1 p_1$. It can then be seen that a "balanced" system (one with no bottlenecks) is one for which

$$\frac{\partial}{\partial \mu_i} A_1\mu_1 p_1 = \frac{\partial}{\partial u_j} A_1\mu_1 p_1 \qquad 1 \leq i, j \leq N \tag{4.104}$$

The reader is referred to [BUZE 71] for further details of this model.

The second model of multiprogramming which we consider is shown in Figure 4.38. This model is due to Gaver and Shedler [GAVE 73a]. It is a two-node cyclic queue in which the CPU is represented as a FCFS exponential service facility and where the (single) peripheral device, the data transmission unit (DTU), is a model for any device that supplies data storage (e.g., drum, disk, tape) and is permitted an arbitrary distribution of service (again FCFS). We assume there are K programs (jobs) in the system, each of which passes through the CPU and the DTU cyclically. We note that neither the Gordon and Newell model nor the Muntz and Baskett model applies here due to the arbitrary distribution of FCFS service in the DTU. Lewis and Shedler [LEWI 71] have solved this system exactly using a rather complex semi-Markov process approach. Here we follow Gaver and Shedler [GAVE 73a] and take advantage of the diffusion approximation to provide a simple approximate way of solving the system; this will demonstrate the power of the diffusion method.

We are particularly interested in the quantity $N_C(t)$, which is defined as the number of customers waiting for or receiving service in the CPU at time t. As in Eq. (2.81) it is clear that $N_C(t) = \alpha_C(t) - \delta_C(t)$ if $N_C(0) = 0$

* This definition differs from that given by Gordon and Newell discussed earlier.

where $\alpha_C(t)$ and $\delta_C(t)$ represent respectively the number of arrivals to and number of departures from the CPU in $(0, t)$. If for the moment we neglect the boundary effects at 0 and K, then $\alpha_C(t)$ and $\delta_C(t)$ will be independent renewal processes. As in Section 2.8, then, we know that they may each be approximated by continuous Gaussian random processes with means $t/\bar{x}_D$ and $t/\bar{x}_C$ and variances $t\sigma_{\bar{x}_D}^2/(\bar{x}_D)^3$ and $t\sigma_{\bar{x}_C}^2/(\bar{x}_C)^3$ respectively as we had seen in Eqs. (2.90) and (2.91); here $\bar{x}_D$ and $\bar{x}_C$ refer to the service time in the DTU and CPU, respectively. We recall that therefore $N_C(t)$ will be approximately normally distributed with infinitesimal mean m and infinitesimal variance σ^2 where

$$m = \frac{1}{\bar{x}_D} - \frac{1}{\bar{x}_C}$$

$$\sigma^2 = \frac{\sigma_{\bar{x}_D}^2}{(\bar{x}_D)^3} + \frac{\sigma_{\bar{x}_C}^2}{(\bar{x}_C)^3}$$

Using our diffusion approximation we then expect $N_C(t)$ to obey the Fokker–Planck equation (2.120) where $F(w, t) = P[N_C(t) \leq w]$. This diffusion equation, of course, must be subject to the boundary conditions $0 \leq N_C(t) \leq K$ which require that $F(w, t) = 0$ for $w < 0$, that $F(0^+, t) \geq 0$, and that $F(K, t) = 1$. When $K = \infty$ we know that Eq. (2.132) gives the closed-form solution to this problem.

For our purposes, however, we seek the equilibrium behavior of this system and therefore obtain the solution as given in Eq. (2.124); this solution, however, did not account for the boundary condition $N_C(t) \leq K$ and if we impose this on the resulting first-order linear differential equation we obtain

$$F(w) = A[1 - Be^{2mw/\sigma^2}] \qquad w \geq 0$$

Upon applying the upper boundary condition we have

$$F(K) = 1 = A[1 - Be^{2mK/\sigma^2}]$$

This determines the constant A and therefore we have

$$F(w) = \frac{1 - Be^{2mw/\sigma^2}}{1 - Be^{2mK/\sigma^2}} \tag{4.105}$$

We must now evaluate the constant B; in [GAVE 73a] a number of suggestions are made, of which we consider only one, namely the case in which B is evaluated so that the solution is an exact fit for the case $K \to \infty$. This is perhaps the case of most practical interest. Since $m < 0$ for a stable system, we will have

$$\lim_{K \to \infty} F(w) = 1 - Be^{2mw/\sigma^2}$$

which then gives

$$\lim_{K \to \infty} F(0^+) = 1 - B$$

However, we know, using simple renewal arguments, that

$$\lim_{t \to \infty} P[N_C(t) = 0] = 1 - \rho$$

for $\rho < 1$ where $\rho = \bar{x}_C/\bar{x}_D$. Therefore we may set $B = \rho$ and substitute back into Eq. (4.105) to find

$$F(w) = \frac{1 - \rho e^{2mw/\sigma^2}}{1 - \rho e^{2mK/\sigma^2}} \qquad (4.106)$$

This is then a simple equation derived from the diffusion approximation; it is a good approximation when K is large and the heavy-traffic condition maintains (m close to but less than 0).

In order to evaluate the quality of this approximation we take the exact results obtained in [LEWI 71] for the CPU utilization (the fraction of time the CPU is busy) and compare it to our approximation. Clearly the CPU utilization, denoted u_C, will merely be $1 - F(0^+)$ and so from Eq. (4.106) we find

$$u_C = \rho \left[\frac{1 - e^{2mK/\sigma^2}}{1 - \rho e^{2mK/\sigma^2}} \right] \qquad (4.107)$$

as our approximation for u_C. In Figure 4.39 we compare the exact results with this approximate equation for two cases, namely, exponential and five-stage Erlangian service time distributions in the DTU. Note the remarkable agreement between the exact and the approximate results. One can also imagine the enormous savings in analytical complexity that this approximation affords us.

In [GAVE 73c] Gaver and Shedler refine the above analysis to permit arbitrary service times in the CPU as well and also to give an approximation that yields superior results, especially when the coefficient of variation of the CPU service time exceeds that of the exponential. This refined approach requires that the exponent $2m/\sigma^2$ used in the diffusion approximation above be replaced by $\log_e B_D{}^*(s_0)$ where $B_D{}^*(s)$ is the Laplace transform of the service time pdf in the DTU and s_0 is the positive root of the characteristic equation $B_C{}^*(-s)B_D{}^*(s) - 1 = 0$ under the condition $m < 0$; $B_C{}^*(s)$ is the corresponding transform for the CPU.

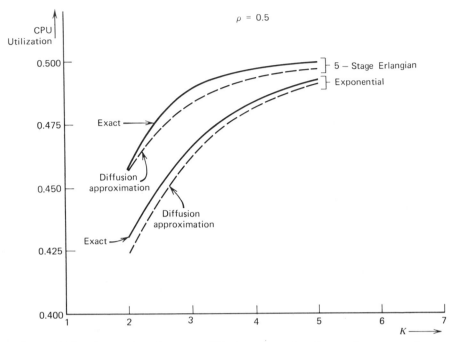

Figure 4.39   Comparison between diffusion approximation and exact results.

Numerous other multiprogramming models are described in [AVI 73 GAVE 67, GAVE 71b]. In [HALA 74] the model of Figure 4.38 is extended to the multiserver system and the diffusion approximation is used with amazing success when compared with simulation and exact results. The interested reader is referred to these papers for further details.

4.14.   REMOTE TERMINAL ACCESS TO COMPUTERS

This section is concerned with the problems of rapidly and efficiently transmitting job requests and data from a remotely located user terminal to a computer system. We consider three models. The first two are concerned with the delays and queues that form at the remote terminals as they are "polled" cyclically by the computer; the approach taken in the first model is to use the diffusion approximation to study this backlog, whereas the second discrete-time model uses a random-walk approach. The third model also considers the use of a single communication line to gather the data from a collection of remote terminals and the use of a finite buffer to aid in this collection; the considerations involve calculating the queueing delays and the probability of exceeding the buffer capacity. Emphasis is placed on the use of multiplexing as a means for transmitting the "concentrated" data to the computer. In [FUCH 70] a number of measurement studies are reported upon that characterize the terminal-to-computer and computer-to-terminal traffic statistics; happily one finds the memoryless distributions in a large number of these statistics justifying our use of such in many of the models in this section and throughout the chapter. These measurements also confirmed the fact that the interactive data communication process is *extremely bursty* in that it has a large peak-to-average load factor. As a result, efficient means for multiplexing many data streams onto a single channel must be found, and we discuss some of these here. This study of remote terminal access to computer systems is of general importance and leads us naturally into Chapters 5 and 6 on computer-communication networks.

Our polling models consider a collection of remote terminals that generate data destined for the computer system and that must reach that computer system over finite-capacity communication channels. Our *first* model is due to Gaver [GAVE 71a]. He assumes that each remote terminal is equipped with a buffer that it fills with data for the computer (CPU); these terminals are connected over a single shared communication channel to the CPU and this channel is the constraining resource in the system. The terminals are polled in a given cyclic order such that if a polled terminal buffer contains information it is emptied and transmitted to the CPU; if there is data at the CPU waiting to be transmitted to the terminal in response to a previous request by that terminal then this transmission takes place next. These transmissions are modeled as random variables due to the possibility of noise contaminating the transmission and requiring a repeated transmission. After a given terminal is so serviced, the next terminal in the cyclic order is polled, and so on. Two types of terminals are distinguished: *input* terminals, in which transmission takes place only from the terminal to the CPU (and no response from the CPU is permitted, except at some later scheduled time, which we neglect); and *response* terminals from which transmissions to the CPU occur and to which, after some time, a response from the CPU is received (in this case service begins with a terminal-to-CPU request and ends when the response from the CPU is received, perhaps after several polling cycles have elapsed). Response terminals service only one customer at a time and are therefore "locked" while awaiting a response from the CPU (as in Section 4.11).

We begin by focusing on a system consisting of a single remote terminal. As usual we are concerned with $N(t)$, which is the number of data units (cards, characters, bytes, or any other appropriate definition) waiting at the terminal for transmission. We intend to use the diffusion approximation and therefore consider the case when $N(t)$ tends to be large. As usual we may relate $N(t)$ to the arrival and departure processes for the queue. If we denote $F(w, t) = P[N(t) \leq w]$ then we know that $F(w, t)$ must satisfy the Fokker–Planck equation (2.120) along with the usual boundary condition at the origin. The parameters in this equation are as follows:

$$m = \lim_{\tau \to 0} \frac{\overline{\alpha(\tau) - \delta(\tau)}}{\tau}$$

and

$$\sigma^2 = \lim_{\tau \to 0} \frac{\sigma^2_{\alpha(\tau)} + \sigma^2_{\delta(\tau)}}{\tau}$$

where $\alpha(\tau)$ and $\delta(\tau)$ represent the number of arrivals and departures, respectively, in an interval of duration τ after the system has been operating for a long time. As usual we require that m be close to but less than zero for the heavy-traffic assumptions. As always we are most interested in the steady-state solution that exists for $m < 0$ and has the usual solution

$$\lim_{t \to \infty} F(w, t) = 1 - e^{2mw/\sigma^2} \qquad (4.108)$$

It behooves us at this point to continue discussing the single terminal system and to study some special cases for which we can easily identify the parameters m and σ^2.

Case 1. In this case we consider a bulk arrival process where the bulks arrive according to a Poisson process at rate λ. We let G be the random bulk size and we let $g_k = P[G = k]$ be the bulk size distribution. Thus the total number of data unit arrivals behaves as a compound Poisson process with infinitesimal mean $\lim_{\tau \to 0} \overline{\alpha(\tau)}/\tau = \lambda \bar{G}$ and infinitesimal variance $\lim_{\tau \to 0} \sigma^2_{\alpha(\tau)}/\tau = \lambda \overline{G^2}$ where this variance is calculated in a fashion similar to that for Eq. (2.121).

Subcase (a). Here we consider a deterministic service (transmission) time of exactly $1/\mu$ sec per data unit. Thus we may immediately calculate the parameters of our diffusion process as

$$m = \lambda \bar{G} - \mu$$

$$\sigma^2 = \lambda \overline{G^2}$$

Subcase (b). Here we assume exponential service time per data unit with mean $1/\mu$ sec. This approximates the need for retransmission due to line noise. Using the heavy-traffic approximation (and therefore assuming that the queue is not empty) we then have that the departure process from the terminal queue is a Poisson process with parameter μ and therefore $\lim_{\tau \to 0} \bar{\delta}(\tau)/\tau = \mu$ and $\lim_{\tau \to 0} \sigma^2_{\delta(\tau)}/\tau = \mu$. Thus the parameters for our diffusion process become

$$m = \lambda \bar{G} - \mu$$

$$\sigma^2 = \lambda \overline{G^2} + \mu$$

These parameters may now be used in Eq. (4.108). Thus we see that the equilibrium distribution of terminal backlog is approximately exponential with the mean length given by

$$\text{subcase (a):} \quad \bar{N} = \frac{\lambda \overline{G^2}}{2(\mu - \lambda \bar{G})} \tag{4.109}$$

$$\text{subcase (b):} \quad \bar{N} = \frac{\lambda \overline{G^2} + \mu}{2(\mu - \lambda \bar{G})} \tag{4.110}$$

Case 2.   In this case we consider an arbitrary renewal process for arrivals where the intervals between arrivals, $t_1, t_2, \ldots, t_n, \ldots$, are independently and identically distributed (represented by $\tilde{t}$); similarly departures form a renewal process with interdeparture times, $x_1, x_2, \ldots, x_n, \ldots$, also independently and identically distributed (represented by $\tilde{x}$). This assumption on the departure process again takes advantage of the fact that the queue is assumed not to empty. The usual renewal theory arguments then show that the number of arrivals and departures in an interval of duration τ are approximately normally distributed such that

$$\lim_{\tau \to 0} \frac{\alpha(\tau)}{\tau} = \frac{1}{\tilde{t}}, \qquad \lim_{\tau \to 0} \frac{\delta(\tau)}{\tau} = \frac{1}{\tilde{x}}$$

$$\lim_{\tau \to 0} \frac{\sigma^2_{\alpha(\tau)}}{\tau} = \frac{\sigma^2_{\tilde{t}}}{(\tilde{t})^3}, \qquad \lim_{\tau \to 0} \frac{\sigma^2_{\delta(\tau)}}{\tau} = \frac{\sigma^2_{\tilde{x}}}{(\tilde{x})^3}$$

where these infinitesimal variance calculations are the same as those which led to Eq. (2.91). Thus the diffusion parameters become

$$m = \frac{1}{\tilde{t}} - \frac{1}{\tilde{x}}$$

$$\sigma^2 = \frac{\sigma^2_{\tilde{t}}}{(\tilde{t})^3} + \frac{\sigma^2_{\tilde{x}}}{(\tilde{x})^3}$$

This then gives the approximate behavior of the queue through Eq. (4.108).

Case 3.   Here we wish to model the equilibrium waiting time rather than the equilibrium queue size. Between the arrival of two adjacent

customers, the increase in waiting time is distributed as the random
variable $\tilde{x} - \tilde{t}$ and so we have immediately

$$m = \bar{x} - \bar{t} \tag{4.111}$$

$$\sigma^2 = \sigma_{\tilde{x}}^2 + \sigma_{\tilde{t}}^2 \tag{4.112}$$

These are the appropriate parameters for the distribution in Eq. (4.108),
which now refers to waiting time.

These three cases describe the behaviour of remote terminal access
when there is only one terminal in the system. We now wish to consider
the case of many terminals. Gaver considers three system configurations
and we follow his treatment here. The first system consists of M input
terminals and no response terminals. We again employ the heavy-traffic
assumption and assume there is a queue (of data units awaiting transmission) at each terminal. Since each terminal behaves in a statistically
identical fashion, we may concentrate on the behavior at terminal 1. We
assume that at time zero a transmission from terminal 1 to the CPU has
just been completed; if we denote by X_i the time to transmit all the data
for a single job at terminal i then we see that the next transmission from
terminal 1 will be completed at the end of a cycle (whose duration we
denote by C) where

$$C = X_2 + X_3 + \cdots + X_M + X_1$$

Thus the average cycle time is

$$\bar{C} = \sum_{i=1}^{M} \bar{X}_i$$

and since these customer service times are assumed to be independent,
we also have

$$\sigma_C^2 = \sum_{i=1}^{M} \sigma_{X_i}^2$$

Here C is the duration of an arbitrary cycle; let $C(j)$ denote the duration
of the jth cycle. We see therefore that transmissions from terminal 1
(representative of any terminal) will take place at the instants $C(1)$,
$C(1) + C(2)$, $C(1) + C(2) + C(3), \ldots$. Our independence assumptions assure us that the set of cycle times forms a sequence of independent and
identically distributed random variables and so the renewal process
defined by the above partial sums will determine how many transmissions
take place. Therefore we see simply that this is an example of Case 2,
where $\tilde{x}$ is to be replaced by the cycle duration $C(j)$. Thus we make use of
our previous results and obtain that the backlog is approximately exponentially distributed with mean given by

$$\bar{N} = \frac{[\sigma_{\tilde{t}}^2/(\bar{t})^3] + [\sigma_C^2/(\bar{C})^3]}{2[(1/\bar{t}) - (1/\bar{C})]} \tag{4.113}$$

A recent generalization of this M-terminal model has been treated by Gaver [GAVE 73b], who permits the terminals to generate a *time-dependent* demand; in particular, the arrival rate of demands from each terminal is $\lambda(t)$ and the system service rate is constant at μ. The treatment uses the diffusion approximation once again. Gaver shows for large M that the system backlog, $N(t)$, may be approximated by the sum of a deterministic portion $Mn(t)$ and a random portion $\sqrt{M}\,r(t)$, that is,

$$N(t) \cong Mn(t) + \sqrt{M}\,r(t)$$

$n(t)$ is given by

$$n(t) = n(0) \exp\left[-\int_0^t \lambda(u)\,du\right] + \int_0^t [\lambda(y) - \mu]^+ \exp\left[\int_y^t \lambda(u)\,du\right] dy$$

where again, $[x]^+ = \max[0, x]$. $r(t)$ turns out to be a normal (Gaussian) diffusion process with mean

$$E[r(t)] = r(0) \exp\left[-\int_0^t \lambda(u)\,du\right]$$

and with variance

$$\text{Var}\,[r(t)] = \int_0^t \exp\left[-2\int_y^t \lambda(u)\,du\right]\left[(1 - n(y))\lambda(y) + \mu\right] dy$$

The reader is referred to [GAVE 73b] for further details.

The second system considered by Gaver again involves M input terminals with the addition of a single response terminal. We are concerned with the response time for this response terminal, which we may recall is the time from when the terminal begins transmission of its request to the CPU and terminates when the response from the CPU has been received; this response time consists of a service time X to transmit the request to the CPU, plus a sequence of cycle times $C(1) + C(2) + \cdots + C(K)$ among the M input terminals during which time the CPU is processing the response terminal's request (we assume that this processing is completed during the Kth such cycle), plus a time X' to transmit the response from the CPU to the terminal plus an additional time θ [which may be interpreted as the time required for the old (satisfied) customer to leave the terminal and for a new customer to take his place]. Thus we see that the response time R for the response terminal may be written as

$$R = X + C' + X' + \theta$$

where C' is the time spent cycling K times among the input terminals. We must now calculate the mean and variance of R so that we may apply our diffusion approximation for predicting the queue size at the response terminal. This queue will consist of additional requests (customers) at the response terminal, none of which can be serviced until the previous request is satisfied. We now assume that the time required by the CPU to process the request of the response terminal is exponentially distributed with parameter β; this time consists of the actual processing time in CPU plus the time spent waiting for this service in the CPU, and so from our heavy-traffic results in Chapter 2 we find that this exponential assumption is not unreasonable. We assume that all M input terminals have identically distributed transmission times, each of which is exponential with parameter γ. We further assume that X and X' are similarly distributed (exponential with parameter γ). In Exercise 4.42 we show that

$$\overline{C'} = \frac{M/\gamma}{1 - [\gamma/(\beta + \gamma)]^M} \tag{4.114}$$

$$\sigma_{C'}^2 = \frac{(M/\gamma)^2}{1 - [\gamma/(\beta + \gamma)]^M} \left[\frac{M+1}{M} + \frac{2[\gamma/(\beta + \gamma)]^M - 1}{1 - [\gamma/(\beta + \gamma)]^M} \right] \tag{4.115}$$

Of course the first two moments of the interchange time θ are assumed known. This completes the analysis of a single response terminal in combination with M input terminals in the sense that we have once again reduced the problem to the situation of Case 2 in which we replace the transmission time $\tilde{x}$ with the response time R; the mean and variance for R are, respectively, $2\bar{X} + \overline{C'} + \bar{\theta}$ and $2\sigma_X^2 + \sigma_{C'}^2 + \sigma_\theta^2$ and may be used in Eqs. (4.111)–(4.112) for the diffusion parameters, which themselves may be used in the distribution of queue size as given in Eq. (4.108).

The third system Gaver considers is for the case of M input terminals and m response terminals. Finding the response time R for a response terminal may be done as an extension of the second system above.

These three systems considered by Gaver provide an approximate method for analyzing the behavior of remote terminal access to computer systems. Admittedly they are approximate and full of assumptions, but one must also admit that they are extremely simple to apply and may be used as a first-cut analysis for this computer-communication problem.

Let us now consider the *second* model. In a series of papers, Konheim and Meister studied a number of remote terminal access systems not unlike those studied by Gaver. Theirs is an exact analysis and while the theory is elegant it is also rather cumbersome. Nevertheless they have studied numerous interesting systems, solved for the transform of the appropriate variables, and have used these transforms to extract the mean and variance of quantities such as queue size and waiting times. Their studies have been very nicely summarized in a paper by Chu and

Konheim [CHU 72] and we wish to describe some of the results below. They concentrate on a class of distributed computer system architectures known as loop or ring systems. Perhaps the first researchers to consider these structures were Farmer and Newhall [FARM 69], who considered a loop system for connecting various peripheral devices (which may be thought of as generalized terminals) to the computer. More recently Pierce [PIER 71] proposed a hierarchy of loops connected together to serve as a computer network for national data transfer. Others have looked at these systems, for example Hayes and Sherman [HAYE 71] and Spragins [SPRA 72].

The results of Konheim and Meister (see for example [KONH 72a] and [KONH 72b]) are perhaps best summarized much as we summarized Gaver's results, by first considering a single input terminal connected to the CPU. The connecting channel is capable of transfering one data unit every time unit (referred to as a slot interval) and is meant to model the transfer of either a character or byte or block, or the like. We assume there is an infinite-capacity buffer at the terminal and the arrival of data units to this terminal is a stationary renewal process just as in Gaver's Case 2. We assume that the number of data units arriving per slot interval has a mean λ, a variance σ^2, and a third central moment $m^{(3)}$. So far we have described the classic G/D/1 queue; however, an additional constraint is placed upon the systems considered by Konheim and Meister, namely, that service may commence only at the beginning of a slot interval on a slotted time axis. Therefore a data unit arriving to an empty terminal in the middle of a slot interval must wait until the beginning of the next slot before its transmission begins; as stated above, the transmission time will be exactly one slot interval in length. This problem is viewed as a gambler's ruin problem (a random walk where the quantity of interest is the time until the first negative excursion takes place). They find that the mean and variance of the number N of data units queued in the buffer just prior to the beginning of a slot interval have the following equilibrium values (this result is also available in [SPRA 72]):

$$\bar{N} = \frac{\sigma^2}{2(1-\lambda)} + \frac{\lambda}{2} \qquad (4.116)$$

$$\sigma_N^2 = \frac{m^{(3)}}{3(1-\lambda)} + \left[\frac{\sigma^2}{2(1-\lambda)} + \frac{1-\lambda}{2}\right]^2 - \frac{(2\lambda-1)(2\lambda-3)}{12} \qquad (4.117)$$

We note that in this case $\rho = \lambda$ and so the stability condition is merely $\lambda < 1$. In the special case when the input is a Poisson process then we find $\bar{N} = (\lambda/[2(1-\lambda)]) + \lambda/2$; the M/D/1 queue gives exactly the same result, which holds for all instants in time [see Eq. (1.83) with $\bar{x^2} = 1$].

Let us now consider a polling system (similar to that considered by Gaver) where, in addition, it is assumed that $r-1$ intervals are wasted in switching from one terminal to the next. We assume again that there are M terminals in the polling sequence and that the arrival process to each is identically distributed. The first equilibrium result gives the average cycle time as

$$\bar{C} = \frac{Mr}{1-M\lambda} \qquad (4.118)$$

with the system utilization equal to $\rho = M\lambda < 1$. The average queue length at an arbitrary terminal in equilibrium is

$$\bar{N} = \frac{\sigma^2}{2(1-M\lambda)} + \frac{Mr\lambda(1-\lambda)}{2(1-M\lambda)} \qquad (4.119)$$

If we now calculate the average time a data unit spends waiting at a terminal we find the result

$$T = \frac{M[\sigma^2 + r(1-\lambda)]}{2(1-M\lambda)} + \frac{1-\lambda}{2} \qquad (4.120)$$

This polling system, as well as the one studied by Gaver, is known as an asynchronous time-division multiplexing (ATDM) system since time is multiplexed in an asynchronous way (that is, it is made available to terminals in relation to the demands placed by that terminal). Another form of polling known as synchronous time-division multiplexing (STDM) is considered next.

In an STDM system each terminal is assigned a fixed proportion of the slot intervals, typically every Mth slot for a given terminal in a set of M terminals. The service therefore is independent of the traffic at the remaining terminals and the analysis may be carried out in a fashion similar to that for the single terminal model if we properly adjust the input process. In particular, let us define λ_M, $\sigma_M{}^2$, and $m_M^{(3)}$ as the mean, variance, and third central moment of the arrival process over M contiguous slot intervals at an arbitrary terminal (recall that all terminals have the same arrival statistics). Having identified these parameters of the input process for M terminals, we may then write down the mean $\bar{N}(M)$ and variance $\sigma_N{}^2(M)$ of the queue size by using Eqs. (4.116) and (4.117) where these new arrival process parameters replace the single terminal parameters. We may also compute delay in such a system. We are interested specifically in the bulk arrival case and so let us assume that at the start of some slot interval we insert k data units; the average delay they experience (defined as the difference between the average time spent in system and average service time) denoted by $W_k(M)$ is given as follows:

$$W_k(M) = \frac{M+1}{2} + k(M-1) + M\left[\bar{N}(M) - \frac{(M+1)\lambda M}{2}\right] \qquad (4.121)$$

One other system that is simply treated using the methods from [CHU 72] is that in which a common buffer is made available to all terminals. We may easily calculate the mean and variance of the stationary queue length in the buffer by using Eqs. (4.116) and (4.117) once again if we reinterpret the arrival process parameters in the obvious way such that we use $\sum_{i=1}^{M} \lambda^{(i)}$, $\sum_{i=1}^{M} \sigma^{2(i)}$, and $\sum_{i=1}^{M} m^{(3)(i)}$ where the superscript i denotes parameters associated with the ith terminal.

We see then that multiplexing (both ATDM and STDM) are incorporated in order to share the use of a communication channel in an efficient way. We note that STDM is an extremely simple system to implement but may be wasteful of channel capacity when terminal traffic is low and slot intervals are assigned to an idle terminal. ATDM makes up for this deficiency by allocating time only on demand but introduces communication and processing overhead since addressing and buffering are required. In the ATDM models described above we have assumed infinite buffer capacity and a single channel per terminal.

For the *third* model, we follow Chu [CHU 69], who considers the G/D/m ATDM system with the additional constraint that the constant service time may be initiated only at the beginning of a slot interval and where the system storage capacity is of maximum size K (where $K \geq m$). We let p_k be the probability that there are exactly k data units in the system (queue plus service) at the end of a slot interval and let a_m represent the equilibrium probability that there are no more than m data units in the system at that time (where m is the number of servers in the system); then we have immediately

$$a_m = \sum_{k=0}^{m} p_k$$

As usual we assume the slot interval is of unit length. Furthermore, let π_k be the probability that k data units are generated from our arrival process during a slot interval (we have above expressed the mean of this distribution as λ); note that we are considering a single input terminal in this analysis but, as above, this may represent the concentration of M terminals. Considering the imbedded Markov chain at the end of a slot interval we may write down the obvious balance equations among these equilibrium probabilities at the termination of adjacent slot intervals as follows. First we have

$$p_0 = a_m \pi_0 \qquad (4.122)$$

This equation merely expresses the fact that at the end of this slot interval the probability of finding no data units in the system equals the probability that no more than m data units were present at the termination of the last slot interval (and therefore each such data unit was placed on one of the transmission channels and is sure to depart) times the probability that no new data units enter the system during the slot interval. In a like fashion we may count the ways in which various patterns of data units may arise in the system and find the following system of equations:

$$p_k = a_m \pi_k + p_{m+1} \pi_{k-1} + p_{m+2} \pi_{k-2} + \cdots$$
$$+ p_{m+k-1} \pi_1 + p_{m+k} \pi_0 \qquad 0 \le k \le K - m \qquad (4.123)$$

$$p_k = a_m \pi_k + p_{m+1} \pi_{k-1} + p_{m+2} \pi_{k-2} + \cdots$$
$$+ p_{K-1} \pi_{k+m+1-K} + p_K \pi_{k+m-K} \qquad K - m < k \le K - 1 \quad (4.124)$$

To complete the set of equations we must add the conservation of probability, namely

$$\sum_{k=0}^{K} p_k = 1$$

Note that when the arrival process is Poisson we have simply $\pi_k = \lambda^k e^{-\lambda} / k!$.

Of interest to this m-server system is P_{of}, the probability of overflow which corresponds to the probability that an arriving data unit is rejected by the buffer; note that $P_{of} \cong P_K$. The system of equations described in the previous paragraph does not have a simple solution; however, in [CHU 69] these equations are solved numerically for the case of Poisson arrivals and P_{of} is shown as a function of K, the buffer size, with the utilization factor (λ/m) as parameter. These numerical curves show that the overflow probability decreases approximately exponentially with K; for example, with $\lambda/m = 0.8$ and with $K = 28$ data units then $P_{of} \cong 10^{-6}$. Moreover if one computes the expected queueing delay as a function of buffer size for various numbers of channels, then one finds that when K exceeds some moderate value then the average delay remains essentially constant independent of increasing buffer size; for example, at $\lambda/m = 0.8$ the critical buffer size is approximately 18 data units, $P_{of} \cong 10^{-4}$ and the queueing delay (for $m = 3$) is approximately 1 slot interval. These delays may be significant as compared to the terminal-to-concentrator delay. In [CHU 69] consideration is also made of the computer-to-terminal traffic; the appropriate equilibrium equations are the same as those above, with the assumption $m = 1$ corresponding to a single computer-to-terminal channel.

In this chapter we have described one of the newest and most successful application areas of queueing theory. The results we have discussed on single resource models are "tidy" in the sense that they are easy to derive and are easy to understand. The multiple-resource models are still under development, and we have given some of the most recent results. The models for multiprogramming and remote terminal access are less understood, and need further refinement for useful application. In this study we have naturally been led into the computer-communication problem and now we generalize that discussion in the obvious way to computer networks in the following two chapters.

REFERENCES

ADIR 69a   Adiri, I. and B. Avi-Itzhak, "A Time-Sharing Queue with a Finite Number of Customers," *Journal of the Association for Computing Machinery.* **16**, 315–323 (1969).

ADIR 69b   Adiri, I., "Computer Time Sharing Queues with Priorities," *Journal of the Association for Computing Machinery,* **16**, 631–645 (1969).

ADIR 69c   Adiri, I., "A Time Sharing Queue," *Management Sciences,* **15**, 639–657 (1969).

ADIR 72   Adiri, I., "Queueing Models for Multiprogrammed Computers," *Proceedings of the International Symposium on Computer-Communication Networks and Teletraffic,* Polytech Press, Brooklyn, N. Y., pp. 441–448, 1972.

ANDE 74   Anderson, H. A. Jr. and R. G. Sargent, "Investigation into Scheduling for an Interactive Computing System," *I.B.M. Journal of Research and Development,* 125–137 (March 1974).

AVI 73   Avi-Itzhak, B. and D. P. Heyman, "Approximate Queueing Models for Multiprogramming Computer Systems," *Operations Research,* **21**, No. 6, 1212–1230 (1973).

BASK 71   Baskett, F., "The Dependence of Computer System Queues upon Processing Time Distribution and Central Processor Scheduling," *Proceedings of the ACM SIGOPS Third Symposium on Operating System Principles,* Stanford University, 109–113, October 1971.

BASK 75   Baskett, F., K. M. Chandy, R. R. Muntz, and F. Palacios-Gomez, "Open, Closed, and Mixed Networks of Queues with Different Classes of Customers," *Journal of the Association for Computing Machinery,* **22**, 248–260 (1975).

BURK 56   Burke, P. J., "The Output of a Queueing System," *Operations Research,* **4**, 699–704 (1956).

BUZE 71   Buzen, J., "Queueing Network Models of Multiprogramming," Ph.D. Thesis, Division of Engineering and Applied Science, Harvard University, Cambridge, Mass. (1971).

BUZE 73   Buzen, J. P., "Computational Algorithms for Closed Queueing Networks with Exponential Servers," *Communications of the Association for Computing Machinery,* **16**, 527–531 (1973).

CHAN 72   Chandy, K. M., "The Analysis and Solutions for General Queueing Networks," *Proceedings of the Sixth Annual Princeton Conference on Information Sciences and Systems*, Princeton University, 224–228, March 1972.

CHNG 70   Chang, W., "Single-Server Queueing Processes in Computing Systems," *IBM Systems Journal*, 36–71 (1970).

CHNG 72   Chang, A. and S. S. Lavenberg, "Work-Rates in Closed Queueing Networks with General Independent Servers," IBM Research Report RJ 989, March 1972.

CHU 69   Chu, W. W., "A Study of Asynchronous Time Division Multiplexing for Time-Sharing Computers," *AFIPS Conference Proceedings*, 1969 Fall Joint Computer Conference, Vol. 35, 669–678, 1969.

CHU 72   Chu, W. W. and A. G. Konheim, "On the Analysis and Modeling of a Class of Computer Communication Systems," *IEEE Transactions on Communications*, **COM-20** Part II, 645–660 (1972).

COFF 66   Coffman, E. G., Jr., "Stochastic Models of Multiple and Time-Shared Computer Operation," Report No. 66-38, Department of Engineering, University of California at Los Angeles, June 1966.

COFF 68a   Coffman, E. G., Jr. and L. Kleinrock, "Computer Scheduling Methods and Their Countermeasures," *AFIPS Conference Proceedings*, 1968 Spring Joint Computer Conference, 11–21, 1968.

COFF 68b   Coffman, E. G., Jr. and L. Kleinrock, "Feedback Queueing Models for Time-Shared Systems," *Journal of the Association for Computing Machinery*, **15**, 549–576 (1968).

COFF 70   Coffman, E. G., Jr., R. R. Muntz, and H. Trotter, "Waiting Time Distribution for Processor-Sharing Systems," *Journal of the Association for Computing Machinery*, **17**, 123–130 (1970).

ESTR 67   Estrin, G. and L. Kleinrock, "Measures, Models and Measurements for Time-Shared Computer Utilities," *Proceedings of the 22nd National Conference of the Association for Computing Machinery*, sponsored by the Association for Computing Machinery, Washington, D.C., 85–96, 1967.

FARM 69   Farmer, W. D. and E. E. Newhall, "An Experimental Distributed Switching System to Handle Bursty Computer Traffic," *Proceedings of the ACM Symposium on Problems in the Optimization of Data Communications Systems*, Pine Mountain, Ga., 1–34, October 1969.

FERD 71   Ferdinand, A. E., "An Analysis of the Machine Interference Model," *IBM Systems Journal*, **10**, No. 2, 129–142 (1971).

FUCH 70   Fuchs, E. and P. E. Jackson, "Estimates of Distributions of Random Variables for Certain Computer Communications Traffic Models," *Communications of the ACM*, **13**, 752–757 (1970).

GAVE 67   Gaver, D. P., "Probability Models for Multiprogramming Computer Systems," *Journal of the Association for Computing Machinery*, **14**, 423–438 (1967).

GAVE 71a   Gaver, D. P., "Analysis of Remote Terminal Backlogs under Heavy Demand Conditions," *Journal of the Association for Computing Machinery*, **18**, 405–415 (1971).

GAVE 71b   Gaver, D. P., "System Service Output, with Application to Multiprogramming," Naval Postgraduate School Report NPS55GV71091A (September 1971).

GAVE 73a   Gaver, D. P. amd G. S. Shedler, "Processor Utilization in Multiprogramming Systems via Diffusion Approximations," *Operations Research*, **21**, 569–576 (1973).

GAVE 73b   Gaver, D. P., "Delays at a Facility with Demand from Many Distinct Sources," Naval Postgraduate School Report NPS 55GV73081A, August 1973.

GAVE 73c   Gaver, D. P. and G. S. Shedler, "Approximate Models for Processor Utilization in Multiprogrammed Computer Systems," *SIAM Journal on Computing*, **2**, 183–192 (1973).

GORD 67   Gordon, W. J. and G. F. Newell, "Closed Queueing Systems with Exponential Servers," *Operations Research*, **15**, 254–265 (1967).

GREE 66   Greenberger, M., "The Priority Problem and Computer Time-Sharing," *Management Science*, **12**, 888–906 (1966).

HALA 74   Halachmi, B. and W. R. Franta, "A Closed Cyclic, Two-Stage Multiprogrammed System Model and its Diffusion Approximation Solution," *Proceedings of the Second Annual ACM Sigmetrics Symposium on Measurement and Evaluation*, Montreal, Canada, 54–64, Sept. 30–Oct. 2, 1974.

HAYE 71   Hayes, F. J. and D. N. Sherman, "Traffic Analysis of a Ring Switched Data Transmission System," *Bell System Technical Journal*, **50**, 2947–2978 (1971).

HSU 71   Hsu, J., "Analysis of a Continuum of Processor-Sharing Models for Time-Shared Computer Systems," UCLA-ENG-7166, School of Engineering and Applied Science, University of California at Los Angeles, October 1971.

JACK 54   Jackson, R. R. P., "Queueing Systems with Phase Type Service," *Operations Research*, **5**, 109–120 (1954).

JACK 57   Jackson, J. R., "Networks of Waiting Lines," *Operations Research*, **5**, 518–521 (1957).

JACK 63   Jackson, J. R., "Jobshop-Like Queueing Systems", *Management Science*, **10**, No. 1, 131–142 (1963).

KLEI 64a   Kleinrock, L., "Analysis of a Time-Shared Processor," *Naval Research Logistics Quarterly*, **11**, 59–73 (1964).

KLEI 64b   Kleinrock, L., *Communication Nets; Stochastic Message Flow and Delay*, McGraw-Hill (New York) 1964. Out of print. Reprinted by Dover Publications (New York) 1972.

KLEI 66a   Kleinrock, L., "Sequential Processing Machines (S.P.M.) Analyzed with a Queueing Theory Model," *Journal of the Association for Computing Machinery*, **13**, 179–193 (1966).

KLEI 66b   Kleinrock, L., "Theory of Queues Applied to Time-Shared Computer Systems," *1966 IEEE Region Six Conference Record*, Tucson, Arizona, 491–500, April 26–28, 1966.

KLEI 67a   Kleinrock, L., "Time-Shared Systems: A Theoretical Treatment," *Journal of the Association for Computing Machinery*, **14**, 242–261 (1967).

KLEI 67b   Kleinrock, L. and E. G. Coffman Jr., "Some Feedback Queueing Models for Time-Shared Systems," *Proceedings of the Fifth International Teletraffic Congress*, 91–92, June 1967.

KLEI 67c   Kleinrock, L. and E. G. Coffman Jr., "Distribution of Attained Service in Time-Shared Systems," *Journal of Computer System Science*, **1**, 287–298 (1967).

KLEI 68a   Kleinrock, L., "Some Recent Results for Time-Shared Processors," *Proceedings of the First Hawaii International Conference on System Sciences*, University of Hawaii, Honolulu, 746–759, January 29–31, 1968.

KLEI 68b   Kleinrock, L., "Certain Analytic Results for Time-Shared Processors," *Proceedings of the International Federation for Information Processing Congress*, 838–845, August 1968.

KLEI 69   Kleinrock, L., "Time-Sharing Systems: Analytical Methods," in *Critical Factors in Data Management*, F. Gruenberger (ed.), Prentice-Hall (Englewood Cliffs, N.J.), 30–32, 1969.

KLEI 70a   Kleinrock, L., "A Continuum of Time-Sharing Scheduling Algorithms," *AFIPS Conference Proceedings*, 1970 Spring Joint Computer Conference 453–458, 1970.

KLEI 70b   Kleinrock, L., "Swap Time Considerations in Time-Shared Systems," *IEEE Transactions on Computers*, **C-19**, 435–540 (1970).

KLEI 70c   Kleinrock, L. and R. R. Muntz, "Multilevel Processor-Sharing Queueing Models for Time-Shared Models," *Proceedings of the Sixth International Teletraffic Congress*, 341/1–341/8, August 1970.

KLEI 71a   Kleinrock, L., R. R. Muntz, and J. Hsu, "Tight Bounds on Average Response Time for Processor-Sharing Models of Time-Shared Computer Systems," *Information Processing* 71, TA-2, 50–58, August 1971.

KLEI 71b   Kleinrock, L., R. R. Muntz, and E. Rodemich, "The Processor-Sharing Queueing Model for Time-Shared Systems with Bulk Arrivals," *Networks Journal*, **1**, 1–13 (1971).

KLEI 72a   Kleinrock, L., "A Selected Menu of Analytical Results for Time-Shared Computer Systems," in *Systemprogrammierung*, R. Oldenburg Verlag (Munich, Germany), 45–73, 1972.

KLEI 72b   Kleinrock, L., "Computer Networks," in *Computer Science*, A. F. Cárdenas, L. Presser, and M. A. Marin (eds.), Wiley Interscience (New York), 241–284, 1972.

KLEI 72c   Kleinrock, L. and R. R. Muntz, "Processor-Sharing Queueing Models of Mixed Scheduling Disciplines for Time-Shared Systems," *Journal of the Association for Computing Machinery*, **19**, 464–482 (1972).

KLEI 73a   Kleinrock, L., "Scheduling, Queueing and Delays in Time-Shared Systems and Computer Networks," in *Computer-Communication Networks*, N. Abramson and F. Kuo (eds.), Prentice-Hall (Englewood Cliffs, N.J.), 95–141, 1973.

KLEI 73b   Kleinrock, L. and J. Hsu, "A Continuum of Computer Processor-Sharing Queueing Models," *Proceedings of the Seventh International Teletraffic Congress*, Stockholm, Sweden, 431/1–431/6, June 1973.

KLEI 75a   Kleinrock, L., *Queueing Systems, Volume I: Theory*, Wiley Interscience (New York), 1975.

KLEI 75b   Kleinrock, L. and A. Nilsson, "On Optimal Scheduling Algorithms for Time-Shared Systems" to appear.

KOBA 74   Kobayashi, H., "Application of the Diffusion Approximation to Queueing Networks, I—Equilibrium Queue Distributions," *Journal of the Association for Computing Machinery*, **21**, 316–328 (1974).

KONH 72a   Konheim, A. G. and B. Meister, "Service in a Loop System," *Journal of the Association for Computing Machinery*, **19**, 92–108 (1972).

KONH 72b   Konheim, A. G., "Service Epochs in a Loop System," *Proceedings of the International Symposium on Computer-Communication Networks and Teletraffic*, Polytech Press, (Brooklyn, N.Y.), 125–143, April 1972.

LEWI 71   Lewis, P. A. W. and G. S. Shedler, "A Cyclic-Queue Model of System Overhead in Multiprogrammed Computer Systems," *Journal of the Association for Computing Machinery*, **18**, 199–220 (1971).

LEE 73   Lee, R. P., "A Comparison of Implicit and Explicit Priorities Scheduling in a Computer System," M.S. thesis, Computer Science Department, School of Engineering and Applied Science, University of California, Los Angeles, June 1973.

MCKI 69   McKinney, J. M., "A Survey of Analytical Time-Sharing Models," *Computing Surveys*, **1**, 105–116 (1969).

MOOR 71   Moore, C. G., III, "Network Models for Large-Scale Time-Sharing Systems," Technical Report No. 71-1, Department of Industrial Engineering, University of Michigan, Ann Arbor, Michigan, April 1971.

MOOR 72   Moore, F. R., "Computational Model of a Closed Queueing Networks with Exponential Servers," *IBM Journal of Research and Development*, 567–572 (1972).

MUNT 72a   Muntz, R. R. and F. Baskett, "Open, Closed, and Mixed Networks of Queues with Different Classes of Customers," Technical Report No. 33, Stanford Electronics Laboratories, Stanford University, August 1972.

MUNT 72b   Muntz, R. R., "Poisson Departure Processes and Queueing Networks," IBM Research Report RC 4145, December 1972.

MUNT 74a   Muntz, R. R. and J. Wong, "Efficient Computational Procedures for Closed Queueing Network Models," *Proceedings of the Seventh*

Hawaii International Conference on System Sciences, Honolulu, Hawaii, 33–36, January 8–10, 1974.

MUNT 74b   Muntz, R. R., "Analytic Models for Computer System Performance Analysis," *Proceedings of the NTG/GI Conference on Computer Architecture and Operating Systems*, Braunschweig, Germany, March 1974.

MUNT 74c   Muntz, R. R. and J. Wong, "Asymptotic Properties of Closed Queueing Network Models," *Proceedings of the Eighth Annual Princeton Conference on Information Sciences and Systems*, Princeton University, March 1974.

ODON 74   O'Donovan, T. M., "Direct Solutions of M/G/1 Processor-Sharing Models," *Operations Research*, **22**, 1232–1235 (1974).

OMAH 73   Omahen, K., "Analytic Models of Multiple Resource Systems," Ph.D. Dissertation, Division of Physical Sciences, The University of Chicago, June 1973.

PIER 71   Pierce, J. R., C. H. Cohen, and W. J. Kropfl, "Network for Block Switching of Data," *IEEE Conference Record*, New York, 222–223, March 1971.

POSN 68   Posner, M. and B. Bernholtz, "Closed Finite Queueing Networks with Time Lags and with Several Classes of Units," *Operations Research*, **16**, 977–985 (1968).

SAKA 69   Sakata, M., S. Noguchi, and J. Oizumi, "Analysis of a Processor-Sharing Queueing Model for Time-Sharing Systems," *Proceedings of the Second Hawaii International Conference on System Sciences*, 625–628, January 1969.

SAKA 71   Sakata, M., S. Noguchi, and J. Oizumi, "An Analysis of the M/G/1 Queue Under Round-Robin Scheduling," *Operations Research*, **19**, 371–385 (1971).

SCHE 67   Scherr, A. A., *An Analysis of Time-Shared Computer Systems*, MIT Press (Cambridge, Mass.) 1967.

SCHR 67   Schrage, L. E., "The Queue M/G/1 with Feedback to Lower Priority Queues," *Management Science*, **13**, 466–471 (1967).

SCHR 69   Schrage, L. E., "The Modelling of Man-Machine Interactive Systems," Report 6942, Center for Mathematical Studies in Business and Economics, University of Chicago, September 1969.

SHEM 67   Shemer, J. E., "Some Mathematical Considerations of Time-Sharing Scheduling Algorithms," *Journal of the Association for Computing Machinery*, **14**, 262–272 (1967).

SPRA 72   Spragins, J. D., "Loops Used for Data Collection," *Proceedings of the International Symposium on Computer-Communication Networks and Teletraffic*, Polytech. Press, (Brooklyn, N.Y.), 59–76, April 1972.

VAND 69   van den Heever, R., "Computer Time-Sharing Priority Systems," Ph.D. Dissertation, University of California at Berkeley, September 1969.

WALL 64   Wallace, V. L., "Queueing Theory in the Analysis of Computer Systems," classnotes for a Michigan Engineering Summer Conference entitled, "Digital Computers in Real Time," University of Michigan, Ann Arbor, Michigan, 1964.

WONG 75   Wong, Johnny W-N., "Queueing Network Models for Computer Systems," UCLA-ENG-7579, School of Engineering and Applied Science, University of California at Los Angeles.

EXERCISES

4.1. Show that Eq. (4.10) is the zero-quantum limit of Eq. (4.9).

4.2. Show that Eqs. (4.9) and (4.11) hold for each priority group in a time-sharing system that discriminates in some fashion based on priority as well as attained service time.

4.3. Consider a round-robin system in discrete time with a finite quantum $q > 0$. Customer arrivals are at an average rate of λ/sec with $P[1$ customer arrival in q sec$] = \lambda q$ and $P[0$ customer arrivals in q sec$] = 1 - \lambda q$ and service is geometrically distributed: $P[\text{service} = nq \text{ sec}] = (1 - \sigma)\sigma^n$ where $0 \le \lambda q \le 1$, $0 \le \sigma \le 1$. Assume that customers arrive just after a quantum begins [thereby joining the queue behind the customer (if any) who was ejected out of service if he requires more service]. Consider those instants just *before* any possible arrivals to the system but just *after* the completion of a quantum q. The number of customers in the system at these instants forms a Markov chain.
Let

$$p_k - P[k \text{ customers in system at the imbedded instants}]$$

- **(a)** Write down the equilibrium equations for p_k.
- **(b)** Find the solution to the equations in (a).
- **(c)** Find $\bar{N}$, the expected number in system.
- **(d)** Let $T(nq) = E[\text{response time for a customer requiring } nq \text{ sec}$ of service]. Let $\bar{N}_i = E[\text{number of customers served between the } (i-1)\text{st and } i\text{th ejection from service for a "tagged" customer}]$. Show that

$$
\bar{N}_i =
\begin{cases}
\bar{N} + 1 & i = 1 \\
\dfrac{1}{1-\rho} - \dfrac{\rho(1-\sigma\alpha)}{1-\rho}\alpha^{i-2} & i = 2, 3, \ldots, n
\end{cases}
$$

where $\rho = \lambda q/(1-\sigma)$ and $\alpha = \sigma + \lambda q$.

(e) From (d) show that

$$T(nq) = \frac{nq}{1-\rho} - \frac{\lambda q^2}{1-\rho}\left[1 + \frac{(1-\sigma\alpha)(1-\alpha^{n-1})}{(1-\sigma)^2(1-\rho)}\right]$$

4.4. Let us take

$$\lim_{\substack{q \to 0 \\ \sigma \to 1}} \frac{q}{1-\sigma} = \frac{1}{\mu}$$

in the discrete-time system of Exercise 4.3.

(a) Show that the limiting arrival process is Poisson at a mean rate λ.

(b) Show that the limiting service time is exponential with mean $1/\mu$ sec.

(c) Let n increase so that

$$\lim_{\substack{q \to 0 \\ n \to \infty}} nq = x$$

Show that

$$\lim T(nq) = T(x) = \frac{x}{1-\rho}$$

as it must for the RR processor-sharing system.

4.5. It can be shown for the system M/G/1 with RR processor-sharing that the quantity $P_k(x)\,dx = P[k$ customers in system with attained service in the interval $(x, x+dx)]$ obeys the following set of differential-difference equations for $x > 0$:

$$\frac{d}{dx}P_k(x) = -k\mu(x)P_k(x) + (k+1)\mu(x)P_{k+1}(x) \qquad k = 0, 1, 2, \ldots$$

where $\mu(x)\,dx = P[$customer with x sec of attained service needs only dx seconds more$]$

(a) Express $\mu(x)$ in terms of $B(x)$.

(b) Assume $B(x) = 1 - e^{-\mu x}$. What now is $\mu(x)$?

(c) Find a differential equation for the attained service density for $n(x)$ with $B(x)$ and $\mu(x)$ as given in part (b).

(d) Solve the differential equation for $n(x)$ in (c) and leave it in terms of $n(0)$.

4.6. Consider the same system as in Exercise 4.5 except we now allow an arbitrary $B(x)$.

(a) Repeat part (c) of Exercise 4.5 for arbitrary $B(x)$.

(b) Repeat part (d) of Exercise 4.5 for arbitrary $B(x)$.

(c) Let

$$P(z, x) = \sum_{k=0}^{\infty} P_k(x)z^k$$

Develop a partial differential equation for $P(z, x)$.

(d) Show that the partial differential equation of (c) is satisfied by

$$P(z, x) = 1 - (1 - z)n(0)[1 - B(x)]$$

(e) From (d) find $P_1(x)$.

4.7. Consider an M/M/1 RR processor-sharing system. Let $w_k(x)$ be the waiting time of a customer who requires x sec of service and who arrives to find k others in the system [that is, his response time is $w_k(x) + x$]. Let $W_k^*(s, x)$ be the Laplace transform of the pdf for $w_k(x)$. Denote $W_k^*(s, x)$ by W_k for simplicity.

(a) Show that

$$\frac{\partial W_k}{\partial x} = \lambda(k + 1)W_{k+1} - [\lambda(k + 1) + (s + \mu)k]W_k + \mu k W_{k-1}$$

where $W_k^*(s, 0) = 1$ for $s \geq 0$ and $k = 0, 1, \ldots$, and $W_{-1} \triangleq 0$.

(b) Let

$$W^*(s, x, z) = \sum_{k=0}^{\infty} W_k^*(s, x)z^k$$

and, for simplicity, let $W^*(s, x, z) = W$. Show that

$$\frac{\partial W}{\partial x} - [\mu z^2 + (s + \mu + \lambda)z + \lambda]\frac{\partial W}{\partial z} = (\mu z - \lambda)W$$

where $W^*(s, 0, z) = 1/(1 - z)$.

(c) It can be shown [COFF 70] that the solution for W in (b) is

$$W = \frac{(1 - \rho r^2)e^{-\lambda x(1-r)}}{(1 - \rho r)(1 - rz) - (1 - r)(z - \rho r)e^{-\mu x(1-\rho r^2)/r}}$$

where r is defined in Eq. (4.20). Show that Eq. (4.19) follows from this expression for W.

4.8. Consider an RR time-shared system with a finite quantum $q > 0$, of which an amount θ is wasted due to swap time. Assume that $B(x) = 1 - e^{-\mu x}$. A customer's last quantum will be less than q sec if he requires less than $(q - \theta)$ sec on his last quantum.

(a) Find the pdf for q, the length of an arbitrary quantum.

(b) Find $\bar{q}$ and $\overline{q^2}$.

(c) Find the distribution for N_q, the number of quanta required by a customer.

(d) Find $E[N_q]$.

(e) Find the Laplace transform of the pdf for X, the total time devoted to a customer, including his swap time.

(f) Find $E[X]$ and $E[X^2]$ in terms only of μ, θ, and q.

(g) Suppose now, that the input process is Poisson at a rate λ. With $\rho = \lambda E[X] < 1$, find the average number of customers in the system.

4.9. In Section 4.4 we showed that $dn(x)/dx = -\mu(x)n(x)$. Apply Eqs. (4.11) and (4.14) directly to this differential equation to obtain the form for $T(x)$ given in Eq. (4.16).

4.10. Consider an M/G/1 processor-sharing system with an arbitrary system of queues and with arrival rate λ and service time distribution $B(x)$.

(a) Express the average number of customers $\bar{N}$ in the system in terms of $n(x)$.

(b) Let us now consider the same system as above but with "swap loss." That is, for every x sec of processing, a fraction ϕ (that is, ϕx sec) is lost in swapping. We may model this by assuming that a job which takes x sec in the system above will take $x/(1-\phi)$ sec in this system, thus giving rise to a new service time distribution, $\hat{B}(x)$. Let $S = E[\text{total time lost in swapping for all those customers still in the system}]$.

  (i) From the results of (a), find an expression for S in terms of ϕ, λ, $\hat{B}(x)$, and $T(x)$.

  (ii) For $B(x) = 1 - e^{-\mu x}$, find S explicitly, when the system is RR.

4.11. From Eq. (4.23) for the preemptive LCFS algorithm, find

(a) the mean response time $T(x)$.

(b) the variance of the response time.

(c) Compare (b) with the response time variance for batch processing.

4.12. **(a)** Prove Eq. (4.27) by using the moment-generating properties of $S^*(s \mid x)$ in Eq. (4.28).

(b) In a similar fashion, find the variance of the response time for FB.

4.13. Consider the M/G/1 two-level foreground background queue described in the footnote on page 172 with parameters λ, $\bar{x}$, $B^*(s)$.

Let the foreground quantum be q and let the background quantum be ∞. We permit the foreground jobs to preempt the background jobs (with no loss); the preempted job resumes service as soon as the foreground queue empties.

(a) Find $W(x)$.

(b) We now permit Poisson arrivals (at a rate λ_0) to enter the background queue as well. These jobs have a service time pdf whose transform will be denoted by $B_0^*(s)$. Find $W(x)$ for jobs that enter at the foreground level.

4.14. Let us consider a bulk arrival process to an M/G/1 queue. That is, at points in time chosen from a Poisson process (of rate λ points/sec), we have *groups* of arrivals to the system. Let us also assume that we serve customers in an RR fashion with quantum size q sec. We assume that all service times are multiples of q and define

$$\sigma_i = P[\text{job which has received } iq \text{ sec of service will require more than } (i+1)q \text{ sec}]$$

$n(iq) = $ average number of jobs in the system with iq sec of *attained* service

$\bar{a} = $ average bulk size of arrivals

$b = $ average number of arrivals entering with a "tagged" job (see below)

We focus attention on a job (the "tagged" job) which requires $x = iq$ sec of service. We assume that, on the average, b jobs arrive with the tagged job. Let us assume that the tagged job is positioned with respect to the rest of his group so that he is last in this group. Let

$T_i = $ average time between the $(i-1)$st and ith quantum of service given to the tagged job

(a) Find an expression for T_1, the average time until the tagged job receives his first quantum.

(b) Show that for $i > 1$,

$$T_i = \sum_{j=0}^{\infty} n(jq)q\sigma_j\sigma_{j+1}\cdots\sigma_{j+i-2}$$
$$+ \sum_{j=1}^{i-1} \lambda\bar{a}T_j q\sigma_0\sigma_1\cdots\sigma_{i-j-1}$$
$$+ q + bq\sigma_0\sigma_1\cdots\sigma_{i-2}$$

(c) Interpret each of the four terms in the equation in (b).

(d) Consider the limit as $q \to 0$. For this limit, let $iq \to \tau$ and $jq \to x$. Note also that $T_i/q \to (d/d\tau)\alpha(\tau)$ where $\alpha(\tau)$ is the average time in the system for jobs requiring τ sec of service. By considering the limit as $q \to 0$ for each of the terms above show that we get Eq. (4.40).

4.15. We wish to establish Eq. (4.43).

    (a) Use the distribution of Eq. (4.42) to replace $B(t)$ in the basic integral equation (4.40).

    (b) Let the operator D denote the differential operator d/dt. Differentiate the result of (a) $d+1$ times to show that

$$Q_0(D)[\alpha^{(1)}(\tau)] + Q_1(D)[\alpha^{(1)}(t_1 - \tau)] = \mu^{d+1}$$

    where Q_0 and Q_1 are given in Eqs. (4.44)–(4.45).

    (c) Show that

$$[Q_0(D)Q_0(-D) - Q_1(D)Q_1(-D)]\alpha^{(1)}(\tau) = [Q_0(0) - Q_1(0)]\mu^{d+1}$$

    (d) Show that the general solution for $\alpha^{(1)}(t)$ is

$$\alpha^{(1)}(\tau) = \frac{1}{1 - \lambda \bar{a} \tau} + \sum_{m=1}^{d+1} (A_m e^{-\beta_m \tau} + B_m e^{\beta_m \tau})$$

    where the roots of $Q_2(y)$ [see Eq. (4.46)] are denoted by $(\beta_i, -\beta_i)$ in pairs, $i = 1, 2, \ldots, d+1$.

    (e) Using the result for $\alpha^{(1)}(\tau)$ in the expression in (b), and matching coefficients, show that

$$A_m = C_m Q_0(\beta_m)$$
$$B_m = -C_m e^{-\beta_m t_1} Q_1(\beta_m) \qquad m = 1, 2, \ldots, d+1$$

    (f) How would you now proceed to determine the coefficients C_m to prove Eq. (4.43)?

4.16. Carry out the complete solution to Eq. (4.43) in the case when $q(\tau) = 1$ (that is, exponential service).

4.17. From the expression for $G(z)$ in the footnote on p. 184, find $G^{(1)}(1)$ and $G^{(2)}(1)$ as well as $\bar{a}$ and b for the M/G/1 bulk arrival RR processor-sharing system.

4.18. Identify the limiting forms for $W^*(s \mid x)$ in Eq. (4.50) and $T(x)$ in Eq. (4.51) for

    (a) $\beta/\alpha = 1$

    (b) $\beta/\alpha = 0$

4.19. From Eq. (4.50), prove that Eq. (4.51) is indeed $T(x)$.

4.20. Let us consider a cascade of SSA systems with a Poisson input at rate λ. In particular, consider an SSA system with parameters α_1 and β_1 corresponding to the original α and β parameters described in the text. Assume now that the service box itself is an SSA system with its own queue box and its own service box with parameters α_2 and β_2. Let us continue this iterated structure in which each service box contains its own queue box and service box down to say N levels.

  (a) Find $W^*(s \mid x)$ for this system explicitly in terms of the α_i and β_i $(i = 1, \ldots, N)$, $B^*(s)$, and $\hat{W}_\lambda^*(s \mid x, N)$, which is the known performance function of the innermost raw scheduling algorithm.

  (b) Discuss any generality offered by this cascade over the simple one-stage SSA system.

4.21. Use the method for deriving Eq. (4.57) to prove Eq. (4.60).

4.22. Derive Eq. (4.65) simply by equating the rate at which jobs enter the dotted box of Figure 4.26 to the rate at which they must leave.

4.23. Consider an M/M/1 processor-sharing system. We observe the arrival of a "tagged" job at a random point in time. Let $m(t)$ be the average number of customers in the system given that the tagged job has attained t sec of service.

  (a) Express $m(t)$ in terms of λ, μ, t, and $T(t)$.

  (b) Place tight upper and lower bounds on $m(t)$ for any conservative algorithm.

4.24. Consider an M/G/1 FB time-sharing system. We wish to derive the average response time as follows [ODON 74]. We consider a tagged job whose service requirement exceeds x sec.

  (a) Find $N_1(x) = E[\text{number of jobs in the system when the "tagged" job arrives which have attained service} \leq x$, but with service requirement $> x]$.

  (b) Find $N_2(x) = E[\text{number of jobs which arrive after the "tagged" job enters, but before the "tagged" job receives } x \text{ sec of service, and which themselves need more than } x \text{ sec of service}]$.

  (c) Clearly $N_1(x) + N_2(x)$ is the expected number of jobs with attained service x when our tagged job has attained service x. By considering the events that occur during the interval of time required to give an additional Δx seconds of service to the tagged job, find an expression relating $dT(x)/dx$ with $N_1(x)$ and $N_2(x)$ and other system parameters.

(d) Prove that the known expression for $T(x)$ for FB satisfies the equation in part (c).

4.25. Consider an M/G/1 time-sharing system with an arrival rate $\lambda = 1$ and the following service time density:

$$b(x) = \begin{cases} 1 & 0 \le x \le 1 \\ 0 & \text{otherwise} \end{cases}$$

(a) Consider a multilevel (ML) scheduling algorithm with two levels as follows: $a_1 = \frac{1}{2}$, $D_1 = \text{RR}$, $D_2 = \text{FCFS}$. Find $T(x)$.

(b) Now consider a selfish scheduling algorithm (SSA) with the raw scheduling algorithm from part (a) and with parameters α, β such that $\beta/\alpha = \frac{1}{2}$. Find $T(x)$ for the SSA.

(c) Give the upper and lower bounds on *any* $W(x)$ function for the $b(x)$ above.

(d) Of the following cases, indicate which are feasible.

    (i) $W(0) = \frac{1}{6}$
    (ii) $W(0) = \frac{1}{2}$
    (iii) $W(\frac{1}{2}) = \frac{1}{10}$
    (iv) $W(\frac{1}{2}) = 1$
    (v) $W(1) = 1$
    (vi) $W(1) = \frac{3}{2}$

(e) Consider $W(x) = \frac{3}{2}x^2$ $(0 \le x \le 1)$. Establish definitely whether $W(x)$ is feasible.

4.26. From Eq. (4.28), show for the FB system that the variance of the response time (for a job requiring x sec of service), which we denote by $\sigma^2(x)$, is given by

$$\sigma^2(x) = \frac{2W_x}{1 - \rho_x} + \frac{\sigma_x^2}{(1 - \rho_x)^2}$$

where σ_x^2 is the variance of the response time in an FCFS system with a truncated service time pdf $b_x(y)$ as in Figure 4.6.

4.27. Consider the FCFS M/G/1 queueing system with input rate λ and with feedback shown below. Each time a customer completes service [whose distribution is $B(x)$], he will reenter service again with probability p, and will depart forever with probability $1 - p$.

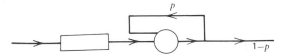

(a) Find $H_1^*(s)$, the Laplace transform of the pdf of a customer's total time in service (including feedback) explicitly in terms of p and $B^*(s)$.

(b) Find $Q_1(z)$, the z-transform of the distribution of the number in system.

(c) Find $W_1^*(s)$, the Laplace transform of the pdf of waiting time in queue.

(d) Find the average wait W_1.

Now consider the FCFS M/G/1 queueing system with feedback shown below. The feedback is now to the tail of the queue instead of directly to service. Here H, Q and W will be as above but will be distinguished by the subscript 2.

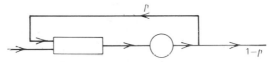

(e) Find $H_2^*(s)$ for this figure.

(f) How is $Q_2(z)$ related to $Q_1(z)$, if at all?

(g) How is $W_2^*(s)$ related to $W_1^*(s)$, if at all?

(h) Find the average wait W_2.

(i) What is the total customer input rate to the queue?

4.28. Let us define a linear cost function $C(W, x)$ which is the "system" cost rate (per average waiting second) of delaying a customer whose total service requirement is x sec by an amount W sec in an M/G/1 time-sharing system; here we have written $W = W(x)$. The total average cost per customer is therefore

$$C = \int_0^\infty C(W, x)W(x)b(x)\, dx$$

We wish to minimize C over all possible scheduling algorithms under the constraint of the conservation law [Eq. (4.58)]. It can be seen [KLEI 75b] that $W(x)$ must satisfy the condition

$$\frac{\partial}{\partial W}[C(W, x)W(x)b(x)] = k[1 - B(x)] \qquad \blacksquare$$

where k is a Lagrange multiplier, that is, some constant.

(a) For each of the following choices of $C(W, x)$ and $b(x)$, find the W which minimizes C and describe a scheduling algorithm which produces such a W.

    (i) $C(W, x) = W$ and   M/M/1

    (ii) $C(W, x) = W/(x + a)$ and   M/M/1

    (iii) $C(W, x) = W/x$ and   M/E$_2$/1

(b) Find the W which minimizes C for the cases below and explain why these solutions are infeasible.

(i) $C(W, x) = W$ and   M/E$_2$/1

(ii) $C(W, x) = W/(x^2 + a)$ and   M/M/1

4.29. Consider the system in Figure 4.26. Let $x = \lambda/\mu$.

(a) For $M \ll M^*$, show that

$$\mu T \cong \frac{1}{1 - (M - 1)x}$$

(b) For $M \gg M^*$ show that

$$\mu T \cong M - M^* + 1$$

4.30. Consider a closed Markovian queueing network with the usual parameters: N, m_i, μ_i, r_{ij}, and K. Now let $m_i = 1$ for $i = 1, 2, \ldots, N-1$ and $m_N = M$. We (may) arbitrarily select $x_N = 1$ in Eq. (4.77). Also, let $k = K - k_N$. Give the simplest form you can for $G(K)$. See [MOOR 72].

4.31. Consider a computer system with two single-server resources under demand. Assume that a Poisson arrival process (at rate λ_i) requires exponential service (at rate μ_i) from resource i ($i = 1, 2$).

(a) Suppose both customer streams must share a single storage space (of capacity K) and any arrivals that find K others in the system are lost. Find the average number of departures per second from each resource.

(b) Suppose we now split the storage so that K_i spaces are reserved for stream i such that $K_1 + K_2 = K$. Find the value of K_1 that maximizes the combined throughput.

4.32. Derive Eq. (4.103).

4.33. Consider the following closed Markovian net with one customer $(K = 1)$:

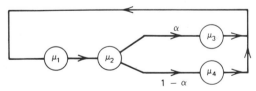

(a) Find $p(k_1, k_2, k_3, k_4)$ *explicitly* in terms of μ_i ($i = 1, 2, 3, 4$) and α.

(b) For $\mu_1 = \mu_2 = \mu_3 = 2$ and $\mu_4 = 1$, find the "saturated" node(s) as α varies over the range $0 \le \alpha \le 1$.

4.34. Consider a closed queueing network with five nodes (service stations), where node i has m_i servers each of which requires $1/\mu_i$ sec to perform a service from an arbitrary distribution. Let $m_5 = M$ and $m_1 = m_2 = m_3 = m_4 = 1$ (we identify node 5 as the "terminal" node). The closed system contains exactly M jobs flowing around. Assume

$$\mathbf{R} = \begin{bmatrix} 0 & \frac{1}{2} & \frac{1}{2} & 0 & 0 \\ 0 & 0 & \frac{1}{2} & \frac{1}{2} & 0 \\ 0 & \frac{1}{2} & 0 & \frac{1}{2} & 0 \\ 0 & 0 & 0 & 0 & 1 \\ 1 & 0 & 0 & 0 & 0 \end{bmatrix}$$

Also assume $\mu_1 = 3$, $\mu_2 = 2$, $\mu_3 = 3$, $\mu_4 = 3$, $\mu_5 = \frac{2}{17}$.
   (a) Draw a fully labeled graph of the network.
   (b) Solve for λ_i/λ_5 for $i = 1, 2, 3, 4$.
   (c) Which is the most saturated node?
   (d) Find M^*, the saturation number.
   (e) Determine the response time T (average time in "rest of the network," that is, from departure to arrival at terminal node) for $M \gg M^*$.
   (f) Repeat part (e) for $M \ll M^*$.
   (g) Show that the two asymptotic curves in parts (e) and (f) intersect at $M = M^*$.
   (h) Draw the (fully labeled) asymptotic behavior on the (T, M) plane.
   (i) Suppose $\mu_2 = 3$ and $\mu_4 = 2$. For parts (a)–(f) above, identify which answers change and which do not.

4.35. Consider the closed Markovian networks for which $m_i = 1$ for $i = 1, 2, \ldots, N$.
   (a) Show that $P[\bar{k}_i \geq k_i] = x_i^{k_i}[G(K - k_i)/G(K)]$.
   (b) From part (a), prove Eqs. (4.80) and (4.81).

4.36. For the same system as in Exercise 4.35, let

$$g(k, n) = \sum_{\mathbf{k} \in B} \prod_{i=1}^{n} x_i^{k_i}$$

where B is that set of vectors $\mathbf{k} = (k_1, k_2, \ldots, k_n)$ for which $k_i \geq 0$ and $\sum_{i=1}^{n} k_i = k$. Note that $g(k, N) = G(k)$ for $k = 0, 1, \ldots, K$.
   (a) Show that $g(k, 1) = x_1^k$ for $k = 0, 1, \ldots, K$.
   (b) Show that $g(0, n) = 1$ for $n = 0, 1, \ldots, N$.

(c) Show that $g(k, n) = g(k, n-1) + x_n g(k-1, n)$ for $n > 1$ and $k > 0$.

(d) The iterative algorithm defined by parts (a), (b), (c) will generate $G(0), G(1), \ldots, G(K)$. Show that the algorithm requires NK multiplications and NK additions, assuming x_i $(i = 1, 2, \ldots, N)$ is given.

4.37. Consider the central server model. We know that

$$A_1 \mu_1 p_1 = A_i \mu_i \qquad i \geq 2$$

Let $A_i^{(j)}$ be the equilibrium probability that there are j or more customers in the ith service station. Thus

$$A_i^{(0)} = 1$$
$$A_i^{(1)} = A_i$$

(a) Find $A_1^{(j)}$.

(b) Find $A_i^{(j)}$ $(i = 2, 3, \ldots, N)$ in terms of $A_1^{(j)}$.

Let Q_i be the average number of customers in the ith station.

(c) Find Q_1 in terms of $A_1^{(j)}$ $(j = 1, 2, \ldots, K)$.

(d) Find Q_1 in terms of $G(k)$ $(k = 0, 1, \ldots, K)$.

(e) Find Q_i in terms of $A_1^{(j)}$ $(j = 1, 2, \ldots, K)$ $(i \geq 2)$.

(f) Find Q_i in terms of $G(k)$ $(k = 0, 1, \ldots, K)$ $(i \geq 2)$.

4.38. Consider the refinement to the multiprogramming model discussed at the end of Section 4.13. Let both the CPU and DTU have exponentially distributed service times.

(a) Find s_0.

(b) Find $F(0)$ from Eq. (4.105) using this refinement.

(c) Solve the two-node closed net by a network of queues model and compare $F(0)$ to part (b).

4.39. Consider an M/M/1 system with parameters λ and μ. A departing customer really departs with probability $(1 - \alpha)$ and returns to the tail of the queue with probability $\alpha < 1$. However, upon his return, he is magically converted into N customers. Let $p_k = P[\text{system contains } k \text{ customers}]$.

(a) Let $N = 1$. Find p_k explicitly. What is the condition for stability?

(b) Let $N = 2$. Find p_k explicitly. What is the condition for stability?

(c) Let $N = 3$. Find p_k explicitly. What is the condition for stability?

4.40. Consider the queueing network shown on the opposite page. Customers arrive according to a Poisson process at rate λ. Upon

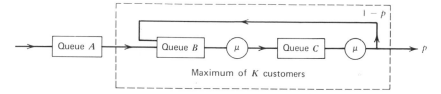

Maximum of K customers

arrival, a customer will enter the dotted box if the number of customers in that box is less than K; otherwise, he will join the FCFS Queue A, which feeds the customer at its head into the dotted box as soon as the number in the box drops below K. Upon entry into the box, a customer proceeds to the FCFS Queue B where he eventually receives service from an exponential server at rate μ. He then proceeds to Queue C, which is served by an independent exponential server at the same rate μ. He then departs the box (and the system) with probability p, or returns to Queue B with probability $1-p$, and so on. Let us assume equilibrium conditions.

(a) Let $K=1$. This is a "familiar" special case.

    (i) Find an expression for $\beta(x)$, which is defined as the pdf of the time a customer spends in the dotted box.

    (ii) In terms of the moments of $\beta(x)$, find the average number, $\bar{N}$, of customers in the entire system (Queue A plus the dotted box).

    (iii) From (ii), find T_1, the average time a customer spends in the entire system.

(b) Let $K=\infty$. This is another familiar special case.

    (i) Find $P(k)=P[k$ in entire system].

    (ii) Find T_∞, the average time a customer spends in the entire system.

(c) Find the conditions for stability in (a) and (b).

(d) For the common range of stability, prove that $T_1 \geq T_\infty$.

4.41. Consider a Markovian queueing network with N nodes and with transition matrix $(i, j = 1, 2, \ldots, N)$

$$
\mathbf{R} =
\begin{bmatrix}
\alpha_1 & 1-\alpha_1 & 0 & 0 & \cdots & 0 \\
\alpha_2 & 0 & 1-\alpha_2 & 0 & & 0 \\
\alpha_3 & 0 & 0 & 1-\alpha_3 & & 0 \\
\cdot & \cdot & \cdot & \cdot & & \cdot \\
\cdot & \cdot & \cdot & \cdot & & \cdot \\
\cdot & \cdot & \cdot & \cdot & & \cdot \\
\alpha_{N-1} & 0 & 0 & 0 & & 1-\alpha_{N-1} \\
\alpha_N & 0 & 0 & 0 & & 0
\end{bmatrix}
$$

where $0 \le \alpha_i < 1$. Let the external input traffic be $\gamma_1 = \gamma$, $\gamma_i = 0$ $(i = 2, 3, \dots, N)$.

(a) Find the nodal traffic λ_i explicitly for $i = 1, 2, \dots, N$.

(b) For $\alpha_N = 0$, what must λ_N be (by physical reasoning)?

(c) From (b), find a simple relationship for the terms in the expression for λ_N from (a) and thereby simplify the answer for all λ_i in part (a).

(d) Let μ_i be the service rate for each (single-server) node. What relationship must hold among the $\{\mu_i\}$ in order that each server have the same utilization?

(e) Suppose we convert this to a *closed* network by letting the last *row* of **R** become

$$[\alpha_N \; \alpha_1 \; \alpha_2 \; \alpha_3 \cdots \alpha_{N-1}].$$

  (i) What must the first *column* sum to?

  (ii) What is the only sensible value for γ?

  (iii) What is the relationship among the relative number of visits to each station?

4.42. Consider Gaver's remote terminal access model with M input terminals and one response terminal. Let $f(x)$ be the pdf [with Laplace transform $F^*(s)$] associated with the generic cycle time $C(j)$.

(a) Show that the number of cycles K, as defined in the text, is geometrically distributed as

$$P[K = k] = (1 - \sigma)\sigma^{k-1}$$

where

$$\sigma = \int_0^\infty e^{-\beta x} f(x) \, dx$$

(b) Show that

$$E[e^{-sC'}] = \frac{F^*(s) - F^*(s + \beta)}{1 - F^*(s + \beta)}$$

(c) From (b), find $\overline{C'}$ and $\sigma_{C'}^2$.

(d) Using the assumption of exponentially distributed transmission times (mean $1/\gamma$), express $F^*(s)$ explicitly and then establish the expressions for $\overline{C'}$ and $\sigma_{C'}^2$ given in the text.

4.43. Compare the average delays in Eqs. (4.120) and (4.121) when $M = r = 1$.

4.44. Consider the system whose solution is given in Eqs. (4.116)–(4.117). Let N_n be the number of queued data units just prior to the nth slot time, and let v_n be the number of data units arriving during the nth slot.

(a) Write down the basic relation among N_{n+1}, N_n, and v_n.

(b) Let $N_n(z) = E[z^{N_n}]$ and $V_n(z) = E[z^{v_n}]$ and also

$$\lim_{n \to \infty} N_n(z) = N(z), \qquad \lim_{n \to \infty} V_n(z) = V(z)$$

Prove that

$$N(z) = (1 - \lambda) \frac{(z-1)V(z)}{z - V(z)}$$

(c) From part (b), prove Eq. (4.116).

(d) From part (b), prove Eq. (4.117).

4.45. Consider the probabilities $\{p_k\}$ as defined in Eqs. (4.122)–(4.124).

  (a) Find α, the average departure rate of data units from the buffer (not counting those which overflow).

  (b) In terms of λ and α, find P_{of}.

  (c) Consider that new data units arrive from a compound Poisson source, that is, bulk arrivals come from a Poisson source (at an average of λ bulks per slot) and the bulks are geometrically distributed with a mean bulk size of $1/\beta$.

  (i) Find the generating function for the number of data units arriving in a slot.

  (ii) Find π_k, the inverse of the generating function in (i).

4.46. Consider a set of data blocks $B_1, B_2, \ldots, B_N$ that is organized as a (one-way) linked list of blocks in a computer memory. That is, the first data block includes a pointer to the address of the second, which itself points to the third, and so on. Assume that independent requests (memory references) for data blocks occur such that

$$p_i \triangleq P[B_i \text{ is requested next}] > 0$$

When B_i is requested, the list is searched sequentially until B_i is found. If the position of the blocks at a given search initiation for B_i is $[B_{a_1}, B_{a_2}, \ldots, B_{a_{n-1}}, B_{a_n}, B_{a_{n+1}}, \ldots, B_{a_N}]$, where $a_n = i$, then after B_i is found, the list is reorganized with B_i at the head of the list; that is, $[B_i, B_{a_1}, B_{a_2}, \ldots, B_{a_{n-1}}, B_{a_{n+1}}, \ldots, B_{a_N}]$. Note that a search of n blocks was required.

  (a) For $j \neq i$, find

$$p_{ij}^{(N)} = P[B_j \text{ is requested at least once between}$$
$$\text{successive requests for } B_i]$$

(b) Let

$E_i^{(N)} = E$[number of unique blocks which are requested between successive requests for (say) B_i when there is a total of N blocks]

    (i) Find $E_i^{(2)}$ in terms of $p_{ij}^{(2)}$ $(j \neq i)$.
    (ii) Find $E_i^{(3)}$ in terms of $p_{ij}^{(3)}$ $(j \neq i)$.
    (iii) Express $E_i^{(N)}$ in terms of $p_{ij}^{(N)}$ $(j \neq i)$.

(c) Using results in parts (a) and (b), find

$$\bar{P}_i = E[\text{position (depth) of } B_i]$$

in the steady state in terms of $p_1, p_2, \ldots, p_N$.

(d) From (b), find $\bar{P}$, the expected depth of a search.

(e) A better way to organize this list would be to keep it unchanged in decreasing order of p_i. For this case, we will have a mean search depth equal to $\bar{Q}$. Show that

$$\bar{Q} \leq \bar{P} \leq 2\bar{Q} - 1$$

4.47. Let us study a time-sharing system with a random swap time $\tilde{\theta}$ which is independent of the service time $\tilde{x}$. We first assume an FCFS M/M/1 system with input rate λ and average service time $\bar{x}$. However, due to the swap time, the total service time $\tilde{y}$ is not exponential and is given by

$$\tilde{y} = \tilde{x} + \tilde{\theta}$$

Assume that $\tilde{\theta}$ is uniformly distributed over the interval $(0, a)$.

(a) Find the pdf for $\tilde{y}$.

(b) Find the (conditional) average response time $T(x)$ explicitly.

Now, assume that service is scheduled according to an RR processor-sharing algorithm. Assume also that once a job is swapped in (to main memory) that it is not swapped out until the job is completed (i.e., only one swap-time cost is incurred per job and therefore the service time for a job is simply $\tilde{y}$ above). Now assume an arbitrary $B(x)$.

(c) Find $T(x)$.

(d) Find a condition on the system parameters that guarantees that the *unconditional* average response time, T, is less for the RR system in part (c) than for the FCFS system in part (b).

Now assume that on the nth visit to service, the swap time $\tilde{\theta}_n$ is $a(\frac{1}{2})^n$ with probability one. Assume an RR processor-shared system where no extra swap time (beyond the first, $\tilde{\theta}_1$) is incurred if a job finds the system empty and leaves before any new job enters; assume an arbitrary service time distribution $B(x)$.

(e) Find the average swap time $\bar{\theta}$ explicitly in terms of a, $\bar{x}$, λ, $B^*(s)$.

(f) What is the relationship among λ, $\bar{x}$, and a to guarantee stability?

Lastly, let us consider an LCFS preemptive algorithm with exponential service (M/M/1) and *no* swap time. The system can hold at most two customers (the newest one in service and the most recently preempted one in the queue); any others are lost.

(g) What is the probability P that a customer is not "lost."

(h) For those jobs that are not lost, find T, the average time in system.

5

Computer-Communication Networks: Analysis and Design

The 1960s was the decade during which time-shared computer systems developed enormously. Toward the end of that decade it was recognized that the special resources and capabilities built up at each of the many separate computer facilities could profit by resource- and load-sharing. (These resources take the form of specialized hardware, software, and data bases.) As a result it was natural* that computer communication should evolve as the next phase in the evolution of information processing systems; thus the 1970s was ushered in as the decade of computer networks!

The structure and sophistication of these networks varies over a considerable range from highly specialized networks designed to handle specific tasks in a carefully controlled environment to more generalized networks that handle a variety of tasks in a highly unstructured environment. The successes and failures of these networks in terms of economy, service, response time, throughput, coverage, reliability, use, and convenience is certainly not uniform from network to network. In this chapter we examine some of the general principles of data network design that have emerged, some of the experience we have gained with respect to a particular generalized network (the ARPANET—an experimental computer network), some advanced concepts, and then lastly, we discuss some of the open problems and challenges that as yet remain in the design of data networks.

The interest in computer networks has grown enormously in the last few years, and the complexity of the issues one faces in creating such

---

* As pointed out by Roberts [ROBE 74], the appearance of sophisticated computer-communication networks coincided with the crossover point (in 1969) when the (processing) cost of dynamically allocating communication resources first became cheaper than the cost of the communication resources being allocated. Thus, for the first time, high-speed networks became cost-effective (see Figure 5.56). This change has come about principally due to the dramatic reduction in the cost of computer hardware.

networks is staggering. The considerations range from the highly technical, mathematical and engineering design questions (many of which are "within reach") to the extremely frustrating political, legal, privacy, social, financial, management, and ethical questions that penetrate the fabric of our society today (most of which are "far out of reach"). The growth in the need for data communications comes from a large number of varied application areas. For example, the finance industry, including banking and insurance firms, has a growing need for remote data processing (such as electronic funds transfer, etc.). In the field of medicine and health there is a need for large information banks with remote access. Educational computing needs currently emphasize interactive use as opposed to routine data entry, retrieval and acquisition. Large government agencies have vast data exchange requirements. Tactical military computer-communication needs are perhaps the largest (global) and most demanding (secure, rapid, reliable). Point of sales terminals by retail organizations is a fast-growing applications field (requiring on-line credit authorization, etc.). Information retrieval is of great importance in the transportation field currently and control of traffic load is a fast growing area. Large corporations now are exchanging, recording, and updating data among their many central and regional offices through integrated corporate data networks. Other industries have a natural need for computer networks (for example, airline reservations systems, travel services, stock market quotation systems, etc.). A vast use is foreseen for access to information processing directly from the consumer's home (shopping, voting, the use of electronic file cabinets, home education, etc.). It is applications such as these and many others that are providing the manpower, time, and money behind the enormous growth of the information processing industry. It is estimated that at the end of 1971 almost a quarter of a million terminals were in use. What is more exciting is that approximately four to five million terminals are projected for 1980 [DATA 73]. Voice input–output for a limited set of commands is already a reality; when voice input–output is available with a large set of commands, then all speech transmitting instruments could become computer terminals, and this will further snowball the need for remote data processing. A similar growth in demand for remote data processing is occurring in many of the industrialized regions of the world. For example, it is estimated [EURO 73, PETE 73] that data traffic in 17 West European countries will grow by a factor of 6 from 1972 to 1980, at which time the demand is estimated at 10^{12} bits/day; in these same countries, the number of terminals will grow from 80,000 in 1972 to an estimate just under a half million in 1980 (and 800,000 by 1985). One major effect of large computer-communication networks may well be a significant reduction in the enormous quantity of paper that is generated by and shuffled among organizations of all types today.

The goal of this chapter is twofold. First, we wish to familiarize the reader with computer-communication networks from a descriptive and operational point of view. (The emphasis on the ARPANET is justified on this ground.) Other descriptions may be found in [ABRA 73b] and [DAVI 73]. Second, we wish to abstract a mathematical model for the performance of the communication subnetwork and then to analyze and heuristically optimize (i.e., design) these networks. We also analyze and optimize some advanced forms of packet-switching systems in a radio environment. In the next chapter we return to our first goal and compare this chapter's analytic predictions with actual measurement, describe the ARPANET flow control procedures, identify trouble spots in these network protocols, and present measurements of the network overhead and throughput.

It is important to emphasize the veritable explosion of activity in this field. This exponential growth manifests itself in the many symposia and conferences devoted to computer communications, which in turn have produced hundreds (perhaps thousands) of papers in the published literature on computer networks. The design of network structures, the measurement of network phenomena, the analysis of sophisticated system models, and the construction of new networks, all add to the growing body of knowledge in a field that still requires considerably more study before we claim to understand its general principles. Consequently, in the face of considerable ignorance compounded with rapid growth and change, the task of writing chapters such as these next two is in some ways impossible and in most ways frustrating. Each new week brings new results to light which modify our thinking in important ways; we are constantly besieged with the obsolescence of system descriptions and network designs. With this apology, we offer the following material. Hopefully, the few general principles we state will be long-lived; perhaps one of the most robust of these is that of *resource sharing*, which we consider in the following section.

5.1. RESOURCE SHARING

Many of the issues involved in the consideration of computer-communication networks deal with the allocation of resources among competitive demands for such resources. These demands may arise, for example, from the huge number of terminals mentioned in the introduction above. These terminals are accessing *resources* of various kinds in the form of processing capacity, storage capacity, and communications capacity. A means for allocating these resources in order to resolve the conflicting demands is one of the most important aspects of today's system design and operation. In fact, resource allocation is at the root of most of the technical (and nontechnical) problems we face today in and beyond the information processing industry. These problems occur in any *multiaccess* system in which the arrival of demands is unpredictable as well as is the size of the demands made upon the resources (as in Chapter 4). It is therefore appropriate that we begin our network discussion with the general concept of resource sharing.

Consider a collection of resources each of which can perform work at some finite rate (i.e., finite-capacity resources). Assume further that there is a collection of users who demand work from these resources in some random fashion; that is, the time when these demands arise and the amount of work required (the size of the demand) are both unpredictable quantities and will vary from user to user. Immediately we are faced with the task of allocating these resources to the users' demands in some effective and efficient fashion, since the quality of service to the users is important to us while at the same time each resource represents an expenditure of capital. We can think of many ways to accomplish this resource allocation, some of which are absolutely terrible! For example, if we merely placed these resources in a common pool and allowed users to seize them as they desired, then some terribly inequitable resource sharing could ensue and very little "sharing" would take place. We depict this situation in Figure 5.1, part (a) of which shows the collection of resources as rectangular boxes, part (b) of which shows the uncontrolled access to these resources by a community of users, and part (c) of which shows the mayhem resulting from such a procedure.

As opposed to this uncontrolled access scheme one could consider the highly controlled resource allocation shown in Figure 5.2 whereby we have provided a *dedicated* resource to each possible user, that is, we have created a collection of private systems. In this way we eliminate the competition among the users for these resources and guarantee an excellent grade of service for each. Unfortunately, this allocation scheme is extremely expensive for two very important reasons: first, it has allocated a resource to a user even during those intervals when he has no need for it; and second, we have not taken advantage of any possible "quantity discount" available with the purchase of large resources. Let us elaborate on the first of these reasons. Specifically, as shown in the measurements of Fuchs, Jackson, and Stubbs [JACK 69, FUCH 70], the behavior of users accessing computing power is extremely *bursty*; that is, the typical user may be characterized as one who makes demands rarely but when he does he requires a high bandwidth in terms of communications as well as in terms of processing. For example, their measurements

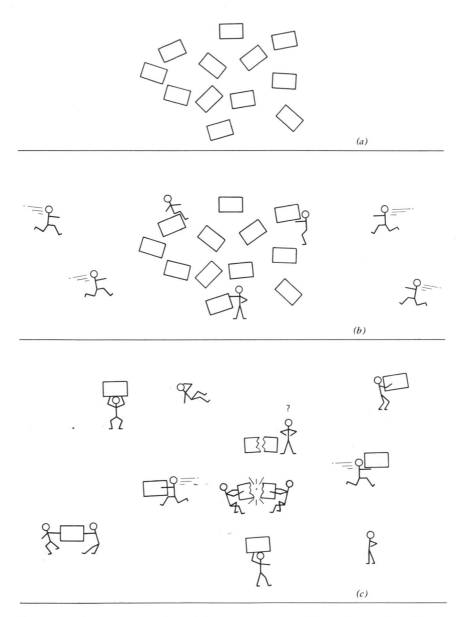

Figure 5.1 Resource sharing? (*a*) The resources; (*b*) the allocation; (*c*) the mayhem.

274

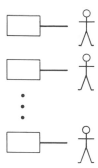

Figure 5.2   Private systems.

show that the communication line from a terminal to the computing system was used, on the average, no more than 5% of the time! Clearly, it is uneconomical to provide the full-time use of a high-capacity resource to such a user; on the other hand, to provide him with a smaller resource would be inadequate for his needs when they do arise. A clear solution to this dilemma is to make available a *pool* of resources which many users share. If done properly this will be an economical means for satisfying the needs of bursty traffic. One of our purposes in this chapter is to espouse that point of view and to point out the gains that are available as the system resources grow in size to satisfy an ever-increasing population of users. That which gives rise to this gain in performance is nothing more sophisticated than the law of large numbers [KLEI 75c, LOEV 63]! As is well known, this law states that the collective demand of a large population of random users is very well approximated by the sum of the *average* demands required by that population. That is, the statistical fluctuations in any individual's demands are smoothed out in the large population case so that the total demand process appears as a more deterministic (i.e., predictable) demand process. Certainly this effect has been used to advantage many times before; perhaps one of the most striking applications has been in the field of life insurance whereby very precise estimates of the mortality rate from a large population can be used to accurately calculate the cost of insuring the lives of individuals in this population; the risk taken by these insurance companies is small indeed!

Let us therefore consider a third allocation scheme in which a pool of resources is made available and each resource is allocated to a user *only when he requires its use*, as shown in Figure 5.3. In this figure, we provide a (fancy) switch that allocates resources upon demand (very much like the switched telephone system). However, we note a remaining difficulty with the shared system of Figure 5.3, namely, when only a few users require access, then the remaining resources lie idle; one would like to put these

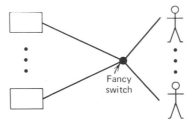

Figure 5.3   Shared system.

idle resources to work. This leads us to the single "large" shared system (for data traffic) of Figure 5.4 in which the *entire* resource is made available to that set of users demanding resources; for example, if only one user needs access then the full capacity is made available to him whereas if more than one need access then the total capacity is allocated among them in some equitable fashion (an example is the set of scheduling algorithms studied in Chapter 4). Not only does this approach eliminate the waste caused by idle resources when work needs to be done, but it also has the potential for taking advantage of any quantity discount available when one purchases large systems (the notorious "economy of scale" possibility).

In all of these systems, of course, we insist that the rate at which the users demand work from the resources is on the average less than the total system capacity for performing work. These large shared systems come in many forms and are known in the computer-communication field by the following names: line-switching, multiplexing, message-switching, store-and-forward systems, packet-switching, and so on. It is such systems that we wish to consider in this chapter.

Let us now consider the quantitative measures of the gains which are available from large resources shared by large populations. The gains we discuss are beyond any effects of "economy or loss due to scale" in the management, maintenance, operation or quantity discounts associated

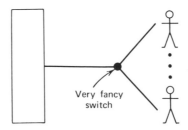

Figure 5.4   Large shared system.

with large systems, but rather arise due to the statistical nature of the demand. We intentionally omit any serious consideration of those other important issues since we now wish to isolate the impact of *scale* on system performance.

The following results have clear applications to computer operating systems construction, terminal network design, and computer-communication networks. In the operating systems environment, one is concerned with providing high performance to a population of users who attempt to share the CPU, main memory, secondary memory devices, printers, plotters, readers, punches, terminals, and other system devices as in Chapter 4. In the terminal network environment, one is anxious to share the data communications capacity required to provide access to and from the terminals and the main processing facility. In our present study of computer-communication networks, we are interested in sharing both the processing facilities that are geographically distributed as well as the data communications required to connect these systems among themselves and among the terminals.

The basic performance parameters of any resource sharing system include the following:

1. the system response time or delay,
2. the throughput,
3. the resource capacity,
4. the resource utilization.

In what follows we take the point of view that there is a stream of job requests accessing the system resource, each of which requires some number of operations from that resource. We let C denote the capacity of the resource in operations per second. We consider an orderly situation in which waiting jobs form a queue. Furthermore, we let $1/\mu$ represent the average number of operations required by a job. Thus we see that the average number of seconds a job requires from the resource is simply* $1/\mu C$. We let λ denote the average number of jobs per second accessing the resource. As usual, the response time of the system is simply the time from when the job arrives until that time when its complete request has been satisfied and is denoted by T. Similarly, the average waiting time for a job (response time minus processing time) is denoted by W; as usual, $T = W + 1/\mu C$. We assume a first-come–first-serve queueing discipline. For a resource of capacity C, under demand by jobs whose input rate is λ per second each of which requires $1/\mu$ operations on the average, then we know that the utilization of such a resource is given simply by $\rho = \lambda/\mu C$.

* In this chapter, we use a different definition for $1/\mu$ than we had used earlier. We do this specifically to introduce the processor (or channel) capacity C.

We are interested in the trading relations among the response time T, the throughput λ, the resource utilization ρ, and the system capacity C. The system structure will affect the relationship among these performance variables in a significant way, and it is our purpose to demonstrate that the simplest structure of all is often superior to some others. We shall also show that large systems give significantly improved performance as compared to smaller ones. Suppose we have a system in operation and find that its average response time T is larger than we desire. We have some options for changing the system parameters in order to reduce T. For example, if we reduce the utilization ρ by either increasing the system capacity C or reducing the system throughput λ, then indeed we will reduce T. This approach is less than satisfactory since we are paying the price of increased system cost [i.e., more capacity (C)] or of reduced throughput (λ) to reduce T; by reducing ρ we are reducing the system efficiency. On the other hand, it is not generally known that this reduction in T can be obtained at *constant* ρ if we merely scale up both the throughput λ and the capacity C. A related tradeoff is that of attempting to increase the throughput of the system. If we simply increase λ then indeed T will degrade (increase). However, we can maintain a constant T as both λ and ρ increase if we permit C to grow less quickly than λ. These effects and the obvious and important tradeoffs among them are investigated below.

Let us now examine some alternative structures for resource allocation and sharing. These structures differ from those in Figures 5.1–5.4 in that we explicitly permit the formation of queues; in our earlier structures we were vague regarding that issue. We begin by considering a collection of m resources each of which has capacity C/m and each of which is individually accessed by a job stream at rate λ/m; this structure is depicted in part (a) of Figure 5.5 and is simply a collection of m G/G/1 queueing systems whose total capacity is C. Our intuition suggests that this system is inefficient since there may be jobs queued up in front of one of the resources when another one is idle; therefore let us consider part (b) of the figure in which we have a single queue accessing the collection of m resources at a total rate equal to λ. This is a G/G/m system. Here we expect an improvement since no job will wait if any resource is idle. Note that both configurations have the same utilization $\rho = \lambda/\mu C$; in part (b) the appropriate interpretation is that ρ is merely the expected fraction of busy resources [see Eq. (1.27)]. This merged queue structure also guarantees that customers will enter service in a first-come–first-serve fashion whereas that is not necessarily the case in part (a); many banks have recently adopted this "imaginative" procedure to everyone's delight. Whereas the merged queue system provides an improvement

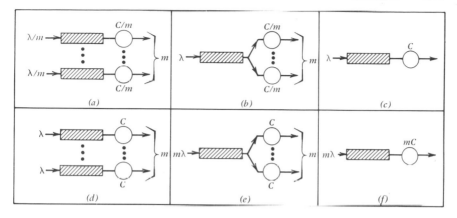

Figure 5.5   Evolution of queueing structures.

over the m separate facilities, there still remains an inefficiency when
there is no queue but less than all the resources are busy as in Figure 5.3;
in this case, some resources remain idle at a time when their services
could be used in speeding the work of the other busy resources. To
overcome this inefficiency we consider part (c) in which we have now
merged the resources as well as the job streams to produce a single G/G/1
queue whose input rate is λ and whose capacity is C much as in Figure
5.4. In part (d) we show a collection of m such single resource systems
similar to that in part (a) but now each has m times the input rate and m
times the resource capacity, and therefore the system is capable of
handling more jobs per second. We have come full circle and are back to
the inefficient system of part (a), which therefore suggests we consider the
system of part (e) in which we have a merged job stream at a rate $m\lambda$
which itself can be improved to the merged-queue merged-resource
system of part (f). That which distinguishes the two single resource
facilities in parts (c) and (f) is that in the latter we have scaled up the
input and the capacity by a factor of m while maintaining a constant load
$\rho = \lambda/\mu C$. Of the six systems shown, all of which have the same value for
ρ, the last is often superior (in the sense of having a smaller response
time T). In fact, if we were to further scale the input and capacity of the
system, we would see yet further improvements. Let us now calculate these
improvements. We begin by discussing the queueing system M/M/m and
following that we consider the system G/G/m.

For the M/M/m system shown in Figure 5.5(b), we know from Section
1.5 with $\rho = \lambda/\mu C$ that p_k, the probability that the system contains k jobs
(counting those in queue as well as those being processed), is simply given

by

$$p_k = \begin{cases} \dfrac{p_0(m\rho)^k}{k!} & k \leq m \\[2ex] \dfrac{p_0\rho^k m^m}{m!} & k \geq m \end{cases}$$

where

$$p_0 = \left[\frac{(m\rho)^m}{(1-\rho)m!} + \sum_{k=0}^{m-1} \frac{(m\rho)^k}{k!} \right]^{-1}$$

From these basic equilibrium probabilities one may easily calculate the average response time T as

$$T = \frac{m}{\mu C} + \frac{P_m}{\mu C(1-\rho)} \qquad\qquad ▬ \quad (5.1)$$

where P_m is simply the probability that the system contains m or more jobs and this probability is given by

$$P_m = \frac{p_0(m\rho)^m}{(1-\rho)m!} \qquad\qquad ▬$$

From these equations we may quantify the relationship among T, λ, C, ρ, and m. In particular it was shown by the author [KLEI 64, Theorem 4.2] that the value of m which minimizes T at constant ρ is $m = 1$ (see Exercise 5.9). In addition, it was demonstrated that T could be reduced at constant ρ by increasing both λ and C. (Results similar to these were discussed by Morse [MORS 58] and Feller [FELL 50]). There are numerous ways to display this tradeoff, some of which we now present.

Perhaps the most striking display of the effect of large systems may be seen in Figure 5.6, where we plot the normalized average response time

$$\frac{\mu C T}{m} = 1 + \frac{P_m}{m(1-\rho)}$$

The normalization is simply the average service time for a job in one of the m servers, and this normalization successfully removes all parameters from the expression leaving only m and ρ ($= \lambda/\mu C$). In this figure, we see that all the curves begin at unity for the value $\rho = 0$ since at this point $P_m = 0$ ($m = 1, 2, \ldots$). As m increases for a given value of ρ, we see that the normalized delay decreases in a dramatic fashion, and as $m \to \infty$ we see that the behavior approaches that of the pure deterministic system (D/D/1) in which no queues form until we exceed the value $\rho = 1$. This figure, however, does not permit one to compare the various structures from Figure 5.5 equitably since we cannot investigate

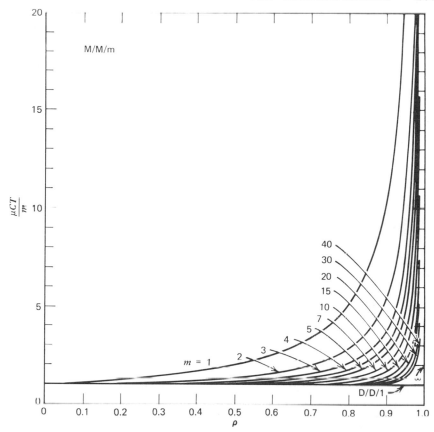

Figure 5.6   Normalized average response time.

the behavior as λ and C vary in some observable way; the difficulty, of course, is that we have normalized the response time and have therefore lost an essential parameter in our performance evaluation. If we return to Eq. (5.1), we see the manner in which the average service time $m/\mu C$ affects the response. In Figure 5.7 we plot Eq. (5.1) under the constant conditions $\mu = \mu_0 = 1$ and $C = C_0 = 1$; thus we assume that the total capacity of our M/M/m system is held constant at $C_0 = 1$ and this is shared equally among the m resources as in Figure 5.5b. In Figure 5.7 we see quite the opposite behavior from that in Figure 5.6, namely that the response time degrades as m increases at constant load ρ. This is simply a demonstration that the system of Figure 5.5c is superior to that of Figure 5.5b. Since μC was held constant in Figure 5.7, we obviously were varying ρ by changing λ. We need not have held μC constant but

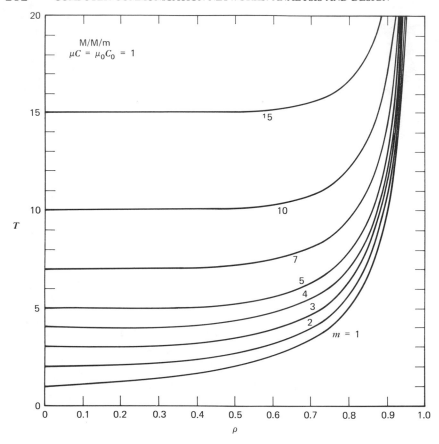

Figure 5.7   Average response time with fixed capacity.

rather could have held λ constant in which case we choose to rewrite Eq. (5.1) as

$$T = \frac{m\rho}{\lambda} + \frac{\rho P_m}{\lambda(1-\rho)} \tag{5.2}$$

In Figure 5.8 we select $\lambda = \lambda_0 = 0.8$ as we permit ρ to change through a variation of μC. Clearly, the response function will now branch out from the origin and again we see the degradation in response time as m increases at a constant ρ; this too demonstrates the superiority of a single resource as opposed to multiple resources at constant total resource capacity.

Let us quantify these graphical observations analytically. We find it convenient to expose the parameters of the response time and so we temporarily write $T = T(m, \lambda, C)$ where once again C denotes the total capacity of the m resource system (throughout we assume that μ is constant). We begin by quoting Theorem 4.2 of [KLEI 64] again, which gives us the important result

$$T(1, \lambda, C) \le T(m, \lambda, C) \qquad m = 1, 2, 3, \ldots \qquad \blacksquare \quad (5.3)$$

demonstrating the improvement due to single resource systems. Furthermore, since P_m and p_0 depend only upon m and ρ (and not upon λ and C separately), then from Eqs. (5.1) and (5.2) we see that scaling λ and C together must give an improvement in T proportional to the scaling factor, namely, for $a > 0$

$$T(m, a\lambda, aC) - \frac{1}{a} T(m, \lambda, C) \qquad \blacksquare \quad (5.4)$$

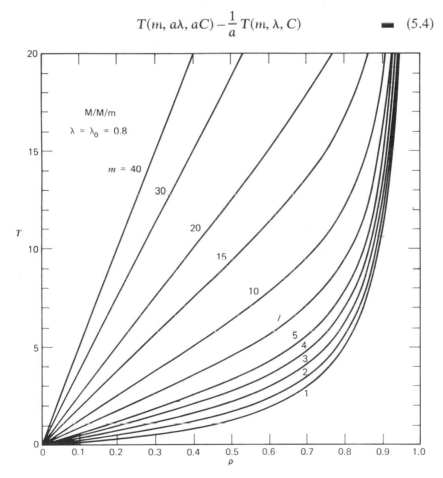

Figure 5.8   Average response time with fixed arrival rate.

Of course, the same is true for the average waiting time in the queue $W = W(m, \lambda, C)$, that is

$$W(m, a\lambda, aC) = \frac{1}{a} W(m, \lambda, C) \qquad \blacksquare \quad (5.5)$$

Thus again we see the significant gains due to scaling up the system parameters. These observations were made by the author [KLEI 64, Section 4.3] where he also generalized the single server case to yield the result

$$T(1, a\lambda, bC) = \frac{(1-\rho)}{b[1-\rho(a/b)]} T(1, \lambda, C)$$

$$W(1, a\lambda, bC) = \frac{a(1-\rho)}{b^2[1-\rho(a/b)]} W(1, \lambda, C)$$

where $\rho = \lambda/\mu C$ and $a \leq \mu Cb/\lambda$. We note that when $a = b$, then these last equations reduce to our earlier results for $m = 1$. Whereas T and W vary as we scale λ and C as just described, we find by using Little's result in Eq. (5.2) that

$$\lambda T(m, \lambda, C) = \bar{N}(m, \lambda, C) = \bar{N}(m, a\lambda, aC) \qquad \blacksquare \quad (5.6)$$

and

$$\lambda W(m, \lambda, C) = \bar{N}_q(m, \lambda, C) = \bar{N}_q(m, a\lambda, aC) \qquad \blacksquare \quad (5.7)$$

where $\bar{N}$ and $\bar{N}_q$ are the average number of jobs in the system and in the queue, respectively. What Eqs. (5.6) and (5.7) tell us is that the average number of jobs in the system or in the queue does not depend upon the scaling parameters, but remains constant for a given value of ρ.

Equations (5.5) and (5.7) give the corresponding results for W as Eqs. (5.4) and (5.6) give for T. It remains for us to find the relationship for W which corresponds to the relationship in Eq. (5.3) and this we do as follows. First we observe from Eq. (5.1) that

$$W(m, \lambda, C) = \frac{P_m}{\mu C(1-\rho)}$$

Our concern is with the behavior of W as m varies. This dependence is contained in the expression P_m, which represents the probability that an M/M/m system contains m jobs or more; that is,

$$P_m = \sum_{k=m}^{\infty} p_k$$

Now we know that ρ is simply the average fraction of busy resources, and this may be written as

$$\rho = \sum_{k=0}^{m-1} \frac{k}{m} p_k + \sum_{k=m}^{\infty} p_k$$

Since the first of these two sums must be non-negative we have immediately that

$$\rho \geq P_m$$

Further we know for the single resource system M/M/1 that

$$W(1, \lambda, C) = \frac{\rho}{\mu C(1-\rho)}$$

From these last two equations and the expression for $W(m, \lambda, C)$ we have the relation we were seeking, namely

$$W(1, \lambda, C) \geq W(m, \lambda, C) \qquad m = 1, 2, 3, \dots \qquad \blacksquare \quad (5.8)$$

This tells us that the average waiting time in an M/M/m system decreases with m, whereas the average response time increases with m. This effect may be seen in Figure 5.9 where we plot T and W at constant values of ρ

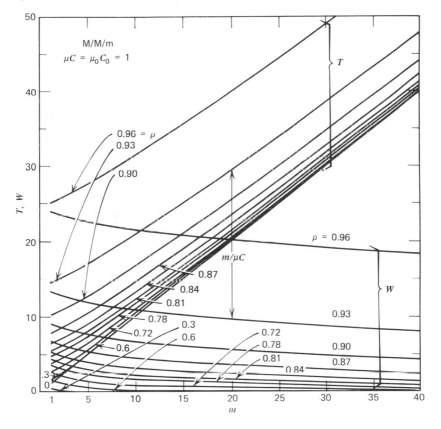

Figure 5.9   Average response time and average wait at constant loads.

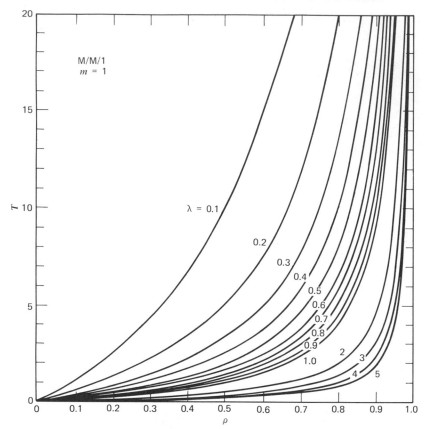

Figure 5.10   Average response time for various input rates for $m = 1$.

as a function of m with $\mu C = 1$. Note that $T(m, \lambda, C) - W(m, \lambda, C) = m/\mu C$ as shown in that figure. Thus, if average queueing time is the performance measure rather than average response time, then partitioned systems are superior; in the teleprocessing systems under study, we take the point of view that the total response time is the appropriate performance measure and so the single resource systems are preferred.

We may display these improvements in yet another way if we focus on the single resource system M/M/1. For example, in Figure 5.10 we show the effect of increasing λ on the average response time for $m = 1$; the appropriate expression for T is

$$T = \frac{\rho/\lambda}{1 - \rho}$$

At constant ρ, we observe the reduction in delay as we increase λ (and therefore also C). In fact, we observe that the average response time improves by a factor of 50 as we pass from $\lambda = 0.1$ to $\lambda = 5$ as anticipated from Eq. (5.4). Similarly, in Figure 5.11 we show the improvement in efficiency (i.e., resource utilization) at constant average response time as the system is scaled up. The function plotted there is simply the solution of this last equation for ρ, namely

$$\rho = \frac{\lambda T}{1 + \lambda T}$$

Thus for the system M/M/m we see that large systems (scaling up the input rate and the system capacity) yield improvements in average response times that are proportional to the scaling factor. For a given scale factor, the single resource system is superior to the multiple resource system.

Computer systems seldom display the simple statistical behavior that we have assumed for the system M/M/m above. Let us therefore consider the more general system G/G/m. We seek to obtain tradeoff relationships similar to those we obtained above. The best we can do here is to work with our known bounds on the system performance and we will show that these bounds predict behavior not unlike that which we were able to obtain above.

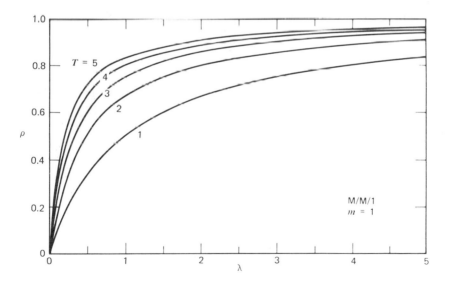

Figure 5.11   Efficiency versus input for $m = 1$ and various response times.

We begin with the single-server system G/G/1. Once again, we assume that the average arrival rate of jobs is given by λ and that the variance of these job's interarrival times is given by σ_a^2. Also the average number of operations per job will again be denoted by $1/\mu$ and with a variance given by σ_p^2. The single resource is assumed to have a processing capacity of C operations per second. Thus the average processing time for a job is again given by $\bar{x} = 1/\mu C$ with a variance σ_b^2. We see that $\sigma_b^2 = \sigma_p^2/C^2$ and also that $\overline{x^2} = \sigma_b^2 + 1/\mu^2 C^2$. For the interarrival time, we have the squared coefficient of variation $C_a^2 = \lambda^2 \sigma_a^2$. For the processing time we have the corresponding quantity

$$C_b^2 = \mu^2 C^2 \sigma_b^2$$
$$= \mu^2 \sigma_p^2$$
$$= C_p^2$$

where C_p^2 is the squared coefficient of variation for the number of operations required by a job. From Eq. (2.22) we have the following upper bound

$$T \le \frac{1}{\mu C} + \frac{\lambda(\sigma_a^2 + \sigma_b^2)}{2(1 - \rho)}$$

We are interested in observing the behavior of this expression as we scale up both λ and C as earlier. In so doing, of course, we will change the interarrival time parameters but will maintain constant all coefficients of variation. In order to see the effects of this scaling, we rewrite the last equation in terms of these coefficients, namely

$$T \le \frac{\rho}{\lambda} + \frac{C_a^2 + \rho^2 C_p^2}{2\lambda(1 - \rho)} \triangleq T_U = T_U(1, \lambda, C)$$

where once again $\rho = \lambda/\mu C$. This last expression is very similar to that given in Eq. (5.2) for $m = 1$; of course, we could just as well have written it in terms of μC. Now we clearly see the effect of scaling λ and C simultaneously, namely, the (bound on the) average response time drops in inverse proportion to this scaling factor. Using our earlier notation and applying it to T_U we have

$$T_U(1, a\lambda, aC) = \frac{1}{a} T_U(1, \lambda, C) \qquad \blacksquare \quad (5.9)$$

For the behavior of the multiple resource system, we now turn our attention to G/G/m. We resort to the known bound derived by Kingman and Brumelle in Eq. (2.66), namely

$$T \le \frac{m}{\mu C} + \frac{\lambda[\sigma_a^2 + (\sigma_b^2/m) + (m - 1)/\mu^2 C^2]}{2(1 - \rho)}$$

If we rewrite this equation in terms of the squared coefficients of variation we then obtain

$$T \le \frac{m\rho}{\lambda} + \frac{C_a^2 + (\rho^2 C_p^2 m) + (m-1)\rho^2}{2\lambda(1-\rho)} = T_U(m, \lambda, C) \qquad (5.10)$$

Again we see the improvement due to scaling at a constant ρ, that is

$$T_U(m, a\lambda, aC) = \frac{1}{a} T_U(m, \lambda, C) \qquad \blacksquare \qquad (5.11)$$

Unfortunately, the bound for G/G/m in Eq. (5.10) is not especially tight and so it is difficult to use it in showing that $m = 1$ is the optimum system as we were able to do for M/M/m. However, it has been shown by Brumelle [BRUM 71] that in the case when $C_p^2 \le 1$ then $m = 1$ is optimum, that is,

$$T(1, \lambda, C) \le T(m, \lambda, C) \qquad \blacksquare$$

(This result extended the results of Stidham [STID 70] for G/M/m, G/D/m, and G/E$_k$/m.) Thus we see that the single large shared resource is superior in terms of response time for a large class of G/G/m systems. However, Brumelle did give an example that showed that G/G/2 can be superior to G/G/1 when $C_p^2 > 1$.

We recall that an improvement for the bound in Eq. (5.10) for G/G/m was conjectured by Kingman (see Eq. 2.64) and takes the form of the following approximation:

$$T \cong \frac{m}{\mu C} + \frac{\lambda [\sigma_a^2 + \sigma_b^2/m^2]}{2(1-\rho)}$$

In fact we recall from Eq. (2.73) that this approximation forms an upper bound for the system G/M/m and further that this Kingman–Köllerström approximation is good as $\rho \to 1$. If once again we express this new upper bound (or approximation) in terms of the coefficients of variation we find

$$T \cong \frac{m\rho}{\lambda} + \frac{C_a^2 + \rho^2 C_p^2}{2\lambda(1-\rho)} \triangleq T_U'$$

or, in terms of μC we have

$$T \cong \frac{m}{\mu C} + \frac{(C_a^2/\rho) + \rho C_p^2}{2\mu C(1-\rho)} = T_U'$$

Once again we have a relationship of the type given in Eq. (5.11) regarding the scaling effect. We note further that the portion of T_U' which corresponds to the waiting time in queue (the second term) is independent of m. T_U' is far superior to the bound given in Eq. (5.10) for G/G/m.

We make the further observation that the response time degrades with increasing m at constant ρ.

Thus in this section we have shown that the concept of large shared single resources is the direction in which we find improvements in the mean response time of the system. Again we comment that these improvements come about basically due to the law of large numbers in that we are smoothing out the relative statistical fluctuations which place demands upon our resource.

5.2. SOME CONTRASTS AND TRADEOFFS

The convenience of network access to information and information processing is possible because of the wedding of two huge but dissimilar industries: the communication and computer industries. (It is perhaps more accurate to describe their union as a "shotgun" wedding with at least one of the partners acting under coercion.) The communication industry may best be characterized as a rather conservative industry that has been around since the beginning of the century. It is quite large (20 billion dollars per year, growing at a rate of approximately 12% per year), is highly regulated, involves a huge physical plant, and contains well-defined problems that are treated by highly trained experts (and is based on a well-understood theory). On the other hand, the computer industry is rapidly changing, is very new, is also large (12 billion dollars a year, growing at an annual rate of 20%), is unregulated, is plagued by an extremely rapid obsolescence of equipment, is rather poorly understood in terms of its basic problems, is not yet a science, has poorly defined goals and problems, and is serviced by perhaps the worst kind of workers (poorly trained, highly paid software "experts"). However, that which drives this conglomerate is the urgent need for information processing. When we try to "marry" these two systems we create some enormous challenges. These systems are large, are expensive, involve exclusive users, are governed by performance measures that are poorly understood, and have significant impact on social, political, and economic facets of our society. Such is the nature of the problem we face.

When we examine the two ends of a remote data processing system, namely, the remotely located terminals and the main computer complex at which processing will take place, we identify a second "incompatibility." Remote terminals come in a variety of types ranging from many low-speed, inexpensive teletypewriters to a few higher-speed, intelligent computer display terminals. Typically, these terminals are located over some distributed geographical region, perhaps in clusters. They tend to operate asynchronously in the sense that the characters they generate are spaced nonuniformly in time. They tend to have a very low duty-cycle and also tend to generate data in bursts (a nasty combination!). Among this multitude of terminal types there is also a multitude of incompatible coded versions of the common alphanumeric symbols. Most of the terminals are relatively cheap and unsophisticated. They operate relatively independently of each other and are unaware of other terminal behavior. On the other hand, a main computer complex typically consists of one, or at most a few, large central processors. These machines are large-scale, high-speed computers that operate synchronously with a high duty-cycle. Usually, a single standard representation for alphanumeric symbols is used. These machines are very expensive and highly sophisticated and, if there is more than one at a given complex, considerable energy may be expended to see that they cooperate in an efficient, symbiotic fashion.

Another difficulty concerns the telephone network, which was originally designed to carry voice traffic, an analog signal whose highly redundant nature combats the various effects of noise present in the telephone network. In order to transmit data traffic over this network, the digital signals generated by computers and terminals must be converted to analog signals which the telephone network* carries on its voice-grade lines; the devices that carry out this conversion are referred to as *modems* (which is a contraction of the word modulator-demodulator) or *data sets*. Since this network is designed to handle voice signals, then one anticipates that adapting such a system for data transmission (whose statistics vary considerably from those of voice signals) is a formidable task. The difference between data and voice traffic lies at the source of many of the difficulties in the design of data communication networks; the principal characteristic of data is that it generates unpredictable demands at unpredictable times, leading to bursty low duty-cycle traffic. This requires buffering, smoothing, multiplexing, concentrating, etc.

Thus we have a complex interaction of partially incompatible systems, and our task is to provide a "message service" that is "invisible" in the sense that the remote terminals may pretend that they are connected directly to the central processor. A high throughput must be maintained in an efficient fashion, and message delivery must be rapid in order to provide acceptable response times to the demands of the terminals. This message service must not place undue burdens on either the terminals or the processors (this, too, is the sense in which it should be invisible). An important consideration is that it should be expandable with ease as the set of terminals shifts or grows and as new computer complexes spring up or enlarge and also as needs change. In the face of all these wonderful

* Digital data networks are currently emerging [DATA 73, GAIN 73].

properties it should also be reliable and relatively economical. This, then, is the environment and the task. Now let us take a more careful look at the network components and structure.

5.3. NETWORK STRUCTURES AND PACKET SWITCHING

A computer-communication network is a collection of nodes at which reside the computing resources [which themselves are connected into the network through nodal switching computers, i.e., the (very) "fancy" switches] which communicate with each other via a set of links (the data communication channels) [DAVI 73, KLEI 64]. Messages in the form of commands, inquiries, file transmissions, and the like, travel through this network over the data transmission lines. At the nodal switching computers, the communications-oriented tasks of relaying messages (with appropriate routing, acknowledging, error and flow controlling, queueing, and so on) and of inserting and removing messages that originate and terminate at the terminals and main processors at that node must be carried out; these tasks are separated from the main computing functions required of the node and are relegated to switching computer which is dedicated to these tasks (e.g., the IMPs in the ARPANET—see below).

Computer-communication networks may therefore be conveniently partitioned into two separate subnetworks: the communication subnetwork

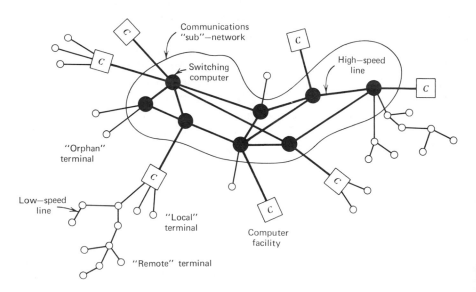

Figure 5.12 The structure of a computer-communication network.

providing the message service, and the collection of computer and terminal resources that forms the "user-resource" subnetwork; see Figure 5.12. In this figure, we show a fairly general structural model of a computer-communication network. First, note the computer facilities denoted as square boxes that carry out the useful processing and storage tasks (so far as the user is concerned). These are connected together by means of the communication subnetwork (which consists of the switching computers and the high-speed data communication channels). As far as the communication subnetwork is concerned, all entry and exit for the network passes through the switching computers. More than one computer facility may be connected through a given switching computer. The terminals may either be local to a computer facility (in which case they may access the network through this facility), or they may be remote (in which case some "remote terminal network" must be provided to connect them to the facility, and then into the high-speed net), or they may not be associated with any computer facility at all, in which case these "orphan" terminals (or even networks of orphan terminals) may gain access to the switching computer directly (e.g., the TIPs in the ARPANET). Most of our attention in this chapter will be directed to the communication subnetwork forming the message service. The function of this message service is to accept messages from any message source (such as a computer or a data terminal), route these messages through the communication network, and then deliver them to their destination in a rapid and reliable fashion.

Communication networks may conveniently be divided into three types: circuit (or line) switching; message switching;* and packet switching.* A *circuit-switching* network provides service by setting up a total path of connected lines from the origin to the destination of the "call" or demand; this complete circuit is set up by a special signaling message that threads its way through the network, seizing channels in the path as it proceeds. After the path is established, a return signal informs the source that data transmission may proceed, and all channels in the path are then used simultaneously. The entire path remains allocated to the transmission (whether or not it is in use), and only when the source releases the circuit, will all these channels be returned to the available pool for use in other paths. Circuit switching is the common method for telephone systems [SYSK 60]. In *message switching*, only one channel is used at a time for a given transmission. The message first travels from its source node to the next node in its path, and when the entire message is received at this node,

---

* Both message and packet switching use a technique known as store-and-forward transmission.

then the next step in its journey is selected; if this selected channel is busy, the message waits in a queue, and finally, when the channel becomes free, transmission begins. Thus the message "hops" from node to node through the network using only one channel at a time, possibly queueing at busy channels, as it is successively stored and forwarded through the network [KLEI 64]. *Packet switching* is basically the same as message switching except that the messages are decomposed into smaller pieces called packets, each of which has a maximum length. These packets are numbered and addressed (as with message switching) and make their way through the net in a packet-switched (store-and-forward) fashion. Thus many packets of the same message may be in transmission simultaneously, thereby giving one of the main advantages of packet switching, namely, the "pipelining" effect; the transmission delay may be considerably reduced (over message switching) as a result (the reduction may be as large as a *factor* proportional to the number of packets into which the message is broken) [BARA 64, DAVI 68, ROBE 67]. In Figure 5.13, we illustrate these three switching modes.* In part (*a*) we show a network transmission path involving four nodes and three transmission lines. For simplicity, we assume that no other traffic in the network interferes with the transmission. In part (*b*), we show the sequence of events (idealized) for circuit switching. Here we include the connection delay at each switch (this is the major component of delay), followed by the transmission of the (assumed zero-length) set-up signal which arrives at switch B after a (speed of light) propagation delay. The cycle repeats along the path, finally generating a return signal which then triggers the data transmission. We note that only one data transmission time is required. The set-up path, in effect, identifies the address for the data. In part (*c*), we show the sequence for message switching. First, there is a small switch processing delay (to select routes and so on), and then the message transmission from *A* to *B* proceeds; in addition to the data, this transmission includes a message header (for identification and routing since the path is not set up as in circuit switching). After *B* fully receives the message, the sequence is repeated from *B* to *C*, and so on. For packet switching, part (*d*), the message is shown as being decomposed into eight packets, each of which requires its own header. The sequence of packets is seen to be pipelining down the chain. Because of header overhead, the number of bits transmitted is least for circuit switching, next larger for message switching, and largest of all for packet switching. However, so long as message lengths are not too long, the network delay for these

---

* In such figures as this, the time axis points down the page and the network nodes are spaced horizontally with the source node to the left and the destination node to the right.

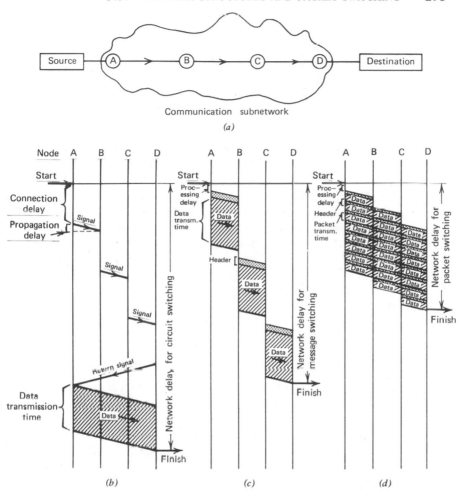

Figure 5.13 Comparison of network delay for circuit, message, and packet switching. (a) The transmission path. (b) Circuit switching. (c) Message switching. (d) Packet switching.

three switching modes favors packet switching as shown. Indeed, the delay for message switching is proportional to the product of the *message* length and the number of hops; for packet switching, the delay is proportional to the product of the *packet* length and the number of hops, plus a term proportional to the message length. Here we have omitted the delay caused by packet-switching control signals, a subject we discuss in detail in Chapter 6. Digital data transmission (especially if it is "bursty") lends itself very nicely to message and packet switching, and we deal only with these modes in this chapter. The decision as to which form of switching to use is a difficult one. Some studies have been conducted in an attempt to compare these switching techniques [PORT 71, CLOS 72, CLOW 73, ITOH 73, MIYA 75, ROSN 75], but to date, no satisfactory comprehensive treatment has been given.* One thing is clear, namely, if there is need for transmitting a long, continuous stream of data, then a leased line (or a circuit-switched connection) makes good sense. On the other hand, if the data flow is bursty (as is typical of computer and terminal data), then some form of resource sharing can be used to great advantage; packet switching is an effective choice here.

There are numerous other properties and issues with regard to packet switching that are important to discuss. For example, since packets are stored as they pass through switching nodes, it is possible to conduct speed, format, and code conversion during the switching process (this is true of any store-and-forward system such as message switching as well); this is not possible with circuit switching, which therefore requires complete end-to-end compatibility in this regard. Further, in a moderately busy network, a set-up signal may find it difficult to locate a complete path of available channels from source to destination, and may return a "busy" signal to the source, i.e., the network is blocked. With packet switching, only the next channel in the path need be available, subject, however, to the ability of the message to initially acquire other kinds of network resources in response to the flow control procedure (again, see Chapter 6). Another key feature of packet switching is its ability to adaptively select good paths for packet journeys as a function of the network congestion (see below and Chapter 6). Besides providing small network delays, packet switching has the desirable feature of rapidly handling small messages in spite of the presence of long messages that may be in transport at the same time; this is because of the decomposition of (long) messages into packets. Another useful property of this decomposition into packets is that the nodal storage requirement is reduced (imagine if we used message switching with message lengths of 10^6 bits!).

In evaluating packet-switching networks, we shall emphasize the following network measures:

- delay
- throughput
- cost
- reliability

* In what follows, we study distributed packet-switched networks. Other packet-switched configurations, known as rings, have been implemented [FARB 72] and proposed [HASS 73, PIER 71].

The first two measures are closely related, but usually are applied as performance criteria to different kinds of traffic. In particular, interactive traffic must be delivered quickly and is usually rather short (in which case, throughput is not a central issue). On the other hand, a long file transfer is not so much concerned with the initial delay in getting the first few bits across the network, but rather, with how many bits per second can be pumped through the network. In Figure 5.14, we show these two cases. We take a very simplistic view here and assume that the network appears to the user as a system that makes him wait (on the average) an amount of time W from when he requests a message delivery until that message *begins* delivery at the other end. The time from when the message begins delivery until delivery is complete will be the "service time," which we denote by x. W is therefore the average time from when the first bit is

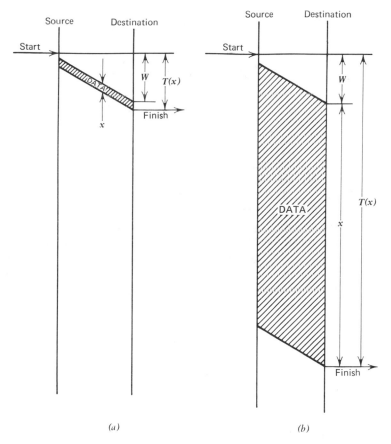

Figure 5.14 Network delay and throughput. (*a*) Short interactive message. (*b*) Long file transfer.

presented to the network until the first bit is delivered; this time is a measure of the (initial) response time of the net. We let $T(x) = W + x$ be the average network delay, which is the principal performance measure. In part (a) of Figure 5.14, we show the case of a short interactive message in which $x \ll W$ and therefore we are mainly concerned with the initial network response time W since it dominates the network delay. In part (b) we show a long file transfer in which $x \gg W$; in this case we are mainly concerned with the network throughput (which we define as γ_{jk} msg/sec between source j and destination k, say) since the network delay behavior is dominated by $x = 1/\gamma_{jk}$. Thus we have

$$T(x) \cong \begin{cases} W & x \ll W \\ x = \dfrac{1}{\gamma_{jk}} & x \gg W \end{cases}$$

A key feature of packet switching is that more than one message (say up to m) is allowed to be in transit through the network at the same time (this is in addition to the packet pipelining for a single message [MAXE 75]). This message multiplexing is possible due both to pipelining along a given path and to alternate routing along many paths. If Z_{jk} is the network delay averaged over all messages passing from j and k (see Section 5.5) then at most we have

$$\gamma_{jk} = \frac{m}{Z_{jk}}$$

If $\bar{b}$ is the average message length (in bits), then the average throughput in bits per second is simply $\bar{b}\gamma_{jk}$ (at most). In addition to handling interactive traffic and file transfers, a network may also be required to handle real-time traffic (such as speech transmission) which demands low delay and high throughput *simultaneously*; such traffic requires different kinds of control procedures for efficient transmission (see Chapter 6).

We have now built up a picture of a packet-switching network as one that pipelines addressed messages along a single path as well as among alternate paths, partitions messages into (pipelined) packets, places headers on packets and messages, and ships them through the network in a store-and-forward fashion. The packets and messages may encounter some unforeseen adventures as they journey through the network, and so for many possible reasons (e.g., errors, blocked storage, time-outs, etc.) the packets and messages may arrive at the destination out of order, or duplicated, or perhaps may even get lost! The network must be prepared to handle these eventualities in an acceptable way (after all, how many users would pay for service on a network that had the habit of accepting a message, losing it, and never informing the user of this loss?). We may summarize the properties of packet-switching networks as follows [CROW 75]:

- random delay
- random throughput
- out-of-order packets and messages
- lost and duplicate packets and messages
- nodal storage
- speed matching between net and attached systems

In order to respond to these properties, the network must provide many of the following functions:

- packetizing
- buffering
- pipelining
- routing procedures
- sequencing and numbering
- error control (noise, duplicate and lost message detection)
- storage (resource) allocation
- flow control

Effective resource sharing in such an environment is critical.

The *analysis* of stochastic flow in store-and-forward networks suffers under many of the combinatorial burdens from network flow theory as well as many of the probabilistic burdens from queueing theory. As a result, the efficient *design* of a computer-communication network is an extremely complex task, and the number of design parameters and operating modes is considerable. In this chapter, we study some of the more important problems which arise in this design process.

In general, a communication (sub)network is made up of (1) the physical network, consisting of the *switching computers* and the *communication channels*; (2) the flow consisting of *messages* (described by their origin, destination, origination time, length, and priority class) that move through the network in a store-and-forward fashion; and (3) the set of *operating rules* for handling the flow of this message traffic.

The synthesis of these networks involves a number of design variables:* the message routing procedure; the flow control procedure; the channel capacity assignment; the priority queueing discipline; and the topological configuration. A *message routing procedure* is a decision rule that determines, according to some algorithm (possibly random), the next node that

---

* For simplicity in the mathematical analysis and design, we omit some of the considerations of buffering strategies, numbering and sequencing, and error control. We return to these important issues in Chapter 6.

a message will visit on its way through the net. The specification of the algorithm gives the routing procedure. The parameters involved in this algorithm may include such things as: origin and destination of the message; priority of the message; availability of certain channels; and congestion (or annihilation) of certain nodes and channels. We define a *fixed routing procedure* as one in which a message's path through the network is uniquely determined from only its origin and destination. When more than one path is allowed, then we refer to this as an *alternate routing procedure*. An alternate routing procedure may choose its alternate paths either deterministically or at random from among the operating links based on the parameter values mentioned above; the former is referred to as *deterministic alternate routing* and the latter as random alternate routing (or more simply as *random routing*) [KLEI 64]. If the routing algorithm bases its decisions on some measure of the observed traffic flow and/or the breakdown of nodes and channels, then we say it is a dynamic or *adaptive routing procedure*.

We see that the routing procedure is obliged to handle all traffic in the network, and in particular, has no direct control over how much traffic is permitted to enter the network. This throttling task is relegated to the *flow control* procedure, which is one of the more critical functions in networks. Basically, the flow control procedure anticipates and prevents congestion by regulating the entry of traffic from the user-resource network into the communication subnetwork. The issues and pitfalls associated with flow control procedures are discussed in Chapter 6.

We have agreed to look only at message- and packet-switching communication networks in which a message (or its packets), upon entering the network, will eventually be given the attention of some communication channel over which it will be transmitted; of course, if the channel is in use when the message requires this service, then the message must join a queue and wait. Note that the service to be offered by this "service facility" is the use of the channel for transmission; the service (transmission) time for a given message is merely its length (say in bits) divided by the capacity of the channel (in bits per second). After a message uses a given channel it then is "received" in the next node (and the channel is released for use by other messages); upon receipt in this node, the routing procedure then assigns it to some outgoing channel from that node along the path to its destination. Again, if that new channel is busy, the message must queue until it is finally given service from the channel, and so on. Eventually the message will be received at its destination; the total time spent in the network is referred to as the *message delay* (or network delay).

The *topological* configuration of the communication net strongly affects

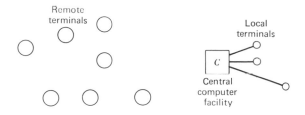

Figure 5.15   The remote terminal connection problem.

its behavior as regards its reliability, message delay, routing, and the like [KLEI 64, FRAN 71a]. The topological design problem is an extremely difficult one and may further be complicated by topological constraints (such as reliability constraints) imposed upon the network. In addition, a *cost function* must be included in the network design, and the form of this cost function strongly affects the structure of the network.

Once a topology is chosen, a *capacity assignment* to each channel must be made. Moreover, the *queueing discipline* that governs the order of service for the various channel queues must be decided upon (we will not discuss this particular issue in the current chapter, since we have already treated it in Chapters 3 and 4).

Before proceeding with the full network problem, let us examine a simpler related one which is important. A common problem in the design of computer communication systems is that of providing access to a *single* central computing facility from a set of remote terminals as depicted in Figure 5.15. In this figure the rectangular box is the central computer facility (which services its own local terminals as well). The resource that one seeks to allocate in this problem is the set of communication lines that will provide access from the remote terminals to the computer. The solution corresponding to the private system of Figure 5.2 is given in Figure 5.16, namely, the star net solution, in which a private line is provided from each remote terminal to the central facility. Such a

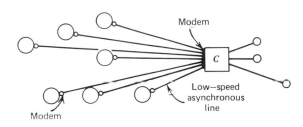

Figure 5.16   The star net solution.

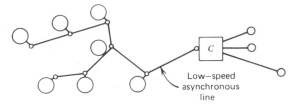

Low—speed
asynchronous
line

Figure 5.17   The minimal spanning tree—the least-cost solution.

solution does indeed avoid conflict for communications capacity, but it is also extremely expensive and, therefore, in an attempt to reduce costs, one is often reduced to low-speed asynchronous communication lines. Moreover, communication gear needed to interface between a digital terminal and an analog communication line (the modem or data set) must be provided at both ends of each line; this is indeed an expensive solution! At the other extreme, we could consider the cheapest possible solution, which consists of connecting these terminals among themselves and the computer using the shortest total line mileage; this is the well-known minimal spanning tree [FRAN 71a] and is easily designed (see Figure 5.17). Unfortunately, one sees that this solution yields, in effect, a single communications facility that must be shared among all terminals, and since it is slow, the interference among terminals is high; we thus have a cheap but slow solution and one that requires a fair amount of control in allocating this resource among the terminals in what is known as a polled or multidrop system [MART 72].

Thus far, we have considered an expensive but private system as well as a least-cost but rather slow solution. Between these two, one could consider the system shown in Figure 5.18, in which we have provided a point of concentration* (known as a multiplexor or concentrator or traffic COP—communications-oriented processor) and which is connected to the central computer facility by a medium- (or even high-) speed synchronous line. It is this line that is expensive, but it may be shared among all the terminals by merging their data streams; this corresponds to the large shared system of Figure 5.4 and if properly designed will enjoy all of its advantages. The multiplexing may be done in any one of a number of ways, but if it is done in a way that assigns a portion of the resource (the medium-speed communication line) to each of the remote terminals in a *permanent* fashion, then we are not much better off than in the private system, since we have really provided a set of private lines from each remote terminal to the computer (although we have taken advantage of

---

* This approach works well when the terminals are clustered (as in an office building) yet remote from the computer (which may be in another city).

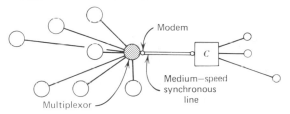

Figure 5.18   The traffic COP solution.

the quantity discount); the real loss is that we have not taken advantage of the averaging effect and this can only be done if the allocation of capacity is performed on a demand basis. One form of demand allocation known as asynchronous time-division multiplexing (ATDM) [CHU 69] was discussed in Chapter 4 (Section 4.14). Indeed, the principle of demand multiplexing of resources is the central issue in the design of efficient packet-switching networks. (The general problem of providing efficient access by many remote terminals to a centralized processing facility is further discussed in [CHOU 73].)

Let us now replace each of the remote terminals previously considered in Figure 5.15 by a computer facility itself, each with its own set of local terminals, and each connected into a sophisticated multiplexor (i.e., a switching computer). We are then considering a collection of computing resources and we are interested now in sharing the capacity of each among all of the terminals in the collection of facilities. In order to accomplish this, we must provide communications (a second type of resource) to permit connections among these computer facilities. Thus we are faced with an allocation of computer *and* communication capacity, an example of which is shown in Figure 5.19 (as well as in Figure 5.12). Here we see a

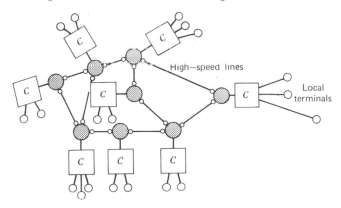

Figure 5.19   A computer-communication network.

computer-communication network in which any local terminal can make use of any computer facility within the network. The object of this chapter is to study such networks and to discuss their behavior. We emphasize, once again, that sophisticated demand multiplexing in the network will provide significant economies of scale.

In the following section we give an operational description of a particular example of a computer-communication network—the ARPANET (later, in Chapter 6, we expand that description considerably and discuss its measured performance). This example will serve as a guide as well as a test bed for many of our results in this and the next chapter. Following that we then identify the class of problems that naturally evolve in network analysis and design. The remainder of the chapter is devoted to the consideration of these various problems and their solutions as well as a discussion of some advanced packet-switching concepts. Throughout we indicate where the basic queueing problems arise, how much progress has so far been made in their solution, and numerous unsolved problems in this field. The fact is that the engineering design of computer-communication networks is at this point fairly well understood and some rather efficient implementations already exist. There are nevertheless some intriguing queueing problems that remain along with numerous topological, optimization and control problems as yet unsolved.

5.4. THE ARPANET—AN OPERATIONAL DESCRIPTION OF AN EXISTING NETWORK

Some years ago Roberts [ROBE 67] proposed an experimental computer network, which was later to become the U.S. Department of Defense Advanced Research Projects Agency (ARPA) Network—the ARPANET. For a number of years before 1967, ARPA had been funding the growth and development of many multiaccess time-shared computer systems at a number of university and industrial research centers across the United States. By 1967, many of these had shown themselves to be valuable computing resources and it was recognized that the Department of Defense as well as the scientific community could benefit greatly if there were to be made available a communication service providing remote access from any terminal to all of these systems. A cost analysis performed at that time indicated that the use of packet switching for the ARPANET would lead to more economical communications and better overall availability and utilization of resources than many other methods. Thus began a serious effort to define the functional details of packet switching. A specification was created for a packet-switched network, a request was sent out for proposal to the technical community, and in early

1969 a contract was awarded for the implementation of the ARPANET to Bolt, Beranek, and Newman (BBN), a Massachusetts-based engineering firm.

In September 1969, the embryonic one-node network (!) came to life when the first packet-switching computer was connected to the Sigma 7 computer at UCLA. Shortly thereafter began the interconnection of many main processors (referred to as HOSTs) at various university, industrial, and government research centers across the United States. In 1970, a series of five papers was presented at the AFIPS Spring Joint Computer Conference in Atlantic City that summarized what we knew about the network at that time [CARR 70, FRAN 70, HEAR 70, KLEI 70, ROBE 70]. The evolution of that network from a small four-node net later in 1969 to a 34-node net in September 1972 is shown in Figure 5.20; Figure 5.21a shows the HOSTs attached to a 39-node network (August 1973) as well as the switching nodes and in Figure 5.21b and c we show the logical and geographical maps, respectively, for a recent 57-node net as of June 1975. Five additional papers summarized our experiences with this network up to 1972 [CROC 72, FRAN 72, ORNS 72, ROBE 72, THOM 72]. Improvements in the operating procedure were reported in [MCQU 72] and [MIMN 73]. A five-year reevaluation was given in [ROBE 73a] and a capsule history was reported in [KARP 73].

The network currently (1975) provides a message service for almost 100 computers geographically distributed across the continental United States and extending by satellite to Hawaii and to a few nodes in Europe. The (HOST) computers are in many ways incompatible with each other, coming from different manufacturers and containing specialized software, data bases, and so on; this in fact presented the challenge of the original network experiment, namely, to provide effective communication among, and utilization of, this collection of incompatible machines. For example: SRI has served as the Network Information Center (NIC); UTAH provides algorithms for the manipulation of figures and for picture processing; the ILLIAC IV at AMES makes available its fantastic parallel processing capability; BBN acts as Network Control Center (NCC); UCLA serves as Network Measurement Center (NMC) and also provides mathematical models and simulation capability for network and time-shared system studies; also at UCLA is the number-crunching power of an IBM 360/91; the MIT Multics time-sharing system is accessible through the net; and many sites provide the services of the PDP-10 TENEX [BOBR 71] time-sharing system.

The topological connection is in the form of a distributed network which provides protection against total line failures by providing at least two

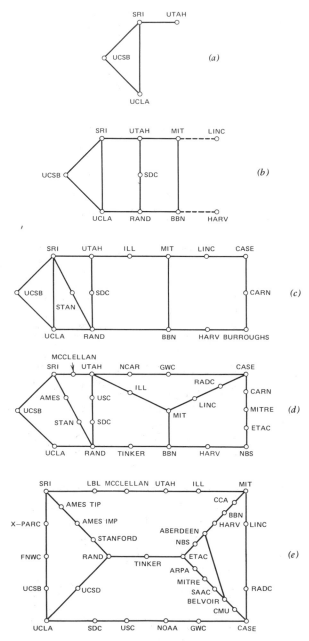

Figure 5.20 Evolution of the ARPANET topology. (*a*) 4-IMP network, 12/1/69. (*b*) 10-IMP network, 7/1/70. (*c*) 15-IMP network, 3/1/71. (*d*) 24-IMP network, 4/1/72. (*e*) 34-IMP network, 9/72.

306

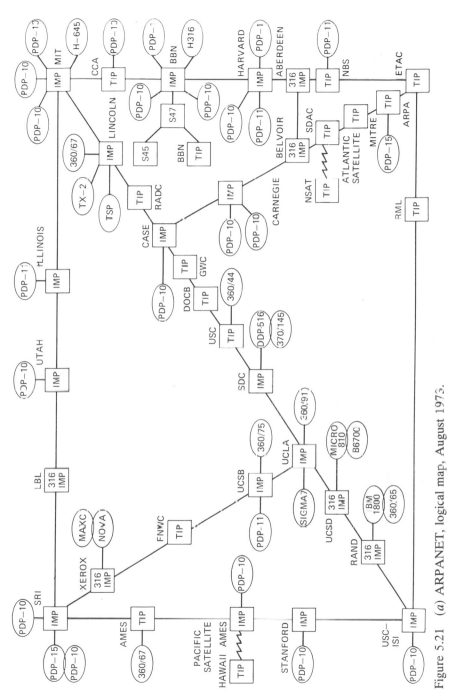

Figure 5.21 (a) ARPANET, logical map, August 1973.

307

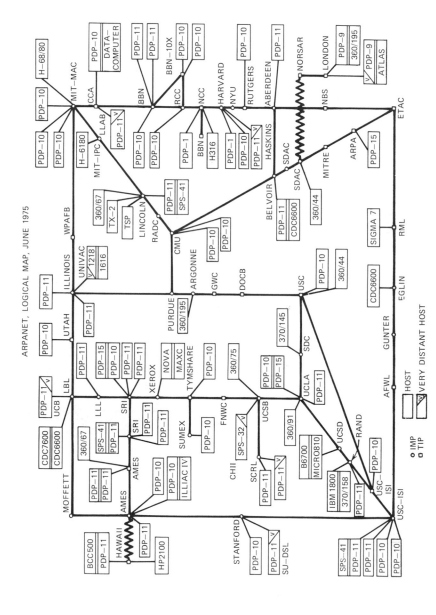

ARPANET, LOGICAL MAP, JUNE 1975

Figure 5.21b ARPANET, logical map, June 1975.

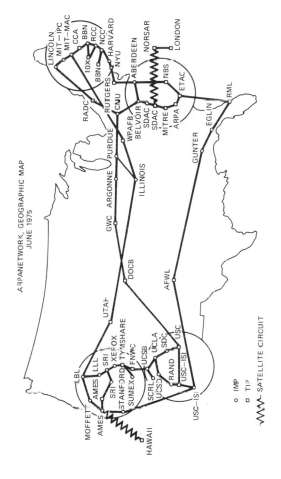

Figure 5.21c ARPANET, geographic map, June 1975.

309

physically separate paths between each pair of nodes. Each HOST is connected [through an asynchronous serial 100-kilobit/sec (KBPS) channel] to the small (local) nodal switching computer called an Interface Message Processor (IMP); the latter are themselves interconnected by leased 50 KBPS full duplex synchronous channels.* The IMPs act as network doorways for the HOST computers. The IMP was introduced to relieve the HOST from many of the message-handling tasks of the communication network. In each HOST, a program referred to as the Network Control Program (NCP) must be implemented and inserted into the operating system (this is major surgery!); the NCP allows HOST computers to communicate with each other according to a HOST–HOST protocol, which is a network-wide standard [CARR 70, MCKE 72a]. In addition, a program known as TELNET acts as a convenient interface between the user and the NCP, allowing him to converse with the network in a more natural way.

In order for a byte stream from (say) a terminal user to be sent to a remote computing system, the user's HOST must package the byte stream into a message stream. This (originating) HOST then delivers each message, including a destination HOST address, to its local IMP. The network IMPs then determine the route, provide error control, handle all message buffering and transmission functions, and finally notify the sender of the eventual receipt of the message at its destination HOST. The collection of HOSTs, IMPs, and channels forms the packet-switched resource-sharing computer-communication network; the IMPs and channels form the message service (i.e., the communication subnet) for the HOST computers. As discussed in the previous section, a dedicated path is not set up between HOST computers that are in communication but rather this communication involves a sequence of message transmissions sharing the communication lines with other messages in transit. The maximum message size is 8063 bits (plus 32 bits used as a HOST–HOST header). Thus a pair of HOSTs will typically communicate over the net via a sequence of transmitted messages. For the reasons discussed in Section 5.3, the IMP program partitions each message into one or more *packets*, each containing at most 1008 bits.† Each packet of a message is transmitted independently to the destination IMP which reassembles the message before shipment to that destination HOST. Following this, an end-to-end

---

* In addition, there currently (June 1975) exist two 230.4 KBPS lines in California, one 50 KBPS satellite channel to Hawaii, one 7.2 KBPS satellite channel to Norway, and one 9.6 KBPS channel from Norway to England; see Figure 5.21b and c.

† The maximum length data message will consist of eight packets of which the first seven will contain 1008 data bits, and the eighth of which will contain 1007 data bits. A message which requires more than one packet is referred to as a multipacket message.

acknowledgement is sent back to the source HOST. In its journey through the net, a packet will "hop" from IMP to neighboring IMP; if the neighbor accepts the packet (i.e., the transmission is found to be valid and there is storage space available), then an IMP-to-IMP packet acknowledgement is returned. If no acknowledgement is received after a time-out period (125 msec), then the packet is retransmitted. This is the ARPANET implementation of a packet-switching network.

The network was designed to achieve both a rapid delivery for the short interactive messages as well as a high throughput for long file transmissions. The first of these goals has been achieved; the response time for short messages varies between 50 and 250 msec even over many hops (excluding satellites, which introduce approximately 0.25 sec just due to propagation delays). As regards the second goal, the network is able to provide a high bandwidth for long messages under light and moderate traffic loads; however, under heavy traffic loads and over many hops the throughput falls somewhat (see Chapter 6). From the beginning, considerable effort has gone into measuring and evaluating the performance of this network. As a result, changes in the IMP program have been made from time to time, with a set of major changes occurring first in the spring of 1972 and later in December, 1974; these changes were introduced, in part, to alleviate some important problems with regard to deadlock and degradation conditions in the network [MCQU 72, OPDE 74] and these are discussed in Chapter 6.

Network routing strategies for this distributed network use a distributed control over the routing decisions that are made in each IMP [FULT 71]. These routing computations are made using information received from neighboring IMPs and local information such as the status of its channels. In practice this approach has worked quite effectively with moderate levels of traffic so far experienced in the network; this has been one of the successes of the ARPANET experiment. The flow control procedure has presented some important challenges that continue to intrigue us. Both routing and flow control are discussed in Chapter 6.

Errors are primarily caused by noise on the communication lines and are handled in the ARPANET by error detection and retransmission between each pair of IMPs along the transmission path; standard cyclic error detection codes have been successfully applied here. Tests have indicated that the raw phone line packet error rates vary from about one in 10^3 to one in 10^5; these (often bursty) errors appear not to affect network performance and, as far as we know, no undetected errors have as yet passed through the network. These failures to detect errors have been designed to occur on the order of years to centuries apart with only a slight additional error control overhead (24 parity bits for error detection per packet in the 50 KBPS lines). The major difficulty with these lines is that occasionally there are extended periods (hours or days) of line outages. The Network Control Center monitors both IMP and line outages continually and their experience [MCKE 72b] is summarized in Chapter 6. It may be concluded that with these error detection precautions, the telephone communication channels do not create a problem in the performance or growth of these networks. When a line or IMP does go down, the network routing procedure automatically adapts to the new condition, thereby preventing congestion.

The original IMP [HEAR 70] was constructed using a Honeywell 516 minicomputer; this is a 16-bit machine with a memory cycle time of 0.96 msec. Configured as an IMP the cost is approximately $100,000. The Honeywell 316 minicomputer (cycle time of 1.6 msec) has also been configured as an IMP costing approximately $50,000; the new IMPs are of this type. An IMP is provided with 16 K of core.* The IMP is responsible for all the processing of packets, which includes decomposition of HOST messages into packets; routing; relaying and receiving store-and-forward packets; acknowledging accepted packets and retransmitting unacknowledged packets; reassembling packets into messages at the destination HOST; generating control messages; and so on. In addition, the IMP program is responsible for gathering statistics, performing on-line testing, and monitoring the line status.

The IMP core storage is divided into 32 segments, each segment consisting of 512 words (16 bits each). Much of this space is devoted to IMP code and tables. The remainder is made available for storing packets in buffers as they pass through the IMP. Currently, a buffer is 74 words (1184 bits) long and is capable of containing one packet (including its overhead); there were approximately 77 buffers available in an IMP towards the end of 1973 and this number dropped to the mid-50 range by the time of the December 1974 changes. These buffers are allocated as follows. Each input line is double buffered, which permits all input traffic to be examined and guarantees that acknowledgements can be processed (which frees buffers within that IMP). The remaining buffers are divided on a limited demand basis between store-and-forward packets and packets being reassembled for delivery to a local HOST. Each output line owns one and is limited to a maximum of eight store-and-forward buffers; originally, a maximum of 20 buffers was made available in a store-and-forward pool from which all lines could request buffers (by mid-1975, this number had been reduced to 9). Toward the end of 1973 there was available a maximum of 66 buffers which could be claimed for reassembly and this number

---

* The original allocation was 12 K.

dropped to 34 by the time of the December 1974 changes (this number is usually chosen as $8k + 2$ where k is some integer; the 8 is for each of k multipacket messages and the $+2$ term is to allow room for two single-packet messages). In terms of performance, the 516 IMP can process approximately 850 KBPS and the 316 IMP can process approximately 700 KBPS (under the most favorable assumption of maximum length messages). The maximum effective capacity of the 50 KBPS line is approximately 40 full-packet messages per second (see Section 6.7). In [MCQU 72] the minimum round-trip delay has been calculated for various message lengths, line speeds, and line distances; for example, using 50 KBPS channels and line lengths of 1000 miles, an 8-packet message traveling between two adjacent IMPs will take approximately 0.25 sec in an otherwise unloaded network; more data on actual message traffic is given in Chapter 6.

As so far described, access to the network comes only from a terminal through a HOST to an IMP. The introduction of a device known as a Terminal IMP (TIP) provides direct terminal access to the network [ORNS 72, MCKE 73]. The TIP performs the dual task of acting as an IMP and as a HOST. It is built around a Honeywell 316 minicomputer with 28 K of core.* Up to 63 terminals (remote or local) of widely differing types may be connected to a given TIP and up to three modem and/or HOST interfaces may be connected (however, this is expandable). That which distinguishes a TIP from an IMP (aside from the additional 12 K of core) is a device known as a multiline controller (MLC), which allows direct connection of terminals to the net. The terminals are handled on a per-character basis with start and stop bits (even on synchronous lines). Data rates (from 75 BPS up to 19.2 KBPS) and character bit lengths may be set for each terminal line by the TIP program itself. For each input and output terminal line, two full characters are buffered—the one currently being assembled or disassembled and one further character to account for memory accessing delays. The MLC contains 256 integrated circuits (MSI and LSI) and is approximately the same complexity as the basic Honeywell 316 computer itself. Each line interface unit contains an additional 31 ICs. A TIP costs approximately $100,000. The additional 12 K memory is required for the special TIP code, tables, buffer storage for terminal messages, and so on. The per-character processing time is about 75 μsec and the overhead per message can be extremely large (a factor of 10 or 20 in bandwidth) when single characters are sent one at a time. Approximately 5% of the TIP is lost in performing as an IMP, even in the absence of IMP traffic. The TIP

---

* Originally 20 K. Those TIPs with a magnetic tape option have 32 K.

bandwidth is approximately 500 KBPS in the absence of terminal traffic (for full-size messages).* The TIP average per-machine down rate is approximately $\frac{1}{2}\%$. A more sophisticated terminal-handling HOST concept has been introduced [BOUK 73] and is known as the ARPA Network Terminal System (ANTS). This HOST is a PDP-11 minicomputer model 20, 40, or 45 that requires connection to an IMP and permits some local processing, editing, and peripheral device attachment. Another front-end terminal system also built around a PDP-11 HOST is the UNIX system [RITC 74]. ANTS, UNIX, and other front end processors provide greater bandwidth than do TIPs for network access.

In October 1972, the first public demonstration of the AR-PANET was conducted in conjunction with the First International Conference on Computer Communications (ICCC) in Washington, D.C. Approximately 30 terminals from various manufacturers were connected to a TIP at the conference site. Instruction booklets were made available to the conference attendees to describe methods for accessing various resources on the ARPANET through these terminals. The procedure (demonstrated there and still in use) that a user goes through in reaching a remote computer facility is as follows. First he sits down, powers up the terminal, and then initiates a simple (login) dialogue with the TIP (or his own HOST if he is IMP-connected). Then he requests the TIP to make a connection to a remote HOST, and when this is accomplished he ignores both the TIP and the net and proceeds to login to the remote HOST. Following this, as has always been the case, the user then ignores the operating system of that HOST and communicates directly with the user process with which he has now been put in contact. During that ICCC demonstration, the true power of the ARPANET became apparent not only to the uninitiated users of the network, but also to the sophisticated and experienced users as they observed peak traffic rates of 60,000 packets per hour passing through the TIP and out into the network. The network traffic has been climbing at a phenomenal rate since 1971 and now sustains a fairly substantial load (see Figure 6.37).

So much for the present description of the ARPANET; more details are given in Chapter 6. Let us now proceed with the broad analytical and synthesis questions associated with a general communication subnet.

5.5. DEFINITIONS, MODEL, AND PROBLEM STATEMENTS

Our object in the remainder of this chapter is to discuss the efficient design of computer-communication networks. In this section we introduce

---

* Specifically, the sum of the HOST, terminal, and channel (modem) traffic cannot exceed 600 KBPS full duplex, and the maximum terminal traffic is roughly 80 KBPS (eight 9.6 KBPS CRT terminals doing output) [BUTT 74].

the appropriate notation and definitions and follow this with the collection of assumptions that define our model of the network. Finally we identify the class of analysis and synthesis problems that confront us in network studies.

Our point of departure is the original M-channel, N-node model for message-switching communication networks described by the author [KLEI 64]. In this model, the M communication channels are assumed to be noiseless, perfectly reliable, and to have a capacity denoted by C_i (bits per second) for the ith channel. The N nodes refer to the message- (or packet-) switching centers (for example, IMPs), which are also considered to be perfectly reliable and in which all of the message-switching functions take place, including such things as message reassembly, routing, buffering, acknowledging, and so on. It is assumed that the nodal processing times are constant with value K (usually K is assumed to be negligible). In addition, there are, of course, the channel queueing and transmission delays. Traffic entering the network from external sources (for example, from the HOSTs) forms a Poisson process* with a mean γ_{jk} (messages per second) for those messages originating at node j and destined for node k. We further define the total external traffic entering (and therefore leaving)† the network by

$$\gamma = \sum_{j=1}^{N} \sum_{k=1}^{N} \gamma_{jk} \qquad (5.12)$$

We observe that the quantity γ_i described in Section 1.6 is merely $\gamma_i - \sum_k \gamma_{ik}$. All messages are assumed to have lengths that are drawn independently from an exponential distribution with mean $1/\mu$ (bits). (For the moment, we ignore the notion of packets). In order to accommodate these messages we assume that all nodes in the network have unlimited storage capacity. For many of the analytical results to follow we assume that messages are directed through the network according to a fixed routing procedure; this therefore implies that a unique path exists through the network for a given origin–destination pair.‡

In high-speed networks spanning large geographical regions it may be important to include the propagation time P_i, which is the time required for the energy representing a single bit to propagate down the length of the ith channel; this energy usually propagates at v (miles per second)—a

---

* See Exercise 5.2 for a discussion of packet arrivals.

† Certain internally generated control traffic, such as network-generated measurement traffic and status reports, may in fact leave the network as well. For now, we ignore such effects.

‡ We could just as well include random routing procedures here as we did in Section 1.6 by the introduction of r_{ij} as the probability that a message currently leaving node i will next be routed to node j. In some practical cases this may introduce a useful degree of freedom.

significant fraction of the speed of light depending upon the particular type of channel used. If the ith channel has a length l_i miles, then we clearly have $P_i = l_i/v$; this term is sometimes neglected but may form a significant portion of delay in nets such as the ARPANET. Thus if a message has a length b bits then the time it occupies the ith channel will be $P_i + (b/C_i)$ sec.* Note that the randomness in the service time comes not from the server (the channel) but rather from the customer (message) in that the message length is a random variable $\bar{b}$. At first glance the network we have so far described is similar to Jackson's open networks described in Section 1.6 [JACK 57, 63]. The careful reader will note a key difference between these two networks, however, and this difference is in the source of the random service time as noted above. In the Jackson networks the service time at each server is an independent random variable whereas, in these computer-communication network models, we see that the service time for a given message at different channels is directly related to the message length $\bar{b}$ and the fixed parameters of the channels; this is anything but independent! Moreover the interarrival time between two successive messages on a given channel can certainly be no less than the service time for the first of these messages on that channel; since the service time for this message on its next channel is directly related to its previous service time (and therefore highly correlated with the interarrival time between the two messages on the first channel), we see that the arrival process of messages to a node due to the internal traffic in the network is *not* independent of the service time these messages receive at that node [KLEI 64]! This is a most discouraging observation since all of our previous calculations in this text and in Volume I have assumed independence between interarrival and service times. In Exercise 5.10 we elaborate upon this dependence. Except for this difficulty, one could apply Jackson's results to these networks immediately. (We continue this discussion in Section 5.6.)

Since each channel in the network is considered to be a separate server, we adopt the notation λ_i as the average number of messages per second which travel over the ith channel. As with the external traffic we define the total traffic within the network by

$$\lambda = \sum_{i=1}^{M} \lambda_i \tag{5.13}$$

We further assume that the cost (say in dollars) of constructing the ith channel with capacity C_i is given by $d_i(C_i)$, an arbitrary function of the

---

*We note, however, that the first bits of the next message may begin transmission while the last bits of the previous message are still in flight, thereby providing some concurrency in channel use.

capacity and of the channel. We let D (dollars) represent the cost of the entire network, which we assume to consist only of the cost for channel construction, and so we have*

$$D = \sum_{i=1}^{M} d_i(C_i) \qquad (5.14)$$

We have earlier defined the message delay as the total time that a message spends in the network. Of most interest is the *average* message delay

$$T = E[\text{message delay}]$$

and we take this to be our basic performance measure. Define the quantity

$$Z_{jk} = E[\text{message delay for a message whose origin is} \\ j \text{ and whose destination is } k]$$

It is clear that these last two quantities are related by

$$T = \sum_{j=1}^{N} \sum_{k=1}^{N} \frac{\gamma_{jk}}{\gamma} Z_{jk} \qquad (5.15)$$

since the fraction γ_{jk}/γ of the total entering message traffic will suffer the delay Z_{jk} on the average. Note that Eq. (5.15) represents a *decomposition* of the network on the basis of origin destination pairs. In Section 5.6 we succeed in a further decomposition down to the single-channel level that is much like the decomposition achieved by Jackson.

Thus, except for the source of service time randomness, we basically have an open queueing network. We will in fact distort this communication network model to correspond exactly to the Jackson model, in which case our analysis is quite trivial and reduces to the calculation of the quantity T; these matters are taken up in Section 5.6.

The communication network problem, however, is not merely one of analysis but also is one of efficient design and, when possible, of optimum design. In any practical network design procedure, a large number of design variables suggest themselves. Among these we include: the selection of channel capacities; the form of routing procedure; the form of flow control procedure; the topological design of the network; the storage capacity at each node; the choice of hardware and software programs to be used for the switching computer; the partitioning of messages into various-sized packets; and so on. Since we are interested mainly in the queueing phenomena for purposes of this text, we discuss neither the

---

* The cost of the nodes may be incorporated in the channel costs directly.

hardware nor many aspects of the software design of the switching computer itself any further. Moreover, since the analysis of realistic routing procedures [FULT 71] is extremely difficult (although it is not hard to invent efficient routing procedures) and since it is also extremely difficult to analyze various practical flow control procedures [DAVI 71, KAHN 71, OPDE 74, PRIC 73] (where here it is hard to invent efficient flow control procedures), we therefore choose to limit discussion on these matters somewhat. Lastly the important topological design of networks is a vast and difficult problem in its own right [FRAN 70, 71a, 72b], and will be discussed briefly in Section 5.10. We wish to focus on three very basic design parameters that we must consider: first is the selection of the channel capacities $\{C_i\}$; second is the selection of the channel flows $\{\lambda_i\}$; and third there is the topology itself. All of these may be varied to improve network performance. We comment here that in any realistic network problem the notion of "optimum design" is an extremely difficult one; however, we may take the liberty of defining a one-dimensional performance criterion, the average message delay T, and attempt to minimize this quantity (thereby optimizing performance). This approach will permit us to make some important qualitative statements about network design and performance. Of course, any optimization problem must be subject to some form of cost constraint, and here we choose the fixed cost constraint given in Eq. (5.14). Thus we have a performance measure T, a cost constraint D, and three variable design "parameters," $\{C_i\}$, $\{\lambda_i\}$, and topology. At this point we must elaborate upon the notion that the set $\{\lambda_i\}$ is a design variable. We recognize the fact that the routing procedure operating upon the message traffic will determine this set of values in any real network. However, we choose to model the routing procedure in a fashion identical to that in Section 1.6 whereby r_{ij} gives the fraction of traffic which leaves node i over the channel connecting node i to node j. (In a fixed routing procedure this fraction will be 0 or 1, depending on the origin and destination of the message traffic.) What we mean, then, is that the optimum selection of the channel traffic $\{\lambda_i\}$ consists of finding those theoretical average message flow rates for each line that will result in a minimum average message delay; we are *not* describing the routing procedure that will in fact *achieve* these channel traffic values (a difficult analytical task, in general).

We may now define four optimization problems that differ only in the set of permissible design variables. In each of these problems it is assumed that we are given the node locations, the external traffic flow requirements γ_{jk}, the channel costs $d_i(C_i)$, the constants D and μ, and also that the flow $\{\lambda_i\}$ we use is feasible (i.e., it satisfies the capacity, conservation, and the external traffic requirement constraints). First, we have the *capacity assignment* (CA) problem:

CA PROBLEM

Given:	Flows $\{\lambda_i\}$ and network topology
Minimize:	T
With respect to:	$\{C_i\}$
Under constraint:	$D = \sum_{i=1}^{M} d_i(C_i)$

Second, we have the *flow assignment* (FA) problem:

FA PROBLEM

Given:	Capacities $\{C_i\}$ and network topology
Minimize:	T
With respect to:	$\{\lambda_i\}$

Third, we have the *capacity and flow assignment* (CFA) problem:

CFA PROBLEM

Given:	Network topology
Minimize:	T
With respect to:	$\{C_i\}$ and $\{\lambda_i\}$
Under constraint:	$D = \sum_{i=1}^{M} d_i(C_i)$

Last, we have the *topology, capacity, and flow assignment* (TCFA) problem:

TCFA PROBLEM

Minimize:	T
With respect to:	Topological design, $\{C_i\}$ and $\{\lambda_i\}$
Under constraint:	$D = \sum_{i=1}^{M} d_i(C_i)$

These four problems are presently solved in various degrees of completeness; that completeness depends very strongly upon the form of the cost functions $d_i(C_i)$. We first consider the simplest case of continuous linear costs, then a manageable case of continuous concave costs, and finally a difficult case of discrete cost functions in which the permissible channel sizes are drawn from a discrete set. We discuss the CA problem in

Section 5.7*, the FA problem in Section 5.8, the CFA problem in Section 5.9*, and the TCFA problem in Section 5.10*. First, however, we must address ourselves to the *analysis* problem, which must be solved before we can attempt any optimization.

5.6. DELAY ANALYSIS

Our goal in this section is to solve for T, the average message delay. We have already expressed this delay in terms of origin–destination paths through the network in Eq. (5.15). We make the assumption of a fixed routing procedure (although these methods may also be used in the case of random and deterministic alternate routing procedures as well—this analysis procedure, however, does *not* extend to the case of adaptive routing procedures).

Let us denote the *path* taken by messages that originate at node j and that are destined for node k (the "j-k traffic") by π_{jk}. We say that the ith channel (of capacity C_i) is included in the path π_{jk} if that channel is traversed by messages following this path; in such a case we use the notation $C_i \in \pi_{jk}$. It is clear therefore that the average rate of message flow, λ_i, on the ith channel must be equal to the sum of the average message flow rates of all paths that traverse this channel, that is

$$\lambda_i = \sum_{\substack{j \\ j,k\,:\,C_i \in \pi_{jk}}} \sum_k \gamma_{jk} \qquad (5.16)$$

Moreover we recognize that Z_{jk} is merely the sum of the average delays encountered by a message in using the various channels along the path π_{jk}. These components are the individual channel delays and so we must define

$$T_i = E[\text{time spent waiting for and using}$$
$$\text{the } i\text{th channel}]$$

Thus T_i is just the average time "in system" where the system is defined as the ith channel (a server) plus the queue of messages in front of that channel; this average system time corresponds to T used in all of our single-node studies, but we now reserve T for the average delay in the network and therefore we subscript the average delay on a given channel within the network. Thus we may write

$$Z_{jk} = \sum_{i\,:\,C_i \in \pi_{jk}} T_i$$

---

* We note that these three problems could be expressed in their dual form, namely to minimize D subject to a fixed (or maximum allowable) T. These dual problems are discussed in Sections 5.9 and 5.10 and are shown to have solutions corresponding to the primal problems stated above.

From Eq. (5.15) we therefore have

$$T = \sum_{j=1}^{N} \sum_{k=1}^{N} \frac{\gamma_{jk}}{\gamma} \sum_{i:C_i \in \pi_{jk}} T_i$$

We now exchange the order of summations and observe that the condition on i becomes the corresponding condition on the pair j, k as is usual when interchanging summations; this yields

$$T = \sum_{i=1}^{M} \frac{T_i}{\gamma} \sum_{j} \sum_{k} \gamma_{jk}$$
$$\scriptstyle j,k:C_i \in \pi_{jk}$$

Using Eq. (5.16) we finally have*

$$T - \sum_{i=1}^{M} \frac{\lambda_i}{\gamma} T_i \qquad\qquad \blacksquare \quad (5.17)$$

We have now *decomposed* the average message delay into its single-channel components, namely the delays T_i; Eq. (5.17) is perfectly general. Our analysis problem therefore reduces simply to the calculation of T_i and it is this approach that we find most fruitful in network problems.

We are now faced with solving for a message's average system time in a single channel that is deeply imbedded within a communication network. As pointed out in the previous section, this problem was faced by Jackson in his network problems, and he was able to establish the remarkable result that this imbedded channel offered a solution identical to that of the same channel acting independently from the network but with Poisson arrivals at a rate equal to that offered by the network. We are not so fortunate here, since as indicated earlier there is dependence among the interarrival and service times.† The author [KLEI 64] discusses this problem at length and comes to the conclusion that not only do we wish this dependence to disappear but also in fact that our wish can approximately be granted. The reader will recall that the phenomenon of messages arriving in sequence on a given channel and departing on some other given channel in the same sequence, gives rise to this dependence; on the other hand, if messages leaving the node on a given channel had entered from distinct channels or if messages entering on the same channel depart on distinct channels then one suspects that this dependence should be reduced. Indeed this is the case, and as shown through

---

* Note that this may also be obtained simply by using Little's result as follows. The average number of messages waiting for or using the ith channel is $\lambda_i T_i$. The average number in the net is merely γT, which is therefore the sum of $\lambda_i T_i$ over all channels. Q.E.D.

† The case of M channels in tandem is treated in [KLEI 64] and limiting results can be obtained when $M \to \infty$ (see Exercise 5.3). Moreover, for the case of constant length messages, the tandem case can be solved exactly (see [RUBI 74] and Exercise 5.4).

numerous simulation results, one may feel confident that the following assumption is reasonable for networks of a moderate connectivity:*

INDEPENDENCE ASSUMPTION: Each time a message is received at a node within the network, a *new length*, $\tilde{b}$, is chosen independently from the pdf

$$p(b) = \mu e^{-\mu b} \qquad b \geq 0 \qquad \blacksquare$$

Thus we are saying that the exponential distribution is used in generating a new length each time a message is received by a node; this is clearly false since messages maintain their length as they pass through the network, but the effect of the assumption on the performance measure T has been shown to be negligible in most interesting networks and so we accept it and proceed with the analysis.

With this independence assumption, then, we find ourselves back to the networks considered by Jackson and we may immediately apply our "isolated-channel" calculations. In our case we see that the ith channel is now representable as an M/M/1 system with Poisson arrivals at a rate λ_i and exponential service times of mean $1/\mu C_i$ sec. What could be simpler! The solution for T_i is given immediately through Eq. (1.60), which yields

$$T_i = \frac{1}{\mu C_i - \lambda_i} \qquad (5.18)$$

and so, from Eq. (5.17) we obtain

$$T = \sum_{i=1}^{M} \frac{\lambda_i}{\gamma} \left[\frac{1}{\mu C_i - \lambda_i} \right] \qquad \blacksquare \quad (5.19)$$

Thus T_i behaves basically as shown in Figure 1.1. This last equation and/or Eq. (5.17) completes the basic part of our analysis of communication networks. We will use these equations in the following sections on optimization.

We have of course neglected a number of factors in generating this basic result; for example, we have assumed $K = 0$ and $P_i = 0$ (recall K = nodal processing time and P_i = propagation delay). When we come to apply this analysis to any realistic network, we must include these variables as well as others. Consider applying these results to our example of the ARPANET. In this case, not only is there message traffic moving through the network, but there is also a certain amount of control traffic. Our basic M/M/1 result gives the message delay of the true message traffic

---

*In particular, we mean that most of the nodes should have more than one channel entering and more than one channel leaving.

for a single channel; of course this delay is composed of two quantities, namely a waiting time on queue and a service time. The service time has an average value related to the average length of the true data traffic; the waiting time, however, is due to the interference of all other traffic in the network and is composed partly of data traffic and partly of control traffic. Therefore, it behooves us to separate these two contributions to delay and to use the appropriate parameters for each. If we let $1/\mu$ denote the average length of a data packet and if we let $1/\mu'$ represent the average length of all packets, then we see that a more accurate expression for T_i in the case of the ARPANET is*

$$T_i = \frac{\lambda_i/\mu'C_i}{\mu'C_i - \lambda_i} + \frac{1}{\mu C_i}$$

Of course, if we set $\mu' = \mu$, this will reduce to Eq. (5.18). If we now account for the nodal processing time K and the channel propagation time P_i we may then write down the following expression for the average message delay in the ARPANET:

$$T = K + \sum_{i=1}^{M} \frac{\lambda_i}{\gamma}\left[\frac{\lambda_i/\mu'C_i}{\mu'C_i - \lambda_i} + \frac{1}{\mu C_i} + P_i + K\right] \quad \blacksquare \quad (5.20)$$

The term in square brackets is just our new expression for T_i and the additional term K comes from the fact that messages pass through one more node than they do channels in their travels through the network. Note that in the case $K = P_i = 0$ and $\mu' = \mu$ the expression is reduced to our basic expression in Eq. (5.19). We shall return to consideration of the delay model in Chapter 6.

From this delay analysis [either Eq. (5.20) or (5.19)] we may predict quantitative as well as phenomenological behavior of the average message delay in networks. In particular, if we assume a relatively homogeneous set of C_i and P_i, then as we increase the load on the network, no individual term in the expression for delay will dominate the summation until the flow in one channel (say channel i_0) approaches the capacity of that channel; this channel corresponds to the network bottleneck. At that point, the term T_{i_0}, and hence T, will grow rapidly. The expression for delay will then be dominated by this term and T will exhibit a threshold behavior; prior to this threshold, T remains relatively constant and as we approach the threshold, T will suddenly grow. Thus we expect an average

* Of course now the line traffic rates λ_i must be adjusted to account for the control traffic as well. Moreover, in the simulation results described below, our early estimates led us to use the values $1/\mu = 560$ bits and $1/\mu' = 350$ bits. Recent measurements indicate that $1/\mu \cong 411$ bits (of which 168 were overhead due to headers, check bits, etc.) and $1/\mu' = 268$ bits (again, including 168 bits of overhead); see Chapter 6.

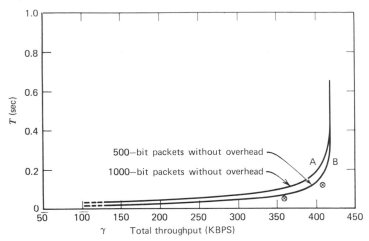

Figure 5.22   Delay versus throughput.

delay in networks that has a much sharper behavior than the average M/M/1 delay shown in Figure 1.1. Let us examine the accuracy of these predictions by plotting the expression given in (5.20) for an $N = 10$ node version of the ARPANET obtained from Figure 5.20(c) by deleting the five rightmost IMPs. The assumptions in this calculation are $K \cong 1$ msec, $v \cong 150{,}000$ miles/sec, and a uniform traffic between all node pairs of $\gamma_{jk} = \gamma/[N(N-1)]$. This results in the curves shown in Figure 5.22 [FRAN 72]. Curve A was obtained using constant-length packets of 1000 bits* while curve B was generated for exponentially distributed packet lengths with an average size of 500 bits; in both these cases all overhead factors were ignored. We observe that the average delay remains relatively small and constant until a total throughput slightly greater than 400 KBPS is reached. The delay then increases rapidly! In this figure we have also illustrated with circled crosses the results of a simulation performed with a realistic routing procedure. This simulation omitted all network over-head and applied to fixed length packets of 1000 bits. It is notable that the delay estimates from the simulation and from the formula are in reasonably close agreement. They both accurately determine the vertical rise in the delay curve in the range just above 400 KBPS; this was

---

* We easily account for the case of nonexponential packet lengths by using the P-K mean value formula [i.e., the first term in the bracket of Eq. (5.20) reduces to half its value shown there for the case of constant length packets—compare Eqs. (1.59) and (1.82)]. Of course, we are taking an extreme liberty here since we have converted the M/M/1 system to an M/G/1 system for which Jackson's result [Eq. (1.76)] no longer applies; this approximation, however, leads us to a model that corresponds very well with simulation, as we shall see.

accomplished through the formula by predicting infinite delay and through the simulation by vigorously rejecting the further input of traffic through its flow control procedure as congestion occurred in the network.

A more detailed simulation has been conducted in which not only was a more realistic message length distribution assumed (thereby leading to an M/G/1 model) but also whereby the internal priority structure of the IMP (in which acknowledgement messages have priority over data traffic) was included using the HOL priority system equations [see Eq. (3.31)]. The results of this simulation are shown in part (b) of Figure 5.23, which applied to the ARPA type 19 node network shown in part (a) [FULT 72]. In part (b) the dots correspond to the simulation data and the curve is calculated from Eq. (5.20) modified to an M/G/1 model (with the same distribution of message length as used for the simulation) with priorities; here the agreement is phenomenally good. The horizontal axis γ corresponds to an arbitrary load scaling factor. Once again we observe the sharp threshold behavior.

This threshold behavior suggests a simplified deterministic (that is, fluid flow) model for delay in computer networks. The essence of the phenomenon is that delay remains essentially constant up to the critical threshold at which point the delay grows in an unbounded fashion. Such a simplification is shown in Figure 5.24. In this figure we represent the delay characteristic in terms of two basic system parameters. The first, T_0, is the constant delay experienced by messages so long as the network load γ is in the stable region ($\gamma < \gamma^*$); the second, γ^* is the network saturation load at which point $T \rightarrow \infty$. T_0 is simply calculated as the "no-load" delay and corresponds to the delay messages would experience in traveling through an unloaded network. For example, were we to use Eq. (5.19) as our model then

$$T_0 = \sum_{i=1}^{M} \frac{\lambda_i}{\gamma} \frac{1}{\mu C_i} \tag{5.21}$$

In this equation we have taken the limit as λ_i, $\gamma \rightarrow 0$ but it is clear (as we now show) that the ratio λ_i/γ has a well-defined limit. A similar result applies to the model in Eq. (5.20). For this limit, we first discuss the notion of *path length*. We define

$$n_{jk} = \text{length of } \pi_{jk}$$

where by length we refer to the number of channels encountered in the path. We are interested in the quantity $\bar{n}$, which we define to be the average path length as given by

$$\bar{n} = \sum_{j=1}^{N} \sum_{k=1}^{N} \frac{\gamma_{jk}}{\gamma} n_{jk}$$

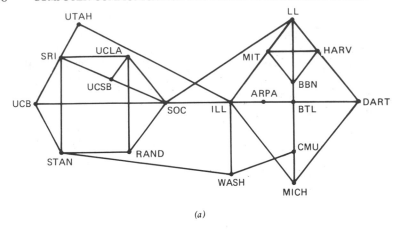

(a)

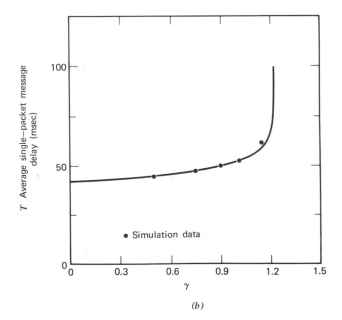

(b)

Figure 5.23   Simulation analysis of delay. (a) The 19-node network simulated. (b) Single-packet message delay as a function of network load.

Clearly, the value of $\bar{n}$ must be invariant to the traffic level if this level is changed by scaling each γ_{jk} by the same factor; the effect of this scaling is to vary the network load γ. Next, consider λ, the total traffic in the network; we observe that the contribution to this total from the j-k traffic will merely be $\gamma_{jk}n_{jk}$ since γ_{jk} messages per second will make n_{jk} "hops" in

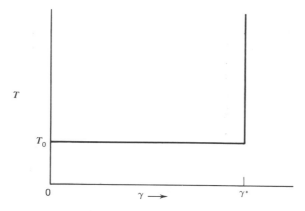

Figure 5.24   The simplified threshold model.

their passage through network channels. Therefore we see

$$\lambda = \sum_{i=1}^{M} \lambda_i = \sum_{j=1}^{N} \sum_{k=1}^{N} \gamma_{jk} n_{jk}$$

These last two equations immediately give that the average path length may easily be expressed as

$$\bar{n} = \frac{\lambda}{\gamma} \qquad\qquad \blacksquare \quad (5.22)$$

This is a general result for all such networks and was established in 1964 [KLEI 64]. Let us now return to Eq. (5.21) and use the average path length to rewrite T_0 as

$$T_0 = \bar{n} \sum_{i=1}^{M} \frac{\lambda_i/\lambda}{\mu C_i} \qquad\qquad \blacksquare$$

Note that the ratio λ_i/λ remains constant (independent of the network load γ) and is a function only of the routing procedure. This then is the method by which we calculate the no-load delay T_0 in our simplified threshold model.

The calculation for γ^*, the network saturation load, is quite straightforward. It corresponds to the smallest value of γ at which some (critical) channel is saturated; in terms of our earlier notation this is the point at which $\lambda_{i_0} = \mu C_{i_0}$ where i_0 is that critical (bottleneck) channel. Therefore, given a fixed routing procedure, one may simply calculate the set $\{\lambda_i\}$ from Eq. (5.16) for any value of γ; one must then examine all the ratios $\lambda_i/\mu C_i$ and identify i_0 as that channel with the largest such ratio. Recognize that we are talking about scaling all terms γ_{jk} (and therefore λ_i) by a common factor (such as γ). The scaling factor is now adjusted to force $\lambda_{i0} = \mu C_{i0}$; the value of throughput γ at which this occurs is precisely γ^*.

This two-parameter simplified threshold model is then a first-cut analysis of the queueing problem. In any realistic computer network, however, the routing procedure will tend to be responsive (i.e., adaptive) to the observed traffic; as a result, if any single channel approaches the saturation point then traffic will be routed through other paths in the network on the way to their destination. If traffic is optimally routed, then γ^* may be calculated as above after one solves for the channel traffic $\{\lambda_i\}$. We note that this method for γ is not equivalent to finding the maximum flow which the network can support between a given pair of nodes since such a calculation involves changing the traffic matrix. However the maximum flow calculation is important and for this we need the well-known Max-Flow, Min-Cut theorem [FRAN 71a]. This theorem states that the maximum flow which a network can support between a source (origin) node s and a termination (destination) node t is simply the value of the minimum s-t cut (defined below). We digress for a moment to elaborate on this procedure. An s-t cut is any collection of branches that, if removed from the network, will stop all traffic flow from origin s to destination t. The capacity of such a cut is the total flow that these removed branches can carry in the direction from origin s to destination t. The minimum s-t cut is that one with smallest capacity. It behooves us at this point to state the well-known Labeling Algorithm for computing the maximum flow between a given origin and destination in a network. For this purpose, we temporarily adopt the notation below.

Let the capacity of the branch (channel) connecting node x and node y be $c(x, y)$ and let the flow actually assigned to this branch be $f(x, y)$ (expressed in bits/sec). Let Γ be the set of branches in the network where (i, j) is a branch directed from node i to node j. Let s be the source (origin) node and t be the termination (destination) node. Initially, let $f(x, y) = 0$ for all branches. The algorithm breaks into two parts:

LABELING ROUTINE

1. Label vertex s by $[s, +, e(s) = \infty]$. s is now labeled and unscanned and all other vertices are unlabeled and unscanned. Go to step 2.
2. Select any labeled and unscanned vertex x.
    (a) For any y such that $(x, y) \in \Gamma$, y is unlabeled, and $c(x, y) > f(x, y)$, then label y by $[x, +, e(y)]$ where $e(y) = \min[e(x), c(x, y) - f(x, y)]$. y is now labeled and unscanned.
    (b) For any y such that $(y, x) \in \Gamma$, y is unlabeled, and $f(y, x) > 0$, then label vertex y by $[x, -, e(y)]$ where $e(y) = \min[e(x), f(y, x)]$. y is now labeled and unscanned.

Repeat (a) and (b) until no more y qualify; then change the label on x by encircling its plus or minus entry. x is now labeled and scanned. Go to step 3.

3. Repeat step 2 until t is labeled (in which case, proceed to the Augmentation Routine) or until no more labels can be assigned. In the latter case the algorithm terminates and the flow currently in the network represents a maximal flow pattern from s to t.

AUGMENTATION ROUTINE

1. Let $z = t$ and go to step 2.
2. If the label on vertex z is $[q, +, e(t)]$ then increase $f(q, z)$ by $e(t)$. If the label on vertex z is $[q, -, e(t)]$ then decrease $f(z, q)$ by $e(t)$. Go to step 3.
3. If $q = s$, erase all vertex labels [but keep all flows $f(x, y)$] and return to step 1 of the Labeling Routine. Otherwise, let $z = q$ and return to step 2 of the Augmentation Routine.

This algorithm will then give the maximum flow between s and t (and therefore the capacity of the minimum s-t cut).

In practice, therefore, and from the analytic and simulation studies of the ARPANET, the average queueing delay is observed to remain small (almost that of an unloaded net) until the traffic within the network approaches the capacity of a cut. The delay then increases rapidly, giving the threshold behavior. Thus a conclusion we can draw is that as long as the traffic is low enough and the routing adaptive enough to avoid the premature saturation of cuts by guiding traffic along paths with excess capacity, then queueing delays will not be significant in networks; this is conditional on an acceptable value of T_0.

So much for the analysis problem.* Our attention will now be directed to the four optimization problems posed in the previous section. We begin with the CA problem.

5.7. THE CAPACITY ASSIGNMENT PROBLEM

One of the more difficult design problems is the optimum selection of capacities from a *finite* set of options. Difficult though it may be, this in fact is a problem commonly faced by network designers. Although there are many heuristic approaches to this problem, exact analytic results are scarce (see [CANT 74] for an algorithm which finds the optimum assignment). It is possible to find reasonable (even optimal) assignments of

---

* We hasten to point out that many intriguing queueing problems remain unsolved [KLEI 64], but we do not pursue them here since our engineering approximations seem to avoid these difficulties in a successful fashion.

discrete capacities for say 200-node networks, but, since these assignments are the results of numerical algorithms very little is known about the relation between such capacity assignments, message delay, and cost. Thus to obtain theoretical properties of optimal capacity assignments we first ignore the constraint that capacities are truly available only in discrete sizes and assume they are available in any continuous size.

With this assumption, we solve the CA problem below in which network topology and traffic flow $\{\lambda_i\}$ are assumed known and fixed. Oberve that λ_i/μ is the average number of bits per second which flow on the ith channel and, therefore, it is clear that any feasible solution to the CA problem must be such that the ith channel is provided with a capacity in excess of this minimal amount. Of course, this is evident from Eqs. (5.19) and (5.20). As far as delay is concerned, it is not critical exactly *how* the excess capacity is assigned as long as we meet the basic condition $C_i > \lambda_i/\mu$.

We begin by considering the case of *linear* capacity costs, namely,

$$d_i(C_i) = d_i C_i \qquad (5.23)$$

Here d_i is the cost incurred for each unit of capacity built into the ith channel.* Note that the cost rate d_i may vary in an arbitrary way with respect to any parameter of the channel except that it must be linear with capacity; for example, d_i is often taken to be proportional to the physical length of the channel (specifically, the distance between its two end points).

We use the form for T given in Eq. (5.19) in obtaining the theoretical properties of the optimal capacity assignment below. To minimize T we proceed by forming the Lagrangian [HILD 65] as follows:

$$G = T + \beta \left[\sum_{i=1}^{M} d_i C_i - D \right]$$

Clearly if we find the minimum value of G with respect to the capacity assignment $\{C_i\}$ then we will have found the solution to the CA problem since the bracketed term is identically equal to zero. (The parameter β is the undetermined multiplier to be evaluated later.) As usual in Lagrangian problems we must satisfy the following set of M equations:

$$\frac{\partial G}{\partial C_i} = 0 \qquad i = 1, 2, \dots, M$$

* In fact, we can just as easily handle the case of a constant plus linear cost, namely, $d_i C_i + d_0$ where d_0 is an extra constant cost for constructing a channel with any capacity as long as $C_i > 0$. We handle this by recognizing through Eq. (5.14) that in this case we really have the constraint $D' = D - M d_0 = \sum_{i=1}^{M} d_i C_i$, which is equivalent to the linear cost function posed in Eq. (5.23) where now D' replaces D.

This gives

$$0 = \left(-\frac{\lambda_i}{\gamma}\right) \frac{\mu}{(\mu C_i - \lambda_i)^2} + \beta \, d_i$$

or

$$C_i = \frac{\lambda_i}{\mu} + \frac{1}{\sqrt{\beta \gamma \mu}} \left(\frac{\lambda_i}{d_i}\right)^{1/2} \qquad i = 1, 2, \ldots, M \qquad (5.24)$$

Once we have evaluated the constant β, this will be our solution; we find β by "forming the constraint" by multiplying this last equation by d_i and summing on i:

$$\sum_{i=1}^{M} d_i C_i = \sum_{i=1}^{M} \frac{\lambda_i d_i}{\mu} + \frac{1}{\sqrt{\beta \gamma \mu}} \sum_{i=1}^{M} \sqrt{\lambda_i \, d_i}$$

We recognize that the left-hand side is just equal to D; solving for the obvious constant we have

$$\frac{1}{\sqrt{\beta \gamma \mu}} = \frac{D - \sum_{i=1}^{M} (\lambda_i \, d_i / \mu)}{\sum_{i=1}^{M} \sqrt{\lambda_i d_i}}$$

Defining the "excess dollars" D_e by

$$D_e \triangleq D - \sum_{i=1}^{M} \frac{\lambda_i \, d_i}{\mu} \qquad (5.25)$$

and using these last two forms in our Eq. (5.24) we come up with the optimal solution to the (linear) CA problem:

$$C_i = \frac{\lambda_i}{\mu} + \left(\frac{D_e}{d_i}\right) \frac{\sqrt{\lambda_i d_i}}{\sum_{j=1}^{M} \sqrt{\lambda_j d_j}} \qquad i = 1, 2, \ldots, M \qquad \blacksquare \quad (5.26)$$

We observe that this assignment allocates capacity such that each channel receives at least λ_i/μ (which is its minimum required amount) and then allocates some additional capacity to each; note that the cost incurred by assigning the minimum capacity to the ith channel is merely $\lambda_i \, d_i / \mu$ dollars and if we sum over all channels we see that the total dollar allocation must exceed this sum if we are to achieve a finite average delay in our network design. The difference between the total dollar allocation and the minimum feasible amount is exactly D_e as given in Eq. (5.25), which explains the name "excess dollars." From Eq. (5.26) we note that these excess dollars are first normalized by the cost rate d_i and then distributed in proportion to the square root of the cost-weighted traffic

$\lambda_i \, d_i$ over all channels; thus we refer to this optimal capacity assignment as the "square root channel capacity assignment" [KLEI 64].

If we substitute the expression for the optimum C_i back into our performance function Eq. (5.19) we obtain the following result:

$$T = \frac{\bar{n}}{\mu \, D_e} \left[\sum_{i=1}^{M} \sqrt{\left(\frac{\lambda_i \, d_i}{\lambda} \right)} \right]^2 \qquad \blacksquare \quad (5.27)$$

This gives the minimum average delay for our network in which capacity has been assigned optimally. We note that D_e plays a critical role here such that as $D_e \to 0$, we find that the average message delay grows without bound. Indeed, as long as $D_e > 0$, the assignment problem is feasible (i.e., $T < \infty$) and this is simply our stability condition; for $D_e \le 0$, the problem is infeasible. These last two equations give the complete solution to the CA problem for the case of linear costs.

A special case of great importance is when $d_i = d$ (the same constant value for each channel); in this case we may, without any loss of generality, assume $d = 1$. This case appears when one considers (stationary) satellite communication channels in which the distance between any two points on earth within the shadow of the satellite is essentially the same regardless of the terrestrial distance between these two points. In this case we note that $D = \sum C_i$, which is merely the sum of all capacities in the network, which we may as well denote by C and whose units we will now take to be bits per second. We see then that our two basic results, the capacity assignment and the performance, give respectively

$$C_i = \frac{\lambda_i}{\mu} + C(1 - \bar{n}\rho) \frac{\sqrt{\lambda_i}}{\sum_{j=1}^{M} \sqrt{\lambda_j}} \qquad i = 1, 2, \ldots, M \qquad \blacksquare \quad (5.28)$$

$$T = \frac{\bar{n} \left(\sum_{i=1}^{M} \sqrt{\lambda_i / \lambda} \right)^2}{\mu C(1 - \bar{n}\rho)} \qquad \blacksquare \quad (5.29)$$

where we have defined

$$\rho \triangleq \frac{\gamma/\mu}{C}$$

This last definition is merely the ratio of γ/μ (the average rate at which bits enter the network from external sources) to C (the total capacity of the network).* The expression for T in this case is quite revealing. First

---

* The use of the symbol ρ is suggestive of a utilization factor, but this is deceptive. The product $\bar{n}\rho = \lambda/\mu C$ is the correct value to consider since $D_e > 0$ corresponds to $\bar{n}\rho < 1$ and this is our criterion for stability and feasibility. ρ is introduced since it is generally a given quantity, not subject to design, whereas $\bar{n}$ varies with topology and routing and should be considered a design variable.

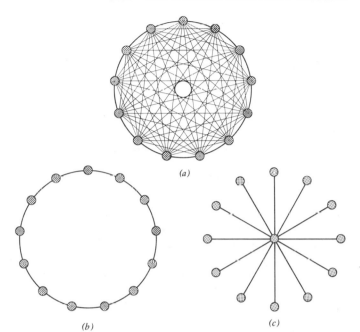

(a)

(b)   (c)

Figure 5.25   Three topologies of interest. (a) The fully connected net. (b) The loop network. (c) The star net.

observe that T is a strictly increasing function of $\bar{n}$ (the average path length). This observation suggests that the topological structure of the network should perhaps be chosen to yield a minimum average path length; this, of course, is achieved in a network that is *fully connected* as shown in Figure 5.25(a) (all pairs of nodes are connected by a communication channel). Furthermore, if one examines the summation in the numerator of T, one finds that it will be minimized over the set $\{\lambda_i/\lambda\}$ if one of these terms is unity and all the rest are zero. Now the numerator summation cannot be minimized to this extent, since then all traffic would exist on one channel with no traffic on any other—this is a highly useless "network." On the other hand, the summation is pointing to the goal of sending lots of traffic on a few large-capacity channels and very little traffic on the others; usually we require at least one channel in and one channel out of each node for communication. A network that achieves this is the simple *loop network* or ring shown in Figure 5.25(b); unfortunately, this network has an average path length proportional to M, depending upon the distribution of γ_{jk}. Another network that also has the one-channel-in, one-channel-out per node for all but the central node is the *star network*, shown in Figure 5.25(c). In this case we do in fact achieve a relatively high concentration of traffic on each line and an average path length $\bar{n} \cong 2$. One must now choose between the fully connected and the star net and all nets "in between" these two. As discussed in [KLEI 64], the particular choice depends on the value of ρ; as $\rho \to 0$ then a star net or a star-like net is most appropriate, whereas when $\rho \to 1$ only the fully connected net will do. Between these two extremes one expects that the addition of some direct channels to the star net will yield the appropriate topology.

The comments above apply to fixed routing procedures, and one wonders if alternate routing procedures might yield an improvement. An alternate routing procedure offers more than one path to the flow of traffic for a given destination and also gives a preferential ordering among these paths; usually the less preferable paths are longer than the primary path. Consequently, we see that alternate routing procedures tend to increase the path length of messages and at the same time tend to disperse traffic among many channels rather than concentrating traffic among a few. Alternate routing violates both of the intuitive rules that we extracted from Eq. (5.29)! In [KLEI 64] these questions are discussed at length and simulations are provided that do in fact demonstrate that the sequence of topologies described above is preferable and that fixed routing appears to be superior to alternate routing procedures. On the other hand, these results were obtained under the assumption that the quantities γ_{jk} are known and constant. If they are either unknown or time-varying or both, then it is clear from Eq. (5.16) that we cannot calculate the channel traffic parameters λ_i and therefore we cannot assign traffic optimally through an equation such as (5.28). Therefore our network will not be optimal for the traffic flow during certain periods of time and if the mismatch is severe, it is then imperative that an (adaptive) alternate routing procedure be provided to seek out paths with excess capacity during these intervals. The conclusion we draw then is that, if the input parameters (specifically γ_{jk}) are known and constant, then under this linear cost assumption we can design optimal networks in the sense of selecting a *capacity assignment that perfectly matches the (fixed routing) network traffic flow;* on the other hand, if the γ_{jk} quantities are unknown or time-varying, then we must introduce an *alternate routing procedure that permits the traffic to adapt to the poor capacity assignment.**

Let us now return to the more general linear cost case with an arbitrary set of $\{d_i\}$. In 1971 [MEIS 71] it was observed that in minimizing T as above, a wide (and perhaps undesirable) variation was possible among the T_i. As a result a new CA problem was posed in which the function to be

---

* The issue of adaptive routing is discussed further in Section 6.1.

minimized was chosen to be

$$T^{(k)} \triangleq \left[\sum_{i=1}^{M} \frac{\lambda_i}{\gamma} (T_i)^k \right]^{1/k}$$

The idea behind this function is that the larger values of T_i raised to large powers will increase this new function much more than previously, so that any minimization procedure with $k > 1$ will reduce the differences among the T_i. If one solves the new CA optimization problem using $T^{(k)}$ in place of T, then the new optimal capacity assignment denoted by $C_i^{(k)}$ is easily shown to be

$$C_i^{'(k)} = \frac{\lambda_i}{\mu} + \frac{D_e}{d_i} \frac{(\lambda_i d_i^k)^{1/(1+k)}}{\sum\limits_{j=1}^{M} (\lambda_j d_j^k)^{1/(1+k)}} \qquad i = 1, 2, \ldots, M \qquad (5.30)$$

and the resulting value of this performance function is

$$T^{(k)} = \frac{(\bar{n})^{1/k}}{\mu D_e} \left[\sum_{i=1}^{M} \left(\frac{\lambda_i d_i^k}{\lambda} \right)^{1/(1+k)} \right]^{(1+k)/k} \qquad (5.31)$$

Note for $k = 1$ that these last results reduce to our former equations (5.26) and (5.27). Of more interest, of course, is the value obtained by the true performance measure T as k changes. It is shown in [MEIS 71] that in fact T deteriorates (increases) only slightly as k increases over a large range while at the same time the variation among the T_i reduces very quickly as k increases, thus achieving the desired result at small additional delay. It is of interest to examine the behavior of these results as $k \to \infty$; here we find

$$\lim_{k \to \infty} C_i^{(k)} = \frac{\lambda_i}{\mu} + \frac{D_e}{\sum\limits_{j=1}^{M} d_j} \qquad (5.32)$$

$$\lim_{k \to \infty} T = \frac{\bar{n}}{\mu D_e} \sum_{j=1}^{M} d_j \qquad (5.33)$$

In this limit we see that the capacity assignment is such as to give each channel its minimum required amount (λ_i / μ) plus a *constant* additional amount. Note that all of the T_i are therefore equal and this corresponds to a mini-max solution to the problem, as might be expected in this severe limit $(k \to \infty)$. For $0 \le k < 1$ this new performance function has the opposite effect, and tends to aggravate (increase) the difference among the T_i. If we examine the limit $k = 0$ we obtain

$$\lim_{k=0} C_i^{(k)} = \frac{\lambda_i}{\mu} + \frac{\lambda_i D_e}{\bar{n} \gamma d_i} \qquad (5.34)$$

$$\lim_{k=0} T = \frac{\bar{n}}{\mu D_e} \sum_{j=1}^{M} d_j \qquad (5.35)$$

We note in this case that the capacity assignment (when all $d_i = 1$) is directly proportional to the traffic λ_i carried by that channel; such a capacity assignment is referred to as the "proportional capacity assignment" and is perhaps a very natural one to consider at first glance.* Of more interest is the fact that the true performance T is identical at both extremes $k = 0, \infty$ although the capacity assignments are quite different!

Let us now generalize beyond the case of linear capacity cost functions. We are still interested in using a continuous approximation to the discretely available channel capacities, but now we seek a better continuous fit to the true discrete costs. In studying the actual tariffs demanded by communications services for high-speed telephone channels, we find that there is effectively a "quantity discount" in that capacity cost (per unit of channel capacity) decreases as the capacity increases. In [KLEI 70] the *logarithmic* cost function

$$D = \sum_{i=1}^{M} d_i \log \alpha C_i \tag{5.36}$$

was studied and it was shown that the CA problem yielded a proportional capacity assignment! However, for the ARPANET it was found [KLEI 70] that a power-law cost function is a better continuous fit to the true values†; thus we are led to consider

$$D = \sum_{i=1}^{M} d_i C_i^{\alpha} \tag{5.37}$$

where $0 \le \alpha \le 1$. If we solve the CA problem using this power-law cost function, we find that the definining equation for C_i is (see Exercise 5.13)

$$C_i - \frac{\lambda_i}{\mu} - g_i C_i^{(1-\alpha)/2} = 0 \tag{5.38}$$

where

$$g_i = \left(\frac{\lambda_i}{\mu \gamma \alpha \beta \, d_i} \right)^{1/2} \tag{5.39}$$

and where β is again the Lagrangian undetermined multiplier that must be adjusted to satisfy the constraint equation (5.37). The equation for C_i may be solved by any convenient iterative algorithm and the results of such a computation for the ARPANET are shown in Figure 5.26; this was the same 19-node network used in Figure 5.23. What we have

---

* In fact, the proportional capacity assignment is not far from optimal [KLEI 64]—see Exercise 5.8.
† It has been observed that an economy of scale seems to be present in many telecommunication resources; the observed value for α was found to have a mean of 0.44 and a standard deviation of approximately 0.10 [ELLI 74].

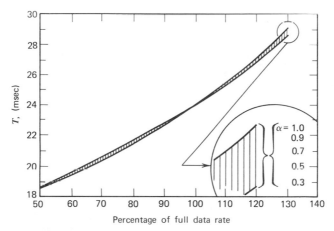

Figure 5.26   Average message delay at fixed cost as a function of data rate for the power loss and linear cost functions.

plotted in Figure 5.26 is the average message delay T versus the load on the network and we have arbitrarily adjusted each of these curves to take on the same value at the full data rate point*; note that this figure gives the result for five values of α. The amazing observation is that T seems to be almost insensitive to the value of α (at least for this version of the ARPANET) and it therefore suggests that an approximate solution for communication networks is to use $\alpha = 1$ [for which we have the exact solutions in Eqs. (5.26) and (5.27)], thereby eliminating the more difficult nonlinear equation (5.38).

In practice, however, as we have mentioned above, the selection of channel capacities must be made from a small finite set. In this case it appears that an approach based upon dynamic programming is perhaps best, and this has yet to be investigated carefully [CANT 74]. However, we wish to point out that the continuous optimization procedures described above do provide a means for selecting among discrete capacities in a suggestive way. For example, in [KLEI 69] the following procedure was investigated. In the early ARPANET, we had $C_i = 50$ KBPS for all channels. For the particular 19-node configuration shown earlier in Figure 5.23, one may calculate the average message delay T (under this fixed-capacity assignment) as well as the true cost, D, of the channels in the network; one may then take this total cost and do a CA optimization with linear costs used as an approximate fit to the actual discrete costs to achieve an optimum assignment and therefore

* "Full" data rate corresponds to an early approximate estimate of the true traffic requirements between all pairs of nodes; thus it is not a "critical" load in any sense.

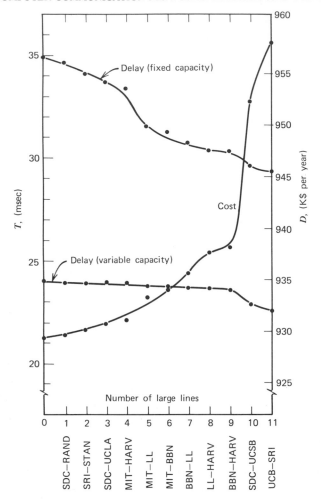

Figure 5.27   Effect of adding large lines of 250 KBPS.

a reduced delay with variable capacity. As a result of this optimization, one will obtain a list of "optimal" capacities for the channels, and
one of these capacities will be largest. Suppose we consider upgrading
this particular "largest" channel in the 50 KBPS network so that it
takes on an increased value of say 250 KBPS; this will of course
increase the cost but will at the same time reduce the delay. We can then
continue this procedure identifying the next most "needy" channel,
upgrading it, increasing the cost, and proceeding; the result of such a
computation is shown in Figure 5.27. In this figure, we note that the
increased capacity lines happened all to be between geographically close

nodes; for example, the distance from SDC to RAND is on the order of 1 mile. If one observes the way in which the fixed*-capacity delay curve drops versus the way in which the cost function increases we see that the introduction of the tenth large-capacity line (SDC-UCSB—roughly 125 miles long) gave a very sharp increase in cost (it was the first channel longer than 20 miles) and only a moderate decrease in delay; thus perhaps that was a poor choice and one should not have increased more than nine channels in this particular example. Such insight and design aids are useful in the practical design of networks. Other procedures of this sort whereby one uses an analytically tractable optimization method to *suggest* directions for improvement in a practical but intractable problem, are extremely worthwhile.

In general, the selection directly from the discrete set of channel capacities is a difficult dynamic programming problem [CANT 74]. However, in the case where the topology of the network is a tree, then a simple and computationally efficient algorithm is known that optimally selects channel capacities for an arbitrary discrete set of costs [FRAN 71b].

Let us now recall the simple threshold approximation for the behavior of networks discussed in the previous section. We wish to compare this to the simple (satellite) channel performance given in Eq. (5.29) for which $d_i = 1$ and $\alpha = 1$. In this equation we observe that as ρ increases, then T grows according to the simple pole at $\rho = 1/\bar{n}$. Thus the performance is much like that given in Figure 1.1 and not at all like the threshold behavior we observed earlier. The reason for this discrepancy lies in the fact that we have optimized the capacities and, as we noted earlier, we guarantee through this optimization to allocate enough capacity to at least equal the average flow a channel must carry; this guarantee ensures that *all channels saturate at exactly the same network load!* Consequently, rather than the case as in the previous section where one channel saturates "prematurely," we now have all channels saturating simultaneously and, therefore, smoothly. However, due to the discrete nature of the capacities and due to the time-varying or unknown nature of the network traffic, it is not generally possible to match channel capacities to the anticipated flows within the channels. Instead, capacities are assigned on the basis of reasonable estimates of average or peak traffic flows and it is the responsibility of the routing procedure to allow the traffic to adapt to the available capacity. Every so often, two HOST computers will engage in heavy communication and thus saturate one or more critical

---

* The variable-capacity delay curve is for the case where the total cost of a given configuration was used with the optimal continuous assignment as given in Eqs. (5.26) and (5.27).

network cuts. In such cases, the routing will not be able to send additional flow across these cuts and the network will therefore experience "premature" saturation in one or a small set of channels leading to the threshold behavior described earlier; this is the more usual case.

Thus the CA optimization problem yields to solution in certain cases only, depending upon the form of the capacity cost function $d_i(C_i)$. For linear costs, we have the exact solution [Eqs. (5.26) and (5.27)] in the form of a square root channel capacity assignment and also the generalized solution for $C_i^{(k)}$ in Eqs. (5.30) and (5.31). For logarithmic costs, we have the proportional channel capacity assignment. For power-law costs, we have the (numerical) solution given in Eq. (5.38). For the more realistic discrete cost functions, we are reduced either to heuristic solutions or to difficult enumeration procedures.

5.8. THE TRAFFIC FLOW ASSIGNMENT PROBLEM

In the CA optimization problem, we required an optimum choice of channel capacities for a given flow configuration $\{\lambda_i\}$. In this section, we are interested in the converse problem, namely, the flow assignment (FA) problem, in which we assume the capacities are given and the flows must be determined so as to minimize the average delay. Previously, we made the comment that alternate routing procedures* that provide more than one path for the j-k traffic were expected to be inferior to fixed routing procedures in which exactly one path existed for this traffic. That comment was based upon the assumption that one could select the capacities to match the traffic requirements. In this section our point of view is that the capacities are given and that the flow of traffic must be adjusted in an optimal way; we therefore expect that it may be necessary to provide more than one path for the j-k traffic since, for example, it is clear if $\gamma_{jk}/\mu > C_i$ (where C_i is the capacity of a channel connecting nodes j and k directly) that more than one path is required to handle the j-k flow.

The basic equation we shall use for our performance function is given in (5.19) and represents the average delay to messages in passing through the network. As a consequence, the "queueing part" of the problem is finished and completely summarized in that equation; what remains is a flow optimization problem, which belongs in the field of network flow theory [FRAN 71a] (this class of problems was discussed in Volume I, Chapter 1, where we labeled them "steady flow through a network of channels"). The Max-Flow, Min-Cut theorem described earlier is the foundation of that theory. Since this material does not properly belong to

---

* The term "bifurcated" routing is sometimes used in place of "alternate" routing.

queueing theory, but yet is significant in the design of computer-communication networks, we devote this short section (and the next) to some of the successful techniques that have been developed in solving these flow problems. In the language of network flow theory we are faced with a multicommodity flow problem with a nonlinear objective function. Each j-k flow requirement, γ_{jk}, forms a unique commodity that must flow from origin j to destination k through the network. As stated in Section 5.5, this multicommodity FA problem requires that we minimize the nonlinear function T with respect to the flows $\{\lambda_i\}$ in such a way that the external flow requirements γ_{jk} are satisfied; this, of course, is under the assumption of a given capacity assignment. In addition, we must obey the usual conservation of flow law at each node on a commodity-by-commodity basis, which merely says that the total flow of j-k traffic into node n must equal the total flow of such traffic out of that node unless $n = j$, in which case the node is our source (origin), or unless $n = k$, in which case the node is our sink (destination). Furthermore, we have the capacity constraint on each channel which requires that the flow (λ_i/μ) on channel i must be (non-negative and) less than the capacity, that is, $(0 \le)\lambda_i/\mu < C_i$. From this last constraint we see that the performance function T has the interesting and obvious property that it will increase without bound whenever any flow approaches the capacity of its channel (that is, whenever the set of flows in the network approaches the upper boundary for these flows as defined by the capacity constraints); in the terminology of mathematical programming we say, therefore, that the performance function T includes the additional capacity constraint as a penalty function. This important property guarantees the feasibility of the solution (with respect to the capacity constraint) during the application of any minimization technique so long as we begin with an initially feasible solution and make "small" steps. Thus if we begin with a feasible solution, then we may *ignore* the capacity constraint and therefore what appeared to be a constrained optimization problem is in fact an unconstrained (and thus simpler) multicommodity flow optimization problem.

The method to be described below is due to Fratta, Gerla, and Kleinrock [FRAT 73] and gives an exact solution to the FA optimization problem which is computationally efficient (in fact, *any* reasonable search procedure will work). Whereas this method is described in terms of the communication network problem, it certainly is applicable to problems of this sort with more general performance functions. A different solution method is described in [CANT 72]. We begin by examining the form of T in Eq. (5.19). We note immediately that this performance measure is separable in the sense that it may be expressed simply as a sum of terms, each of which depends only on the flow in a single channel.* Furthermore, we observe from Eq. (5.19) that

$$\frac{\partial T}{\partial(\lambda_i/\mu)} = \frac{C_i}{\gamma\,[C_i-(\lambda_i/\mu)]^2} \qquad i=1,2,\ldots,M$$

We see that $\partial T/\partial(\lambda_i/\mu)>0$ for all i and a similar computation shows that $\partial^2 T/\partial(\lambda_i/\mu)^2>0$ (both as long as the capacity constraints are satisfied); therefore we may immediately conclude that T is a *convex* function of the flows. Furthermore, the set of feasible flows is itself a convex polyhedron. Thus we may conclude that if the problem is feasible at all, then any local minimum is a global minimum for T. Any method we have for finding a local minimum, therefore, will solve the global problem.

The method we investigate here is the "flow deviation" (FD) method [GERL 73], which locates this global minimum. In order to understand the FD method we must first explain the key motion of a *shortest-path flow*. Assume that we have a network where each channel is labeled with some length l_i. For such a network we might naturally ask for the shortest path between an origin node j and a destination node k, and we would then be tempted to send whatever flow requirement we had, say γ_{jk}, along that path (assuming no capacity or delay considerations). If we do likewise for all pairs j, k, then the resulting flow is referred to as the shortest-path flow. There are numerous well-known and efficient algorithms for finding the set of j-k shortest paths; let us describe one such algorithm published by Floyd [FLOY 62]. Consider a network with N nodes. Let $\mathbf{D}_0=(d_{jk})$ be an $N\times N$ matrix whose j, k component d_{jk} gives the length (a given quantity for this calculation) of the channel *directly* connecting node j to node k; if no such channel exists, then this entry is infinity (also $d_{jj}=0$). Furthermore, we assume that there are no cycles such that the total length around that cycle is negative. (Observe that we are using a two-subscript notation for the length of a channel in this algorithm, whereas previously we had used the single subscripted symbol l_i to denote the length of the ith channel; this is only for convenience of stating this algorithm.) If we now consider any path π_{jk} that connects node j to node k, then we denote the length of this path (the sum of the channel lengths) by $l(\pi_{jk})$. Our task is to calculate the $N\times N$ matrix $\mathbf{H}=(h_{jk})$ where h_{jk} is the length of the shortest path connecting node j and node k; thus, $\mathbf{H}$ is the matrix of shortest paths that we seek. Floyd's shortest-path algorithm begins with the distance matrix $\mathbf{D}_0$ and iteratively modifies it through a sequence of N matrices, which at the nth step is denoted by the matrix $\mathbf{D}_n$, eventually terminating in the matrix of shortest paths $\mathbf{D}_N=\mathbf{H}$. Starting with $n=0$ and $d_{jk}(0)=d_{jk}$,

---

* This separability is a critical property and is exploited in the work of Everett [EVER 63] and Whitney [WHIT 72].

we calculate $\mathbf{D}_{n+1}$ from $\mathbf{D}_n$ by the iterative operation

$$d_{jk}(n+1) = \min[d_{jk}(n), d_{j,n+1}(n) + d_{n+1,k}(n)]$$

After calculating $\mathbf{D}_1$, we see that $d_{jk}(1)$ (the j, k element in $\mathbf{D}_1$) gives the length of the shortest path from node j to node k restricted to paths with node 1 as the only intermediate node. Similarly, by the time we have calculated $\mathbf{D}_n$ at the nth iteration, then $d_{jk}(n)$ will give the shortest distance from node j to node k in which the intermediate nodes are restricted to belong to the set $\{1, 2, \ldots, n\}$. Thus by the time we reach the Nth iteration we have $d_{jk}(N) = h_{jk}$, the desired result. The total algorithm requires approximately N^3 addition-subtractions and N^3 comparisons. Many other shortest-path algorithms exist.

Now, returning to our original problem, the key to the FD method is to associate a "length" with the ith branch whose value is given by

$$l_i \triangleq \frac{\partial T}{\partial(\lambda_i/\mu)} = \frac{C_i}{\gamma[C_i - (\lambda_i/\mu)]^2} \qquad \blacksquare \quad (5.40)$$

when the flow on that channel is λ_i/μ. This, of course, is the linear rate at which T increases with an infinitesimal increase in the flow on the ith channel. These "lengths" or "cost rates" may then be used to pose a shortest-route flow problem, and the resulting paths will represent the "cheapest" (i.e., marginally best for reducing T) paths to which some of the flow can be deviated; the question is *how much* of the original flow should be deviated to these new paths. Once this is determined, the process may be repeated by recalculating new lengths using the updated flows, solving a new shortest-route flow problem, finding the correct flow deviation quantity, and so on. This iterative procedure continues until an acceptable performance tolerance is reached.

Let us now phrase these ideas in the form of a specific algorithm. To do so we introduce the vector notation

$$\mathbf{f}^{(n)} = \left(\frac{\lambda_1^{(n)}}{\mu}, \frac{\lambda_2^{(n)}}{\mu}, \ldots, \frac{\lambda_M^{(n)}}{\mu}\right) \qquad (5.41)$$

where $\mathbf{f}^{(n)}$ is the flow vector at the nth iteration of the algorithm whose ith component $\lambda_i^{(n)}/\mu$ represents the total flow on the ith channel at the nth iteration. Temporarily we assume that we have a feasible starting flow $\mathbf{f}^{(0)}$. We may then state the following:

FD OPTIMAL ROUTING ALGORITHM:

Step 1. Let $n = 0$

Step 2. For each $i = 1, 2, \ldots, M$ compute

$$l_i = \frac{C_i}{\gamma[C_i - (\lambda_i^{(n)}/\mu)]^2}$$

Step 3.   Compute β_n, the incremental cost rate at this flow where

$$\beta_n = \sum_{i=1}^{M} l_i \frac{\lambda_i^{(n)}}{\mu}$$

Step 4.   Solve the shortest-route flow problem using the lengths l_i. Let ϕ_i be the resultant flow on the ith channel that would result if all flow were routed via these shortest paths. Let the vector of these flows be denoted by

$$\boldsymbol{\phi} = (\phi_1, \phi_2, \ldots, \phi_M)$$

Step 5.   Compute b_n, the incremental cost rate for the shortest-route flow where

$$b_n = \sum_{i=1}^{M} l_i \phi_i$$

Step 6.   (Stopping rule) If $\beta_n - b_n < \varepsilon$ where $\varepsilon > 0$ is a properly chosen acceptance tolerance, then STOP. Otherwise, go to Step 7.

Step 7.   Find that value of α in the range $0 \le \alpha \le 1$ such that the flow $(1-\alpha)\mathbf{f}^{(n)} + \alpha\boldsymbol{\phi}$ minimizes T. Let this optimum value be denoted by a. This optimum value may be located by any convenient search method (such as a Fibonacci search [ZANG 67]).

Step 8.   (Flow deviation) Let $\mathbf{f}^{(n+1)} = (1-a)\mathbf{f}^{(n)} + a\boldsymbol{\phi}$.

Step 9.   Let $n = n+1$. Go to Step 2.

We note that the critical steps in this algorithm are: Step 2, the length calculation; Step 4, the shortest-route flow calculation; Step 6, the stopping test; Step 7, the portion of flow to be deviated; and finally Step 8, the flow deviation itself. We note that the flow deviation is moving in a direction of maximum decrease for T. In general, it results in a deterministic alternate routing procedure.

Let us now address the question of finding a feasible starting flow $\mathbf{f}^{(0)}$. We assume once again that the total external flow requirement is γ [see Eq. (5.12)]. We introduce a scaling factor h such that $h\gamma$ is the amount of flow we handle for a given value of h. The algorithm for finding a feasible starting flow is as follows:

Step 1.   Set $h_0 = 1$ and let $\mathbf{f}^{(0)}$ be the solution to the shortest-route flow for a network whose lengths are $l_i = 1/\gamma C_i$ [note that this is the length in Eq. (5.40) at 0 flow]. This step will route a total flow $h_0\gamma$ through the network and we denote the flow in the ith channel at this stage by $\lambda_i^{(0)}/\mu$. Let $n = 0$.

Step 2. Let

$$\sigma_n = \max_i \left(\frac{\lambda_i^{(n)}}{\mu C_i} \right)$$

If $\sigma_n/h_n < 1$ then set $\mathbf{f}^{(0)} = \mathbf{f}^{(n)}/h_n$; STOP (this is a feasible starting flow). If on the other hand, $\sigma_n/h_n \geq 1$, then let $h_{n+1} = h_n[1 - \varepsilon_1(1 - \sigma_n)]/\sigma_n$ where ε_1 is an appropriate precision parameter such that $0 < \varepsilon_1 < 1$.

Step 3. Let $\mathbf{g}^{(n+1)} = (h_{n+1}/h_n)\mathbf{f}^{(n)}$. This is a feasible multicommodity flow that routes a total traffic in an amount equal to $h_{n+1}\gamma < 1$.

Step 4. Now form a flow deviation operation on the flow $\mathbf{g}^{(n+1)}$; that is, perform Steps 2, 4, 7, and 8 of the FD algorithm to find $\boldsymbol{\phi}$ (the shortest-route flow with lengths based on the flow $\mathbf{g}^{(n+1)}$), and the optimum value of α, namely a, such that the flow $\mathbf{f}^{(n+1)} = (1 - a)\mathbf{g}^{(n+1)} + a\boldsymbol{\phi}$ minimizes T. If $n = 0$, go to Step 6; otherwise go to Step 5.

Step 5. If

$$\left| \sum_{i=1}^{M} l_i(\phi_i - g_i^{(n+1)}) \right| < \theta \quad and \quad |h_{n+1} - h_n| < \delta$$

(where θ and δ are properly chosen positive tolerances), then STOP; the problem is infeasible within the tolerances θ and δ. Otherwise, proceed to Step 6.

Step 6. Let $n = n + 1$ and go to Step 2.

This procedure will either find a feasible starting flow or will declare the problem infeasible within the tolerances chosen.

Whereas the FD method produces an optimum routing of traffic in the network and whereas it is relatively efficient in terms of computation, it turns out that there is a simpler suboptimal method that produces a fixed routing procedure and often yields extremely good results using much less computation. This suboptimal method avoids the task of deciding how much flow to deviate; it merely decides to deviate all or none of the flow for each γ_{jk}. The approximation is based on the observation made in the previous section that fixed routing procedures have the correct properties in terms of short average path lengths and highly concentrated traffic. The class of networks for which this fixed routing algorithm is effective is for the case of "large and balanced networks." A network is said to be large if it has a large number of nodes; it is said to be balanced if the elements γ_{jk} do not differ from each other by large percentages. We now make these notions quantitative in an ad hoc, but intuitive fashion. In particular, let

$$\gamma_0 \triangleq \frac{\gamma}{N(N-1)} \tag{5.42}$$

be the average flow requirement per ordered pair of nodes (i.e., traffic flow in one direction), and let

$$m \triangleq \max_{j,k} \frac{\gamma_{jk}}{\gamma_0} \tag{5.43}$$

be the ratio of the largest to the average node pair flow value (many other appropriate definitions of m are possible here). We have $m \geq 1$ and we observe that the case $m = 1$ is the uniform traffic case for which $\gamma_{jk} = \gamma_0$ for all j, k. A network is said to be balanced if m is close to unity. We further let the branch-to-node density for a network be denoted by $K = M/N$ and denote by $\bar{n}_0$ the average path length when traffic is routed according to the shortest path and where the length of each channel in this computation is taken as unity. We then consider the parameter

$$\eta = \frac{Km}{(N-1)\bar{n}_0} \tag{5.44}$$

We say a network is "large and balanced" if $\eta \ll 1$. From this definition it can be shown that in a large and balanced network, on the average the contribution of the j-k traffic in any channel can be considered negligible as compared to the total flow in that channel. This being the case, the reader can see why the consideration of fixed routing procedures (deviate all or none) leads to a good approximation. Let us now give the suboptimal fixed routing flow algorithm. Again we assume that we have a feasible fixed routing starting flow $\mathbf{f}^{(0)}$.

FIXED ROUTING FLOW ALGORITHM

Step 1. Let $n = 0$.

Step 2. Using the flow $\mathbf{f}^{(n)}$, find the set of shortest routes [under the metric l_i defined in Eq. (5.40)].

Step 3. Let $\mathbf{g} = \mathbf{f}^{(n)}$. For each flow requirement γ_{jk}:

Step 3a. Let $\mathbf{v}$ be the flow obtained from $\mathbf{g}$ by deviating *all* of the flow γ_{jk} from its path in the flow $\mathbf{f}^{(n)}$ to the shortest j-k path.

Step 3b. If both $\mathbf{v}$ is feasible and if T using $\mathbf{v}$ is strictly less than T using $\mathbf{g}$, then go to Step 3c. Otherwise, go to Step 3d.

Step 3c. $\mathbf{g} = \mathbf{v}$.

Step 3d. If all flows γ_{jk} have been processed, go to Step 4. Otherwise, select any unprocessed γ_{jk} and go to Step 3a.

Step 4. If $\mathbf{g} = \mathbf{f}^{(n)}$ STOP; the method cannot improve the fixed routing flow any further. Otherwise, let $\mathbf{f}^{(n+1)} = \mathbf{g}$, set $n = n + 1$ and go to Step 2.

This algorithm will converge in a finite number of steps since there are only a finite number of fixed routing flows to be considered, and we know that the same flow may not be considered twice due to the stopping condition. The feasible fixed routing starting flow $\mathbf{f}^{(0)}$ may be found by a method similar to that used for the FD algorithm.

Let us apply the FD flow algorithm as well as the fixed routing flow algorithm to the case of the ARPANET. We consider a 21-node AR-PANET* with a uniform requirement in the sense that for all distinct ordered pairs j, k ($j \neq k$) we have $\gamma_{jk} = \gamma_0 = 1.187$ KBPS; for this particular network the maximum possible value of γ_0 is 1.25 (corresponding to $\gamma^* = 525$ KBPS) and so we are asking that our algorithms find a flow which is within 95% of the maximum possible flow the network can sustain. For this network we have the parameters $N = 21$, $M = 52$ (26 full-duplex channels), $m = 1$, and so

$$\eta = \frac{52/21}{20\bar{n}_0} < 0.1 \ll 1$$

Thus the large and balanced network condition is satisfied; we therefore expect the fixed routing flow algorithm to give us results comparable to those from the optimum FD flow algorithm.

The result of applying the optimum FD algorithm yields a minimum value for T equal to $T_{min} = 0.2406$ sec, which was obtained after 80 shortest-route flow computations to an accuracy of 10^{-4} for T. The execution time for this algorithm was 30 sec (programmed in FORTRAN and run on an IBM 360/91). On the other hand, the fixed routing flow algorithm yielded $T_{min} = 0.2438$ sec after 12 shortest-route flow computations and required an execution time of only 4 sec (again in FORTRAN on an IBM 360/91). In fact, both these times could be improved if the code were optimized. We note that the much faster suboptimal algorithm gave a result well within 2% of the optimum algorithm, showing the power of the fixed routing flow algorithm for large and balanced networks. In Figure 5.28 we show the result for the fixed routing flow algorithm. Curve C_0 corresponds to the search for a feasible starting flow in which flow is routed along the shortest paths at $h_0 = 0$. The saturation level for this particular flow is at $h_{0,\text{SAT}} = 0.85$, and so $h = 1$ is infeasible and we must continue to search for a feasible starting flow. Curve C_1

---

* This is the network shown in Figure 5.20(d) without the following three nodes: MCCLEL-LAN, USC, and RADC. Also, an additional channel between SRI and UCLA was added.

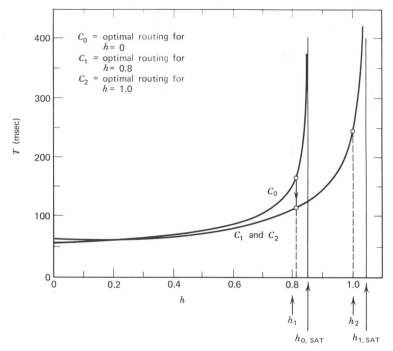

Figure 5.28   Average delay T versus normalized traffic h, using optimal routing.

corresponds to the next iteration in the search for a feasible flow, and in this case we find saturation at $h_{1,SAT} = 1.05$, thereby showing that $h = 1$ is feasible and we may now go to the main fixed routing flow algorithm. Curve C_2 gives the resulting performance of this flow algorithm and it practically coincides with curve C_1 giving the result for $T_{min} = 0.2438$ sec at $h = 1$. Note that C_0 is indeed the optimum flow at very small values for h, as expected.

Thus we see that the FD flow algorithm is an optimum procedure for solving the FA problem and that it results in an alternate routing flow. Furthermore, we see that in the large and balanced network case, we have a faster and rather good algorithm of a fixed routing type.

5.9. THE CAPACITY AND FLOW ASSIGNMENT PROBLEM

In the two previous sections we have given optimum solutions to the CA problem and the FA problem. When we combine these to form the CFA problem we are no longer able to give globally optimal solutions, but rather we describe procedures that find local minima for T.

Since we are once again in a position to select C_i, it seems that the fixed routing algorithms for flow should also be optimum. In Exercise 5.17 we show this to be the case for the linear cost function in which $d_i = 1$. It has also been shown [FRAT 72] for the more general linear cost case with arbitrary d_i that once again the fixed routing flow is optimum; this comes about due to a certain concavity property for T as expressed in its optimum form in Eq. (5.27). Furthermore, it can be shown that the local minima for T are obtained with shortest-route flows (a subclass of the fixed routing flows) since the minima must occur at the corners of the convex polyhedron of the feasible flow set [GERL 73].

Our approach to finding these local minima, then, is to begin with a feasible starting flow, calculate the optimum capacity assignment under linearized costs, carry out the FD algorithm to find the optimum flows, repeat the CA problem for these new flows, and continue to iterate between the CA solution and the FA solution until we find a (local) minimum. We note in particular that the FD algorithm will be especially simple in this case since the step size α will always be equal to unity (fixed routing flow). Thus our algorithm is as follows under the assumption of a feasible starting flow $\mathbf{f}^{(0)}$:

CFA SUBOPTIMAL ALGORITHM

Step 1. Let $n = 0$.
Step 2. Carry out the CA algorithm for the flow $\mathbf{f}^{(n)}$ to find an optimal set of capacities, using linearized costs.
Step 3. Using the lengths $l_i = \partial T/\partial(\lambda_i/\mu)$ carry out an FD algorithm in which $\alpha = 1$ at each step. Let the optimum flow so obtained be denoted by $\mathbf{f}^{(n+1)}$.
Step 4. If T for the flow $\mathbf{f}^{(n+1)}$ is greater than or equal to T for the flow $\mathbf{f}^{(n)}$, then STOP; the flow $\mathbf{f}^{(n)}$ is the local minimum obtained. Otherwise, let $n = n + 1$ and go to Step 2.

The algorithm will certainly converge, since there are only a finite number of shortest route flows.

At the completion of Step 2 (the CA algorithm) we know that C_i will be given by Eq. (5.26) and the performance measure, T, will be given by Eq. (5.27), where we interpret λ_i to mean $\lambda_i^{(n)}$. For these values, we then see from Eq. (5.40), that the form for l_i is given by

$$l_i \triangleq \frac{\partial T}{\partial(\lambda_i/\mu)} = \frac{\bar{n} \sum_{j=1}^{M} (\lambda_j d_j/\lambda)^{1/2}}{D_e} \left[\left(\frac{d_i}{\lambda \lambda_i} \right)^{1/2} + \frac{d_i}{\mu D_e} \sum_{j=1}^{M} \left(\frac{\lambda_j d_j}{\lambda} \right)^{1/2} \right] \blacksquare \quad (5.45)$$

We see, therefore, that $l_i \geq 0$, and so negative loops cannot exist (as required by the shortest-path algorithm). We also observe that $\lim_{\lambda_r \to 0} l_i = \infty$, which indicates that whenever the flow (and therefore the capacity) of a channel is reduced to zero at the end of an iteration, then both flow and capacity will remain at zero for subsequent iterations since the incremental cost of inserting flow is then infinity!

We are now faced with two problems. First, we must find a feasible starting flow $\mathbf{f}^{(0)}$ and secondly, we must search over many local minima in an attempt to find a global minimum. We solve both of these difficulties by repeating the CFA algorithm for many different starting feasible flows. Each starting feasible flow is found by assigning initial lengths to the channels in a random way. For each assignment we then do a shortest-route flow algorithm using the lengths so chosen and test for the case $D_e > 0$ for this flow; if the case is satisfied, we have found a feasible starting flow and we can begin the CFA algorithm; otherwise, the initial flow is rejected and we try another set of random lengths.

As mentioned at the end of Section 5.5, the CFA problem may also be posed in its dual form, namely

CFA PROBLEM (DUAL FORM)

Given: Network topology

Minimize: $D = \sum_i d_i(C_i)$

With respect to: $\{C_i\}$ and $\{\lambda_i\}$

Under constraint: $T = \sum_i \frac{\lambda_i}{\gamma} T_i \leq T_{\text{MAX}}$

Here we choose to minimize the cost D under the basic constraint of a fixed maximum average delay T_{MAX}. The CFA suboptimal algorithm described above will essentially work for this dual form of the problem if we replace the definition of length given there as $l_i = \partial T/\partial(\lambda_i/\mu)$ with the new definition $l_i = \partial D/\partial(\lambda_i/\mu)$ in the obvious way. In the case when the capacity costs are linear, that is, $d_i(C_i) = d_i C_i$, then it is easy to show [see Exercise (5.15)] that the solution to the CA problem at the end of Step 2 will be

$$C_i = \frac{\lambda_i}{\mu} + \left(\frac{\lambda_i}{\mu \gamma T_{\text{MAX}}}\right) \frac{\sum_{j=1}^{M} \sqrt{\lambda_j d_j}}{\sqrt{\lambda_i d_i}} \qquad \blacksquare \quad (5.46)$$

and the minimum cost that satisfies the maximum average delay constraint is simply

$$D = \sum_{i=1}^{M} \frac{\lambda_i d_i}{\mu} + \frac{1}{\gamma T_{\text{MAX}}}\left[\sum_{i=1}^{M} \left(\frac{\lambda_i d_i}{\mu}\right)^{1/2}\right]^2 \qquad \blacksquare \quad (5.47)$$

It can be shown that this function is concave with respect to the flows $\{\lambda_i\}$; moreover, it can be shown that any flow corresponding to a local minimum of the dual CFA problem is also a shortest-route flow and that the step size α of the FD algorithm is always equal to unity [GERL 73]. (These three properties are also true for *any* concave cost capacity function under the constraint $T \leq T_{MAX}$.) The expression for length corresponding to Eq. (5.45) for the dual problem is

$$l_i \triangleq \frac{\partial D}{\partial(\lambda_i/\mu)} = d_i \left(1 + \frac{\sum_{j=1}^{M} \sqrt{\lambda_j d_j}}{\gamma T_{MAX}\sqrt{\lambda_i d_i}} \right) \quad \blacksquare \quad (5.48)$$

where d_i is the slope of the cost capacity curve for the ith channel linearized about the current value of capacity. Equations (5.46), (5.47), and (5.48) correspond in the dual problem to Eqs. (5.26), (5.27), and (5.45) in the primal problem. Once again we notice that all lengths are non-negative and therefore negative loops cannot occur. In addition, we note that the effective length of a channel goes to infinity as the flow on that channel drops to zero as earlier. Here again we find local minima and therefore must randomize our search in order to find several of these minima.

Since the CFA algorithm eliminates certain channels as it iterates to a (local) minimum, we see that it has possibilities as an aid in the topological design of networks (i.e., the TCFA problem). This we study in the next section.

5.10. SOME TOPOLOGICAL CONSIDERATIONS—APPLICATIONS TO THE ARPANET

We have played around long enough with our three subproblems (CA, FA, CFA). Let us now face the true design problem for networks, the TCFA problem, which we consider in its dual form, namely, to minimize the cost D at a fixed maximum average delay T_{MAX}. As stated in Section 5.5, we are given the nodal locations (as, for example, in Figure 5.29 below) and an external source-to-destination flow requirement, γ_{jk}. In real networks, we are also immediately faced with (i.e., given) the difficult case of a set of *discrete* channel capacities, an example of which is given in Table 5.1 (these were the options available to the ARPANET). If we make no approximations to these discrete costs, then the direct solution to the TCFA problem involves an integer problem with $N(N-1)/2$ variables (i.e., the channel capacities, assuming full-duplex operation) as well as a multicommodity flow problem; this is far too complex for networks the

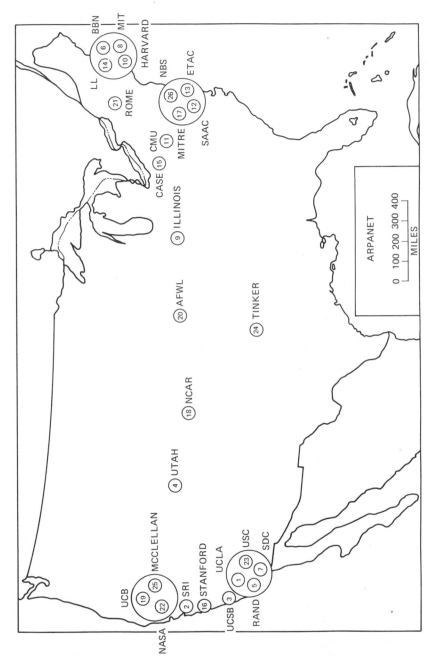

Figure 5.29 26 ARPA sites.

Table 5.1

Channel Capacities and Corresponding Costs Used in the Optimization

CAPACITY (KBPS)	TERMINATION COST ($/month)	LINE COST ($/month/mile)
9.6	650	0.40
19.2 $(2 \times 9.6)^a$	1300	0.80
19.2	850	2.50
50	850	5.00
100 $(2 \times 50)^a$	1700	10.00
230.4	1350	30.00

Note: The total cost per month of a channel is given by:
total cost = termination cost + (line cost) × (length in miles)
[a] Options obtained by using lower capacities in parallel.

size of the ARPANET! We will settle for approximate (heuristic) solutions which we have reason to believe come within 5–10% of the optimum at a greatly reduced computational burden (after all, since we seldom can predict the γ_{jk} very accurately in the first place, this compromise seems to be a good engineering decision).

Our (heuristic) solution to the TCFA problem is simply an iterative form of the CFA solution, and takes advantage of the fact that channels may be eliminated (and therefore we observe topological changes) as the CFA algorithm itself proceeds. This branch (channel) elimination occurs since, as we mentioned in the previous section [and proved in Exercise (5.15)], the cost D as expressed in Eq. (5.47) is concave with respect to the flows $\{\lambda_i\}$. For this reason, we refer to our iterative algorithm as the *concave branch elimination method* (CBE) [GERL 73]. The algorithm proceeds as follows:

CBE SUBOPTIMAL ALGORITHM

Step 1. Select an initial topology (this selection process is discussed below; the fully-connected net is often a good choice in practice).

Step 2. For each channel in the topology, make a power-law approximation* as in Eq. (5.37). For each iteration in Step 3 below, use a linearized value for capacity about the value of flow for that channel.

---

* An alternate choice is to use a piecewise linear approximation here.

Step 3.   Carry out the CFA algorithm. If, at any iteration, a connectivity constraint* is violated, then stop the optimization and proceed to Step 4; otherwise, let the CFA algorithm run to completion and then proceed to Step 4.

Step 4.   From the CFA suboptimal solution, discretize the continuous capacities obtained. For example, the continuous capacity may be "rounded" to the nearest allowable $(\lambda_i/\mu < C_i)$ discrete value such that we continue to satisfy $T \leq T_{MAX}$. This will tend to change the total investment D.

Step 5.   Conduct a final flow optimization by an application of the FD algorithm (and even perform some "tuning up" of capacities and flows if desired).

Step 6.   Repeat Steps 3–5 for a number of feasible random starting flows (by selecting random initial lengths with a shortest-route flow as discussed in Section 5.9).

Step 7.   Repeat Steps 1–6 for a number of initial topologies.

The number of repeats in Steps 6 and 7 depends upon how many dollars one is willing to spend on finding the solution. Experience with the ARPANET indicates that 20 to 30 repeats at Step 6 and a few (roughly 5) initial topologies (of which one is fully connected and the others are highly connected) at Step 7 yield good results.

Shortly, we shall apply the CBE method to the ARPANET. First let us discuss some other approaches to the TCFA problem. An excellent review of some previously studied techniques is given by Frank and Chou [FRAN 72b]. One class of techniques is known as the *branch X-change method*. Starting from an arbitrary feasible topology, a class of local transformations (the branch X-changes) is defined in which one branch is removed and some new branch is added, such that feasibility (e.g., 2-connectivity) is preserved. Then a (simple-minded) CFA problem is solved using a minimum fit procedure [FRAN 70]; if this results in an improvement, then the transformation is kept (otherwise it is rejected). This procedure is carried out until the set of local transformations has been exhausted. An extension of the branch X-change method is the *cut-saturation method* [GERL 74]. This method reduces the set of local transformations to those that are good candidates for improving the throughput-cost performance. It begins with a tree (or other low-connected topology) and identifies the critical cut-set resulting from an

---

* For example, as mentioned in Section 5.4, the ARPANET has a 2-connectivity constraint, i.e., the net must provide at least two independent paths (no IMPs or channels in common except at each end of the path where there must be an IMP in common) between every pair of IMPs.

FA solution. It then adds a channel across this cut (or increases the capacity of a channel in this cut). This is repeated until a feasible topology is obtained. The procedure then continues, but, in addition, at each step the least utilized channel in the network is removed (if feasible). After a given number of iterations, the algorithm terminates. The *key* to all of these heuristic algorithms is the availability of efficient methods for evaluating topologies along with the random generation of several starting topologies (to search for many local minima).

Let us now apply the CBE algorithm to solve the TCFA problem for one version of the ARPANET. In Figure 5.29, we show a collection of 26 nodes that we wish to connect into a network. We choose the cost rates d_i to be proportional to the channel lengths (since we see from Table 5.1 that this is the form of tariff used by the telephone company) and for each channel we make a power-law approximation to the capacity for use in Step 2. We will further assume that the traffic requirement is uniform such that $\gamma_{jk} = \gamma_0 = 1.0$ KBPS, resulting in $\gamma = 650$ KBPS.

We begin by discussing CBE solutions without the 2-connectivity constraint. The best CBE solution in this case depends upon the exponents (α) used in the power-law fit to the true capacity costs. A number of properties emerge when one studies the solutions obtained from many starting topologies; these may be summarized as follows. As α decreases, corresponding to a stronger economy of scale [see Eq. (5.37)], then topologies with fewer channels, each of which has large capacities, are superior. In fact, for $\alpha \leq 0.6$, tree topologies (25 branches) were always the best ones found. Furthermore, as α decreases, the variation in the number of channels decreases, whereas the variation in the cost of the networks increases, these statements applying to the set of local minima found in the search; this is because the number of local minima increases and the costs of such local minima are widely diversified, due to the rapidly changing objective function in this vicinity. Consequently, if one begins with a highly connected topology (say, more than 40 channels in this example), then whereas it is more likely to contain the optimal topology than would a low connected topology (26 to 29 channels), such a topology will contain a much larger number of local minima and "good" minima may never be found. However for α in the range $0.8 \leq \alpha \leq 1.0$, it appears that the highly connected topologies do offer the probability of containing good, if not optimal, solutions since the number of local minima are relatively small and the values of these minima are relatively close. Conversely, the low connected topologies force one to look at a smaller set of solutions, none of which may be near the optimum. However, for smaller values of α, the number of local minima is huge and their values change so quickly that the highly connected topologies will probably

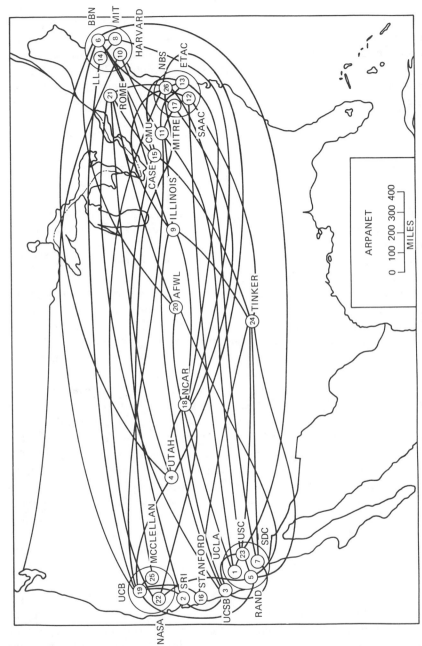

Figure 5.30   Best solution: monthly cost = $89,580, $\gamma_0 = 1.05$.

settle on poor local minima; it is perhaps better in this case to use carefully chosen low connected starting topologies since we expect the optimum topology itself to be of this type.

Thus we may conclude that for α in the range $0.8 \leq \alpha \leq 1.0$, the CBE method is effective and the choice of the initial (highly connected) topology is not all that critical; further, the exploration of a few local minima is probably sufficient. For moderate values of α ($0.5 \leq \alpha \leq 0.8$), the CBE method will only work well with carefully chosen (low con-nected) starting topologies and with a search over many local minima. For smaller values of α the CBE method is probably not very effective, and the methods such as the branch X-change or the cut-saturation are probably more effective.

If we now require the two-connectivity condition in the ARPANET, then, as stated in the algorithm, we may still use the CBE method in which we force the CFA algorithm to stop whenever we attempt to eliminate a branch that renders the network one-connected. Our statements above show that when we already have solutions with a fair degree of connectivity (as in the case when $0.8 \leq \alpha \leq 1.0$) then the two-connectivity constraint imposes few restrictions on us, and so the CBE method works well in these cases. The 26-node net shown in Figure 5.29 was studied with the two-connectivity requirement and with the constraint $T_{MAX} = 0.20$ sec. Approximately 30 initial topologies were studied, and for each, 30 local minima were found using the CBE method. Each of the 30 topologies consumed between 30 and 60 sec on an IBM 360/91 (i.e., 1–2 sec/minimum). In all cases, the capacity cost functions were fitted with power-law expressions and after the best solution was found for a given topology, the capacities were discretized as mentioned above; in this process of discretizing the capacities, the total cost usually increases and so does the total throughput. Surprisingly, the best discretized solution was obtained by starting with a fully-connected topology; it resulted in an unusual network of 61 channels, as shown in Figure 5.30! In Table 5.2 we show the way in which various discrete capacities were used as a function of channel length for this 61-channel solution. We note that the short lines used the higher capacities and that the longer lines used the smaller capacities, as might be expected. The cost of this optimum was $89,580 per month. Interestingly, the second best solution (shown in Figure 5.31) came in at a cost of $94,288 per month, and it was obtained from a rather low connected topology that contained 29 channels in its initial and final forms; it uses higher-speed lines (50 and 100 KBPS as marked in the figure) on the medium and long distances and 230 KBPS on the very short lines. It turns out that an important part of this procedure (in terms of its impact on cost and

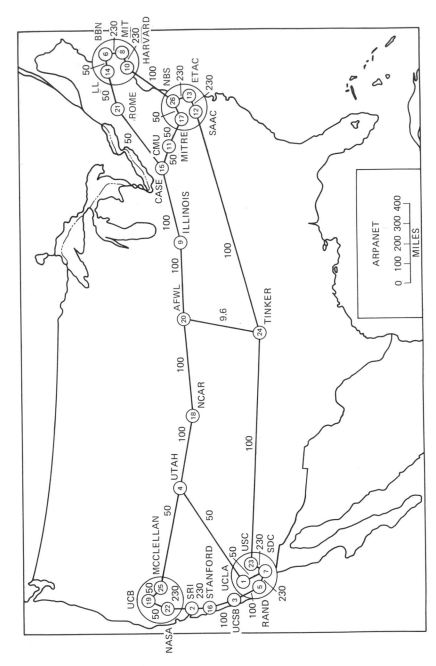

Figure 5.31 Second best solution: monthly cost = $\gamma_0 = 1.00$.

Table 5.2

Distribution of Capacities Versus Link Lengths for the 61-Channel Topology of Figure 5.30

CAPACITY (KBPS)	LINK LENGTH (miles)			
	< 100	100–500	500–1000	> 1000
9.6	0	3	8	20
19.2	1	8	2	6
50	11	2	0	0

Note: each entry represents the number of links having the specified capacity and lying within the specified length range.

performance) is the process of discretizing the continuous capacities, and this aspect of the problem has yet to be perfected.

If we use the continuous solutions as approximate lower bounds on performance, then we obtain the curve shown dashed in Figure 5.32. This figure displays what is perhaps the most meaningful information for network design, namely, the *cost-performance profile*. We show the per-node throughput $\gamma_0 = \gamma_{jk}$ (a uniform traffic requirement) as a function of

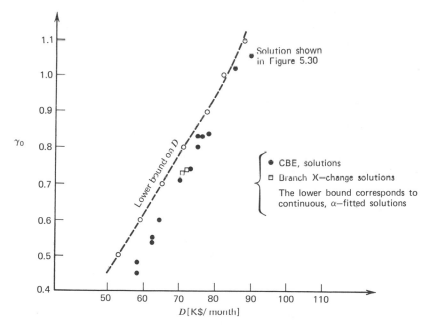

Figure 5.32 Cost-performance profile for computer network.

the monthly network cost D. The solid circles correspond to other CBE solutions that were best among those that we studied for different values of γ_0; these suboptimal solutions were all derived from fully-connected initial topologies, since, for the ARPANET, α is not far from unity, thereby presenting only a moderate economy of scale. (Also shown are the results of two branch X-change solutions.) We note that this range of solutions lies within 5–10% of the lower bound, yielding good discrete solutions over a rather broad range of throughputs and costs; these solutions were obtained with only a moderate amount of computation. The cost effectiveness of large shared systems is evident from this figure since the throughput is increasing faster than the network cost! Indeed, we see that the (γ_0, D) relationship is approximately linear, that is,

$$\gamma_0 \cong mD + b \qquad \blacksquare$$

where, from Figure 5.32 we have roughly $m = 0.02$, $b = -0.7$. Thus for a small γ_0, the cost per unit of throughput is larger (e.g., $\gamma_0 = 0.5$ gives $D/\gamma_0 \cong 120$) than for a scaled-up network (e.g., $\gamma_0 = 1.0$ gives $D/\gamma_0 \cong 87$). The slope m is an important system parameter since under the above assumption of linearity, $\lim_{\gamma_0 \to \infty} D/\gamma_0 = 1/m$; for the example above, the limiting cost is approximately \$50/(msg/sec)/node-pair/month. Thus we see that the CBE method is a fairly effective topological design tool if one is willing to come within 5–10% of the optimum.*

These methods represent the state-of-the-art in the subnet design of computer networks. In the following two sections, we discuss the use of radio as a communication medium for networks.

5.11. SATELLITE PACKET SWITCHING

Packet switching networks, such as the ARPANET, came into existence in response to the need for rapid, efficient, and economical data communications. As such networks grow in size and coverage, the need to provide inexpensive, long-haul, high-capacity communication channels becomes more pressing. The technology of satellite communications offers promise as a solution for such data transmission, and in this section we wish to discuss the use of packet switching over a broadband satellite channel. In the next section, we address ourselves to a problem at the opposite extreme, namely, that of providing inexpensive communications from the users' terminals into the high-level network itself; this will lead us into a discussion of ground radio packet switching.

---

* This and other heuristic topological design procedures including bounding techniques are discussed further in [GERL 73, 75].

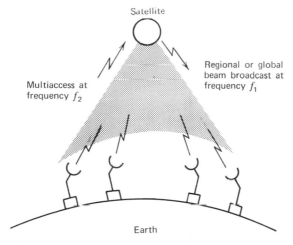

Figure 5.33   Packet switch in the sky.

There are a number of characteristics of a (stationary) earth satellite that are of importance in its use for packet switching (see Figure 5.33). First, and perhaps most striking, is the *long propagation delay* in a round-trip transmission (up and down) to a satellite transponder that is in a synchronous orbit roughly 36,000 km above the earth; this delay is approximately 0.24 sec if the satellite is directly overhead, and is roughly 0.27 sec if the satellite is near the horizon. Second, a single voice channel provides a *broadband* data communications capability (50 KBPS as currently used in the ARPANET with the INTELSAT IV satellite link from Hawaii to California). Third, the satellite transponder can retransmit back down to earth in a *broadcast* mode to all earth stations in its "shadow." Fourth, each transmitter can listen to his own transmission since he, too, is in the broadcast shadow; if errors due to random noise can be neglected (as we assume in this and the following section), then we have "perfect feedback" that gives us *automatic acknowledgements.* These are the principal technical considerations of interest to us in studying delay and throughput, but of course there are many other important features (which we do not discuss) such as vulnerability, cost, access to terrestrial communication channels, distance-independent operation, and so on.

The existence of the inherent quarter-second propagation delay suggests that we introduce access schemes that differ radically from land-based communications. In particular, packet switching permits us to take advantage of the long delay, the broadband, the broadcast, and the automatic acknowledgment of these channels. (The case of ground radio packet switching in the next section takes advantage of the fact that the propagation delays of interest are *small* compared to a packet transmission time.)

There are many ways to use a given satellite channel for data communications. For example, one could make permanent subchannel assignments to private users as in Figure 5.2; as we know, this can be very wasteful in a bursty user environment. Also, one could permit a dial-up procedure for sharing the set of subchannels as in Figure 5.3. Alternatively, one could provide the entire channel capacity to users on a demand basis (with some form of polling control) as in Figure 5.4; this requires buffering at the source. As another alternative, one could permit "random" access to the full capacity of the channel in a packet-switching mode; this too is in the spirit of the ARPANET philosophy, and we wish to study the performance of some such random access schemes.

Of interest, then, is the consideration of satellite channels for packet switching. As earlier, a packet is defined merely as a package of data that has been prepared by one user for transmission to some other user in the system. The satellite is characterized as a high-capacity channel with a fixed propagation delay that is large compared to the packet transmission time. We consider a transmission scheme wherein a given transmitter forms his packet and then *bursts* it out rapidly on the channel at full capacity. Many users operating in this fashion *automatically multiplex* their transmissions on a demand basis. The (stationary) satellite acts as a pure transponder repeating whatever it receives and beaming this transmission back down to earth*; this broadcasted transmission can be heard by every user of the system and in particular a user can listen to his own transmission on its way back down (this is a crucial property!). Since the satellite is merely transponding, then whenever a portion of one user's transmission reaches the satellite while another user's transmission is being transponded, the two collide and "destroy" each other. The problem we are then faced with is how to control the allocation of time at the satellite in a fashion that produces an acceptable level of performance.

The ideal situation would be for the users to agree collectively when each could transmit. The difficulty is that the means for communication available to these geographically distributed users is the satellite channel itself, and we are faced with attempting to control a channel that must carry its own control information. There are essentially three decentralized approaches to the solution of this packet-switching problem. The first has come to be known as a "pure ALOHA" system [ABRA 73a] in

---

* All users transmit packets on the same transmission frequency (i.e., they share the entire channel) and all receive on a second common frequency. A given receiver picks out packets addressed only to itself.

which users transmit any time they desire.* If, after one propagation delay, they hear their successful transmission, then they assume that no conflict occurred (i.e., they have a positive "acknowledgement"); otherwise, they know a collision (or perhaps some other source of noise) did occur and they must retransmit (i.e., they assume a negative "acknowledgement"). If all users retransmit immediately upon hearing a conflict, then they are sure to conflict again, and so some scheme must be devised for introducing a random retransmission delay to spread these conflicting packets over time.

The second method for using the satellite channel is to "slot" time into segments whose duration is exactly equal to the transmission time of a single packet (we assume constant length packets). If we now require all packets to begin their transmission only at the beginning of a slot (where time is always referenced to the satellite), then we enjoy a gain in efficiency, since collisions are now restricted to a single slot duration; such a scheme is referred to as a "slotted ALOHA" system.

The third method for using these channels is to attempt to schedule their use in some direct fashion; this introduces the notion of a reservation system in which time slots are reserved on a fixed or demand basis for specific users' transmissions.

Thus we are faced with a finite-capacity communication channel subject to unpredictable and conflicting demands. When these demands collide, we "lose" some of the effective capacity of the channel and we must characterize the effect of that conflict. Note that it is possible to use the channel up to its full rated capacity when only a single user is demanding service; this is true since a user will never conflict with himself (he has the capability to schedule his own use).

The remainder of this section is organized as follows. We begin with the calculation of *throughput* in pure and slotted ALOHA channels. Then we inquire about delay in these channels and develop the *throughput-delay* tradeoffs. Next, we consider the inherent instability of these channels and discuss the *stability-throughput-delay* tradeoffs. We also develop *control* methods for obtaining stability in slotted ALOHA channels. We then discuss the behavior of a satellite *reservation* channel. Last we compare slotted ALOHA with a fixed allocation scheme.

Let us now consider the transmission of a packet in a pure ALOHA system, as shown in Figure 5.34. If the packet transmission period is P sec, as shown, then this packet is vulnerable during a period of $2P$ sec

---

* In fact, the ALOHA system implemented at the University of Hawaii [ABRA 70] is a ground radio system using this access method; ground radio systems are discussed in the following section.

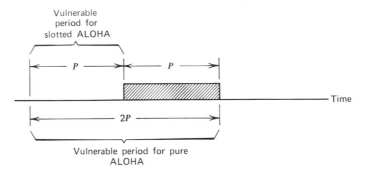

Figure 5.34   The vulnerable period for ALOHA.

in the sense that any other packet transmission generated and initiated in the vulnerable period will transmit, collide with, and destroy our transmitted packet. Let S denote the throughput of the satellite channel (average number of successful transmissions per transmission period P), and let G denote the average channel traffic (again measured in number of packet transmissions attempted per P sec). If we assume that the total traffic (G) entering the channel is an independent process generated by an *infinite* population of users (each generating channel traffic at an infinitesimally small rate summing to G), then we obviously have

$$S = Gp_0 \qquad \blacksquare$$

where $p_0 = P$ [no additional packets are generated during the vulnerable period]. If we assume that the channel traffic is Poisson, then we have $p_0 = e^{-2G}$, which gives

$$S = Ge^{-2G} \qquad \blacksquare \quad (5.49)$$

This equation was first obtained by Abramson [ABRA 70] for the case of a radio channel (of course, it applies equally well to satellite channels). It is easily seen that the maximum throughput one can obtain from a pure ALOHA channel occurs at a value $G = \frac{1}{2}$ and yields a (low) maximum efficiency of $1/2e \cong 0.184$.

As discussed, a variation to pure ALOHA is the slotted ALOHA system, in which it is assumed that time at the channel is slotted into segments exactly equal to a packet transmission time P, and it is further assumed that all users are synchronized to these slot intervals. Whenever a user generates a packet, he must delay his transmission so that it will synchronize exactly with the next possible channel slot. As a result, the vulnerable period for slotted ALOHA is only P sec long (see Figure 5.34); conflicts occur only when more than one user generates a request during the same slot period (and the conflict is total, rather than partial as in pure ALOHA). As a result, again assuming Poisson channel traffic, we have $p_0 = e^{-G}$, and so for slotted ALOHA we have

$$S = Ge^{-G} \qquad \blacksquare \qquad (5.50)$$

The notion of slotting was first introduced by Roberts [ROBE 72], and he showed the significant gain in efficiency that could be achieved by optimizing G; that is, when $G = 1$, we obtain the maximum efficiency for a slotted ALOHA system which is $1/e \cong 0.368$ (twice that of a pure ALOHA system).

Let us temporarily neglect delay (and stability) considerations and study only the permissible values of the throughput S and the channel traffic G. In Figure 5.35, we plot the S, G relationship for pure and slotted ALOHA [Eqs. (5.49) and (5.50)]. (Note that $G > 1$ is possible since more than one packet/slot may attempt a transmission.) These curves give the performance in the case of an infinite population of users whose collective rate of new packet generations is S. Let us now consider a *finite* population model [ABRA 73a] with a total of M independent

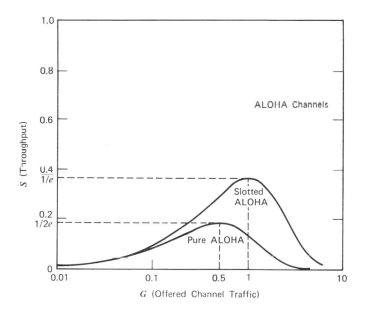

Figure 5.35   Throughput for pure and slotted ALOHA.

users in a slotted ALOHA system. We model each user's packet transmissions* as a sequence of independent Bernoulli trials, that is, let

$G_m = P$ [mth user transmits a packet in any given slot] $m = 1, 2, \ldots, M$

The average traffic (per slot) due to the mth user is therefore G_m. Thus the total average channel traffic is $G = \sum_{m=1}^{M} G_m$ packets per slot. Let

$S_m = P$ [mth user's packet is successful] $m = 1, 2, \ldots, M$

Clearly, the average throughput (per slot) due to the mth user is S_m and the average total throughput is $S = \sum_{m=1}^{M} S_m$. The fraction of the mth user's traffic that results in successful transmissions is merely the probability that no other user transmits when he does, that is,

$$S_m = G_m \prod_{i \neq m} (1 - G_i) \qquad m = 1, 2, \ldots, M \qquad (5.51)$$

This set of M equations has a solution set, $\{S_m\}$, which defines the allowable mixtures of source rates which this channel can, at most, support. We now consider a simple example.

Example 1

Let us assume that all users are statistically identical. Then $S_m = S/M$ and $G_m = G/M$. Equation (5.51) then gives

$$S = G\left(1 - \frac{G}{M}\right)^{M-1} \qquad \blacksquare \quad (5.52)$$

Further, for $M \to \infty$, we have $S = Ge^{-G}$, which corresponds to the case of an infinite population Poisson source as in Eq. (5.50). The expression in Eq. (5.52) rapidly decays to the Poisson throughput as M increases. It is noteworthy for $M = 1$ that $S = G$ with a maximum value of $S = 1$. We shall return to this later.

Returning now to Eq. (5.51), we define

$$g = \prod_{i=1}^{M} (1 - G_i) \qquad (5.53)$$

and we then have

$$S_m = \frac{G_m}{1 - G_m} g \qquad m = 1, 2, \ldots, M \qquad (5.54)$$

Although we are neglecting delay and stability questions, we certainly must invoke the average load constraint as always; this is, the channel has

* These may be old retransmissions or new packet transmissions.

a (finite) capacity and we must restrict our traffic so that the rate of new packet generation does not exceed that capacity (for example, in the pure ALOHA channel, $S \le 1/2e$). If we attempt to drive the channel beyond this maximum value, the channel throughput will lag behind the new packet generation. We will then be faced with the awful consequences of unbounded queues and delays, and of vanishingly small throughput (in fact, as we shall see later, this situation will occur in an ALOHA channel with an infinite input population with probability one at *any* value of $S > 0$ unless we adopt some form of control). This being the case, it is useful for us to determine the allowable values of throughput, and the boundaries of that region. Equation (5.54) defines the set of achievable throughputs $\{S_m\}$ in terms of the set of channel traffic values $\{G_m\}$. This equation defines a region in the M-dimensional space whose coordinates are the new packet generation rates $S_1, S_2, \ldots, S_M$. The boundary to this region which defines the maximum throughput for S_m when all other S_i ($i \ne m$) are held fixed (at allowable values) is found (for all m simultaneously) by setting the Jacobian (determinant) $J = J(S_1, S_2, \ldots, S_M; G_1, G_2, \ldots, G_M)$ equal to zero.* From Eq. (5.54) we see that

$$\frac{\partial S_i}{\partial G_i} = \begin{cases} \dfrac{g}{1 - G_i} & i = j \\[2ex] \dfrac{-gG_j}{(1 - G_j)(1 - G_i)} & i \ne j \end{cases} \tag{5.55}$$

Thus, after performing a bit of algebra, we find

$$J = g^{M-2} \begin{vmatrix} (1-G_1) & -G_1 & -G_1 & \cdots & -G_1 \\ -G_2 & (1-G_2) & -G_2 & \cdots & -G_2 \\ -G_3 & -G_3 & (1-G_3) & \cdots & -G_3 \\ \cdot & \cdot & \cdot & \cdot & \cdot \\ \cdot & \cdot & \cdot & \cdot & \cdot \\ \cdot & \cdot & \cdot & \cdot & \cdot \\ -G_M & -G_M & -G_M & \cdots & (1-G_M) \end{vmatrix}$$

$$= g^{M-2}(1 - G_1 - G_2 - \cdots - G_M) \tag{5.56}$$

Applying the condition $J = 0$ yields the following general condition on the set of channel traffic rates that achieves the set of maximum throughput values:

$$G = \sum_{m=1}^{M} G_i = 1 \qquad \blacksquare \tag{5.57}$$

* The i, j element of this determinant is $\partial S_j / \partial G_i$. This condition ($J = 0$) is discussed at length in [BEVE 70, Section 3.2].

This last equation gives us our throughput performance contours, which we now discuss through some examples.

Example 2

Let $M = 1$. We then have $S = G = 1$. This, of course, is the best we can ever do (one successful packet per slot time), and is explained by the fact that a user never (collides with and) destroys himself; that is, whereas we permit incest (see Chapter 6) in the ARPANET, we do not permit suicide!

Example 3

Let us assume that the M users form two groups such that m_1 of them have a rate S_1 and $m_2 = M - m_1$ of them have a rate S_2. Thus $S = m_1 S_1 + m_2 S_2$ and $G = m_1 G_1 + m_2 G_2$. The M equations (5.54) reduce to two equations:

$$S_1 = G_1(1 - G_1)^{m_1 - 1}(1 - G_2)^{m_2}$$
$$S_2 = G_2(1 - G_2)^{m_2 - 1}(1 - G_1)^{m_1} \tag{5.58}$$

which define the set of permitted rates $\{S_1, S_2\}$ as a function of the traffic rates $\{G_1, G_2\}$. If we now impose the optimization condition $G = 1$, we may then solve for S_1 and S_2 at any given value of G_1 in the range $0 \le G_1 \le 1/m_1$ [note that $G_2 = (1 - m_1 G_1)/m_2$]. The results of this calculation are shown in Figures 5.36 and 5.37 for various (m_1, m_2) pairs [ABRA 73a]. Due to the grouping, we are able to reduce the M-dimensional contour to two dimensions, namely, $m_1 S_1$ and $m_2 S_2$. Here, again, we observe throughputs S that exceed $1/e$. In fact, when $S_1 = 0$ and $S = MS_2$, then the optimum contour takes on the value $S = (1 - 1/M)^{M-1}$ as in Example 1 (where we first observed the case $S > 1/e$). For $m_1 = m_2 = 1$, we have

$$S_2 = 1 - 2\sqrt{S_1} + S_1 \tag{5.59}$$

Also, for $m_2 = \infty$, we obtain

$$S_1 = G_1(1 - G_1)^{m_1 - 1}e^{-(1 - m_1 G_1)}$$
$$m_2 S_2 = (1 - m_1 G_1)(1 - G_1)^{m_1}e^{-(1 - m_1 G_1)} \tag{5.60}$$

Lastly, for $m_1 = m_2 = \infty$, we obtain

$$\lim_{m_1, m_2 \to \infty} (m_1 S_1 + m_2 S_2) = \frac{1}{e} \tag{5.61}$$

where, recall, $S = m_1 S_1 + m_2 S_2$.

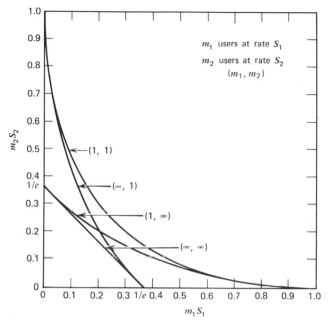

Figure 5.36 Allowable source rates for slotted ALOHA.

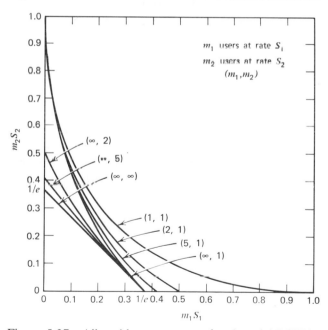

Figure 5.37 Allowable source rates for slotted ALOHA.

The only way a slotted ALOHA channel can achieve $S = 1$ is for the case of exactly one user active in the system! We note a gain in throughput as the source rates become unbalanced; this is discussed in more detail next in our delay model.

The throughput limits discussed above pay no regard to the delay that packets experience. We now focus upon that issue in slotted ALOHA systems [KLEI 73, LAM 74]. We begin with the model in which we assume an infinite number of users* who collectively form a source that generates packets independent of conditions on the channel. This source generates V packets per slot from the distribution $v_k = P[V = k]$ with a mean of S packets per slot.†

Again, we assume that each packet is of constant length requiring P sec for transmission; in the numerical studies presented below we assume that the capacity of the channel is 50 KBPS and that the packets are each 1125 bits in length, yielding $P = 22.5$ msec. Note that $S' = S/P$ is the average number of packets arriving per second from the source. Let d be the maximum round-trip propagation delay that we assume each user experiences and let $R = d/P$ be the number of slots that can fit into one round-trip propagation time; for our numerical results we assume $d = 270$ msec and so $R = 12$ slots. R slots after a transmission, a user will either hear that he was successful or know that he was destroyed. In the latter case, if he now retransmits during the next slot interval and if all other users behave likewise, then *for sure they will collide again*; consequently, we shall assume that each user transmits a previously collided packet at random during one of the next K slots (each slot being chosen with probability $1/K$). Thus retransmission will take place either $R + 1$, $R + 2, \ldots$, or $R + K$ slots after the initial transmission. As a result, traffic introduced to the channel from our collection of users will now consist of new packets and previously blocked packets, the total number adding up to L packets transmitted per slot where $p_k = P[L = k]$ with a mean traffic of G packets per slot. We assume that each user in the infinite population will have *at most one packet* requiring transmission at any time (including any previously blocked packets). Of interest to us is a description of the maximum throughput rate S and the average packet delay T as a function of the channel traffic G. It is clear that S/G is merely the probability of a successful transmission and G/S is the average number of times a packet must be transmitted until success. From Eq. (5.50) we already know that the peak throughput is equal to $1/e$, which

---

* These will be referred to as the "small" users.

† Below we shall be forced to distinguish between the throughput rate and the new packet generation rate. For now, we assume they are equal.

occurs when $G = 1$. However, this throughput yields an infinite value for T since $G = 1$ corresponds to $\rho = 1$. Let us assume that the channel traffic is an independent process (simulation experiments show that this is an excellent assumption [LAM 74]). We define $P(z) = \sum_k p_k z^k$ and $V(z) = \sum_k v_k z^k$ as the generating function for the number of packets transmitted per slot and the number of packets generated per slot, respectively. Then, as shown in Exercise 5.22 we have [LAM 74]

$$P(z) = \left[\frac{p_1}{K}(1-z) + P\left(1 - \frac{1-z}{K}\right) \right]^K V(z) \tag{5.62}$$

If, further, the source is an independent process and is Poisson distributed, then $V(z) = e^{-S(1-z)}$ [see Eq. (1.52)], and then we see immediately that,

$$\lim_{K \to \infty} P(z) = e^{-G(1-z)}$$

This shows that our assumption of Poisson channel traffic is quite reasonable.

We have so far defined the following critical system parameters: S, G, K, T, and R. In the ensuing analysis, we shall distinguish packets transmitting in a given slot as being either newly generated or ones that have in the past collided with other packets. This leads to an approximation since we do not distinguish how many times a packet has met with a collision. Simulation shows that the correlation of traffic in different slots is negligible, except at shifts of $R+1$, $R+2$, ..., $R+K$; this exactly supports our approximation since we concern ourselves with the most recent collision. We require the following two additional definitions:

$q = P$ [newly generated packet is successfully transmitted]

$q_t = P$ [previously blocked packet is successfully transmitted]

Our principal concern is to investigate the tradeoff between the average delay T and the throughput S, where

T = average time (in slots) until a packet is successfully received

Our point of departure is Eq. (5.50). We must alter this equation to account for the effect of the random retransmission delay parameter K. The result of this is (see Exercise 5.23),

$$S = G \frac{q_t}{q_t + 1 - q} \tag{5.63}$$

where

$$q = \left[e^{-G/K} + \frac{G}{K} e^{-G} \right]^K e^{-S} \tag{5.64}$$

and

$$q_t = \left[\frac{e^{-G/K} - e^{-G}}{1 - e^{-G}} \right] \left[e^{-G/K} + \frac{G}{K} e^{-G} \right]^{K-1} e^{-S} \tag{5.65}$$

The considerations which led to Eq. (5.63) were inspired by Roberts who developed an approximation for Eq. (5.65) of the form

$$q_t \cong \frac{K-1}{K} e^{-G} \tag{5.66}$$

This is a reasonably good approximation. Equations (5.63)–(5.65) form a set of nonlinear simultaneous equations for S, q, and q_t that must be solved to obtain an explicit expression for S in terms of the system parameters G and K. In general, this cannot be accomplished. However, we note that as K approaches infinity, these three equations reduce simply to

$$\lim_{K \to \infty} \frac{S}{G} = \lim_{K \to \infty} q = \lim_{K \to \infty} q_t = e^{-G} \tag{5.67}$$

Thus we see that Eq. (5.50) is the correct expression for the throughput S only when K approaches infinity, which corresponds to the case of infinite average delay. Note that the large K case avoids the large delay problem only if P is small (very high-speed channels).

The numerical solution to Eqs. (5.63)–(5.65) is given in Figure 5.38 where we plot the throughput S as a function of the channel traffic G for various values of K. We note that the maximum throughput at any given K occurs when $G = 1$. The throughput improves as K increases, finally yielding a maximum value of $S = 1/e \cong 0.368$ for $G = 1$, $K = \infty$. Thus we have the unfortunate situation that the ultimate capacity of this channel supporting a large number of small users is less than 37% of its theoretical maximum (of 1). We note that the efficiency rapidly approaches this limiting value (of $1/e$) as K increases and that for $K = 15$ we are almost there. The figure also shows some delay contours that we discuss below.

The average packet delay is given by (see Exercise 5.25)

$$T = R + 1 + \frac{1-q}{q_t} \left[R + 1 + \frac{K-1}{2} \right] \qquad \blacksquare \tag{5.68}$$

If we use Little's result along with Eqs. (5.63) and (5.68), we get $\bar{N}$, the average number of packets in the system:

$$\bar{N} = ST = G \left[R + 1 + \frac{K-1}{2} \right] - S \left[\frac{K-1}{2} \right] \tag{5.69}$$

In Figure 5.38 we have also plotted the loci of constant delay in the (S, G) plane. Note the way these loci bend over sharply as K increases,

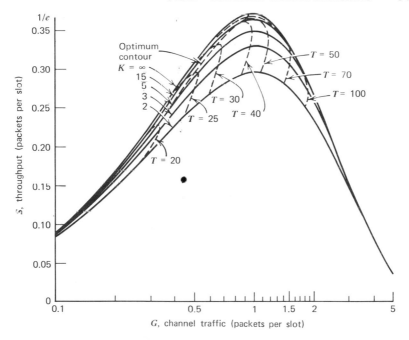

Figure 5.38   Throughput as a function of channel traffic.

defining a maximum throughput $S_{max}(T)$ for any given value of T; observe the cost in throughput if we wish to limit the average delay. This effect is clearly seen in Figure 5.39, which is the fundamental display of the tradeoff between delay and throughput; this figure shows the delay-throughput contours for constant values of K. We also give the minimum envelope of these contours, which defines the optimum performance curve for this system (a similar optimum curve is shown in Figure 5.38). Note how sharply the delay increases near the maximum throughput $S = 1/e$; it is clear that an extreme price in delay must be paid if one wishes to push the channel throughput much above 0.360 and the incremental gain in throughput here is infinitesimal. Note that, as S approaches zero, T approaches $R+1$. Also shown are the constant G contours. Thus this figure and Figure 5.38 are two alternate ways of displaying the relationship among the four critical system quantities S, G, K, and T.

We now wish to study the throughput-delay tradeoff for a special case of the finite population model described earlier. In particular, we let $m_1 = \infty$ and $m_2 = 1$. This gives us an infinite population of small or "background" users who behave as in Figures 5.38 and 5.39; we denote

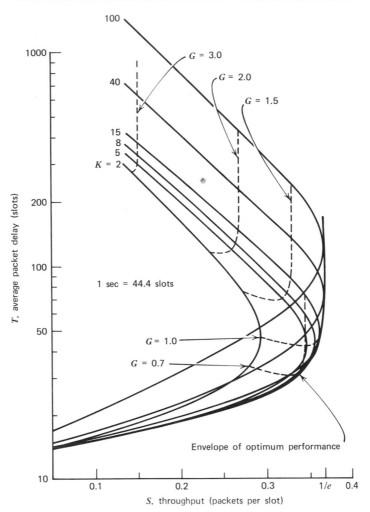

Figure 5.39   Delay-throughput tradeoff.

their total throughput by $S_\infty = \lim_{m_1 \to \infty} m_1 S_1$ and their total channel traffic by $G_\infty = \lim_{m_1 \to \infty} m_1 G_1$. Also, we have one *large* user with throughput S_2 and channel traffic G_2. This large user generates packets according to an independent Poisson process (at rate S_2 packets per P sec) that he may place in a queue to enable him to schedule transmission of these packets without mutual interference among his packets. (The details regarding how he schedules his transmissions to share the channel with the small users are given in [KLEI 73, LAM 74].) Of course, we have the total throughput $S = S_\infty + S_2$ and the total channel traffic $G = G_\infty + G_2$. As earlier, we know at a constant value of S_∞, that S_2 will be maximized when $G = G_\infty + G_2 = 1$. Expressions similar to those in Eqs. (5.63)–(5.68) for throughput and delay have been developed for the small and large users. The optimum throughput-delay performance is given in Figure 5.40, in which we display a family of optimum delay curves for various values of S_∞. The delay T is the weighted average of the background user's delay (T_∞) and the large user's delay T_2; that is, $T = (S_\infty T_\infty + S_2 T_2)/S$. On any constant S_∞ curve, we have $S_2 = S - S_\infty$, and for any S_∞ we define the maximum total throughput as S_{max}. By restricting S_∞ to be less than $1/e$, we are able to achieve an increase in total throughput S by introducing the additional large-user traffic. This additional capacity increases slowly as S_∞ drops; however, when S_∞ falls below 0.1, we begin to pick up significant gains for S_2. Also observe that each of the constant curves "peels off" from a common curve at a value of $S = S_\infty$. The collection of peel-off points generates the optimum performance envelope that had been given in Figure 5.39. At $S_\infty = 0$, we have only the large user operating with no collisions; this reduces to the classical queueing system with Poisson input and constant service time (M/D/1) and represents the *absolute optimum performance* contour for any method of using the satellite channel when the input is Poisson (for other input distributions,

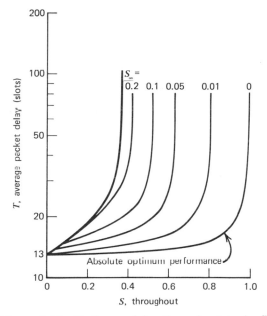

Figure 5.40   Optimum delay-throughput tradeoffs.

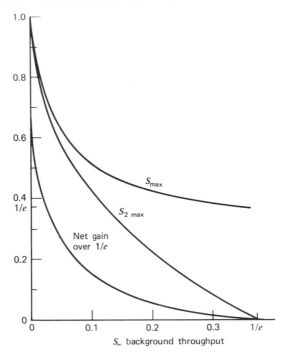

Figure 5.41   Throughput contours.

we may use our G/D/1 queueing results to calculate this absolute optimum performance contour).

In Figure 5.41, we show the throughput tradeoffs between the background and large users. The upper curve shows the absolute maximum S
at each value of S_∞; this is a clear display of the significant gain in S_2 that
we can achieve if we are willing to reduce the background throughput.
The middle curve [also shown in Figure 5.36 as the $(\infty, 1)$ curve] shows
the absolute maximum value for S_2 at each value of S_∞. The lowest curve
shows the net gain in system capacity (S) as S_∞ is reduced from its maximum
possible value of $1/e$.

So far, we have obtained steady-state performance results for throughput and delay under the assumption of equilibrium conditions. Often this
assumption is not satisfied, in which case the aforementioned performance applies only for a (possibly small) interval of time. We now investigate the effect of this phenomenon [KLEI 74a, LAM 74]. The slotted
ALOHA channel is now assumed to support a large (but finite) number
(M) of active terminals. Each terminal has buffer space for exactly one
(fixed length) message packet. Only when a terminal's buffer is empty

may a new message packet be generated (by the terminal's external source) and this will occur with probability σ in a slot. We let $N(t)$ be a random variable representing the total number of nonempty (i.e., busy) terminals (and this number will be referred to as the channel backlog) and let $S(t)$ be the combined input rate of packets into all terminals at time t. The vector $[N(t), S(t)]$ is defined to be the *channel state vector* or *channel load* at time t. For $N(t) = n$, $[N(t), S(t)] = [n, (M-n)\sigma]$, giving rise to what we call the linear feedback model. Note that the channel input rate $S(t)$ decreases linearly as $N(t)$ increases due to the finite population. Packets that collide are retransmitted (after a round-trip propagation delay of R slots) during one of the next K slots, each such being chosen at random with probability $1/K$. Thus retransmission takes place on the average $R + (K+1)/2$ slots after the previous transmission. This re-transmission scheme is difficult to analyze, and so we assume the simpler scheme* whereby every backlogged packet independently retransmits with probability p. This is an excellent approximation for the slotted satellite channel above as shown by simulation for moderate-to-large K when we choose

$$p = \frac{1}{R + (K+1)/2} \tag{5.70}$$

As above, we assume $R = 12$ and we will express our numerical results in terms of K [using Eq. (5.70)] rather than p.

We define S_{out} to be the delivered output (throughput)† rate of the channel, which is the probability of exactly one (successful) packet transmission in a channel slot. For the model above, if $[N(t), S(t)] = [n, (M-n)\sigma]$ then

$$S_{\text{out}} = (1-p)^n (M-n)\sigma(1-\sigma)^{M-n-1}$$
$$+ np(1-p)^{n-1}(1-\sigma)^{M-n} \tag{5.71}$$

In the limit when $M \to \infty$ and $\sigma \to 0$ (such that $M\sigma = S < \infty$) we have the *infinite population model* in which new packets are generated for transmission over the channel at the constant Poisson rate S. In this case, Eq. (5.71) reduces to

$$S_{\text{out}} = (1-p)^n S e^{-S} + np(1-p)^{n-1} e^{-S} \tag{5.72}$$

This expression is very accurate even for finite M if $\sigma \ll 1$, and if we replace $S = M\sigma$ by $S = (M-n)\sigma$.

---

* Such a scheme is possible also for ground radio packet switching systems; see Section 5.12.
† For the first time, we distinguish between the input rate S and the throughput rate S_{out} (which may be more or less than S).

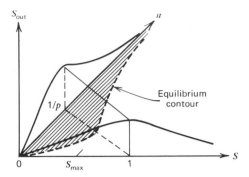

Figure 5.42   Channel throughput rate as a function of load and backlog.

In Figure 5.42, for a fixed K, we show the qualitative behavior of S_{out} as a function of the channel load $[n, S]$ as expressed in Eq. (5.72). Note that there is an equilibrium contour in the (n, S) "phase plane" on which the channel input rate S is equal to the channel throughput rate S_{out}. In the shaded region enclosed by the equilibrium contour, S_{out} exceeds S; elsewhere $S > S_{out}$ (the system capacity is exceeded!). The area of the shaded region

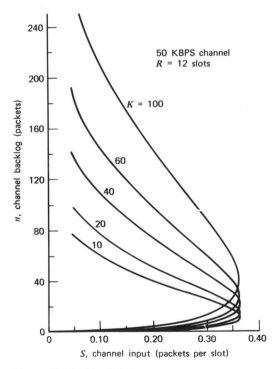

Figure 5.43   Equilibrium contours.

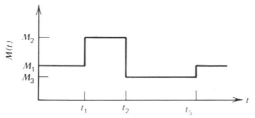

Figure 5.44   Time-varying source rates.

may be increased by increasing K as shown in Figure 5.43 where a family of equilibrium contours are displayed; these contours are similar to the constant-K loci of Figure 5.39.

Consider the example in which $M = M(t)$ as shown in Figure 5.44. We will use the fluid approximation for the trajectory of the channel state vector $[N(t), S(t)]$ on the (n, S) plane as sketched in Figure 5.45; we show

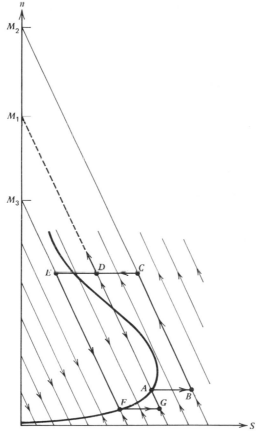

Figure 5.45   Fluid approximation trajectories.

two possible cases corresponding to Figure 5.44 for different values of M_3. (The arrows indicate the directions of "fluid" flow except along the horizontal portions of the trajectory. We note that this fluid approximation claims that the trajectory must lie on the line $n = M - (S/\sigma)$ whenever $M(t) = M$; the motion along this line will be up if $S > S_{out}$ and down if $S < S_{out}$.) The first case represents the trajectory ABCDEFGA that returns to the original equilibrium point A on the equilibrium contour despite the input pulse. The second case shows the unhappy situation in which the decrease in channel input at time t_2 is not sufficient to bring the trajectory back into the "safe" region. Here we show the situation where $M_3 = M_1$ and the trajectory goes from A at times prior to t_1 to B at t_1 to C by t_2 and to D at t_2. After t_2, the trajectory rises along the dashed line until it intersects the equilibrium contour (this intersection is not shown); eventually, the channel will be paralyzed as a result of an increasing backlog and a small channel throughput rate.

We have demonstrated channel "instability" due to a time-varying input. Now let us study the conditions under which the channel with a *stationary* input is "unstable." (When both M and σ are constant in time, we have a "stationary" input.) For this case, we define the *channel load line* in the (n, S) plane as the line $n = M - (S/\sigma)$. *A channel is said to be "stable" when its load line intersects (nontangentially) the equilibrium contour in exactly one place.* In Figure 5.46(a) we show an example of a *stable* channel and its *operating point.* If M is finite, a "stable" channel can always be achieved by using a sufficiently large K (see Figure 5.43). Of course, a large K implies large average packet delays.

In Figure 5.46(b) we show an example of an *unstable* channel. (Note that a load line which misses or is only tangential to the equilibrium contour is also unstable by our definition.) The point (n_0, S_0) is the desired

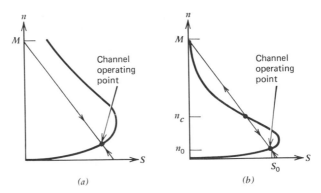

Figure 5.46   Channel stability. (a) A stable channel. (b) An unstable channel.

operating point since it yields the largest channel throughput and the smallest packet delay. The channel, however, cannot maintain equilibrium at this operating point indefinitely since $N(t)$ is a random process; that is, with probability one, the backlog $N(t)$ crosses the "critical" value n_c in a finite time and as soon as it does, S exceeds S_{out}. Under this condition, although there is a small probability that $N(t)$ may return below n_c, all our simulations show that the channel state vector $[N(t), S(t)]$ accelerates up the channel load line producing an increasing backlog and a vanishing throughput rate. In this state, the channel is disabled and external intervention is necessary to restore proper channel operation (see below). A number of studies have shown that the uncontrolled infinite population model is always unstable [FAYO 74, FERG 75, LAM 74].

From the above discussion and referring to Figure 5.46(b), we divide the channel load line into two regions: the "safe" region consisting of the channel states $\{[n, S] : n \le n_c\}$ and the "unsafe" region consisting of the channel states $\{[n, S] : n > n_c\}$. A good "stability" measure (for these unstable channels!) is the average time to exit into the unsafe region starting from a safe channel state. To be exact, we define FET to be the average *first exit time* into the unsafe region starting from an initially empty (zero backlog) channel. The FET will be used as our measure of channel stability. Its derivation and an efficient computational procedure for calculating its value are given in [LAM 74].

As earlier, our numerical computations assume a 50 KBPS satellite channel with 1125 bit packets and a round-trip propagation delay of 0.27 sec (giving $R = 12$ and giving 44.4 slots per second).

In Figure 5.47 we have shown the FET as a function of K for the infinite population model and for fixed values of the channel throughput rate S_0 at the channel operating point (if it remained there). The infinite population model results give us the worst case estimates, as shown in Figure 5.48 where we display the FET as a function of M for $K = 10$ and four values of S_0. The channel FET increases as M decreases and there is a critical M below which the channel is always stable in the sense of Figure 5.46(a). In Figure 5.47, we see that the channel stability (FET value) can be improved either by decreasing the channel throughput rate or by increasing K (which in turn increases the average packet delay). For example, if we limit S_0 below 0.25 and use $K = 10$, then the channel enters the unsafe region at most only once every two days on the average.

In Figure 5.49, we show two sets of throughput-delay performance curves with guaranteed FET values. (For comparison, we have also shown as a lower bound, the optimum performance curve that was displayed in Figure 5.39 without regard to channel stability.) The first set consists of three solid curves corresponding to an infinite population model with

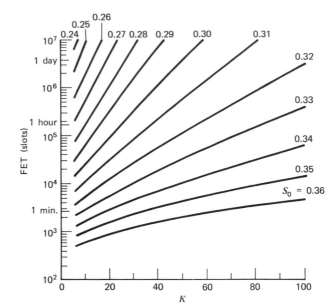

Figure 5.47   FET values for the infinite population model.

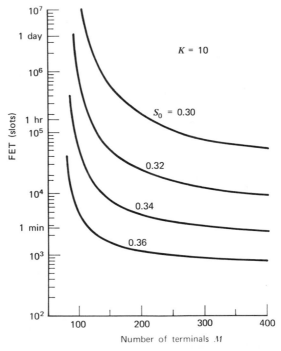

Figure 5.48   FET as a function of M for the finite population model.

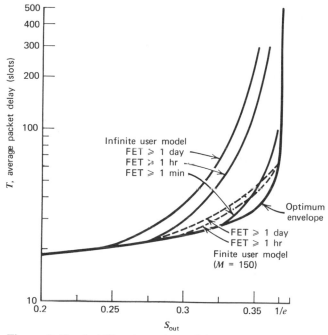

Figure 5.49   Stability-throughput-delay tradeoff.

channel FET ≥ 1 day, 1 hr, or 1 min. The second set consists of two dashed curves corresponding to a finite population of $M = 150$ terminals with channel FET ≥ 1 day or 1 hr. This figure displays the tradeoff among channel stability, throughput, and delay.

Thus we see that for an unstable channel, any throughput-delay performance results obtained under steady state assumptions (as given earlier) will be achievable *only for finite periods of time.*

So far, we have studied the ultimate throughput contours (which neglected the delay characteristics), the delay-throughput tradeoff (which neglected stability), and the basic instability of the uncontrolled satellite packet-switching channel. It remains for us to examine control methods for stabilizing the channel behavior; this we do below.

We wish to achieve optimum delay-throughput performance including a control procedure that guarantees channel stability. We shall find this control procedure under the assumptions (to be relaxed later) of zero propagation delay $(R \rightarrow 0)$ and of terminals that know the channel state [that is, the backlog $N(t)$] exactly. We deal with the linear feedback model (finite population of size M) such that the channel input rate $S(t) = [M - N(t)]\sigma$. Further, we assume that each backlogged packet retransmits in any given slot with probability p. We consider only those control policies [LAM 74] that permit each of the backlogged packets to be retransmitted independently with one of two probabilities: $p = p_0$ or $p = p_1$ ($p_0 > p_1$). The control policy will decide between p_0 or p_1 (the same value for all terminals at any instant) based only on the current backlog $N(t)$. The selection of p_0 and p_1 is really equivalent to the selection of retransmission delay parameters K_0 and K_1, respectively, since from Eq. (5.70) we assume $p = \{R + (K+1)/2\}^{-1}$. Thus we exercise control of the channel by increasing the average retransmission delay when the backlog gets large, and decreasing it when the backlog gets small. The problem, then, is to determine the value of p to use for each value of $N(t)$; we wish to find the optimal decision rule such that we either maximize throughput S_{out} under some constraint on average delay T, or we minimize T subject to some constraint on S_{out}. Fortunately, it can be shown [LAM 74, 75] that a stationary control policy always exists that simultaneously minimizes T and maximizes S_{out} over the class of all policies under consideration. This policy can be found by applying Howard's policy-iteration method [HOWA 60] and the computational cost of applying this method can be reduced considerably by the use of clever algorithms. This procedure, in all of the cases considered [LAM 74], seems to result in a stationary policy of the control limit type. That is, a critical value of backlog, n^*, exists such that

$$p = \begin{cases} p_0 & \text{when } N(t) < n^* \\ p_1 & \text{when } N(t) \geq n^* \end{cases} \tag{5.73}$$

The performance of the channel under this stabilizing control limit procedure is shown in Figure 5.50 where we show the analytic prediction (solid) and simulation (circles). Here we see the simultaneous minimization of T and maximization of S_{out} at the critical value of the control limit. More striking, perhaps, is the lovely insensitivity (for reasonable loads) of the channel performance with respect to variations of the control limit value! Another way to display these results is as in Figure 5.51. Here we see the optimum delay-throughput performance curves. These curves move out to the right as M decreases. Each point on these optimum curves corresponds to an optimum n^*, and, as shown, any increase or decrease from n^* results in degraded performance in both T and S_{out}; this shows the simultaneous optimization of T and S_{out} as discussed above. Also shown is the optimum lower bound for the infinite population case ($M = \infty$).

We have assumed that each terminal knows the channel state $N(t)$ exactly. Of course, this is unrealistic. However, it is possible to estimate $N(t)$ by observing the number of idle slots over an appropriate time interval into the past (on the order of one round-trip propagation delay).

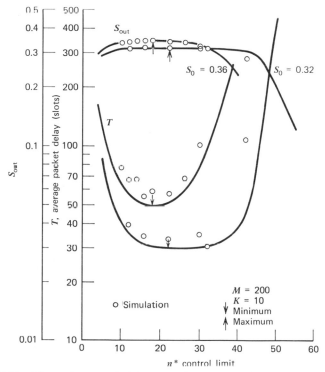

Figure 5.50   Throughput and delay as a function of the control limit value.

Simulation results show that algorithms based on such estimates operate rather well due to the flatness of the performance curves in Figure 5.50 (that is, they achieve near-optimum delay and throughput while guaranteeing stability) [LAM 74].

Let us now consider the third packet-switched channel access method— a *dynamic reservation system*. Thus far, we have studied the behavior of ALOHA channels that permit open (and largely uncontrolled) competition for time on the channel. As a result, collisions and retransmissions were common. Of course, there exist other means for sharing (multiplexing) the channel, and we wish to close this section with a brief discussion of such methods. We draw this material mainly from [ROBE 73b].

Currently, the typical allocation of a satellite channel is on a pair-by-pair basis, using Frequency Division Multiplexing (FDM) with roughly a 2 KHz minimum guard band per channel; this is clearly a "private system" allocation. These guard bands can be eliminated if (Synchronous) Time Division Multiplexing (STDM) is used; this too is a private

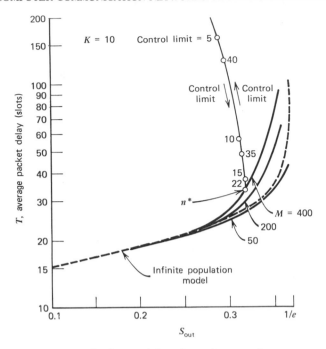

Figure 5.51   Optimum delay-throughput performance.

system. We stated in Section 5.1 that large shared systems provide many advantages, principal among these being the more efficient use of capacity as compared to private, low duty-cycle, bursty behavior. On the other hand, an STDM system* will be highly efficient if each derived channel is permitted to buffer data in order to provide a smooth, high duty-cycle supply of traffic to the channel; this is possible, for example, if each derived channel is handling a merged traffic stream already (as in the ARPANET). The point is that these a priori reservations of time slots guarantee no collisions! A similar effect is possible if we permit *dynamic* slot assignment also, and it is such a reservation system we wish now to discuss.

The dynamic (slotted) reservation system we have in mind contains both reserved and unreserved (ALOHA) slots interleaved with each other (see below). The (smaller) ALOHA slots are used by stations to place their reservations for as many reserved slots as they need (up to some maximum). These reservation requests may collide with others, as always with ALOHA. However, if a request is successful, then each station adds

_____
* Also known as Time Division Multiple Access (TDMA).

to a count, J, the number of new slots that have just been successfully reserved. There is, in effect, a single "queue in the sky" of reserved slots whose total number (J) is known to all users. The particular location on this queue that belongs to a given user need be known only by him; others need only know the total queue length so that they will properly select future slots for their own reservations.* Since it is critical that users correctly receive reservation requests, a good strategy is to transmit each such request three times;† this gives excellent noise immunity. Whenever no slots are reserved, all slots become ALOHA slots; however, when the first request is successful, r slots (each of duration P sec) are used for reserved slots followed by one P-sec interval in which V (triplicated) ALOHA slots may fit ($V \cong 6$). This pattern repeats as long as some slots are reserved. A good value for r is $r = AV\bar{m}/2e$ where $A = (K-1)/K$ [see Eq. (5.66)], $K \cong 2.3\sqrt{R}$, $\bar{m} = E$[packets per message] ($\bar{m} \cong 4.5$); this gives $r = 5$ (rounded up to the nearest integer).‡ The value of r can also be adjusted dynamically. Using these numbers, and a model for delay, it is found [ROBE 73b] that TDMA, reservation, and slotted ALOHA compare as shown in Figure 5.52. This figure assumes that half of the newly (Poisson-) generated messages are single packets and half are multipackets of eight packets each. Note that for small throughput, slotted ALOHA is superior. However, for increased throughput, the reservation method is superior, and, under heavily loaded channels ($S \to 1$), both TDMA and the reservation method are comparable. The reader is referred to [ROBE 73b] for further comparisons among these methods.§

The studies and experiments described in this section have led to the development of a new IMP—the satellite IMP (SIMP)—for use in the ARPANET. The SIMP is an IMP with additions and modifications to the software and hardware [BUTT 74]. For example, 32 additional buffers are provided for satellite packet storage. The hardware permits accurate timing and detection information for purposes of slotting. The SIMP provides for random retransmission delay by use of the p parameter [see Eq. (5.70)]. Routing updates, routing procedures, and acknowledgements have been adapted to the peculiarities of the satellite channel. The

---

* That is, from a given user's point of view, only two kinds of reserved slots exist: those that are his, and those that belong to others in which he is not permitted to transmit.
† Even with this triplication, many requests can fit into the P-sec slots since requests are much smaller than packets [ROBE 73b].
‡ Roberts uses 1350-bit packets, giving $R = 10$.
§ Another form of dynamic reservation assignment is reported by Binder [BIND 75a], who superimposes a dynamic RR assignment over a TDMA system.

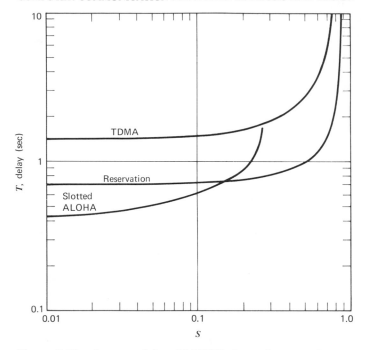

Figure 5.52   Average delay, 50 KBPS channel, ten stations.

predicted performance of the SIMP is upper-bounded as follows: $(7C/M) + C \leq 1.2$ MBPS (megabits per second) where C is the effective channel capacity and M is the number of SIMPs sharing the channel. An experimental SIMP network was initiated in the Summer of 1975.

It is worthwhile at this point to compare these random multiaccess resource-sharing schemes with fixed allocation from yet another point of view [KLEI 75a]. For this purpose, we shall compare slotted ALOHA to Frequency Division Multiple Access (FDMA).* The fixed channel assignment in FDMA is effective in preventing channel collisions but succeeds in this at the expense of possibly poor utilization of each channel since the smoothing effect of a large population is absent. To analyze FDMA, we adopt the following assumptions: (a) an assumed finite (but large) population of M users; (b) each user generates a new fixed length packet (of b bits) according to a Poisson process at a rate λ per second; (c) the total channel has a bandwidth of H hertz modulated at 1 bit/sec/hertz (giving a channel capacity of H bits/sec). Thus, with M users in this FDMA mode, each is assigned a channel of H/M bits/sec. Each such channel behaves as

* We let $R = 0$ for purposes of this comparison.

an M/D/1 queueing system giving an average time in system T (waiting plus transmission) from Eq. (1.82)

$$T = \frac{(\rho/\lambda)(1 - \rho/2)}{1 - \rho} \qquad (5.74)$$

where $\rho = \lambda M b / H$.

We are assuming that queueing is permitted at each terminal. However, the analysis for slotted ALOHA assumes an infinite population of users with an aggregate input rate of $M\lambda$ packets per second and this produces an upper bound on delay. (We note that a finite population model with M users each at rate λ and with queueing permitted will produce fewer collisions than the infinite population would since each terminal will avoid conflicts among its own packets.)

Equation (5.74) for FDMA is compared with the results for delay in slotted ALOHA with an infinite population (see Figure 5.39) as follows. We consider the (M, λ) plane in Figure 5.53, in which we represent constant T contours. Comparing the delay performance of the two

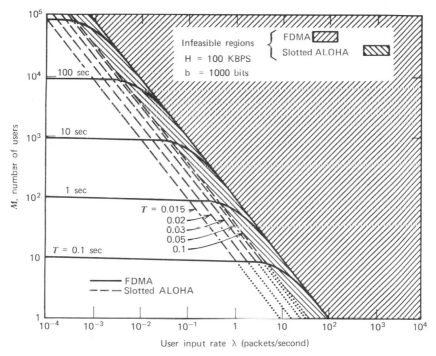

Figure 5.53   FDMA and slotted ALOHA random access: performance with 100 KBPS bandwidth.

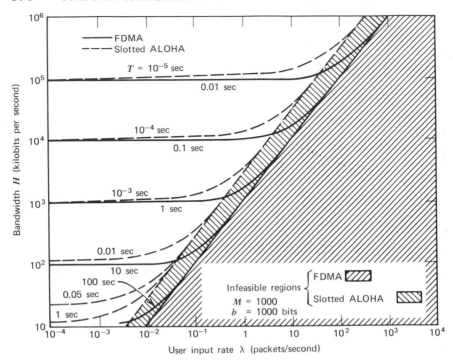

Figure 5.54   FDMA and slotted ALOHA random access: bandwidth requirements for 1000 terminals.

systems, we note that when we are in the presence of bursty users (small λ), slotted ALOHA can support many more users than FDMA, for the same packet delay. For example, at $T = 0.1$ sec, slotted ALOHA can support a number of users which is well over three orders of magnitude greater than the number that FDMA can support when $\lambda = 10^{-3}$ packet/sec; as λ increases (i.e., as the burstiness decreases), this difference reduces until at $\lambda \cong 5$ the two systems can support roughly an equal number of users. Beyond this point, FDMA is superior. This crossover point clearly depends upon the value of T examined. In fact, slotted ALOHA can support a total traffic only in the range $\rho < 1/e \cong 0.37$ and beyond that, FDMA will always be superior until it too saturates at $\rho = 1$; this tradeoff was also clearly evident in the curves of Figure 5.52.

The above result can also be presented in the following manner. Let M be some large number, say 1000. Figure 5.54 shows constant T contours in the (H, λ) plane. Again we note that if we are in the presence of bursty users, in order to achieve the same small delay, FDMA requires a bandwidth larger than slotted ALOHA by as much as three orders of magnitude. This factor is exactly equal to M as $\lambda \to 0$ since then the queueing effects are insignificant; in this limit, the delay T is simply the packet transmission time (observe the flatness of the curves in Figures 5.53 and 5.54), which for FDMA is $T = Mb/H$ and for slotted ALOHA is $T = b/H$. It is also obvious here ($\lambda \to 0$), for the same total bandwidth H, that FDMA will give M times the delay as compared to slotted ALOHA. This gain diminishes as λ increases, until finally as $\rho \to 1/e$ the situation reverses as mentioned above.

Last, let us fix λ and consider the delay contours in the (H, M) plane. Figure 5.55 corresponds to $\lambda = 10^{-1}$ packets per second. Such input rates correspond to bursty users. We note again that in order to support a large number of users, FDMA requires a larger bandwidth for the same delay performance as compared to slotted ALOHA.

It is all too evident from the above comparison that random access is by far superior to FDMA (or TDMA) when the environment consists of a large population of bursty users. However, we have noted that slotted ALOHA itself does not use the channel as efficiently as we might hope and

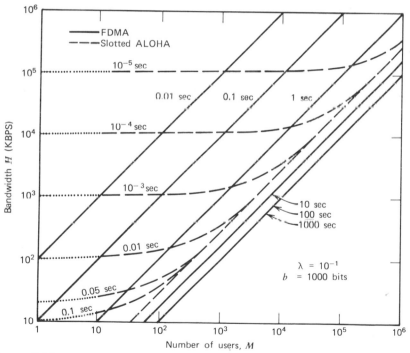

Figure 5.55 FDMA and slotted ALOHA random access: performance for $\lambda = 10^{-1}$ packets per second.

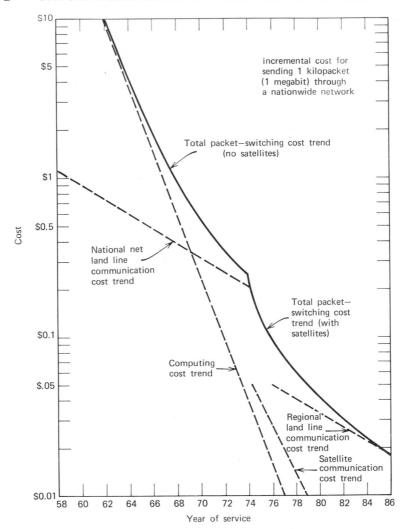

Figure 5.56 Incremental cost for sending 1 kilopacket (1 megabit) through a nationwide network.

this prompted us to study reservation schemes such as those described earlier; in fact, FDMA and TDMA are reservation schemes with permanent reservations. In the next section, we analyze another access protocol which gives superior performance than that of slotted ALOHA in an environment different from satellites.

The effect of satellite communications will be quite significant, as discussed, for example, by Roberts [ROBE 74]. This is the case since the cost of land-based communication services has been dropping at a rate of only 11% per year (a factor of 10 in 22 years) whereas a cost-performance study of computing machines (of various sizes) has shown their cost to be reducing at an annual rate in excess of 50%. As mentioned in the first footnote of this chapter, the (processing) cost of allocating communication resources first fell below the cost of the communication resource being allocated in 1969; the cost for the communication portion of computer-communication networks has been growing relative to the processing cost ever since. This is displayed in Figure 5.56 [ROBE 74] above where we plot the incremental cost of sending 10^6 bits (1 kilopacket) through a nationwide network as a function of the year when that transmission would have taken place. The assumption is that this data must travel a distance of 1200 miles (roughly half the cross country distance in the United States, based on uniformly distributed traffic requirements) in four hops; therefore, the average length of a leased line is assumed to be 300 miles (average distance between switching computers in a triply connected network consisting of approximately 100 nodes). Thus the two cost factors included in handling the average packet are (1) the communications cost for four 300-mile leased lines, and (2) the processing required in five switching computers. We plot the communications and computing (switching) components of cost separately, as well as their sum in a packet-switching network. Note that prior to 1969, the computing portion dominated. By 1974, the communications portion was roughly 85% of the cost. However, with the introduction of domestic (U.S.) satellites in 1974, only regional access to the satellite stations need be made by land-line communications; the sum of switching, satellite, and land-based communication costs are shown projected in the figure beyond 1974, and we see that the costs are significantly reduced (and, in the future, appear to be dominated by land-line costs again). In this way, satellite packet switching will make its impact on computer-communication networks in the near future.

5.12. GROUND RADIO PACKET SWITCHING

In this section, we concern ourselves with one method for providing communications among a large number of local user terminals. This local communication system could provide, for example, the means by which these terminals gain access to a high-level network such as the AR-PANET. In 1972, Roberts [ROBE 72] outlined the design for a hand-held terminal that could make use of a radio channel in a packet-switched mode. This led to a series of studies in ground radio packet switching as a communication system for local (possibly mobile) terminal interconnection under the guidance of R. E. Kahn [Kahn 75].

Mobile radio systems have been available for many years now (automobile, aircraft, marine, etc.). However, the principal mode of communication has been through the use of voice grade lines with a telephone-oriented philosophy. This corresponds to the shared system of Figure 5.3. We wish to study the suitability of the highly multiplexed random access packet-switching technology for this application.

Of course, the basic principles involved are quite similar to the satellite packet-switching studies reported in the previous section. We consider a collection of terminals, each of which is attempting to transmit to a central station. We have a broadcast channel in which two packet transmissions that overlap in part (or completely) will destroy each other. The fundamental difference is that with ground radio, the round-trip propagation delay is small compared to a packet transmission time (that is, $R \ll 1$). For example, 1000-bit packets traveling over 100 KBPS channels yield a transmission period of $P = 10$ msec and, if the maximum transmission distance is 10 miles, then the (speed of light) packet propagation delay $d = 0.054$ msec, giving $R = d/P = 0.005 \ll 1$. (With satellites, we had the case $R \gg 1$ and this led us to the consideration of pure ALOHA, slotted ALOHA, and reservation methods.) The small value for R suggests yet a fourth method for using the packet-switched channel, namely, the *Carrier Sense Multiple Access* mode (CSMA). In CSMA, we permit the terminal to listen to ("sense") the channel, and, if the carrier signal is heard, then the terminal realizes that the channel is in use by some other terminal and will politely postpone its transmission until the channel is sensed to be idle. (This information is useless with satellites since the sensed channel state provides information about the channel that is "ancient" history—that is, R packets into the past.)

The particular rules (that is, the protocol) for deciding when a terminal may transmit determines a maximum capacity for the channel. For example, the (nonsense!) pure ALOHA protocol achieves a maximum throughput of $1/2e \cong 0.184$ packets per P sec, whereas (nonsense again) slotted ALOHA gives $1/e \cong 0.368$. Two "persistent" CSMA protocols (known as nonpersistent CSMA and p-persistent CSMA) have been studied and we consider these below [KLEI 74b, KLEI 75a, KLEI 75b, TOBA 74, TOBA 75].

(Before proceeding with the analysis, we hasten to point out that, in addition to questions of access protocols, there are many other important aspects to ground radio packet switching. For example, the topological organization of a packet radio network is plagued with fascinating problems. Also, radio propagation, with the usual multipath, barriers, and noise effects, gives rise to serious considerations. Add to this the mobility of the terminals, and one finds that the use of packet switching in such an environment leads to many new and interesting problems. See [BIND 75b, BURC 75, FRAL 75a, FRAL 75b, FRAN 75, KAHN 75, KLEI 75a]).

The various protocols considered below differ from one another by the action (as regards packet transmission) that a terminal takes after sensing the channel. In all cases, when a terminal determines (by the absence of a positive acknowledgement from the central station) that its previous transmission was unsuccessful, then it reschedules the retransmission of the packet according to a randomly distributed retransmission delay. At this new point in time, the terminal senses the channel and repeats the algorithm dictated by the protocol. At any instant a terminal is called a *ready terminal* if it has a packet ready for transmission at this instant (either a new packet just generated or a previously conflicted packet rescheduled for retransmission at this instant).

A terminal may, at any one time, either be transmitting or receiving (but not both simultaneously). However, the delay incurred to switch from one mode to the other is assumed to be negligible. All packets have a constant length of b bits and are transmitted over an assumed noiseless channel (i.e., the errors in packet reception caused by random noise are not considered to be a serious problem and are neglected in comparison with errors caused by overlap interference). The system assumes noncapture (i.e., the overlap of any fraction of two packets results in destructive interference and both packets must be retransmitted). We further simplify the problem by assuming that the one-way propagation delay $d/2$ is identical[*] for all source-destination pairs.

Let us now discuss two protocols. We first consider the *nonpersistent CSMA*. Here, the idea is to limit the interference among packets by always rescheduling any packet which finds the channel busy upon its arrival. Specifically, a ready terminal senses the channel and operates as follows:

- If the channel is sensed idle, it transmits the packet.
- If the channel is sensed busy, then the terminal schedules the retransmission of the packet to some later time according to the retransmission delay distribution. At this new point in time, it senses the channel again and repeats the algorithm described.

[A slotted version of the nonpersistent CSMA can be considered in which the time axis is slotted with the slot size equal to $d/2$ seconds (the one-way propagation delay). All terminals are synchronized[†] and are forced to

---

[*] By considering this constant propagation delay to be equal to the largest possible, one gets lower (i.e., pessimistic) bounds on performance.

[†] We do not discuss the practical problems involved in synchronizing terminals.

start transmission only at the beginning of a slot. When a packet's arrival occurs during a slot, the terminal senses the channel at the beginning of the next slot and operates according to the protocol described above.]

We next consider the *p-persistent CSMA* protocol. However, before treating the general case (arbitrary *p*), we introduce the special case of $p = 1$.

The *1-persistent CSMA* protocol is devised in order to (presumably) achieve acceptable throughput by never letting the channel go idle if some ready terminal is available. More precisely, a ready terminal senses the channel and operates as follows:

- If the channel is sensed idle, it transmits the packet with probability one.
- If the channel is sensed busy, it waits until the channel goes idle (i.e., persisting on sensing) and only then transmits the packet (with probability one—hence, the name 1-persistent).

(A slotted version of this 1-persistent CSMA can also be considered by slotting the time axis and synchronizing the transmission of packets in much the same way as for the previous protocol.)

The above 1-persistent and nonpersistent protocols differ by the probability (one or zero) of not rescheduling a packet which upon arrival finds the channel busy. In the case of 1-persistent CSMA, we note that whenever two or more terminals become ready during another terminals' transmission, they wait for the channel to become idle (at the end of that transmission) and then they all transmit with probability one. A conflict will also occur with probability one! The idea of randomizing the starting time of transmission of packets accumulating at the end of a transmission suggests itself for intereference reduction and throughput improvement and brings us to the *p-persistent CSMA* protocol which is a generalization of the 1-persistent CSMA protocol. This scheme consists of including an additional parameter *p* which is the probability that a ready packet persists ($1 - p$ is the probability of delaying transmission by $d/2$ seconds). The parameter *p* will be chosen so as to reduce the level of interference while keeping the idle periods between any two consecutive nonoverlapped transmissions as small as possible. Specifically, the protocol consists of the following: the time axis is slotted where the slot size is $d/2$ seconds. For simplicity of analysis, we consider the system to be synchronized such that all packets begin their transmission at the beginning of a slot.

Consider a ready terminal:

- If this terminal senses the channel to be idle, then
  —with probability *p*, the terminal transmits the packet

—with probability $1-p$, the terminal delays the transmission of the packet by $d/2$ seconds (i.e., one slot). If at this new point in time, the channel is still detected idle, the same process above is repeated; otherwise, some packet must have started transmission, and our terminal schedules the retransmission of the packet according to the retransmission delay distribution (i.e., it acts as if it had conflicted and learned about the conflict).

- If the ready terminal senses the channel busy, it waits until the channel becomes idle (at the end of the current transmission) and then operates as above.

We assume a Poisson traffic source consisting of an infinite population of users each generating channel traffic at an infinitesimally small rate, summing to an average channel traffic of G packets per P sec; the average channel throughput is S packets per P sec. Each packet is of constant length, requiring P sec for transmission. If we were able to schedule the packets perfectly into the available channel space with absolutely no overlap or space between the packets, then we would have $S = 1$; therefore, as earlier S is the *channel utilization*, or *throughput*. The maximum achievable throughput for an access mode is called the *capacity* of the channel under that mode. Each ready terminal has exactly one (and not more than one) packet awaiting transmission. All source-destination pairs are assumed to have a constant (maximum) one-way normalized propagation delay denoted by a [$= (d/2)/P = R/2$]. The radio channel is assumed to be noiseless and all terminals are within line-of-sight (i.e., all terminals can hear each other in a mutually broadcast mode). The channel for acknowledgements from the central station is assumed to be separate from the channel we are studying (i.e., acknowledgements arrive reliably and at no cost).

We wish to solve for the channel capacity of the system for all of the access protocols described above. This we do by expressing S in terms of G (as well as other system parameters). The channel capacity is obtained by maximizing S with respect to G. Note that S/G is merely the probability of a successful transmission and that G/S is the average number of times a packet must be transmitted or scheduled until success. We derive results only for the nonpersistent CSMA protocol and refer the reader to the references listed earlier for the other protocols (the spirit of the derivation in all cases is similar).

For the nonpersistent CSMA protocol, we consider a packet (say, packet 0) that is newly generated at a terminal at time 0 when the channel is idle. This terminal will surely sense the channel as idle and transmit the packet. Now comes the race! If no other terminal transmits a packet

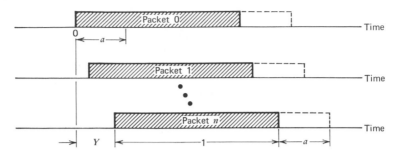

Figure 5.57   Interference in nonpersistent CSMA.

during the next $d/2$ sec (the vulnerable period), then packet 0 will be successful (since after $d/2$ sec all other terminals sense the channel busy due to packet 0 and will avoid interfering transmissions); the probability that no terminal transmits during these $d/2$ sec is $e^{-dG/2P} = e^{-aG}$. It is convenient, as earlier, to normalize time to units of P sec, and we use this scaled time axis in Figure 5.57. In this figure, we show the case where n other packets enter during packet 0's vulnerable period (if $n \geq 1$, then certainly they all get wiped out!). Let us define Y as the arrival time of the last packet that collides with packet 0, where, of course $0 \leq Y \leq a$; if $n = 0$, then we say $Y = 0$. In Exercise 5.27, we show that $P[Y \leq y] = e^{-G(a-y)}$, which gives

$$\bar{Y} = a - \frac{1 - e^{-aG}}{G} \qquad (5.75)$$

Now, the average duration of a busy interval is $\bar{Y} + 1 + a$; that is, packet n arrives, on the average $\bar{Y}$ units after the busy period begins, spends 1 unit in transmission, and clears the channel a units later (normalized propagation delay). The average duration of an idle period is simply $1/G$ (Poisson arrivals!). Thus an average cycle is $\bar{Y} + 1 + a + (1/G)$ units in duration. The average duration in a busy interval that the channel is used without conflicts (i.e., a successful transmission of duration 1) is e^{-aG}; this divided by the average cycle time is therefore S, the average throughput in packets per unit time (P sec). Thus using Eq. (5.75), we have

Nonpersistent CSMA

$$S = \frac{Ge^{-aG}}{G(1 + 2a) + e^{-aG}} \qquad \blacksquare \quad (5.76)$$

Note that whereas e^{-aG} is the success probability for packet 0, the success probability for an arbitrary packet is S/G, which clearly must be less than

e^{-aG}. Note further that

$$\lim_{a \to 0} S = \frac{G}{1+G} \qquad (5.77)$$

Similar throughput equations can be derived for the other protocols and we now merely quote the results:

Slotted nonpersistent CSMA

$$S = \frac{aGe^{-aG}}{1 - e^{-aG} + a} \qquad \blacksquare \quad (5.78)$$

1-Persistent CSMA

$$S = \frac{G[1 + G + aG(1 + G + aG/2)]e^{-G(1+2a)}}{G(1+2a) - (1 - e^{-aG}) + (1 + aG)e^{-G(1+a)}} \qquad \blacksquare \quad (5.79)$$

Slotted 1-persistent CSMA

$$S = \frac{Ge^{-G(1+a)}[1 + a - e^{-aG}]}{(1+a)(1 - e^{-aG}) + ae^{-G(1+a)}} \qquad \blacksquare \quad (5.80)$$

p-Persistent CSMA

$$S(G, p, a) = \frac{(1 - e^{-aG})[P'_s\pi_0 + P_s(1 - \pi_0)]}{(1 - e^{-aG})[a\bar{t}'\pi_0 + a\bar{t}(1 - \pi_0) + 1 + a] + a\pi_0}$$

$$\blacksquare \quad (5.81)$$

where P'_s, P_s, $\bar{t}'$, $\bar{t}$ and π_0 are defined in [KLEI 75a, TOBA 74]. We note that

$$S(G, p \to 0, a = 0) \to \frac{G}{G + e^{-G}}$$

In Figure 5.58 for $a = 0.01$, we plot S versus G for the various random access modes introduced so far and show the relative performance of each.

While the capacity of an ALOHA channel does not depend on the propagation delay, the capacity of a CSMA channel does. An increase in a increases the "vulnerable" period of a packet and reduces the channel capacity. This also results in "older" channel state information from sensing. In Figure 5.59 we plot, versus a, the channel capacity for all of the above random access modes. For large a, we note that slotted ALOHA (and even pure ALOHA) is superior to any CSMA mode since decisions based on partially obsolete data are deleterious; this effect is due in part to our assumption about the constant propagation delay.

Let us now discuss the average delay T. We must now identify the delay incurred when a packet is rescheduled for later retransmission (due either to a conflict or a sensed carrier). As with satellite channels, we

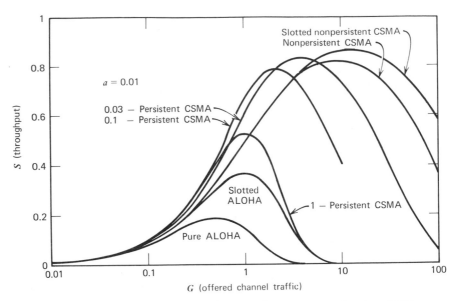

Figure 5.58   Throughput for the various random access modes ($a = 0.01$).

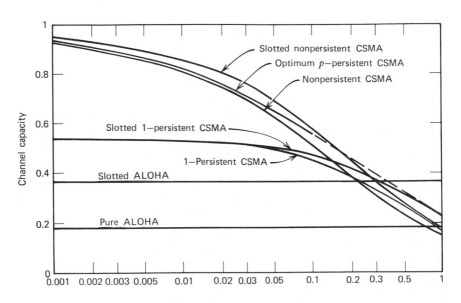

Figure 5.59   Effect of propagation delay on channel capacity.

400

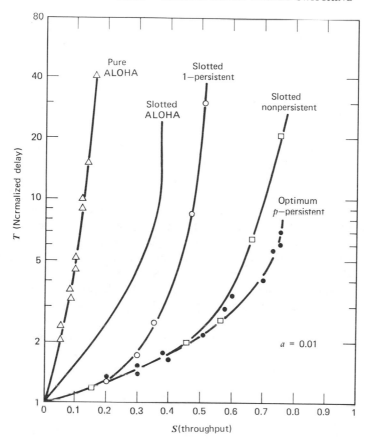

Figure 5.60   CSMA and ALOHA: Throughput-delay tradeoffs from simulation
$(a - 0.01)$.

assume a uniformly distributed retransmission delay with a mean of $\bar{X}$
time units* (normalized with respect to P). In Exercise 5.28, we show for
nonpersistent CSMA that

$$T = \left(\frac{G}{S}-1\right)[2a +1+\beta +\bar{X}]+1+a \qquad \blacksquare \quad (5.82)$$

where β is the normalized time to receive an acknowledgement and where
we assume each blocked (carrier-sensed) or conflicted packet suffers an
acknowledgement delay. Note that $\bar{X}$ should be optimized with respect to
S. The delay-throughput performance as obtained by simulation for the
two ALOHA modes and three CSMA modes is shown in Figure 5.60.

* $\bar{X}$ may depend on S for design purposes.

The throughput-delay performance described above was based on the (strong) assumption that all terminals were in line-of-sight and within range of each other. There are many instances where this is not the case, forcing us to relax that assumption. Terminals can be within range of the station (computer center, gateway to a network, SIMP, etc.) but out-of-range of each other, or they can be separated by some physical obstacle opaque to UHF radio signals. This gives rise to what is called the "hidden terminal" effect. It is evident that the existence of hidden terminals in an environment affects (degrades) the performance of CSMA. In [TOBA 74, 75] the effect of this degradation is quantified. For example, if we have the case of N identical independent groups (each an infinite population) each of which is hidden from the other (but within a given group, all terminals are able to sense each other), then the channel capacity behaves as shown in Figure 5.61. Note that the channel capacity experiences a drastic decrease between the two cases $N = 1$ (no hidden terminals) and $N = 2$. For $N \geq 2$, slotted ALOHA performs better than CSMA. This decrease is more critical for the nonpersistent CSMA than for the 1-persistent CSMA as shown in the figure. For $N > 2$, the channel capacity is rather insensitive to N and approaches pure ALOHA for large N.

The previous example did not show the effect of a small fraction of the population being hidden from the rest. Let us now consider an example in

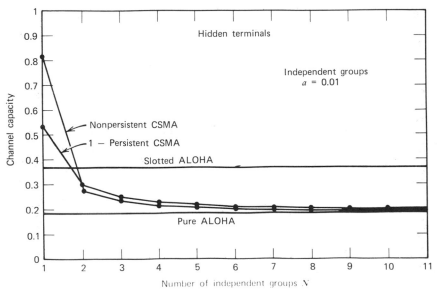

Figure 5.61   Independent group case: channel capacity versus the number of groups.

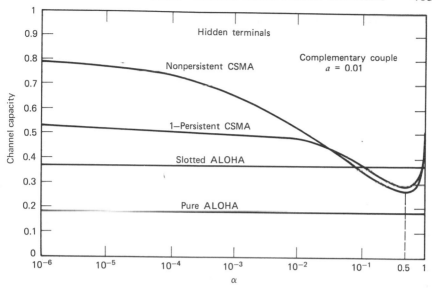

Figure 5.62 Two-group couple configuration: channel capacity versus a.

which the population consists of two independent groups ($N = 2$) of unequal sizes such that

$$S_1 = \alpha S$$

$$S_2 = (1 - \alpha)S$$

The channel capacity is plotted versus α for two CSMA protocols in Figure 5.62. Here again we note that the capacity decreases rapidly as α increases from 0. This decrease is much more ciritical for the nonpersistent than for the 1-persistent protocol. As soon as $\alpha = 10^{-2}$, the capacity of nonpersistent CSMA is approximately only 0.5, as compared to 0.82 when $\alpha = 0$. In addition, CSMA performs (in regard to capacity) only as well as slotted ALOHA when $\alpha = 0.08$ for the nonpersistent protocol and when $\alpha = 0.1$ for the 1-persistent protocol. In both cases, we note that the minimum capacity is obtained for $\alpha = 0.5$; this corresponds to the case $N = 2$ in the previous example. Another way to see the effect of the group split (α) is shown in Figure 5.63. Here we see the tradeoff among S_1, S_2, $S_1 + S_2$, and α.

Let us consider a solution to this hidden terminal problem which we call the Busy-Tone Multiple Access mode (BTMA). The operation of BTMA rests on the assumption that the station is, by definition, within range and in line-of-sight of all terminals. The total available bandwidth is to be divided into two channels: a message channel and a busy tone (BT)

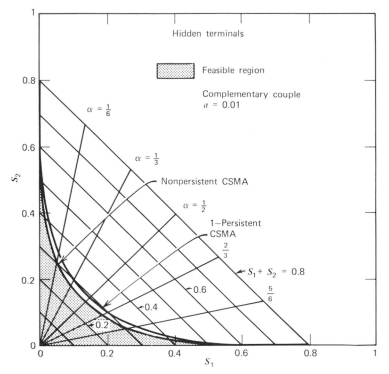

Figure 5.63   Two-group configuration: feasible regions.

channel. As long as the station senses a (terminal) carrier on the incoming message channel it transmits a (sine wave) busy-tone signal on the busy-tone channel. It is by sensing a carrier on the busy-tone channel that terminals determine when the message channel is busy. The action regarding the transmission of a packet that a terminal takes (again) is prescribed by the particular protocol being used. In CSMA, the difficulty of detecting the presence of a signal on the message channel when this message uses the *entire* bandwidth is minor and therefore has been neglected. It is not so when we are concerned with the (statistical) detection of the (sine wave) busy tone signal on the *narrow band* BT channel. The detection time, denoted by t_d, is no longer negligible and must be accounted for. Let us consider only the nonpersistent BTMA protocol which is similar to the nonpersistent CSMA protocol and corresponds to the following. Whenever a terminal has a packet ready for transmission, it senses the busy-tone channel for t_d seconds (the detection time), at the end of which it decides whether the BT signal is present or absent. (t_d is a system parameter to be optimized.) If the terminal decides that the BT signal is absent then it transmits the packet, otherwise it reschedules the packet for transmission at some later time incurring a random rescheduling delay; at this new point in time, it senses the BT channel and repeats the algorithm. In the event of a conflict, which the terminal learns about by failing to receive an acknowledgement from the station, the terminal again reschedules the transmission of the packet for some later time, and repeats the above process. This system is discussed in [KLEI 75b, TOBA 74, TOBA 75]. The result of optimization of various system parameters (including F, the probability of falsely detecting the channel as busy) is shown in the simulation curves of Figure 5.64. We see, in particular, that the cost in delay and throughput is really quite small due to hidden terminals when we use the BTMA solution.

In Table 5.3 we summarize the results for the many access protocols we have presented. From the table, we see that most of the throughput which is lost with the ALOHA systems may be regained with various of the CSMA systems.

As with satellite packet switching, one may introduce dynamic reservation systems as, for example, in [TOBA 74, 76] where it is shown that such schemes are efficient over a wide and interesting range of parameters.

As with most "contention" systems, these random multiaccess broadcast channels (ALOHA, CSMA, BTMA) are characterized by the fact

Table 5.3

Capacity for various access protocols $(a = 0.1)$

PROTOCOL	CAPACITY C
Pure ALOHA	0.184
Slotted ALOHA	0.368
1-Persistent CSMA	0.529
Slotted 1-persistent CSMA	0.531
Nonpersistent BTMA	
H = 100 KHz	0.680
H = 1000 KHz	0.720
0.1-Persistent CSMA	0.791
Nonpersistent CSMA	0.815
0.03-Persistent CSMA	0.827
Slotted nonpersistent CSMA	0.857
Perfect scheduling	1.000

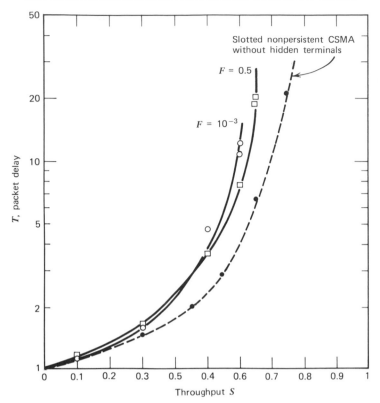

Figure 5.64 Slotted nonpersistent BTMA: throughput-delay tradeoff ($a =$ 0.01).

that the throughput goes to zero when the load grows too large, thus determining a limited system capacity. This and the throughput-delay performance have been discussed here in terms of a steady-state analysis under the assumption of equilibrium conditions. These channels also present unstable behavior at most input loads as discussed in Section 5.11 for satellite packet switching, and so here too special control procedures must be adopted.

The study of ground radio packet switching is in its early stages. The analysis poses many interesting problems and the design and optimization considerations are extremely complex. Nevertheless, these systems offer what appear to be promising access methods which are relatively efficient for the bursty case of terminal-to-computer communication.

This brings to a close our discussion of the analysis and design of computer-communication networks. We have identified some intriguing unsolved problems, but in general have identified effective analysis and design procedures which produce cost-effective packet-switched networks. In the next chapter, we turn our attention to the observed behavior of the ARPANET and discuss the importance of flow control and the traps to which it can lead.

REFERENCES

ABRA 70   Abramson, N., "The ALOHA System—Another Alternative for Computer Communications," *AFIPS Conference Proceedings*, 1970 Fall Joint Computer Conference, **37**, 281–285.

ABRA 73a   Abramson, N., "Packet Switching with Satellites," 1973 National Computer Conference, *AFIPS Conference Proceedings*, **42**, 695–702.

ABRA 73b   Abramson, N. and F. Kuo, *Computer-Communication Networks*, Prentice-Hall (Englewood Cliffs, New Jersey), 1973.

BARA 64   Baran, P., "On Distributed Communications," RAND Series Reports, August 1964.

BEVE 70   Beveridge, G. S. and R. S. Schecter, *Optimization: Theory and Practice*, McGraw-Hill (New York), 1970.

BIND 75a   Binder, R., "A Dynamic Packet-Switching System for Satellite Broadcast Channels." *Proceedings of the IEEE International Conference on Communications*, San Francisco, California, 41-1 to 41-5, June 16–18, 1975.

BIND 75b   Binder, R., N. Abramson, F. F. Kuo, A. Okinaka, and D. Wax, "ALOHA Packet Broadcasting—A Retrospect," *AFIPS Conference Proceedings*, 1975 National Computer Conference, **44**, 203–215.

BOBR 71   Bobrow, D. G., J. D. Burchfiel, D. L. Murphy, and R. S. Tomlinson, "TENEX, A Paged Time Sharing System for the PDP-10," paper presented at the Third ACM Symposium on Operating System Principles, October 18–20, 1971, published in *Communications of the Association for Computing Machinery*, **15**, 135–143 (1972).

BOUK 73   Bouknight, W. J., G. R. Grossman, and D. M. Grothe, "The ARPA Network Terminal System: A New Approach to Network Access," *Proceedings of the Third IEEE Symposium on Data Networks Analysis and Design*, St. Petersburg, Florida, 73–79, November 13–15, 1973.

BRUM 71   Brumelle, S. L., "Some Inequalities for Parallel Service Queues," *Operations Research*, **19**, 402–413 (1971).

BURC 75   Burchfiel, J., R. Tomlinson, and M. Beeler, "Functions and Structure of a Packet Radio Station," *AFIPS Conference Proceedings*, 1975 National Computer Conference, **44**, 245–251.

BUTT 74   Butterfield, S., R. Rettberg, and D. Walden, "The Satellite IMP for the ARPA Network," *Proceedings of the Seventh Hawaii*

International Conference on System Science, University of Hawaii, Honolulu, Hawaii, 70–73, January 8–10, 1974.

CANT 72   Cantor, D. G. and M. Gerla, "The Optimal Routing of Messages in a Computer Network via Mathematical Programming," *IEEE Computer Conference Proceedings,* San Francisco, California, 167–170, September 1972.

CANT 74   Cantor, D. G. and M. Gerla, "Capacity Allocation in Distributed Computer Networks," *Proceedings of the Seventh Hawaii International Conference on System Sciences,* University of Hawaii, Honolulu, Hawaii, 115–117, January 8–10, 1974.

CARR 70   Carr, C. S., S. D. Crocker, and V. G. Cerf, "HOST-HOST Communication Protocol in the ARPA Network," *AFIPS Conference Proceedings,* 1970 Spring Joint Computer Conference, **36,** 589–597.

CHOU 73   Chou, W. and A. Kershenbaum, "A unified Alogrithm for Designing Multidrop Teleprocessing Networks," *Proceedings of the Third IEEE Symposium on Data Networks Analysis and Design,* St. Peterburg, Florida, 148–156, November 13–15, 1973.

CHU 69   Chu, W. W., "A Study of the Technique of Asynchronous Time Division Multiplexing for Time-Sharing Computer Communications," *Proceedings of the Second Hawaii International Conference on System Science,* University of Hawaii, Honolulu, Hawaii, 607–610, January 1969.

CLOS 72   Closs, F., "Time Delays and Trunk Capacity Requirements in Line-Switched and Message-Switched Networks," *ISS Record,* 428–433 (1972).

CLOS 73   Closs, F., "Packet Arrival and Buffer Statistics in a Packet Switching Node," *Proceedings of the Third IEEE Data Communications Symposium on Data Networks Analysis and Design,* St. Petersburg. Florida, 12–17 November 13–15, 1973.

CLOW 73   Clowes, G. J. and C. S. Jayasuriya, "Traffic Considerations in Switched Data Networks," *Proceedings of the Third IEEE Symposium on Data Networks Analysis and Design,* St. Petersburg, Florida, 18–22, November 13–15, 1973.

CROC 72   Crocker, S. D., J. F. Heafner, R. M. Metcalfe, and J. B. Postel, "Function-Oriented Protocols for the ARPA Computer Network," *AFIPS Conference Proceedings,* 1972 Spring Joint Computer Conference, **40,** 271–279.

CROW 75   Crowther, W. R., F. E. Heart, A. A. McKenzie, J. M. McQuillan, and D. C. Walden, "Issues in Packet-Switching Network Design," *AFIPS Conference Proceedings,* 1975 National Computer Conference, **44,** 161–175.

DATA 73   *Data Channels,* Vol. 1, No. 3 (Verona, New Jersey), November 1973.

DAVI 68   Davies, D. W., "The Principles of a Data Communication Network for Computers and Remote Peripherals," *Proceedings of the IFIP Congress 68,* Edinburgh, Scotland, Hardware, 709–714.

DAVI 71   Davies, D. W., "The Control of Congestion in Packet Switching Networks," *Proceedings of the Second ACM IEEE Symposium in the Optimization of Data Communications Systems,* Palo Alto, California, 46–49, October 1971.

DAVI 73   Davies, D. W. and D. Barber, *Communication Networks for Computers,* John Wiley, (New York) 1973.

ELLI 74   Ellis, L. W., "The Law of the Economies of Scale Applied to Computer-Communication System Design," *Proceedings of the International Conference on Computer Communication,* Stockholm, Sweden, 299–306, 1974.

EURO 73   "Eurodata—A Market Study on Data Communications in Europe, 1972–1985," study sponsored by the European Conference of Postal and Telecommunications Administrations (1973).

EVER 63   Everett, H., III, "Generalized LaGrange Multipliers Method for Solving Problems of Optimal Allocation of Resources," *Operations Research,* **11,** 399–418 (1963).

FARB 72   Farber, D. J. and K. C. Larson, "The Structure of the Distributed Computer System—The Communications System," *Proceedings of the International Symposium on Computer-Communication Networks and Teletraffic,* Polytech Press, (Brooklyn, N.Y.), 539–545, April 1972.

FAYO 74   Fayolle, G., E. Gelenbe, J. Labetoulle, and D. Bastin, "The Stability Problem of Broadcast Packet Switching Computer Networks," in *Computer Architectures and Networks,* E. Gelenbe and R. Mahl (eds.), North-Holland (Amsterdam), 135–140, 1974.

FELL 50   Feller, W., *An Introduction to Probability Theory and Its Applications,* John Wiley (New York) 1950.

FERG 75   Ferguson, M. J., "On the Control, Stability, and Waiting Time in a Slotted ALOHA Random Access System," *Proceedings of the IEEE International Conference on Communications,* San Francisco, California, 41-6 to 41-9, June 16–18, 1975.

FLOY 62   Floyd, R., "Algorithm 97, Shortest Path," *Communications of the ACM,* **5,** 345 (1962).

FRAL 75a Fralick, S. C. and J. C. Garrett, "Technological Considerations for Packet Radio Networks," *AFIPS Conference Proceedings,* 1975 National Computer Conference, **44,** 233–243.

FRAL 75b Fralick, S. C., D. H. Brandin, F. Kuo, and C. Harrison, "Digital Terminals for Packet Broadcasting," *AFIPS Conference Proceedings* 1975 National Computer Conference, **44,** 253–261.

FRAN 70   Frank, H., I. T. Frisch, and W. Chou, "Topological Considerations in the Design of the ARPA Network," *AFIPS Conference Proceedings,* 1970 Spring Joint Computer-Conference **36,** 381–587.

FRAN 71a Frank, H., and I. T. Frisch, *Communication, Transmission, and Transportation Networks,* Addison-Wesley (Reading, Mass.), 1971.

FRAN 71b Frank, H., I. T. Frisch, W. Chou, and R. Van Slyke, "Optimal

Design of Centralized Computer Networks," *Networks*, Vol. 1, No. 1, Wiley (New York), 43–57, 1971.

FRAN 72a   Frank, H., R. E. Kahn, and L. Kleinrock, "Computer Communication Network Design—Experience with Theory and Practice," *AFIPS Conference Proceedings*, 1972 Spring Joint Computer Conference, **40,** 255–270.

FRAN 72b   Frank, H. and W. Chou, "Topological Optimization of Computer Networks," *Proceedings of IEEE*, **60,** 1385–1397, (1972).

FRAN 75   Frank, H., I. Gitman, and R. van Slyke, "Packet Radio System— Network Considerations," *AFIPS Conference Proceedings*, 1975 National Computer Conference, **44,** 217–231.

FRAT 72   Fratta, L. and M. Gerla, "The Synthesis of Computer Networks: Properties of the Optimum Solution," *ACM-International Computing Symposium*, Venice, Italy, April 1972.

FRAT 73   Fratta, L., M. Gerla, and L. Kleinrock, "The Flow Deviation Method—An Approach to Store-and-Forward Communication Network Design," *Networks*, **3,** 97–133 (1973).

FUCH 70   Fuchs, E. and P. E. Jackson, "Estimates of Distributions of Random Variables for Certain Computer Communications Traffic Models," *Communications of the ACM,* **13,** 752–757 (1970).

FULT 71   Fultz, G. L. and L. Kleinrock, "Adaptive Routing Techniques for Store-and-Forward Computer-Communication Networks," *Proceedings of the IEEE International Conference on Communications*, 39-1 to 39-8, June 14–16, 1971.

FULT 72   Fultz, G. L., "Adaptive Routing Techniques for Message Switching Computer-Communication Networks," School of Engineering and Applied Science, University of California, Los Angeles, Engineering Report UCLA-ENG-7252, July 1972.

GAIN 73   Gaines, E. C., "Specialized Common Carriers—Competition and Alternative," *Telecommunications*, **7,** No. 9, 15–26 (1973).

GERL 73   Gerla, M., "The Design of Store-and Forward (S/F) Networks for Computer Communications," University of California, Los Angeles, School of Engineering and Applied Science, Engineering Report UCLA-ENG-7319, 1973.

GERL 74   Gerla, M., H. Frank, W. Chou, and J. Eckl, "A Cut-Saturation Algorithm for Topological Design of Packet Switched Communications Networks," *Proceedings of the National Telecommunications Conference*, San Deigo, California, December 1974.

GERL 75   Gerla, M., "Approximations and Bounds for the Topological Design of Distributed Computer Networks," *Proceedings for the Fourth Data Communications Symposium*, Quebec, Canada, 4-7 to 4-15, October 1975.

HASS 73   Hassing, T. E., R. M. Hampton, G. W. Bailey, and R. S. Gardella, "A Loop Network for General Purpose Data Communications in a Heterogeneous World," *Proceedings of the Third IEEE Symposium on Data Networks Analysis and Design*, St. Petersburg, Florida, 88–96, November 13–15, 1973.

HEAR 70   Heart, F. W., R. E. Kahn, S. M. Ornstein, W. R. Crowther, and D. C. Walden, "The Interface Message Processor for the ARPA Computer Network," *AFIPS Conference Proceedings*, 1970 Spring Joint Computer Conference, **36,** 551–567.

HILD 65   Hildebrand, F. B., *Methods of Applied Mathematics* (2nd ed.), Prentice-Hall (Englewood Cliffs, New Jersey), 1965.

HOWA 60   Howard, R. A., *Dynamic Programming and Markov Processes*, M.I.T. Press (Cambridge, Mass.), 1960.

ITOH 73   Itoh, K. and T. Kato, "An Analysis of Traffic Handling Capacity of Packet Switched and Circuit Switched Networks," *Proceedings of the Third IEEE Symposium on Data Networks Analysis and Design*, St. Petersburg, Florida, 29–37, November 13–15, 1973.

JACK 57   Jackson, J. R., "Networks of Waiting Lines," *Operations Research*, **5,** 518–521 (1957).

JACK 63   Jackson, J. R., "Jobshop-Like Queueing Systems," *Management Science* **10,** No. 1, 131–142 (1963).

JACK 69   Jackson, P. E. and C. D. Stubbs, "A Study of Multi-access Computer Communications, *AFIPS Conference Proceedings*, 1969 Spring Joint Computer Conference, **34,** 491–504.

KAHN 71   Kahn, R. E. and W. R. Crowther, "Flow Control in a Resource Sharing Computer Network," *Proceedings of the Second ACM IEEE Symposium on Problems in the Optimization of Data Communications Systems*, Palo Alto, California, 108–116, October 1971.

KAHN 75   Kahn, R. E., "The Organization of Computer Resources into a Packet Radio Network," *AFIPS Conference Proceedings*, 1975 National Computer Conference, **44,** 177–186.

KARP 73   Karp, P. M., "Origin, Development and Current Status of the ARPA Network," CompCon 73, *Proceedings of the Seventh Annual IEEE Computer Society International Conference*, 49–52.

KLEI 64   Kleinrock, L., *Communication Nets; Stochastic Message Flow and Delay*, McGraw-Hill (New York), 1964. Out of print. Reprinted by Dover Publications, 1972.

KLEI 69   Kleinrock, L., "Models for Computer Networks," *Proceedings of the IEEE International Conference on Communications*, Boulder, Colorado, 21-9 to 21-16, June 9–11, 1969.

KLEI 70   Kleinrock, L., "Analytic and Simulation Methods in Computer Network Design," *AFIPS Conference Proceedings*, 1970 Spring Joint Computer Conference, **36,** 569–579.

KLEI 73   Kleinrock, L. and S. S. Lam, "Packet Switching in a Slotted Satellite Channel," *AFIPS Conference Proceedings*, 1973 National Computer Conference, **42,** 703–710.

KLEI 74a   Kleinrock, L. and S. S. Lam, "On Stability of Packet Switching in a Random Multi-access Broadcast Channel," *Proceedings of the Special Subconference on Computer Nets, Seventh Hawaii International Conference on System Sciences*, University of Hawaii, Honolulu, Hawaii, January, 1974.

KLEI 74b   Kleinrock, L. and F. A. Tobagi, "Carrier Sense Multiple Access for Packet Switched Radio Channels" *Proceedings of the IEEE International Conference on Communications*, Minneapolis, Minnesota, 21B-1 to 21B-7, June 1974.

KLEI 75a   Kleinrock, L. and F. A. Tobagi, "Random Access Techniques for Data Transmission over Packet Switched Radio Channels," *AFIPS Conference Proceedings*, 1975 National Computer Conference, **44,** 187–201.

KLEI 75b   Kleinrock, L. and F. A. Tobagi, "Packet Switching in Radio Channels: Part I—Carrier Sense Multiple-Access Modes and their Throughput-Delay Characteristics," *IEEE Transactions on Communications*, **Com-23,** 1400–1416 (1975).

KLEI 75c   Kleinrock, L., *Queueing Systems, Vol. I: Theory*, Wiley Interscience (New York), 1975.

LAM 74   Lam, S. S., "Packet Switching in a Multi-access Broadcast Channel with Applications to Satellite Communication in a Computer Network," Computer Science Department, School of Engineering and Applied Science, Engineering Report UCLA-ENG-7429, March 1974.

LAM 75   Lam, S. S., and L. Kleinrock, "Dynamic Control Schemes for a Packet Switched Multi-access Broadcast Channel," *AFIPS Conference Proceedings*, 1975 National Computer Conference, **44,** 143–153.

LOEV 63   Loeve, M., *Probability Theory*, Van Nostrand (Princeton, New Jersey), 1963.

MART 72   Martin, J., *Systems Analysis for Data Transmission*, Prentice-Hall, (Englewood Cliffs, New Jersey), 1972.

MAXE 75   Maxemchuk, N. F., "Dispersity Routing," *Proceedings of the IEEE International Conference on Communications*, San Francisco, California, 41-10 to 41-13, June 16–18, 1975.

MCKE 72a   McKenzie, A. A., "HOST/HOST Protocol for the ARPA Network" (NIC 8246), ARPA Network Information Center, Stanford Research Institute, Menlo Park, California (January 1972).

MCKE 72b   McKenzie, A. A., B. P. Cossell, J. M. McQuillan, and M. J. Thrope, "The Network Control Center for the ARPA Network," *ICCC Proceedings*, 185–191, October 1972.

MCKE 73   McKenzie, A. A., "Status Report on the Terminal IMP," *ARPANET News*, Issue 9 (NIC 19720), 3–11, November 1973.

MCQU 72   McQuillan, J. M., W. R. Crowther, P. P. Cossell, D. C. Walden, and F. E. Heart, "Improvements in the Design and Performance of the ARPA Network," *AFIPS Conference Proceedings*, 1972 Fall Joint Computer Conference, **41,** 741–754.

MEIS 71   Meister, B., H. Muller, and H. Rudin, "New Optimization Criteria for Message-Switching Networks," *IEEE Transactions on Communication Technology*, **Com-19,** 256–260 (1971).

MIMN 73   Mimno, N. W., B. P. Cossell, D. C. Walden, S. C. Butterfield, and J. B. Leven, "Terminal Access to the ARPA Network: Experience and Improvements," CompCon 73, *Proceedings of the Seventh Annual IEEE Computer Society International Conference*, 39–43, February 1973.

MIYA 75   Miyahara, H., T. Hasegawa, and Y. Teshigawara, "A Comparative Analysis of Switching Methods in Computer Communication Networks," *Proceedings of the IEEE International Conference on Communications*, San Francisco, California, 6-6 to 6-10, June 16–18, 1975.

MORS 58   Morse, P. M., *Queues, Inventories and Maintenance*, John Wiley (New York), 1958.

OPDE 74   Opderbeck, H. and L. Kleinrock, "The Influence of Control Procedures on the Performance of Packet-Switched Networks," *Proceedings of the National Telecommunications Conference*, San Diego, California, December 1974.

ORNS 72   Ornstein, S. M., F. E. Heart, W. R. Crowther, H. K. Rising, S. B. Russell, and A. Michel, "The Terminal IMP for the ARPA Computer Network," *AFIPS Conference Proceedings*, 1972 Spring Joint Computer Conference, **40**, 243–254.

PETE 73   Peters, R. A. and K. M. Simpson, "Eurodata: Data Communications in Europe 1972–1985," *Datamation*, **19**, No. 12, 76–80 (1973).

PIER 71   Pierce, J. R., "Network for Block Switching of Data," *IEEE Convention Records*, New York, 222–223, March 1971.

PORT 71   Port, E., and F. Closs, "Comparison of Switched Data Networks on the Basis of Waiting Times," *IBM Zurich Research Laboratory Report*, April 28, 1971.

PRIC 73   Price, W. L., "Simulation of Packet-Switching Networks Controlled on Isarithmic Principles," *Proceedings of the Third IEEE Symposium on Data Networks Analysis and Design*, St. Petersburg, Florida, 44–49, November 13–15, 1973.

RITC 74   Ritchie, D. M. and K. Thompson, "the UNIX Time-Sharing System," *Communications of the Association for Computing Machinery*, **17**, 365–375, (1974).

ROBE 67   Roberts, L. G., "Multiple Computer Networks and Inter-Computer Communications," *Proceedings of the ACM Symposium on Operating Systems*, Gatlinburg, Tennessee, 1967.

ROBE 70   Roberts, L. G. and B. D. Wessler, "Computer Network Development to Achieve Resource Sharing," *AFIPS Conference Proceedings*, 1970 Spring Joint Computer Conference **36**, 543–549.

ROBE 72   Roberts, L. G., "Extensions of Packet Communication Technology to a Hand Held Personal Terminal," *AFIPS Conference Proceedings*, 1972 Spring Joint Computer Conference **40**, 295–298.

ROBE 73a   Roberts, L. G., "Network Rationale: A 5-Year Re-evaluation," CompCon 73, *Seventh Annual IEEE Computer Society International Conference*, 3–5, February 1973.

ROBE 73b   Roberts, L. G., "Dynamic Allocation of Satellite Capacity through Packet Reservation," *AFIPS Conference Proceedings*, 1973 National Computer Conference, **42**, 711–716.

ROBE 74   (Roberts, L. G., "Data by the Packet," *IEEE Spectrum*, 46–51, February 1974.

ROSN 75   Rosner, R. D. and B. Springer, "A Cost and Performance Tradeoff Study of Circuit and Packet Switching of Data," to appear in *IEEE Transactions on Communications*, 1975.

RUBI 74   Rubin, I., "Communication Networks: Message Path Delays," *IEEE Transactions of the Professional Group on Information Theory* **IT-20,** 738–745 (1974).

STID 70   Stidham, S. Jr., "On the Optimality of Single-Server Queueing Systems," *Operations Research*, **18,** No. 4, 708–732 (1970).

SYSK 60   Syski, R., *Introduction to Congestion in Telephone Systems*, Oliver & Boyd (Edinburgh and London), 1960.

THOM 72   Thomas, R. H. and D. A. Henderson, "McROSS—A Multi-Computer Programming System," *AFIPS Conference Proceedings*, 1972 Spring Joint Computer Conference **40,** 281–293.

TOBA 74   Tobagi, F. A., "Random Access Techniques for Data Transmission Over Packet Switched Radio Networks," School of Engineering and Applied Science, University of California, Los Angeles, UCLA-ENG 7499, December 1974.

TOBA 75   Tobagi, F. A. and L. Kleinrock, "Packet Switching in Radio Channels: Part II—The Hidden Terminal Problem in Carrier Sense Multiple-Access and the Busy Tone Solution," *IEEE Transactions on Communications*, **COM-23,** 1417–1433 (1975).

TOBA 76   Tobagi, F. A. and L. Kleinrock, "Packet Switching in Radio Channels; Part III: Split Channel Reservation Multiple Access," to appear in *IEEE Transactions on Communications*.

WHIT 72   Whitney, V. K. M., "LaGrangian Optimization of Stochastic Communication Systems Models," *Proceedings of the International Symposium on Computer-Communication Networks and Teletraffic*, Polytechnic Press, (Brooklyn, N.Y.), 385–395 April 1972.

ZANG 67   Zangwill, W. I., *Nonlinear Programming, A Unified Approach*, Prentice-Hall (Englewood Cliffs, New Jersey) 1967.

EXERCISES

5.1. Consider a G/M/m system. Using the upper and lower bounds on W as given in Eq. (2.73), show that the results given in Eqs. (5.3) and (5.8) for M/M/m also hold for G/M/m, namely

$$T(1, \lambda, C) \le T(m, \lambda, C)$$
$$W(1, \lambda, C) \ge W(m, \lambda, C)$$

5.2. [CLOS 73] Here we seek the pdf for the interarrival time of packets. Assume that messages are generated from a very large number of Poisson sources whose collective rate is λ. Each message is decomposed into an integer number of packets; the mean number of packets per message is α, and the fixed length of each packet is $\bar{b}$ bits. Upon the initiation of a given message, its packets begin to appear at intervals of $\bar{b}/C$ sec (where C is the capacity of a channel feeding these packets to a central switch). Packets from many sources may be arriving simultaneously, and we denote by $\tilde{t}$, the interarrival time between adjacent packet arrivals.

(a) Show that

$$P[\tilde{t} \leq t] = \begin{cases} 1 - e^{-\lambda \alpha t} & 0 \leq t < \dfrac{\bar{b}}{C} \\[2ex] 1 - \left[\dfrac{1}{\alpha} e^{-\lambda \bar{b}(\alpha-1)/C} \right] e^{-\lambda t} & \dfrac{\bar{b}}{C} \leq t \end{cases}$$

(b) Find $E[\tilde{t}]$.
(c) Find $\sigma_{\tilde{t}}^2$.

5.3. In this and the next problem, we investigate the behavior of message traffic in a chain of M tandem channels as shown below. We have $N = M + 1$ nodes

Thus $\gamma_{1,M+1} = \gamma$ ($\gamma_{jk} = 0$ otherwise) and $\lambda_i = \gamma$ ($i = 1, 2, \ldots, M$). All traffic enters node 1 (Poisson traffic at a rate γ) and departs in node $M+1$. Assume that message lengths are chosen from an arbitrary distribution with a mean of $\bar{b}$ bits per message. Note that once a message arrives at node 1, its entire future behavior can be calculated deterministically.

(a) Let $U_i(t)$ be the unfinished work in the ith node at time t. Assume a sequence of arrivals to node 1 at the times $\{0, 1, 3, 8, 10, 11, 14.5, 18, 19, 20, 22, 23, 24, 27, \ldots\}$ and a corresponding sequence of message lengths $\{2, 3, 1, 3, 1, 2, 2, 2, 1, 2, 1, 2, 2, 1, \ldots\}$. Assume the entire network is empty prior to $t = 0$. Accurately sketch $U_i(t)$ for $i = 1, 2, 3$, and 4 to show the behavior of these 14 messages, under the assumption that $C_1 = C_2 = \cdots = C_M = 1$ bit per second.

(b) From the results of (a), discuss the behavior of the message traffic in the Nth node as $N \to \infty$ [KLEI 64]. In particular, for node N, discuss:
    (i) the system time (queue plus service) of a message in (say) the nth busy period;

      (ii) the system time for messages in the nth busy period as compared to the $(n-1)$st busy period;

     (iii) the interval separating the $(n-1)$st and nth busy period.

5.4. Consider the chain of M tandem channels as in the previous problem. Here we assume that all messages have a *constant* length of $\bar{b}$ bits and permit arbitrary channel capacities C_i [RUBI 74]. Thus the service time in the ith channel is exactly $\bar{b}/C_i$ sec, which we shall denote by $x(i)$. The average message delay is clearly $T = \sum_{i=1}^{M} T_i$ where $T_i = W_i + x(i)$ and W_i is the average wait for the ith channel. For $M = 1$, we have a simple M/D/1 system whose complete solution may be obtained from Section 1.7.

(a) Show that

$$w_{n+1}^{(i)} = [w_n^{(i)} + u_n^{(i)}]^+$$

where $w_n^{(i)}$ is the waiting time (on queue) for the nth message in node i and

$$u_n^{(i)} = \begin{cases} x(i) - x(i-1) & \text{if } w_{n+1}^{(i-1)} > 0 \\ x(i) - x(i-1) - I_{n+1}^{(i-1)} & \text{if } w_{n+1}^{(i-1)} = 0 \end{cases}$$

and $I_n^{(i)}$ is the duration of that idle period in node i which is terminated by the nth message.

(b) Show that $w_n^{(i)} = 0$ if $x(i) \leq x(i-1)$ for $i \geq 2$.

(c) From (b) we see that a natural set of nodes to "eliminate" are those for which $w_n^{(i)} = 0$. Thus we define a set of (ladder) indices $\{k_j\}$ such that $k_1 = 1$ and

$$k_j = \min\{i : k_{j-1} < i \leq M, x(i) > x(k_{j-1})\}$$

For the sequence $\{C_1 = 1, C_2 = \frac{1}{2}, C_3 = 2, C_4 = 2, C_5 = \frac{1}{10}, C_6 = 1\}$, find $\{k_j\}$.

(d) Let $W(M) = \sum_{i=1}^{M} W_i$ and $T = T(M)$. Thus

$$T(M) = W(M) + \sum_{i=1}^{M} x(i)$$

Let $w_n = \sum_{i=1}^{M} w_n^{(i)}$. Show that

$$w_{n+1} = [w_n + \max(x(1), x(2), \ldots, x(M)) - t_{n+1}]^+$$

where t_n is the interarrival time at node 1 [see Eq. (1.2)].

(e) Compare the expression for w_{n+1} in (d) above with Eq. (1.120). From this, show that the distribution of w_n is the same as that for an M/D/1 system with mean arrival rate γ and service time $\bar{x} = \max(x(1), x(2), \ldots, x(M))$.

(f) Prove that the stability condition for the tandem queue is merely that $\rho^* = \gamma x(k^*) < 1$ where k^* is the largest (ladder) index.

(g) Prove that

$$W(M) = \frac{1}{2} \frac{\rho^* x(k^*)}{1 - \rho^*}$$

(h) From (g), solve for W_i.

(i) How does a permutation of the N channels affect $W(M)$?

5.5. Prove that the minimum value for the numerator sum in Eq. (5.29) is given by

$$\lambda_i = \begin{cases} \lambda & i - i_0 \\ 0 & i \neq i_0 \end{cases}$$

for any choice of i_0.

5.6. Suppose we place a set of *lower* bounds on the channel flows, that is,

$$\lambda_i \geq a_i$$

For the linear CFA problem with $d_i = d$ for all $i = 1, 2, \ldots, M$, what now is the optimum C_i which minimizes T subject to a fixed total cost of D dollars and a total traffic λ.

5.7. Suppose we have N independent M/M/1 facilities indexed by $i = 1, 2, \ldots, N$. Let the parameters λ_i and C_i for the ith facility be such that

$$\lambda = \sum_{i=1}^{N} \lambda_i$$

$$C = \sum_{i=1}^{N} C_i$$

Let

$$T = \sum_{i=1}^{N} \frac{\lambda_i}{\lambda} \left(\frac{1}{\mu C_i - \lambda_i} \right)$$

where $1/\mu$ is the average message length in bits.

(a) For a given set $\{\lambda_i\}$, what is the optimum selection for C_i ($i = 1, 2, \ldots, N$) in the sense of minimizing T?

(b) Using the optimum $\{C_i\}$ what choice of λ_i ($i = 1, 2, \ldots, N$) will permit the smallest value for T?

5.8. Let T_p be the value for T in Eq. (5.19) when $C_i = (\lambda_i/\lambda)C$ (proportional to traffic). Let T_s be the solution given in Eq. (5.29).

    (a) Find T_p/T_s.

    (b) Prove that $T_p/T_s \geq 1$. Under what conditions will $T_p = T_s$?

5.9. Consider an M/M/m system with $\bar{x} = m/\mu C$. Prove that $m = 1$ gives a minimum $\bar{F}$.

5.10. Consider a simple communication network consisting of two nodes in tandem as shown below.

Assume that all messages originate from a Poisson source (at rate λ) and arrive to node 1 where they must queue while awaiting transmission on channel 1 (capacity C_1) after which they queue for transmission over channel 2 (capacity C_2). Let $C_1 = C_2 = 1$. All messages are exponentially distributed with a mean of $1/\mu$ bits. Let x_n be the message length of the nth message (in bits) and t_{in} be the time between arrivals of the $(n-1)$st and nth message at node i.

    (a) Find the pdf for t_{1n} and for t_{2n}.

    (b) Find the joint density $p(x_n, t_{2n})$ under the assumption that messages maintain their original length (that is, *no* independence assumption). Note that this proves the dependence of x_n and t_{2n}.

5.11. **(a)** Derive Eqs. (5.30)–(5.35).

    (b) Compare the delay T when capacity is optimized for $k = 0$ and $k = 1$.

5.12. Consider a computer-communication network with M channels with a logarithmic cost function as given in Eq. (5.36). *Derive* the optimal channel capacity assignment explicitly in terms of $\{\lambda_i\}$, $\{d_i\}$, μ, α, D, M, and the Lagrangian multiplier.

5.13. Consider a computer-communication network with M channels with a power-law cost function as given in Eq. (5.37). Show that the optimum solution for C_i is as defined in Eqs. (5.38)–(5.39).

5.14. Consider a network that has a nonlinear cost function $d_i(C_i)$ that is continuous and strictly increasing. Let $T^{(\infty)}$ be the performance measure we wish to minimize.

    (a) Find the optimum C_i in terms of $T^{(\infty)}$ and the cost function.

    (b) Find the (single) nonlinear equation that defines $T^{(\infty)}$; assume the total cost is to be D.

5.15. **(a)** Prove Eq. (5.45).

    (b) Solve the CFA problem in dual form for linear costs $d_i(C_i) = d_i C_i$ and prove Eqs. (5.46) and (5.47).

(c) Show that D as given in Eq. (5.47) is concave with respect to $\{\lambda_i\}$.

(d) Prove Eq. (5.48).

5.16. Consider an idealized communication network with $N=4$ nodes for which our simple M/M/1 model from Eq. (5.19) holds. Assume $\mu = 1$ and that we have the following traffic matrix (γ_{jk}) and the following routing matrix (r_{ij}), where $r_{ij} =$ number of the next node to visit if the message is currently in node j and has node i as its final destination:

$$(\gamma_{jk}) = \begin{bmatrix} 0 & 2 & 1 & 2 \\ 1 & 0 & 1 & 1 \\ 4 & 1 & 0 & 1 \\ 1 & 1 & 0 & 0 \end{bmatrix}, \quad (r_{ij}) = \begin{bmatrix} - & 3 & 1 & 3 \\ 2 & - & 1 & 2 \\ 2 & 3 & - & 3 \\ 4 & 3 & 1 & - \end{bmatrix}$$

(a) Find the average path length $\bar{n}$.

(b) Suppose we are given a total capacity $C = \sum_i C_i = 34$. Find that assignment of capacity $\{C_i\}$ which minimizes the average message delay T.

(c) Draw the net and label each branch with an arrow and the pair (λ_i, C_i).

(d) Evaluate T.

(e) Define the "length" of a channel as in Eq. (5.40).

  (i) Find the length of all paths from node 1 to node 3 and identify the shortest path.

  (ii) Is this the path taken by the γ_{13} traffic? Will your answer always be the same, in optimized networks?

(f) Now suppose we make a *proportional* capacity assignment. Find $\{C_i\}$ and T, and compare to part (d).

(g) Now suppose we change the routing matrix as follows:

$$(r_{ij}) = \begin{bmatrix} - & 1 & 1 & 1 \\ 2 & - & 2 & 2 \\ 3 & 3 & - & 3 \\ 4 & 4 & 4 & - \end{bmatrix}$$

If we have the same (γ_{jk}), μ, and C, and we make an optimal channel capacity assignment, what will be the value of T?

5.17. Consider a computer-communication network with a linear cost function $D = C = \sum_i C_i$ ($d_i = 1$). Also assume that fixed routing is used *except* that between node n_1 and n_2, there are two equal-length alternative paths (of length L) such that of the traffic going from n_1 to n_2, an amount λ_1 goes over the first path (path *a*) and an amount λ_2 goes over the second path (path *b*), as shown below.

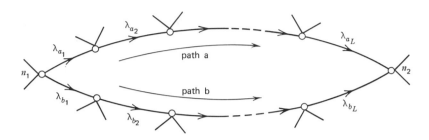

λ_{a_i} (λ_{b_i}) is the total channel traffic on the ith branch of path a (path b). If capacities are always assigned optimally, show that the overall network delay T can always be reduced by routing *all* the $n_1 \rightarrow n_2$ traffic $(\lambda_1 + \lambda_2)$ either over path a or path b, but not both.

5.18. Find **H,** the matrix of shortest paths for the distance matrix $\mathbf{D}_0$ below:

$$\mathbf{D}_0 = \begin{bmatrix} 0 & 2 & 2 & \infty & \infty \\ 2 & 0 & 1 & \infty & 3 \\ 2 & 1 & 0 & 3 & 1 \\ \infty & \infty & 3 & 0 & 1 \\ \infty & 3 & 1 & 1 & 0 \end{bmatrix}$$

5.19. The linear feedback channel (see p. 377) has a backlog $N(t)$ that is a Markov process with stationary transition probabilities p_{ij}. Find p_{ij} in terms of M, σ, and p.

5.20. Prove Eq. (5.56).

5.21. Prove Eqs. (5.59)–(5.61).

5.22. Prove Eq. (5.62).

5.23. Prove Eqs. (5.63)–(5.65).

5.24. From Eqs. (5.63)–(5.65), prove that

$$\lim_{K \to \infty} \frac{S}{G} = \lim_{K \to \infty} q = \lim_{K \to \infty} q_t = e^{-G}$$

5.25. Prove Eq. (5.68).

5.26. From Eq. (5.68), show that

$$\lim_{K \to \infty} \frac{\partial T}{\partial K} = \frac{1 - e^{-G}}{2e^{-G}}$$

5.27. Consider the random variable Y in Figure 5.57.
   (a) Show that

$$P[Y \le y] = \begin{cases} 0 & y < 0 \\ e^{-G(a-y)} & 0 \le y \le a \\ 1 & a < y \end{cases}$$

   (b) From this, show that $\bar{Y} = a \quad (1 - e^{-aG})/G$.

5.28. Prove Eq. (5.82) by first providing an interpretation for the quantity $(G/S) - 1$.

6

Computer-Communication Networks:
Measurement, Flow Control, and
ARPANET Traps

"Without measurement, it is difficult to have a science."* The truth of this statement derives from the fact that our models of the real world are only that—just models. In order to validate these models and to improve upon them one must go out into the real world and make measurements to determine the true fashion in which nature behaves. It is against such measurements that our mathematical models must stand up. So far, in the previous chapter we have basically been concerned with theoretical modeling, analysis, and optimization of computer-communication networks. In this chapter, we explore the validity of these tools in terms of simulation and measurement experiments conducted on the ARPANET. Happily, we find that the models are fairly robust under this test. These experiments also exposed some deadlocks, degradations, and traps due to the subtleties of the flow control procedure; these, too, we discuss below. Indeed, by its very nature, the easiest and most "dramatic" role of network measurement (and therefore also of the Network Measurement Center at UCLA) is to identify the weak points of system operation. As a result, in this chapter we tend to dwell on the shortcomings (rather than the strong points) of the ARPANET. The reader is urged not to take this as a criticism of the ARPANET, but merely as an attempt to emphasize the lessons to be learned from the ARPANET experiment. It is only through such objective introspection that deep understanding can develop and eventually lead to mature system development. In fact, the AR-PANET experiment has been a huge success and functions beautifully as an advanced communications service for most of its users most of the time. It is no surprise that through these experiments, we identify a large

* A favorite quote of Richard W. Hamming (Bell Telephone Laboratories, Murray Hill, New Jersey).

422

number of new phenomena and new problems that require analysis and understanding and for which research is currently being conducted.

6.1.   SIMULATION AND ROUTING

We have already seen some comparison between our basic network queueing model and *simulation* results for the ARPANET (see Figures 5.22 and 5.23). There we were basically concerned with single-packet message delay as a function of the load (i.e., throughput) on the network. We found the agreement between simulation and theory exceedingly good and were able to identify and explore the threshold behavior due to "premature" saturation. The reader will recall that in the ARPANET, messages up to 8063 data bits are acceptable as one message which are further decomposed into packets (with a maximum size of 1008 bits). A single-packet message, of course, is one that requires no more than one packet; on the other hand, recall that a multipacket message is one that requires between two and eight packets. In [FULT 72] a multipacket model has been developed and a comparison between those analytical results and simulation results is shown in Figure 6.1 (see Section 6.6 for

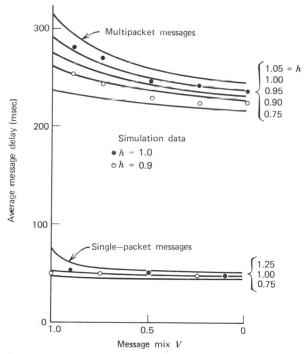

Figure 6.1   Message delay for the 19-node net versus message mix.

further details of the multipacket model). For this simulation it was assumed that single-packet messages had priority over multipacket messages when competing for a channel. In the figure we plot message delay as a function of V, the percentage of traffic that is single packet while maintaining a constant average bit rate into the network; when $V = 1.0$ we have all single packets, whereas $V = 0$ corresponds to all multipackets. The multipacket model is similar to Eq. (5.20) and provides the solid theoretical curves in this figure. The dots represent simulation results. (Recall that h is the scaling factor on network load.) From this figure it is clear that we have generated a suitable model for predicting average message delay in such networks.

We may put our simulation experiment to much better use by attacking problems that do not lend themselves to mathematical analysis. Such a case is the evaluation of adaptive network routing procedures in which routing decisions are based upon the dynamics of the network on an instantaneous basis. It is clear that an efficient message routing procedure is an essential ingredient for the successful operation of a computer network. As we have said, the function of a routing procedure is to direct message traffic along paths within the network in a fashion which avoids congestion. (Also, in conjunction with the routing procedure one must provide methods for controlling the flow of traffic entering the network which would otherwise congest the system; this we have defined earlier as a flow control procedure and we shall consider it below.) The basic requirements for a good routing procedure are as follows:

1. It should ensure rapid and reliable delivery of messages.
2. It should adapt to changes in the network topology resulting from nodal and channel failures.
3. It should adapt to varying source-destination traffic loads.
4. It should route packets away from temporarily congested nodes within the network.
5. It should determine the connectivity of the network.
6. It should allow easy and "automatic" insertion and deletion of IMPs.

In [FULT 71] a number of routing algorithms have been defined and classified. Using the FD method, it has been possible to generate approximate lower bounds on the message delay as a function of relative load (h) for various of these algorithms, and in Figure 6.2 this lower envelope is shown. This figure corresponds to the 19-node network shown in Figure 5.23. The performance of a routing procedure similar to the one *originally* used in the ARPANET, namely a periodic update algorithm (PUA1) is shown as the upper curve and we see that its performance

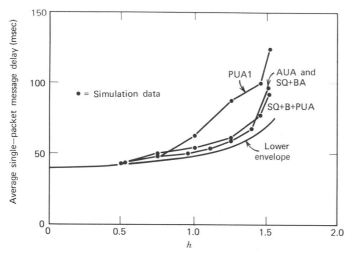

Figure 6.2 Comparison of routing algorithm performance as a function of network load.

deteriorates and departs from the optimum as the network load grows. The principle feature of this algorithm is that it is a *distributed* control algorithm whereby no overall decision-making authority is vested in some particular location. All nodes make *local* routing decisions in a dynamic fashion as follows. Each node (say node n) keeps a routing table, which is simply a directory with one entry per destination in the net. This entry (say, for destination k) gives the name of the node to which node n will route *all* traffic it receives which is destined for node k. In Figure 6.3(a), we show the case where node n has four neighbor nodes (4, 7, 8, 17). The routing table in Figure 6.3(c) indicates that all traffic passing through node n and destined for node k will be routed next to node 7. This entry in the routing table is, in effect, arrived at as follows. Every so often (periodically), each neighbor node will deliver a column to the delay table in node n which gives that neighbor's current best estimate of his minimum delay to all destinations; to each entry in this column, node n will add an amount $4+q$, where q is the number of packets currently on the output queue from node n to this neighbor (the quantity 4 is an arbitrary value to represent the delay in passing over one hop). These entries are assembled into the delay table [see Figure 6.3(b)]. Node n then selects the minimum across each row to form the routing table; in addition to the next node, the routing table contains the value of the best (minimum) estimated delay. In fact, the entire delay table is no longer kept in the ARPANET IMPs since the periodic updating (which actually occurs every 640 msec on a heavily loaded 50 KBPS channel) is randomly phased

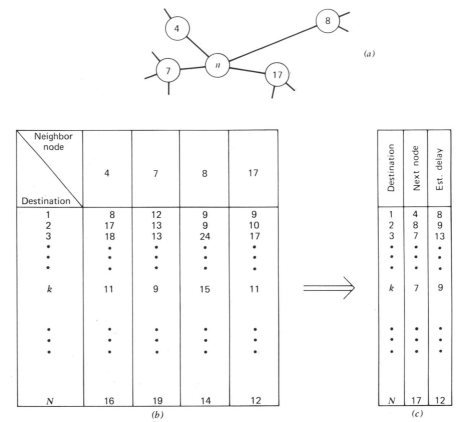

Figure 6.3   Formation of the routing table.

among nodes. Therefore, whenever an update (column) comes in from a neighbor (and $4+q$ is added to each entry), each delay entry is directly compared with that in the routing table; if the new delay for some k is smaller than the old estimate, then that entry is replaced and the route to k will be changed (if this update came from a neighbor that was not the next-node entry originally). Currently, some refinements to this procedure are being considered. We note that this neighbor-to-neighbor updating is really carrying global routing information. One of the successes of the ARPANET has been to demonstrate that such a distributed routing control procedure is basically stable and can converge to fairly efficient routes. It is reasonably responsive to network nodal and channel failures (see below), but more important, it can automatically become aware of a new node as soon as it is connected (or repaired and returned) into the network; this growth capability is an excellent engineering feature of the ARPANET.

Returning to Figure 6.2, we see that other procedures using an asynchronous update algorithm (AUA)* or the shortest queue plus bias algorithm (SQ+BA)* are superior to PUA1, although they are slightly more costly in terms of the processing load that they place upon the IMP. A good candidate for a routing algorithm is the shortest queue plus bias plus periodic update algorithm (SQ+B+PUA),* which is reasonably simple to implement and not too costly in processing load [FULT 72]. One of the important problems in the consideration of routing procedures is to eliminate loops; a loop is caused when the routing procedure sends a message around a portion of the network and eventually returns that message to a previously visited node; clearly, if such a loop persists, messages get trapped in the network and the average delays grow to infinity. Any effective routing procedure will either avoid loops or destroy them shortly after they are formed. In the PUA1 algorithm it is not uncommon to observe short-lived loops of length 2 ("ping-ponging" between two nodes). Looping has become an important issue in the ARPANET [KLEI 75] and loop-free routing algorithms are under investigation [NAYL 75].

Another effective use of the simulation program has been to observe the *adaptability* of routing procedures in the presence of unreliable channels. For the 19-node network mentioned above, we simulated the transient response to a line failure followed later by the repair of that line. In Figure 6.4, we show the smoothed† average delay T as a function of time for the case where a critical (cross-country) channel (channel 4-17) is declared "broken" at time $t = T_F$ and is later declared repaired at time $t = T_R$. We also show the behavior when a noncritical channel (channel 8-11) is broken at T_F and repaired at T_R. For reference, we plot the case of no failures as well. In Figure 6.4(a) these three cases are shown for the PUA1 algorithm; we see a poor response to failure of the critical channel (delays get very large and little, if any, adaptation seems

---

* Whereas in the PUA1 algorithm, routing information is passed between neighboring IMPs roughly twice a second, permitting a rather frequent updating of the routing tables, with AUA updating takes place between neighbors only when critical delay thresholds are crossed. The SQ+BA algorithm does not exchange any data between neighbors, but merely makes routing decisions based upon internal queue lengths (SQ) and shortest path lengths (bias, B). Finally, the SQ+B+PUA algorithm adds to SQ+BA a periodic update between neighbors to announce channel and IMP failures (this updates the path length values).

† The value for T is a geometrically averaged (smoothed) value for delay over its present and past values. At $t = T_R$, the smoothing is reinitialized.

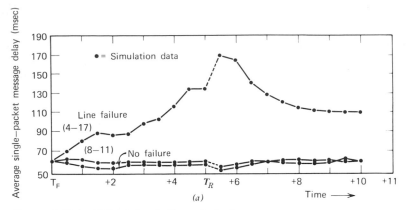

Figure 6.4(a)   PUA1 (periodic updating).

to take place) and slow (incomplete) recovery after repair at T_R. In part (b), we show the response for the AUA algorithm; we see a much improved behavior, both in adapting to the failure and in recovering after repair. In part (c), we see the excellent performance of the SQ+B+ PUA algorithm. The rapidity with which a routing algorithm adapts to these conditions is a most important measure of that algorithm's performance. Currently, the routing procedure in the ARPANET is undergoing modifications to accommodate the increased number of nodes and higher traffic levels [MCQU 74a, 74b]. One modification currently allows for variable-rate updating of routing tables where the rate increases with the channel's capacity and decreases with the channel's utilization. A routing

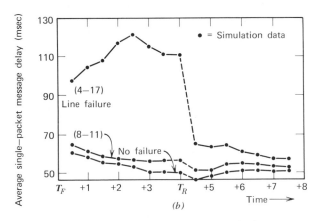

Figure 6.4(b)   AUA (asynchronous updating).

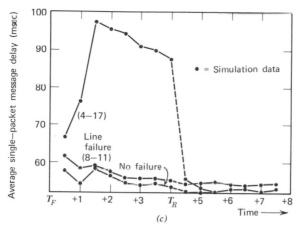

Figure 6.4(c) SQ+B+PUA (shortest queue plus bias plus periodic updating).

algorithm that searches for high-bandwidth paths rather than paths of low delay as at present is also under consideration; this point of view is motivated in part by the anticipated growth in file transfers and remote job entry. Additional considerations address themselves to the peculiar needs of satellite channels and to the concept of regional routing procedures that look attractive for large network design [KAMO 76].

6.2. EARLY ARPANET MEASUREMENTS

But simulation is not enough, and when one has the opportunity (as we do with the ARPANET), one should perform direct measurements on the system. Early in the design of this network, it was understood that the ability to make direct measurements was of paramount importance and would form an essential part of the experiment of creating the network itself. Consequently, a fairly sophisticated measurement capability has been built into the network both within the IMPs and within certain of the HOST computers (especially at the UCLA Network Measurement Center—NMC). These measurements provide information regarding the traffic delays, throughput, congestion and deadlocks in the network, provide means for testing the limits of its traffic-carrying capacity, provide a method of validating and improving the theoretical models, and lastly, eventually may provide an on-line mechanism for controlling network traffic en route.

The measurement tools provide six capabilities: trace; accumulated statistics; snapshots; artificial message generation; control, collection, and analysis; and status reports. The first four of these are statistics (or traffic) that are collected (or generated) within the network IMPs only when one (or more) of these tools have been enabled; these IMPs then create measurement packets and send them to any HOST that has requested them to gather these statistics (typically, the HOST is the UCLA Network Measurement Center). The fifth is a collection of programs written and used by the UCLA-NMC for control and analysis of the measurement experiments. The sixth is a continual sequence of IMP-generated reports that are sent to the Network Control Center (NCC) at BBN. The details regarding these various measurements follows.

Trace is a mechanism whereby messages may be "traced" as they pass through a sequence of IMPs. Those IMPs whose trace parameter has been set will generate one trace block for each marked packet (i.e., a packet with its trace bit set) that passes through that particular IMP. A trace block contains four time stamps that occur when (1) the last bit of the packet arrives; (2) the packet is put on an output queue; (3) the packet starts transmission; and (4) transmission is completed correctly to a neighboring IMP (i.e., acknowledgement received) or to a HOST. Also contained in the trace block are the length of the packet, an address indicating where the packet was sent, and the IMP header. The IMP header consists of the source and destination addresses and several other pieces of control information. An "auto trace" facility exists by which every nth message of the γ_{jk} traffic may be stamped for tracing; also, a new feature of "packet trace," will trace every nth packet passing through a designated IMP.

The *accumulated statistics* message consists of several tables of data summarizing activity at a network node over an interval of time that is under program control. First is a summary of the HOST activity. This consists of message-size statistics for messages leaving and entering the set of HOSTS connected to that IMP. (Individual HOST activity is not available; only summary data is collected.) The message-size statistics include a histogram of message lengths (in packets) for multipacket messages and a log (base 2) histogram of packet lengths (in words) for all last packets (i.e., a count is recorded of those packets whose data content is from 0 to 1, 2 to 3, 4 to 7, 8 to 15, 16 to 31, or 32 to 63 IMP words in length). Also included are the round-trip statistics. These contain the number of round trips (message sent and RFNM returned*) sent from the probed site to each other site and the total time recorded for those round trips. These statistics are listed for each possible destination from the probed site. Second are the probed site's channel statistics, which consist

---

* The RFNM is an end-to-end message acknowledgement which means "request-for-next-message". See Section 6.3 for the details.

of (1) the number of "hellos" sent per channel (channel test signals, which really are the routing update messages), (2) the number of data words sent per channel, (3) the number of inputs received per channel (all inputs: data packets, control packets, acknowledgements, etc.), (4) the number of errors detected per channel, (5) the number of "I-heard-you" packets received per channel (response to hello), (6) the number of times the free buffer list was empty, and (7) log histograms of packet length, in data words (one histogram per channel).

Snapshots give an instantaneous peek at an IMP. The snapshot records HOST and channel queue lengths. Also included is information about storage allocation: the length of the free storage list, the number of buffers in use for store-and-forward packets, the number of buffers in use for reassembly of messages, and the number of buffers allocated to reassembly (but not yet in use—see the reservation schemes below). Snapshots also include the IMP routing table. Entry k in the routing table contains the channel address indicating where to send a packet destined for site k, the minimum number of hops to site k and the delay estimate to site k.

In addition to the above instrumentation package built into each IMP, we have a capability for *artificial message generation*. The message generator in any IMP can send fixed-length messages to one destination at fixed or RFNM-driven interdeparture times. Together with the generation facility, there exists the capability to discard data at "fake" HOSTs in each IMP. (Several message generator–acceptor pairs have also been implemented for a subset of the HOSTs on the network as well; these are extremely useful for experimentation.) The above-mentioned measurement and message generation facilities are turned on and off by sending messages to the "parameter change" background program in the IMPs. A set of programs at UCLA automatically formats and sends the correct parameter change messages to initiate experiments.

Control, collection, and analysis of the data by the NMC is accomplished by supplying specific subroutines for a general driver program; the data analysis is currently done on the UCLA 360/91.

In addition to the above tools, which are mainly for experimental use, the NCC has built into the IMPs a monitoring function called *status reports*. Each IMP sends a status report to the NCC HOST once a minute (actually, every 52 sec) and also whenever exceptional conditions occur. Contained in the status report are the following: (1) the up–down status of the HOSTs and channels; (2) for each channel, a count of the number of hello messages that failed to arrive (during the last minute); (3) for each channel, a count of the number of packets (transmitted in the last minute) for which acknowledgements were received; and (4) a count of the number of packets entering the IMP from each HOST. These status reports are continually received at the NCC and this data is processed by a minicomputer that advises the operator of failures in the network and creates summary statistics.

Numerous measurements of round-trip message delay, throughput, utilization, traffic flow and HOST behavior have been conducted at UCLA, and we wish now to compare some of these measurement results with some of the theoretical and simulation results we have so far discussed.

In [COLE 71], it was shown (by accounting for processing, transmission, and propagation times) that at that time, the expected round-trip delay between two nodes in the ARPANET (SRI-UTAH) could easily be calculated to give $20 + 7.68b$ msec, where b is the length of the message (in bits) traveling this round trip; this equation assumes that no other traffic exists in the network and, therefore, provides a strict lower bound on such delays. In Figure 6.5, we give the measured results for this round-trip delay and also show the theoretical equation just described; the calculations are shown to be accurate and the delays above the minimum are due to interfering traffic within the network. A second revealing measurement was made early (also on the original system before certain improvements such as acknowledgement piggybacking

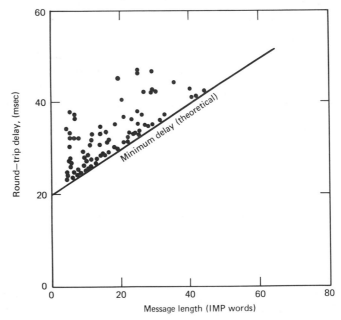

Figure 6.5   Round-trip delay measurements of the 1971 SRI-UTAH traffic.

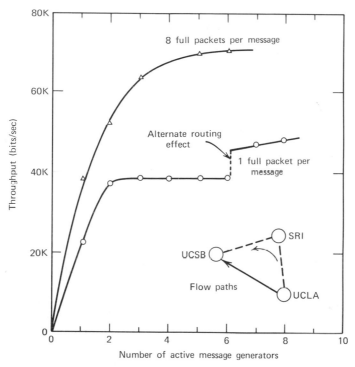

Figure 6.6 Throughput measurements between UCLA and the neighboring UCSB IMP.

were introduced—see below) to determine the throughput handling capability and this is shown in Figure 6.6. Here we see the flow of traffic from UCLA to UCSB. This test was made using an artificial traffic generator that has the capability of creating more than one "conversation" between a pair of sites. The original network flow control procedure constrained each conversation (a logical link between a sending and a receiving process) to have *at most one* message in the network at any time; only when that message was delivered could the next message be generated in this conversation. That is, when a message reached its destination, the flow control procedure sent a control message from the destination IMP back to the source IMP, which signaled its readiness to accept a new message from the source HOST into the network; this is the end-to-end message acknowledgement, that is, the RFNM (request-for-next-message). In Figure 6.6 we see the effect of this early control procedure. The lower curve corresponds to single-packet messages, and we see the growth in throughput as the number of generators (conversations) increases, reaching a plateau when two or more generators are in operation; this plateau occurs at approximately 38 KBPS, which corresponds to the effective raw data rate permissible on a single 50 KBPS line when overhead considerations in the original system are accounted for. This level of traffic is not attainable with a single message generator due to the enforced delays while awaiting the RFNM. Note that the routing procedure permitted alternate routing to take place when more than six generators were active; at this point a large enough queue is formed at UCLA so that at the next routing table update, alternate routing is triggered via SRI. The upper curve in this figure corresponds to full length messages composed of eight packets each. Here we do not expect the RFNM to constrain the throughput in a significant way, since it comes only once in every eight packets, and we note that a single message generator immediately achieves the "plateau" value of throughput seen earlier. In this multipacket mode we observe that its plateau is reached at about 70 KBPS; this is below the expected 76 KBPS for the two paths in use since the alternate routing procedure could not keep both paths busy and due to the finite storage capacity of the IMP. This and other early measurements indicated that one of the most critical resources in the network was the amount of storage in an IMP used to reassemble multipacket messages before they are delivered to the destination HOST. In the original IMP operating system, when a multipacket message made its way through the network as separate packets (and had to be collected and reassembled at the final IMP before delivery), eight buffers were first allocated at the destination IMP (for a maximum size message of eight packets) only when the first of these packets arrived at that IMP (rather than reserving this space ahead of time, as is done currently—see below). The original flow control procedure provided enough reassembly space in an IMP to handle only three multipacket messages simultaneously; this critical resource was a determining factor in throughput and delay for multipacket messages and was partially remedied with the additional 4K of IMP core memory, most of which was used for reassembly.

The throughput measurement described in Figure 6.6 was carried out in the presence of no interfering traffic from other sources. A more interesting experiment is to observe the mutual interference among competing conversations within the network. This was done for a network similar to that in Figure 5.20(d)* in which UCLA was attempting to send

---

* The measured network differed as follows: NCAR and GWC were not yet connected and so there was no link from UTAH to CASE and McCLELLAN, USC, and TINKER were not yet inserted in the lines connecting their neighbors; furthermore, in series with the MIT-BBN link there was an additional node (BBN-TIP), and in series with the SRI-AMES link was another node (AMES-TIP).

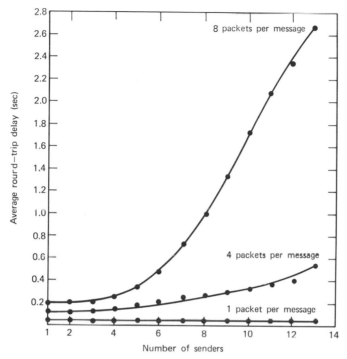

Figure 6.7   Measured round-trip times.

message traffic to RAND. Measurements were taken at UCLA as, one by one, at 6-min intervals, additional nodes were instructed to send interfering traffic to RAND (using each IMP's artificial traffic generator). The order in which new IMPs entered the sequence is as follows: UCLA, BBN, SDC, STAN, HARV, UTAH, AMES, MIT, ILL, BBN-TIP, CASE, LINC, and finally, CARN.

In Figure 6.7 we show the way in which the average round-trip delay from UCLA to RAND (and back for the RFNM) varies with the number of interfering users. Three curves are shown, each with different message lengths; these are the lengths used for all messages generated within the network for this experiment. We note for the single-packet messages that essentially no interference occurred as the number of users increased;* however, with four and, in particular, with eight packets per message the interference caused significant round-trip delays to the message traffic. A large part of this delay was due to the competition for reassembly storage space at the RAND IMP. In Figure 6.8 we see a similar plot for the

* This result is conditional on the fact that UCLA is directly connected to RAND.

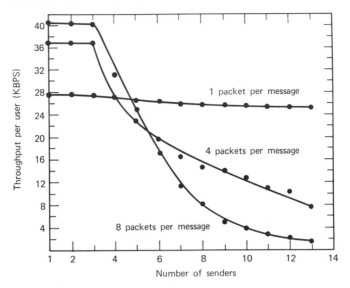

Figure 6.8   Measured throughput per user.

throughput per user (i.e., UCLA-RAND). As expected, the single-packet message flow is maintained at an essentially constant data rate. When we go to four packets per message and then to eight packets per message we observe that the data rate increases with message size as long as the number of senders is less than four. This increase in throughput is due to the large penalty paid by small (single-packet) messages for the flow control procedure with its use of RFNMs. The delay due to a RFNM comes once per packet for single-packet messages but only once per eight packets for eight-packet messages and this accounts for the increased throughput for a small number of interfering users. The reason that the multi-packet message throughput begins to deteriorate seriously when four or more senders are active is due to the finite storage capacity at each IMP that is set aside for the reassembly of multipacket messages as mentioned above. From Figures 6.7 and 6.8 we see the effect of this interference on the increase in delay and the decrease in throughput. However, one may argue that the *total* throughput in the network may be increasing as the number of conversations increases. In Figure 6.8 we displayed the throughput *per user* as a function of the number of users. We may *approximate* the total throughput as a function of the number of users by taking each measured point in Figure 6.8 and multiplying by the number of users to which that point corresponds; this, of course, is a fiction, since the measured points from Figure 6.8 correspond to the throughput seen

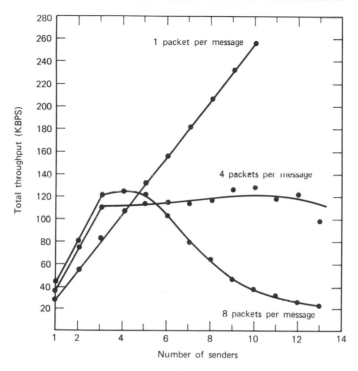

Figure 6.9 Estimated total throughput.

in the UCLA-RAND transmission only. However, this derived curve will provide some insight as to what is happening with regard to total throughput, and this we show in Figure 6.9. We see that the single-packet-per-message case yields a linearly increasing total throughput that is part of the fiction since, clearly, this function can rise only until it reaches the maximum throughput capability of the IMP itself. In the case of four packets per message, we see that the throughput levels off when three or more senders are active, and this saturation level occurs somewhat above 100 KBPS. The more interesting (albeit unfortunate) case is for eight packets per message in which the total throughput peaks at around four users and then begins to fall off in a disastrous fashion* as the network goes into what is known as *reassembly lockup* [KAHN 71]; indeed, for more than 13 interfering users we were able to reduce the throughput to zero! Reassembly lockup, the most famous of the ARPANET deadlock

---

* This is similar to the fundamental diagram of road traffic [HAIG 63] for which strong analogies may be drawn.

conditions, was due to a logical flaw in the original flow control procedure. It occurred when partially reassembled messages could not be completely reassembled since the congested network prevented their remaining packets from reaching the destination; that is, each of the destination's neighbors had given *all* of their store-and-forward buffers to additional messages heading for that same destination (at which no unassigned reassembly buffers were available). Thus the urgently needed remaining packets could not pass through the barrier of blocked IMPs surrounding the destination.

The importance of this reassembly lockup problem indicated the need for a new flow control procedure within the ARPANET which we describe in the following section; basically, it guarantees that reassembly space is available for all messages in the net. The general behavior of a network in which nodes become "blocked" (due to lack of sufficient buffer space, etc.) appears to be a relatively important effect in networks. In [ZEIG 71] a theoretical study is described in which an approximation for the behavior of networks of queues with blocking is formulated. The success of that particular model in describing one form of network blocking is surprisingly good.

6.3. FLOW CONTROL

As we promised in Section 5.4, we now discuss the philosophy of flow control and describe some of the various control procedures that have been used in the ARPANET. In later sections, we identify some of the traps (deadlocks and degradations) they have caused and calculate the overhead required to implement these procedures [OPDE 74, KLEI 75].

Whenever two information processing systems exchange data, carefully designed control procedures are necessary to ensure safe and correct transfer of the information. Internal network control procedures (such as the routing procedure) deal with packets that have already been accepted by the network. However, there is an equally important decision to be made concerning the admission of packets to the network in the first place. Networks cannot afford to accept all the traffic that is offered without some control. This would inevitably lead to heavy congestion (and possibly lockup!) of the entire network. There must be rules to govern the acceptance of traffic from the outside. These rules are commonly known as *flow control procedures.*

There are two types of flow control: local flow control and global flow control. Local flow control is an important characteristic of any packet-switched network. It is a direct consequence of the limited buffer space in each node. Whenever this buffer space is used up, a node has to stop further input from outside the net and from its neighbors. To avoid congestion, the input is usually stopped even before all the buffers are occupied. For example, there may only be a limited number of packets allowed on each output queue. Packets that would have to join a full output queue are rejected. There may also be a limit on the total number of "store-and-forward" buffers such that when all store-and-forward buffers are occupied, that node will accept only those packets which leave the network at that node (e.g., "reassembly packets") [HEAR 70, MCQU 72].

The existence of local flow control procedures implies that packets may experience an *admission delay* that will be nonzero in case of (local) congestion. It is important to realize that this admission delay contributes to the total delay in the same way as the queueing delay in the output queues.

Local flow control alone is not sufficient to avoid congestion in a packet-switched network. There also needs to be some limitation on the total number of packets that can be handled by the network simultaneously. Procedures that achieve this limitation are called global flow control procedures. If the global flow control works properly, further input to the communication network is stopped well before all the buffer space in the net is occupied. There are two methods of global flow control that have been investigated: end-to-end flow control (ARPA-like [MCQU 72]) and isarithmic flow control (NPL-like [DAVI 71]).

When the flow control procedure is not designed properly, it may lead to lockup or deadlock conditions, which are among the most serious system malfunctions that can occur in a computer system or network. Communication protocols have to be designed very carefully to avoid the occurrence of these lockups. The common characteristic of lockups is that they occur only under unusual circumstances that were not foreseen or deemed too unlikely to occur by the protocol designers. (However, these designers often are not the ones in a position to evaluate such likelihoods quantitatively.) We have already seen in the previous section how reassembly lockup could occur in the subnet when reassembly space was unavailable to store incoming multipacket messages. Lockups were predicted and discussed in [KAHN 71].

Direct store-and-forward lockup is another example of a lockup that can occur in a packet-switched network if no proper precautions are taken [KAHN 71]. Let us assume that all store-and-forward buffers in some IMP A are filled with packets headed to some destination IMP C through a neighboring IMP B and that all store-and-forward buffers in IMP B are filled with packets headed to some destination IMP D through

IMP *A*. Since there is no store-and-forward buffer space available in either IMP *A* or *B*, no packet can be successfully transmitted between these two IMPs and a deadlock situation results. There is, of course, an easy way to remedy this situation. One has to make sure that not all of the store-and-forward buffers can reside on a single output queue. (In the ARPANET, only eight of the 20 pooled store-and-forward buffers can be placed on a single output queue.) *Indirect store-and-forward lockup* can occur when all the store-and-forward buffers in a loop of IMPs become filled with packets which all travel in the same direction (clockwise or counterclockwise) [KAHN 71]. Although such a highly structured traffic pattern is very unlikely to occur, it can be shown that, for the lockup to establish itself, this undesirable packet flow need only persist for about 1 sec in the ARPANET.

The characteristic of the original ARPANET flow control procedure was to accept only one message at a time between any two end-to-end communicating processes. Specifically, a new message would not be accepted at the source IMP until the end-to-end message acknowledgement (RFNM) was sent from the destination IMP back to the source IMP. This guaranteed the proper sequencing of process-to-process messages. Of course, this permitted many different pairs of communicating processes to send messages simultaneously. In Figure 6.10(*a*) we show the details of a single-packet transmission along a 3-hop, 4-IMP path (Source IMP $A \rightarrow$ IMP $B \rightarrow$ IMP $C \rightarrow$ Destination IMP D). At time *a*, the packet (message number 1) arrives at *A*. After a queueing delay (w_A) the packet is transmitted to *B* and an ACK is returned to *A*. After another queueing delay (w_B) the packet is transmitted to *C*. We assume that line noise corrupts this transmission causing IMP *C* to detect an error and to discard this packet. After a 125-msec timeout for the ACK, *B* transmits the packet to *C* again; this time *C* receives it correctly, and indicates this to *B* with the ACK.* The last hop is from *C* to *D*; destination IMP *D* first acknowledges the *C* to *D* transmission and then generates a RFNM that makes its way back through the net (causing ACKs along the way) reaching source IMP *A* at time *c*. In Figure 6.10(*b*) we show the simplified notation to represent the arrival of message 1 at time *a*, its reception at the destination at time *b* and the delivery of the RFNM at time *c*. Whereas this simple RFNM control appeared innocent enough, it had the unfortunate property of permitting demands (messages) to enter the net for which all the necessary resources had not been allocated (i.e., reassembly buffers), and this led to the catastrophic reassembly deadlock

---

* This IMP-to-IMP acknowledgement procedure has remained basically unchanged up to the present time (except for the feature of permitting ACKs to "piggyback" on packets travelling toward the waiting IMP; see below).

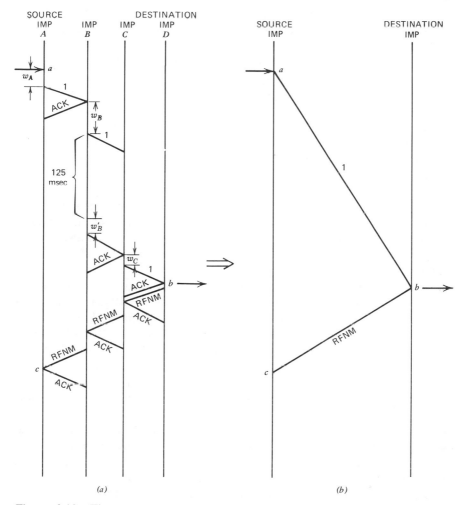

Figure 6.10 The anatomy of a single-packet transmission. (a) Details showing ACKs, queueing, and retransmission. (b) Equivalent diagram without details.

mentioned earlier. Aside from this deadlock, the ARPANET has crashed occasionally because of hardware failures. Although transmission line errors are handled by cyclic error detecting codes that are hardware generated and checked by the line modems, no protection against hardware errors in the IMP was provided originally. This led to some amusing network crashes. For example, there was the case of an IMP that improperly generated routing update messages claiming it had zero-delay paths to all destinations in the net! This IMP became an absorbing node for an unlimited amount of traffic, finally bringing the network to its knees. In another case a certain amount of chaos was caused when an IMP claimed it was the UCLA IMP (which it was not!). These hardware errors are now detected by the inclusion of a (16-bit) software checksum that accompanies all packets in their journey through the net.

The difficulties with the original flow control procedure (which we shall refer to as version 1) led to some major changes in the spring of 1972 [MCQU 72], which we shall refer to as version 2. (Additional changes were made in December 1974 which produced version 3, to be described in Section 6.8. The principal features of these three versions are listed in Table 6.8 in Section 6.8.) Let us now describe version 2.

In version 2 of the ARPANET flow control was implemented by setting a maximum number of four messages that could be outstanding between any pair of source-destination IMPs. Note that this was a significant departure from version 1, which placed a flow constraint on individual communication links between a given pair of HOST processes. This scheme has two major advantages: (1) It is easy to implement (whenever four RFNMs are outstanding, further input from *any* of the HOSTs attached at the source end of the pair is stopped); (2) The input to the net at the source is stopped rapidly when the destination IMP or HOST goes down. Unfortunately, it requires different HOSTs attached to the same IMP to compete for access to these four "tokens" and therefore to influence each other's performance. Furthermore, it forces one to invent methods that guarantee the proper sequencing of messages.

As mentioned, the flow control mechanism in the ARPANET was modified in some significant ways to avoid the occurrence of reassembly lockup. Specifically, in version 2, no multipacket message is allowed to enter the network until storage for that message has been allocated at the destination IMP. As soon as the source IMP takes in the first packet of a multipacket message, it sends a small control message to the destination IMP requesting that reassembly storage (i.e., eight buffers) be reserved. It does not take in further packets from the HOST until it receives an allocation message in reply. The *high bandwidth for long sequences* of messages is obtained by requiring only the first message in a sequence of multipacket messages to go through the reservation procedure (i.e., sending a request for an eight-buffer allocation and awaiting the allocation); from that point on, the reservation is held for the message sequence as long as more data is delivered to the source IMP within 250 msec* after the RFNM is received (see below). We obtain *rapid response for*

---

* When this version (2) of the IMP operating system was first installed in the spring of 1972, this timeout was originally 125 msec. It was changed to 250 msec in 1974.

short messages (single-packet messages) by transmitting them directly to their destination while maintaining a copy at the source IMP. If there is buffer space at the destination, then the message is accepted and passed on to the HOST. At this point, the RFNM is generated and finds its way back to the source IMP (which then discards its copy of the message). If no space is available at the destination, this single-packet message is then discarded and this transmission is then considered to be a request for space; when space becomes available, the source IMP is so informed, at which point the (stored) single-packet message may be retransmitted. This permits immediate delivery of short messages when space is available at the destination. Note that version 2 requires storage of the first packet of a message at the source IMP.

In the following description of version 2, we shall refer to Figure 6.11 to illustrate the sequence of events in handling the transmission of a data

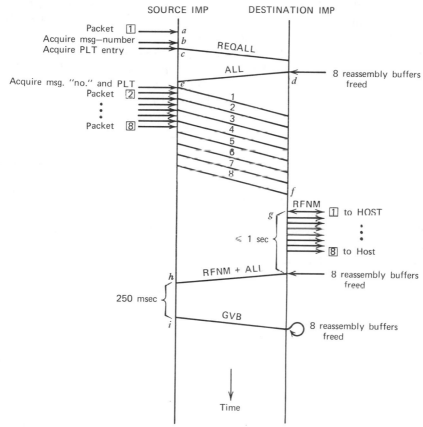

Figure 6.11 The sequence of events for a multipacket message transmission.

stream which requires one full (i.e., eight-packet) multipacket message. A message is treated as a multipacket message if the HOST-IMP interface has not received an end-of-message indication after the input of the first packet is completed (shown at point *a* in the figure). At this time, transmission of the remaining packets of this message from the HOST is temporarily stopped until the message acquires some network resources which we now describe. First, the multipacket message must acquire a message number (from the IMP), which is used for message sequencing (point *b*); recall that all messages originating at this IMP and heading to the same destination IMP share a common number space so that at most four such numbers may be assigned to messages in the net. Next, an entry in the *pending leader table* (PLT) must be obtained as shown at point *c*. The PLT contains a copy of the leader of all multipacket messages that are currently being handled by the source IMP. Among other things, the function of the PLT is to construct the packet headers for the successive packets of the multipacket message. Such an entry is deleted when the RFNM is received from the destination IMP. The PLT is shared by messages from all HOSTS attached to the same IMP and used for all possible destinations; the PLT can hold at most only six entries.

After the PLT entry has been obtained by the multipacket message, a table is interrogated to find out whether there are eight reassembly buffers reserved for this source IMP at the desired destination IMP. If this is not the case, a control message (REQALL) which requests an allocation of these buffers is generated and sent from the source IMP (also shown at point *c*) to the destination IMP. This REQALL steals the acquired message number and PLT entry for its own use at this point. This request is honored by the destination IMP as soon as it has eight buffers available (point *d*). To report this fact, a subnet control message (ALL) is returned to the source IMP thus delivering the eight buffer allocation. Since the previously acquired message number and PLT entry have been used, a new message number and a new PLT entry may have to be obtained for the multipacket message itself. (Had eight reassembly buffers been reserved in the first place, this would have shown at the source IMP by the presence of an unassigned ALL.) Only when all these events have taken place can the first packet begin its journey to the destination IMP and the input of the remaining packets be initiated, as shown at point *e*.

When all packets of the multipacket message have been received by the destination IMP (point *f*), the message is put on the IMP-to-HOST output queue. After the transmission of the first packet to the HOST (point *g*), the RFNM for this message is generated at the destination IMP (also point *g*) to be returned to the source IMP. This RFNM prefers to carry a "piggybacked" ALL (an implicit reservation of eight buffers for the next multipacket message) if the necessary buffer space is available. If not, the RFNM will wait for at most 1 sec for this buffer space. (In case the necessary eight reassembly buffers do not become available within this second, the RFNM is then sent without a piggybacked ALL.) The source IMP will wait roughly 30 sec for the RFNM before generating a control message to inquire about the message.

After the reception of the RFNM at the source IMP (point h), the message number and the PLT entry for this message is freed and the source HOST is informed of the correct message delivery. In case the RFNM carries a piggybacked ALL, the allocate counter for the proper destination IMP is incremented. This implicit reservation of buffers is returned to the destination IMP if some HOST attached to the source IMP does not make use of it within the next 250 msec (shown at point i); the cancellation is implemented as a "giveback" control message (GVB) that is generated at the source IMP. If, however, the next multipacket message to the same destination IMP is received from *any* source HOST within 250 msec, this message will capture the ALL and need only acquire a message number and a PLT entry before it can be sent to the destination IMP.

Thus we see that three separate resources must be obtained by each multipacket message prior to its transmission through the net: a message number, a PLT entry, and an ALL.

The system under consideration extracts a price for the implementation of its control functions. This price is paid for in the form of overhead in the packets as they are transmitted over the communication channels, in the form of buffers as the packets are stored in IMP, in control messages (IMP-IMP, IMP-HOST, HOST-HOST), in measurement and monitoring, and so on. Let us now state some of the overhead costs for version 2. The shortest "data" packet permitted is 168 bits long in its transmitted form (consisting of no data bits,* 96 bits of software, and 72 bits of hardware overhead), and is 176 bits long in its stored form (no data bits and 176 bits of software overhead); the longest transmitted packet is 1176 bits long (consisting of 1008 data bits plus the 168 bits of overhead). In its transmitted form, a routing update is 1160 bits whereas most other subnet control messages are 152 bits. Another feature introduced in version 2 is the piggybacking of IMP-to-IMP ACKs on packet traffic traveling to the waiting IMP; this requires an additional 16 bits in the packet header. The details behind these numbers and how they affect network throughput will be found below in Section 6.7; for now, let us

---

* Actually, because of the word-padding requirements of the IMP, there must be at least one padding bit in the data and this would require one more (16-bit) IMP word for "data."

identify some unusual network phenomena that were found to arise in version 2.

6.4. LOCKUPS, DEGRADATIONS, AND TRAPS

The message sequence numbers mentioned above for version 2 are used to guarantee that messages leave the destination IMP in the same order as they entered the source IMP. This sequencing of messages has the potential of introducing deadlock conditions. The reason for this is that any message, say the $(n+1)$st, which is out of order (and therefore cannot be delivered to its destination HOST) may use up resources in the destination IMP that are required by the nth message which must be delivered next. Therefore, the nth message may not be able to reach its destination IMP, which, in turn, prevents the other messages ($n+1$, $n+2$, etc.) that are out of order from being delivered to their destination HOST(s). For this reason one has to be very careful not to allocate too many resources (e.g., buffers) to messages that are out of order.

To avoid lockup conditions, version 2 of the ARPANET flow control procedure takes the two following precautions:

1. Requests for buffer allocation (REQALLs) are always serviced in order of message number; that is, no ALL is returned for the $(n+1)$st message if the nth message (or request for buffer allocation for the nth message) has not yet been received and serviced.
2. Single packet messages that arrive at the destination IMP out of order are not accepted unless they were retransmitted in response to a previous buffer allocation. These messages are rather treated as a request for the allocation of one buffer (according to precaution 1 above) and the message text is discarded.

With these two precautions the occurrence of deadlock conditions appears to be impossible, but nevertheless can occur! Indeed, whenever one introduces conditions on the message flow, there exists the danger that these conditions cannot be met and then the message flow will cease. Reassembly and sequencing are examples of such conditions. Let us now describe some of the deadlocks and degradations that have been observed in version 2 of the ARPANET.

In December, 1973, a dormant lockup condition was brought to life and named the "*Christmas lockup.*" This lockup was exposed by collecting snapshot measurement messages at UCLA from all sites simultaneously. The Christmas lockup occurred when snapshot messages arrived at the UCLA IMP for which reassembly storage had been allocated but for which no reassembly *blocks* were free. (A reassembly block is a piece of storage used in the actual process of reassembling packets back into messages.) These messages had no way to locate their allocated buffers since the pointer to an allocated buffer is part of the reassembly block; as a consequence, allocated buffers could never be used and could never be freed! The difficulty was caused by the system first allocating buffers before it was assured that a reassembly block was available. To avoid this kind of lockup, reassembly blocks are now allocated along with the reassembly buffers for each multipacket message.

Another lockup condition recently identified (but, as far as we know, it has never been observed) is *"piggyback lockup"* [OPDE 74]. This lockup is due to message sequencing, and comes about because of the ALL which piggybacks on a RFNM (see Figure 6.11). The piggybacked ALL represents a buffer allocation for the next multipacket message, and *not* for the next message in sequence. Thus, if the next message in sequence is a single-packet message, the piggybacked ALL in effect allocates buffers to a message that is out of order. Let us see how this situation can lead to a deadlock condition. Assume there is a maximum of eight reassembly buffers in each IMP; the choice of eight is for simplicity, but the argument works for any value. Let IMP A continually transmit eight-packet messages to the same destination IMP B such that all eight reassembly buffers in IMP B are used up by this transmission of multipacket messages. If now, in the stream of eight-packet messages, IMP A sends a single-packet message it will generally not be accepted by destination IMP B since there is no reassembly buffer space available. (There may be a free reassembly buffer if the single-packet message just happens to arrive during the time when one of the eight-packet messages is being transmitted to its HOST). The single-packet message will therefore be treated as a request for buffer allocation. This request will not be serviced before the RFNM for the previous multipacket message has been sent. At this time, however, all the free reassembly buffers will have been immediately allocated to the next multipacket message via the piggybacked ALL mechanism. In this case, the eight-packet message from IMP A that arrives later at IMP B cannot be delivered to its destination HOST because it is out of order. The single-packet message that should be delivered next, however, will never reach the destination IMP since there is no reassembly space available and therefore its REQALL will never be serviced. Deadlock! Once we recognize a deadlock such as this, it is usually easy to prevent its future occurrence (the trick, however, is to expurgate all deadlocks from the control mechanism ahead of time). A minor modification removes the piggyback lockup as follows. The described deadlock can only occur because single- and multipacket messages use the same pool of reassembly buffers. If we set aside a single reassembly buffer (or one for each destination HOST) that can be used only by single-packet messages, this lockup condition which is due to message sequencing cannot occur.

The ARPANET was originally designed to handle only two kinds of traffic: interactive traffic characterized by short, bursty transmissions that require a small network delay; and file transfers, which generally are characterized by long sequences of multipacket messages that require a large network throughput (see Figure 5.14). However, a third type, namely real-time traffic, has recently been recognized as a potential candidate for transmission through a packet-switched network. The throughput and delay requirements for real-time traffic are quite different from the throughput and delay requirements for interactive use or file transfers. For the transmission of digitized speech [FORG 75], for example, it is necessary to achieve a relatively high throughput for small messages since long messages result in long source delays (unacceptable for speech) to fill the large buffers. We realize that originally little effort was put forth to optimize the transmission of real-time traffic in the ARPANET. It was nevertheless surprising to find that the observed throughput for single-packet messages was in many cases only about one-fourth of what one would expect. Let us explain why this happens. As mentioned before, single-packet messages are not accepted by the destination IMP if they arrive out of order. Rather, they are treated as a request for the allocation of one reassembly buffer. The corresponding ALL is then sent back to the source IMP only after the RFNM for the previous message has been processed. Consider Figure 6.12. In part (a) we show the (good) case in which a sequence of single-packet messages passes through the network; the basic throughput limitation is that only four messages may be accepted at a time,* and so we observe waves of four messages flowing through the net per round-trip time. Upon arrival of message 1, a RFNM is returned to the source IMP, which permits this IMP to destroy its copy of message 1 and to release the next message (5), and so on. In part (b), we show the case where (say) the first message gets delayed (perhaps due to a transmission error) and arrives at the destination IMP later than

---

* There is a peculiarity in the HOST-to-HOST protocol in this regard which is a carryover from version 1. Specifically, the HOST-to-HOST protocol continues to be RFNM-driven and to restrict any logical process-to-process communication link to only one message at a time. A single HOST may use the full limit of four outstanding messages by multiplexing more than one logical link onto the net. Similarly, a set of HOSTs can (indeed, must) share the limit of four if they are attached to the same source IMP. However, we are describing a real-time traffic source, which is itself capturing the full set of four "tokens"; this is only possible by causing the communicating HOSTs to violate the standard protocol, which is the assumption in this paragraph.

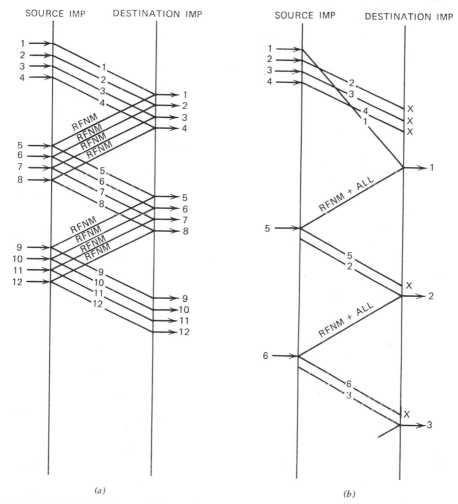

Figure 6.12   Real-time traffic throughput degradation (*a*) The good case. (*b*) The bad case.

messages 2, 3, and 4. Since these three are received out of order, they are not accepted at the destination IMP (i.e., the packets are discarded; this is denoted by *X*), but are considered to be requests for a single-buffer allocation. When message 1 finally arrives, it is delivered, a RFNM is generated *and* an ALL for message 2 is piggybacked and returned to the source (i.e., message 2 is now *in* order and may be released). When the RFNM+ALL arrive at the source, the RFNM releases the next message (5), and the ALL (for message 2) causes the source IMP to retransmit message 2. When message 5 arrives at the destination, it too is discarded (X) and queued up (behind messages 3 and 4) as a request for allocation. When message 2 arrives, it is now in order and so it is accepted and delivered; this generates a RFNM for message 2 and a piggybacked ALL for message 3, and so on. The most interesting fact about this sequence of events is that the arrival of message 2 before message 1 at the destination IMP causes not only message 2, but all future messages to be retransmitted! We see that the round-trip time for all future messages is more than four times as large as it would be without these undesirable retransmissions. It is also noteworthy that, once this retransmission pattern has established itself, there is almost no way the system can recover from this *throughput degradation* condition other than interrupting the input stream at the source IMP. In the case of speech transmission, however, this might not occur for some time. Therefore speech transmission systems would in many cases have to work with only one-fourth of the expected single-packet bandwidth. Since this is clearly an unacceptable condition, we now describe a method that could be used to avoid the undesirable retransmission of messages. Recall that a single-packet message is rejected at the destination IMP and later retransmitted if the RFNM for the preceding message has not yet been sent to the source IMP. This is mainly done to prevent the occurrence of reassembly lockup conditions. Therefore the problem cannot be solved by simply accepting all single-packet messages without additional measures to prevent deadlocks. This could lead to a reassembly lockup if a large number of single-packet messages from several source IMPs arrives at their common destination IMP out of order. In this case the destination IMP might not be able to accept those messages that *are* in order because of the lack of reassembly buffers. As a result, the system could become deadlocked. Any solution of the throughput degradation problem must guarantee that all messages that arrive in order can be accepted by the destination IMP. Suppose all single-packet messages are initially accepted (or stored). Let us take a closer look at the situation where all single-packet messages are accepted (or stored) such that there is no reassembly buffer available for messages that have to be delivered to their HOSTs next. This is not really a lockup condition because the source IMPs have kept a copy of all single-packet messages for which a RFNM has not yet been received. Therefore any single-packet message that arrived out of order but was accepted by the destination IMP can nevertheless be deleted later without the message being lost. The destination IMP only has to send an ALL for each deleted single-packet message to the corresponding source IMP when reassembly buffer space is available. This can also be considered as a deferred rejection. But now a retransmission is only necessary if the destination IMP is really running out of reassembly buffers. In this case, the physical limitations of the system are reached and we cannot hope to gain large throughput increases by means of protocol changes.

So much for lockups and degradations in version 2. We next consider the measured throughput for this version.

6.5. NETWORK THROUGHPUT

Let us now describe the limitations to and the measurement of the throughput that can be achieved between a pair of HOSTs in the ARPANET using the version 2 flow control procedure [KLEI 75]. First consider the limitations imposed by the hardware. The line capacity represents the most obvious and important throughput limitation. Since a HOST is connected to an IMP via a single 100 KBPS transmission line, the throughput can never exceed 100 KBPS. If there is no alternate routing in the subnet, the throughput is further limited by the 50 KBPS line capacity of the subnet communication channels. (The issue of alternate routing is discussed below.)

Recall that the processing bandwidth of the IMP allows for a throughput of about 700 and 850 KBPS for the 316 and 516 IMP, respectively [MCQU 72]. Therefore the IMPs can handle several 50 KBPS lines simultaneously. To avoid throughput degradations because of a CPU-limited HOST computer for our throughput experiments, we used a PDP 11/45 minicomputer at UCLA whose only task it was to generate eight-packet messages.

Let us now discuss what throughput limitations are imposed on the system by the subnet *control* procedures. As discussed above, there are two kinds of resources a message must acquire for transmission: buffers and control blocks (specifically message numbers and table entries). Naturally, there is only a finite number of each of these resources available. Moreover, most of the buffers and control blocks must be shared with messages from other HOSTs. The lack of any one of the resources can create a bottleneck that limits the throughput for a single HOST. Let us now discuss how many units of each resource are available and comment on the likelihood that it becomes a bottleneck.

A packet is allowed to enter the source IMP only if that IMP has at least four free buffers available. In October, 1974 (when the measurements described below were made), the total number of packet buffers in an IMP with and without the VDH (very distant HOST) software was, respectively, 30 and 51. This meant that an interruption of message input because of buffer shortage could only occur in the unlikely event that the

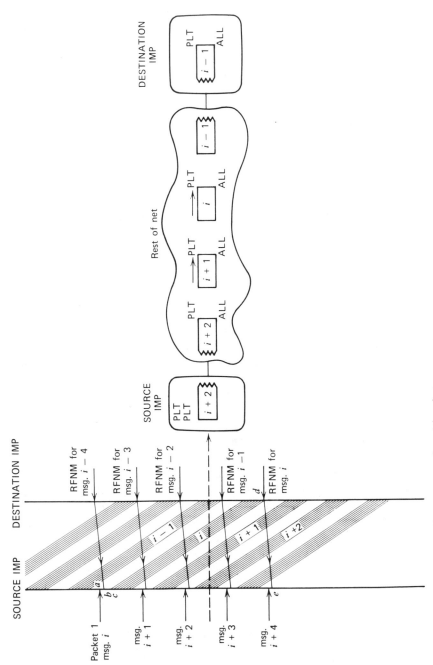

Figure 6.13   The normal sequence of multipacket messages.

source IMP was heavily engaged in handling store-and-forward as well as reassembly traffic.

The next resources the message had to obtain were the message number and the PLT entry. Recall that there is a limitation of only four message sequence numbers allocated per source IMP–destination IMP pair. This means that all source HOSTs at some source IMP A communicating with any of the destination HOSTs at some destination IMP B share the same stream of message numbers from IMP A to IMP B. This possible interference between HOSTs and the fact that there are only four message numbers that can be used in parallel means that the message number allocation can become a serious bottleneck in cases where the source and destination IMP are several hops apart. (This was the major reason for the recent change to the message processing procedure which has recently been implemented to produce version 3; see Section 6.8.)

After a message number is obtained, the multipacket message must acquire one of the PLT entries of which there is a shared pool of six. Since the PLT is shared by all HOSTs that are attached to the source IMP and used for all possible destinations, it also represents a potential bottleneck. This bottleneck can easily be removed by increasing the number of entries permitted in the PLT. However, the PLT also serves as a flow control device that limits the total number of multipacket messages that can be handled by the subnet simultaneously. Therefore, removal of the throttling effect due to the small-size PLT may introduce other congestion or stability problems. A corresponding consideration applies to the message number allocation.

The number of simultaneously unacknowledged eight-packet messages is further limited by the finite reassembly space. In October 1974, a maximum of 34 buffers (for IMPs without the VDH software) was available for reassembly in the destination IMP.* This means that at most four eight-packet messages can be reassembled at the same time. The reassembly space must of course be shared with all other HOSTs that are sending messages to the same destination IMP. It may therefore become another serious throughput bottleneck. We know that even if there is no interference from other HOSTs there cannot be more than four messages in transmission between any pair of HOSTs because of the message number limitation. This restriction decreases the achievable throughput in the event that the line bandwidth times the round-trip time is larger than four times the message length. Figure 6.13 depicts this case. The input of the first packet of message i is initiated at time a after the last packet of

---

* Whenever more IMP core memory is required for new functions, the number of available buffers decreases; this has been a continuing trend in the ARPANET.

message $i-1$ has been processed in the source IMP. After the input of this first packet is complete, the source IMP waits until time b when the RFNM for message $i-4$ arrives (herein lies the bottleneck). Shortly after this RFNM has been processed (at time c), the transmission of the first packet over the first hop commences and the input of the remaining packets from the HOST is initiated. At time d, all packets have been reassembled in the destination IMP, the first packet has been transmitted to the destination HOST and the eight reassembly buffers have been acquired by the RFNM, which is then sent to the source IMP. The RFNM reaches the source IMP at time e and thereby allows the transmission of message $i+4$ to proceed. In this figure we also show a snapshot of the net at the time slice indicated by the dashed arrow. We show four messages (each with their own ALL and PLT): $i+2$ is leaving the source IMP, $i+1$ and i are in flight, and $i-1$ is entering the destination IMP. We also see the two unused PLT entries in the source IMP. The possible gaps in successive message transmissions represents a loss in throughput and can be caused by the limitation of four messages outstanding per IMP pair; this manifests itself in the next (fifth) message awaiting the return of a RFNM.

We have not yet mentioned the interference due other store-and-forward packets that can significantly decrease the HOST-to-HOST throughput. This interference causes larger queueing times and, possibly, rejection by IMPs along the path. Such a rejection occurs if either there are 20 packets allocated for store-and-forward in the neighbor IMP or if the output queue for the next hop is full. (Recall that a maximum of eight packets is allowed on each output queue.) A rejected packet is retransmitted if no acknowledgement has been received after a timeout of 125 msec.

We now turn to a brief discussion of alternate routing and its impact on our throughput experiments. It turns out in the ARPANET that alternate paths are rarely used if they are longer than the primary path by more than two hops. The reason for this comes from the way the delay estimate is calculated and updated and from the way the output queues are managed. Each hop on the path from source to destination contributes four (arbitrary) delay units to the delay estimate. Each packet in an output queue between source and destination contributes one delay unit to the delay estimate. Since the length of the output queues is limited to eight packets, one hop can therefore increase the delay estimate by at most 12 units. Thus the minimum and maximum delay estimates over a path of n hops are, respectively, $4n$ and $12n$ delay units. Packets are always sent over the path with the smallest current delay estimate as described in Section 6.1. From this it follows that an alternate path is never used if it is three or more times longer (in terms of hops) than the primary path. Thus, for a primary path at length n, alternate routing is only possible over paths of length less than $3n$ hops. Let us assume that all the channels along the primary and alternate path have the same capacity and that there is no interfering traffic. If we send as many packets as possible over the primary path, these packets usually will not encounter large queueing delays because this stream is fairly deterministic as it proceeds down the chain (see Exercises 5.3 and 5.4). This means that the delay estimate increases only slightly, although all of the bandwidth is used up. Therefore a switch to an alternate path occurs only if it is only slightly longer than the primary path. In the case of interfering traffic, the output queues will grow in size and therefore a switch is more likely to occur. Such a switch to an alternate path may therefore help to regain some of the bandwidth that was lost to the interfering traffic. It has already been pointed out in [MCQU 73] that, even if primary and secondary paths are equally long, at most a 30% increase in throughput can be achieved. This is due to the restriction of a maximum of eight packets on an output queue and the

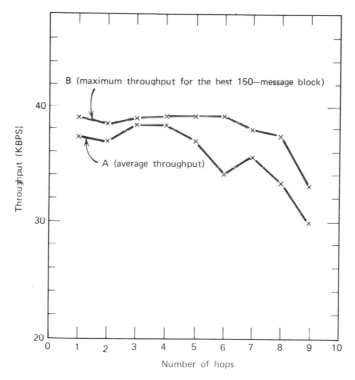

Figure 6.14   Throughput in the ARPANET (October 1974).

fact that the frequency of switching between lines is limited to once every 640 msec (for heavily loaded 50 KBPS lines). Thus the backlogged queue of eight packets on the old path will provide overlapped transmission for only 8×23.5 msec $= 188$ msec of the total of 640 msec between updates (the only times when alternate paths may be selected). The relatively slow propagation of routing information further reduces the frequency of switching between primary and alternate paths. This discussion shows that alternate paths have only a small effect on the maximum throughput that can be achieved. However, alternate paths are of great importance for the reliability of the network as discussed earlier.

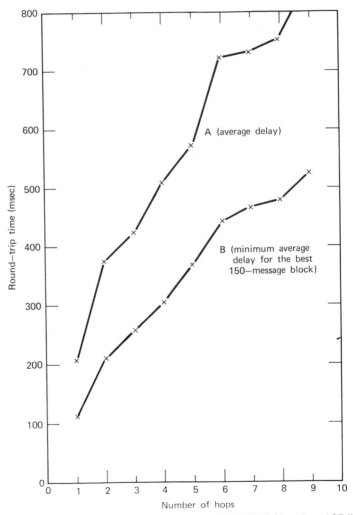

Figure 6.15   Round-trip delay in the ARPANET (October 1974).

The October 1974 throughput experiments for version 2 (KLEI 75] produced the results shown in Figure 6.14. Here we show the throughput (in KBPS) as a function of the number of hops between source and destination. Curve A is for the throughput averaged over the entire 10-minute experiment; curve B is the throughput for the best block of 150 successive messages. Note that we are able to pump an average of roughly 37–38.5 KBPS out to five hops; it drops beyond that, falling to 30 KBPS at nine hops largely because of the four-message limitation. Also, the best 150 message throughput is not much better than the overall average. In Figure 6.15, we show the corresponding curves for the average round-trip delays as a function of source-destination hop distance. Note that the average delay for n hops may be approximated by $200 + 75n$ (msec). The measured histogram for delay is given in Figure 6.16 for hop distances of 1, 5, and 9. Of further interest is the correlation coefficient of round-trip delay for successive messages in the network; this

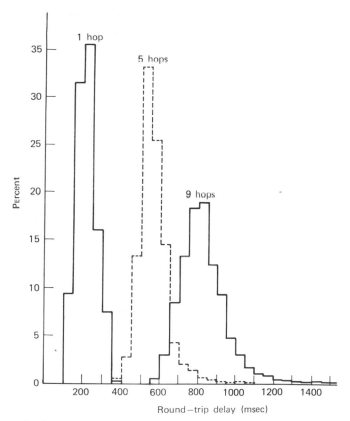

Figure 6.16   Histogram of round-trip delay (msec).

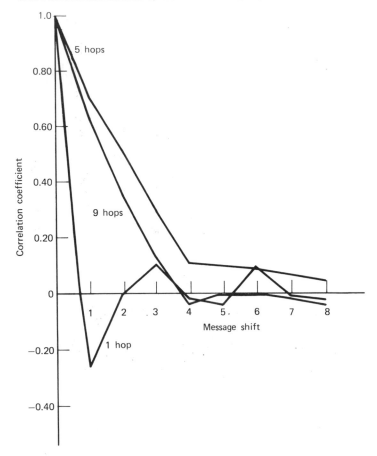

Figure 6.17   Correlation coefficient for message delay.

is shown in Figure 6.17. Note that message delay is correlated out to about three or four successive messages.

Let us now study a week-long measurement of the ARPANET.

6.6. ONE WEEK OF ARPANET DATA

In August 1973, a week-long measurement was made on all traffic carried by the (version 2) 39-node ARPANET shown in Figure 5.21(a). The purpose of this experiment was to observe the traffic characteristics of the operating network [KLEI 74]. These characteristics include (1) message and packet size distributions, (2) mean round-trip delay, (3) mean traffic-weighted path length, (4) "incest" (the flow of traffic to and from HOSTs at the same local site), (5) "most popular" sites and channels, (6) "favoritism" (that property which a site demonstrates by sending many of its messages to one or a small number of sites), and (7) channel utilization. This data has more than just historical significance—it also permits us to evaluate our design methodology. In particular, there are several network design parameters whose values were chosen prior to the actual network implementation and that deserve to be reevaluated as a result of these measurements. Among these parameters are packet (and therefore buffer) size, number of buffers, channel capacity, and single-multipacket message and reassembly philosophy.

To observe these traffic characteristics, data was gathered over a (nearly) continuous seven-day period from 8:37 AM on August 1, 1973 through 5:06 PM on August 7, 1973. Accumulated statistics were sent to UCLA-NMC from each site in the network at periodic intervals of approximately 7 min. The data was subsequently processed, giving the following results.

During the "seven" days,* a total of some 6.31 billion bits of HOST traffic were carried through the network by some 26.0 million messages (an average of 4.1 million messages per day). This means that, on the average, the entire network was accepting $\gamma = 47.4$ messages per second and carrying an average of 11,495 bits per second among HOST computers. The messages were distributed in length as shown in Figure 6.18 for messages entering the network from all HOSTs. From this histogram, we find that there are, on the average, only 1.12 packets per HOST message! Moreover, the measured mean length of a message is only 243 bits of data! These facts indicate that not only are there very few multipacket messages,† but also most single-packet messages are quite short (the measured value is 218 data bits). This latter fact is borne out in the $\log_2$ histogram of packet length for packets entering the network from the HOSTs, as shown in Figure 6.19. This small message size has an impact on the efficiency of storage utilization which we now consider. We may define the buffer utilization efficiency η as

$$\eta = \frac{\bar{b}_p}{L + H}$$

where $\bar{b}_p = E$ [packet length], $L =$ (maximum) packet length, and $H =$ length of packet storage overhead; that is, the fixed length of an IMP buffer for packet storage is $L + H$ bits. The seven-day measurements show that $\bar{b}_p = 218$ bits. Using this value, along with $L = 1008$ and $H = 176$, we have a buffer utilization efficiency of only $\eta = 0.184$! In

* Actually, a total of 548,951 sec.

† This suggests that the philosophy of multipacket messages should be reexamined.

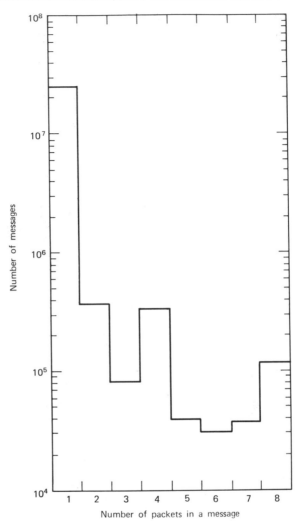

Figure 6.18   Histogram of HOST message length in packets.

Exercise 6.5 [COLE 71], we assume exponentially distributed messages, with a mean of $\bar{b}$ bits, and show* that in this case $\bar{b}_p = \bar{b}[1 - e^{-L/\bar{b}}]$. When we use the measured value $\bar{b} = 243$ bits in this last relationship, we find $\bar{b}_p = 239$, which is roughly 10% larger than the measured value ($\bar{b}_p = 218$); this means that there are more short messages than predicted by the exponential distribution. Even using $\bar{b}_p = 239$, we find $\eta = 0.202$ (again, a

---

* The effect of *message* truncation at 8063 bits has no noticeable effect on these results.

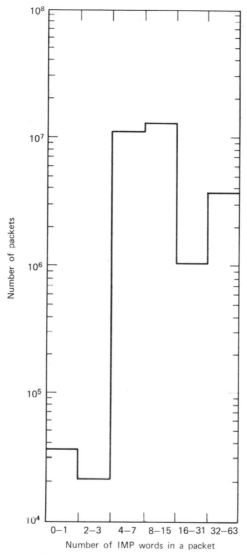

Figure 6.19   Histogram of packet size in words.

small value). From Exercise 6.5, we solve for the optimum packet length L_o, which, for $\bar{b} = 243$ and $H - 176$, gives us $L_o = 244$ bits (assuming an exponential distribution again) and a (maximum possible) efficiency of $\eta = 0.366$ ($\bar{b}_p = 154$ bits).* This value of efficiency (roughly 37%) is

_____

* The only way to increase efficiency beyond this is to reduce H.

considerably higher than the measured values. Thus, based on this particular week-long measurement (and supported by previous and more recent measurements), the use of packets four times shorter than those currently allowed is indicated if one wishes to maximize buffer utilization. On the other hand, the major cause of these very short messages is the way in which the common PDP-10 time-shared operating system (TENEX) behaves. Specifically, TENEX [BOBR 71] is a character-oriented system that performs character echoing! This means that for each character typed at a user terminal, an eight-bit message is created (to which is added all the overhead) for passage through the net (this generates IMP-to-IMP acknowledgements and a RFNM); after TENEX handles the interrupt caused by this character, it generates another eight-bit message (the echo!) for the net. Of course, this creates an enormous percentage of overhead (to be discussed below), but for this discussion, we note that this procedure greatly reduces the average network message length, especially when we see how many PDP-10 systems are attached to the ARPANET. Currently, TENEX is being modified to improve the efficiency of its message processing (more than character-at-a-time transmission and local character echoing).

The measured mean round-trip* message delay for the seven-day period was 93 msec. Indeed, the network was meeting its design goal of less than 200 msec for single-packet messages. Thus, as desired, the communication subnet is essentially transparent to the user, as far as delay is concerned. The principal source of delay (occasionally many seconds) seen during a user interaction comes both from his local HOST and from the destination HOST on which he is being served (and almost never from the communication subnet except for those occasions when looping occurs). Major contributors to this small network message delay are the small message size discussed above and the fact that a significant number of messages traverse very short paths in the network. Below, we shall continue the discussion regarding delay. For now, let us investigate the traffic distribution and explain the presence of short paths, incest, favoritism, and the like. From Eq. (5.22), we know that the average path length $\bar{n}$ (measured in hops) may be calculated easily by merely measuring the entering message rate (γ) and the internal message rate λ, that is, $\bar{n} = \lambda/\gamma$. The seven-day measurements produced a value of $\bar{n} = 3.31$. It is possible to obtain a lower bound on this average path length by merely measuring

---

* Round-trip delay is measured by the IMPs and is the time from when a message enters the network until the network's end-to-end acknowledgement in the form of a RFNM is returned.

how much traffic passes between pairs of HOSTs and assuming that this traffic travels by the shortest network path; if one makes such a calculation from the measured data, one finds that the minimum (traffic-weighted) path length is 3.24 hops, which, when compared to $n = 3.31$, shows that much of the traffic currently traveling in the network is indeed going by the shortest path! On the other hand, if one calculates the shortest path length between node-pairs for the topology shown in Figure 5.21(a), and averages this over the entire network, then one finds that two sites are, on the average, separated by 5.32 hops. This is quite a bit different from the measured traffic-weighted value of 3.31! One wonders about such a significant difference between the two measures of path length since it suggests that network users are communicating with sites that are nearby. This is surprising since distance in the network should be invisible to the users! The answers can be found by examining in more detail where the traffic is going. In fact, one finds that a significant fraction of the traffic is going over a distance of *zero* hops! This corresponds to traffic traveling from one HOST at a given site through its own IMP and then back to a second HOST at the same site (after all, the IMP is a very convenient interface between *local* machines also). Such traffic we naturally refer to as "incest." For example, we find that 70% of SRI's traffic is incestuous, similarly, 34% of MIT's traffic, 34% of Lincoln Laboratory's traffic, 54% of USC's traffic, 33% of ARPA's traffic, 46% of CCA's traffic, and 36% of XEROX's traffic is also incestuous. All in all, averaged over the entire network, approximately 22% of the traffic is incestuous. Moreover, the AMES' IMP and the AMES' TIP (at a hop distance of one, but colocated at AMES) seem to have a relationship (which is incestuous in spirit) whereby the former sends 85% of its traffic to the latter and in the reverse direction, we find 58% traveling (this accounts for 13% of the total network traffic); in fact, a total of 16% of the network traffic travels a total hop distance of one. We can examine this phenomenon further by observing what percentage of traffic is traveling over various hop distances in the network. This is shown in Figure 6.20, where we have plotted the percentage of traffic as a function of the hop distance traveled; we see, for example, that 22% travels a distance zero (incest) and approximately 16% travels a distance of one, and so on. One might argue that these percentages should be modified by the number of node-pairs at a given hop distance; therefore, in Figure 6.20 we also show the number of node-pairs at a given distance for the ARPANET [of Figure 5.21(a)]. For this topology, it can be seen that the following list of ordered pairs (x, y) provides the distribution of node-pair minimum distances as plotted in Figure 6.20 (where x is the hop distance

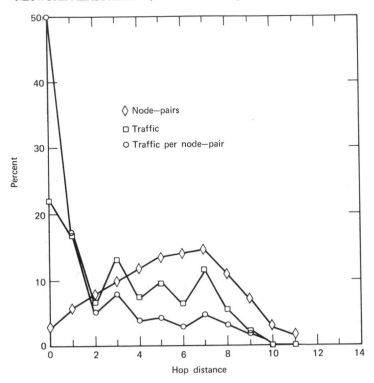

Figure 6.20   Distance-dependence of traffic (incest, etc.).

and y is the number* of node-pairs at this distance): $(0, 39)$, $(1, 86)$, $(2, 118)$, $(3, 148)$, $(4, 176)$, $(5, 204)$, $(6, 210)$, $(7, 218)$, $(8, 160)$, $(9, 102)$, $(10, 40)$, and $(11, 20)$. No nodes are more than 11 hops apart. Note that more nodes are at a distance of 7 than any other distance (with the average distance equal to 5.32, as mentioned above). In a network with N nodes and $M/2$ full-duplex channels, the first two entries on the list must always be $(0, N)$, $(1, M)$. Dividing the measured percentage traffic by the number of node-pairs at a given distance, we come up with the third curve in Figure 6.20, which shows the percentage of traffic per node-pair at a given distance; aside from the distance zero (incest) and distance one traffic, we find that this normalized traffic is approximately constant (at least for hop distances of $2, 3, \ldots, 9$). [The last effect that contributes to

* We consider node-pairs as *ordered* pairs; thus the pair (MIT, UCLA) is distinct from (UCLA, MIT). This is natural since traffic flow is not necessarily symmetrical. The (important) special case of (node i, node i) counts as one "pair."

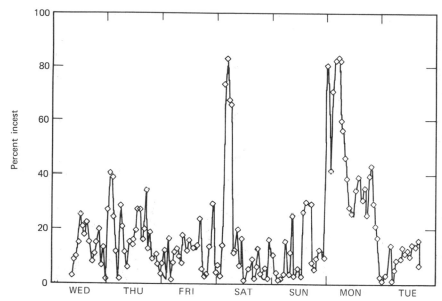

Figure 6.21   Hourly incest in the ARPANET.

the remaining nonuniformity is the existence of large users (e.g., University of ILLINOIS) and large servers (e.g., ISI).] If the traffic were uniformly distributed in the network, then the traffic curve in Figure 6.20 would be a horizontal line at the value 8.25%. In Figure 6.21, we display the percent of incest in the network during each hour* of the experiment. Note that incest accounts for over 80% of the traffic during certain hours (the weekly average is 22%), peaking in the wee hours of the morning.

A further illustration of the nonuniformity of the traffic is seen in Figure 6.22. Here, we have plotted the fraction of messages sent from the n most busy sources as a function of n. Notice that over 80% of the traffic is generated by the busiest one-third of the sites. A similar effect is true of the busiest (most popular) destinations.

Another phenomenon regarding the movement of traffic among HOSTs is the occurrence of "favorite sites" in the network. HOST A's favorite site over a given time interval is defined to be that HOST to which it sends more messages than to any other during that interval. For example, averaged over the seven days, SDC sent 70% of its traffic to

---

* This, and the other "hourly" plots below, actually show points that are separated by 56 min (related to the IMP clock speed which provided 7-minute statistics, eight of which gave the 56 minute "hour"). The separation between the days on the horizontal axes occurs at midnight.

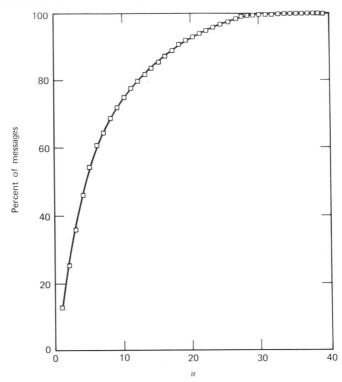

Figure 6.22   The busy-source effect.

BBN, ILLINOIS sent 50% of its traffic to ISI, BBN-TIP sent 59% of its traffic to BBN, and ARPA sent 47% of its traffic to ISI. We have already seen that some HOSTs send more to themselves than to any other. Accounting for all the favorite-site traffic, we find that 44% of the network traffic is going to the favorite site! In Figure 6.23, we show (summed over all sources) the percent of traffic traveling to a source's n most favored destinations as a function of n. If these orderings and percentages remained invariant over time (i.e., a stationary traffic matrix), then one could use this information in the topological design; however, it can be shown [FRAN 72, GERL 73] that both the network design and performance are relatively insensitive to changes in the traffic matrix (and so a uniform requirement is usually assumed for design calculations as we have done in Section 5.10). (A uniform traffic matrix would have given the most favorite site a percentage of only $100/N = 2.56\%$.) Note that by the time we have asked for traffic going up to and including the ninth most favorite site, we have accounted for over 90% of the traffic in the

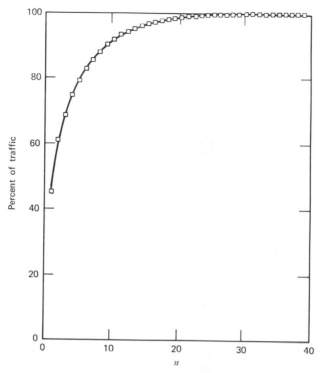

Figure 6.23   The favorite-site effect.

network. However, it is important to realize that this involves more than nine sites, since each source need not have the same set of nine most favorites (in fact, 33 unique destinations were involved). This favorite site effect is more dramatically displayed in Figure 6.24, which shows the percentage of traffic to the most favored site of all sources on an *hourly* basis. Most of the traffic (a minimum of 40% and an average of 61%) was caused by conversations between the N sources and their favorites. There are N^2 pairs in total; thus, on a weekly basis, the N favorites accounted for $0.44N = 17$ times the traffic they would have generated if the traffic matrix had been uniform (on an hourly basis it is $0.61N = 24$ times). Note that the favorite-site effect must increase as we shrink the time interval over which "favorite" is defined; in fact, if we choose an interval comparable to a packet transmission time, then the most favorite sites will account for almost 100% of the traffic, since the name of each source's favorite site will change dynamically to equal the name of the destination site for this source's traffic of the moment. Thus the amount of traffic due to favorite sites has an interpretation that changes as the time interval

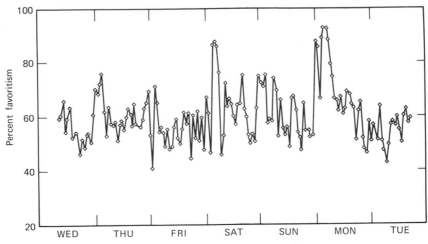

Figure 6.24   Hourly behavior of the favorite-site effect.

changes. The weekly value of 44% has two possible interpretations. The first is that there exists a true phenomenon of favoritism due, perhaps, to the existence of a few useful "server" systems.* The second interpretation is that network users are lazy; once a user becomes familiar with some destination HOST, he continues to favor (and possibly encourage others to favor) that HOST in the future rather than experimenting with other systems as well. A further explanation for this phenomenon is that it is not especially easy to use a foreign HOST at this stage in the evolution of networks. Except insofar as legislative procedures (such as accounting) inhibit freedom, we expect this trend to diminish in the future as we learn to make remote processing power more readily accessible to users in a meaningful way.

Related to Figure 6.24 is Figure 6.25, in which we have plotted the number, K, of favored destinations necessary to sum to 90% of the overall traffic on an hourly basis. This means that in any hour, 90% of the messages were sent between at most NK of the total $N^2 = 1521$ pairs in the network. Notice that K has a maximum hourly value of 7 (this is less than the weekly average of $K = 9$ due to the smaller averaging interval as discussed above). Therefore, for any hour, it requires at most 18% of the node-pairs to send 90% of the messages (in the most extreme case, $K = 1$, and so for those hours, at most $1/N$ or 2.56% of the node-pairs sent 90% of the messages).

* Of the 39 weekly favorites, only 17 unique destinations were involved.

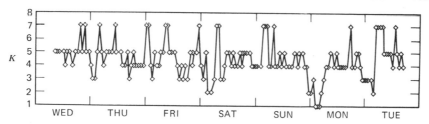

Figure 6.25   Number of favored destinations required to achieve 90% of traffic.

The last (traffic matrix) phenomenon we wish to discuss is also related to the favorite-site effect. If we list the most popular site, we find it to be SRI; 13% of the total network traffic was destined for this site. Second on the list is ISI, to which 12.3% of the traffic was directed. Similarly, 10.8% of the traffic was directed to the third most active site, namely, the AMES TIP. "Most" traffic here is defined in terms of total number of messages. However, if we examine these numbers more closely, we find that of the 13% to SRI, more than two-thirds of it (9%) was self-directed and less than one-third (4%) really came from remote HOSTs. In the case of ISI, less than $\frac{1}{100}$ was self-directed, and so in a real sense ISI was the most popular site in the network. Another measure might be the average number of bits per second directed toward a site (in which case ISI comes up with 842). Based on this measure, the most popular site was CCA, which received an average of 1931 bits per second; however, CCA was merely sixth based on the number of *messages* per second receiving 5.4% of the message traffic (of which almost half of this was self-directed!). Thus we see that the definition of "most popular" is somewhat evasive. Nevertheless, it is also clear that by incestuous traffic one can easily make oneself the most popular site on the network!

Let us now examine the behavior of γ, λ, and other similar global network measures as a function of time. In Figure 6.26, we show the average rate at which HOST messages were generated (per second) on an hourly basis; this gives us an indication as to when the "work" was done on the network. There are no real surprises here: the curve shows a predominance of traffic during daylight hours and on weekdays. It is interesting that Monday had noticeably heavier traffic than the other weekdays (were the users manifesting feelings of guilt or anxiety for having slowed down during the weekend?). Observe that a truly world-wide network with its many time zones could perhaps take advantage of these hourly and daily slow periods.

Figure 6.27 illustrates the change in network use as a function of time by showing the behavior of the mean number of packets per message. The peaks are associated with those hours during which file transfers

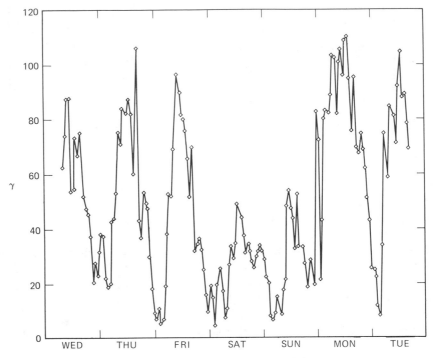

Figure 6.26   HOST message arrival rate (γ).

dominated the interactive traffic. These peaks in general occur during off-shifts hours (as with incest). Perhaps users feel that they get better data rates, reliability, or HOST service late at night; or, perhaps the background of file transfers is continually present, but is noticed only when the interactive users are asleep.

The internal traffic on channels is one measure of the effectiveness of the network design and use. In Figure 6.28, we show the channel utilization averaged over the entire network on an hourly basis both with and without overhead (due to routing updates, etc.). The low utilization (whose weekly average was 0.071 if overhead is included or 0.0077 neglecting overhead) suggests that the lines in the network have a great deal of excess capacity on the average (this excess capacity is desirable for peak loads). The maximum hourly line load (averaged over all channels) was approximately 13.4% (occurring 5 hours before the end of the measurement), and corresponded to an internal network flow in excess of 600 KBPS; without overhead the maximum hourly average utilization was approximately 2.8% (129 KBPS internal flow). It is interesting to observe the heaviest loaded line during each hour; this we plot

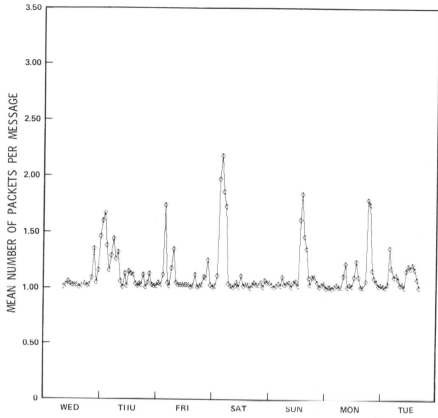

Figure 6.27   Hourly variation of the number of packets/message.

in Figure 6.29 both without (part a) and with (part b) overhead. Note from part (b) that the busiest line of any hour (HARVARD to ABERDEEN) had a utilization of 0.48 for that hour; without overhead, the busiest line (GWC to CASE) had a utilization of 0.225 for its busiest hour. Over the seven days, these channels themselves had hourly load histories as shown in Figure 6.30. Note how bursty the traffic was on these lines (even averaged over an hour). Another interesting line is that one which had the maximum load averaged over the week. Neglecting routing updates and all other overhead, the channel from ISI to RML had the largest weekly load (0.017), and its hourly behavior is shown in Figure 6.31(a); again, we see bursty behavior. If we include overhead, then the satellite channel to Norway (SDAC to NSAT) had the largest utilization averaged over the week since it is only a 7.2 KBPS channel and therefore, all traffic placed almost seven (50/7.2) times the load on it (in this

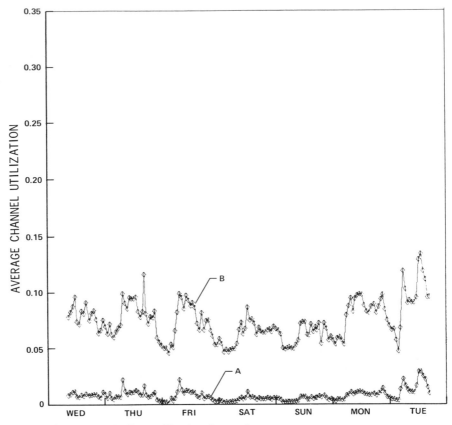

Figure 6.28   Hourly line utilization for entire net.

case, roughly 2 KBPS, or 28% of the line, is used for routing updates alone). The hourly history for this channel is shown in Figure 6.31(b). Also on this figure we have shown the UP/DOWN status of this line (in both directions).* Note that the channel was operational in both directions for a small fraction of the measurement (mainly on Monday) and only during this time was it carrying its own routing updates as well as responses to the NSAT to SDAC channel's routing updates in the form of "I heard you's"; this gives the 28% overhead mentioned above. This

---

* Our measurements actually give the UP/DOWN status of the IMPs as seen by the NMC. When NSAT could not be reached (i.e., no "I heard you's"), we have displayed the NSAT to SDAC channel as being down in Figure 6.31(b), and similarly, when SDAC is declared down, we have shown the SDAC to NSAT (and the NSAT to SDAC) channel down; in fact, when a full-duplex channel (or the node at one end) goes down, it does so in both directions simultaneously, but the routing update attempts (which are the implicit "hello's") continue to inquire if the channel is up.

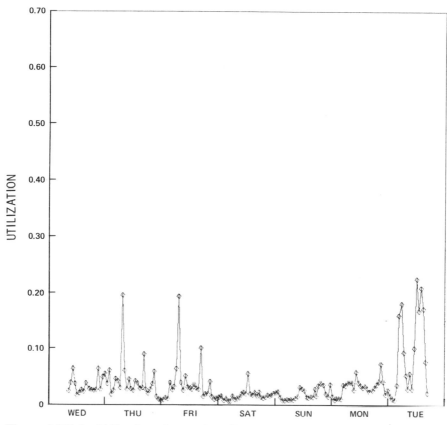

Figure 6.29(*a*) Utilization of most heavily used channel in each hour (without overhead).

channel was down for a large part of Friday during which time it carried no traffic. For the rest (most) of the week, the NSAT to SDAC channel was down and so no "I heard you" traffic was recorded on the SDAC to NSAT channel, as can be seen in Figure 6.31(*b*).

With few exceptions, the channels in the network are fairly reliable. The average packet error rate was one error in 12,880 packets transmitted. Of the 86 channels in the network, 14 reported no errors during the seven days, while six channels had packet error rates greater than 1 in 1000. The worst case was 1 in 340 packets for the channel from the RADC-TIP to Lincoln. These errors do not impair the network performance since we include error detection and retransmission as mentioned earlier. In Figure 6.32, we show the error behavior of these lines for the seven-day measurement. (Over the past four years, averaged over all

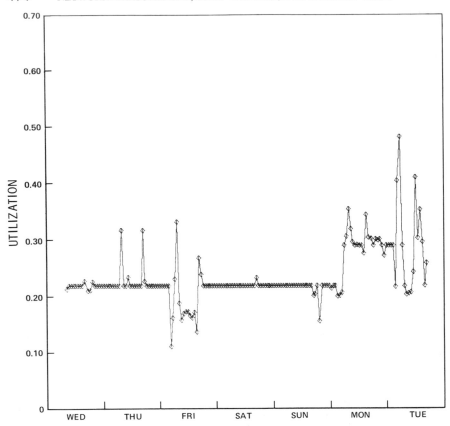

Figure 6.29(*b*)   Utilization of the most heavily used channel in each hour (with overhead).

channels, the average down time per channel has been very close to one percent.) The failure rate of the IMPs should also be included here, but a seven-day measurement is clearly insufficient for this purpose. Consequently, in Figure 6.33, we show the performance of the IMPs over a 19-month interval from June 1972 through December 1973 [MCKE 75]. The average IMP down rate was 1.64%, with the worst case being 9.13%.

Let us now return to considerations of average message delay. In Figure 6.34, we plot the hourly behavior of this measured delay. The minimum value is approximately 20 msec (occurring when the incest exceeded 80%—see Figure 6.21) and the maximum is roughly 325 msec;*

* This does *not* violate our design goal of 200 msec for interactive traffic since we can see from Figure 6.27 that multipacket messages were dominating the traffic flow at this time.

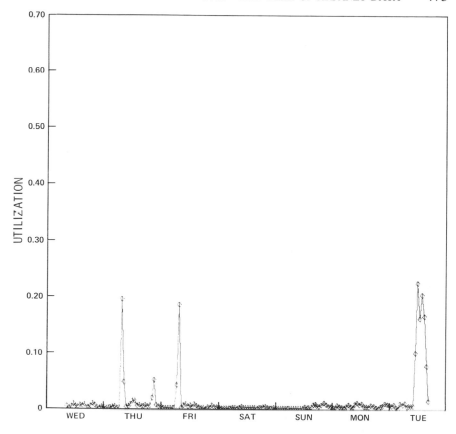

Figure 6.30(a)   Utilization of the channel (GWC to CASE) with the highest hourly average (without overhead).

recall that the weekly average was 93 msec. One wonders how well the network delay model we created in Section 5.6 predicts the behavior shown in Figure 6.34. Let us extend this model to fit the specific implementation of the ARPANET. Following this model extension, we will present a comparison between the predicted and measured delay.

Our point of departure is Eq. (5.20). As earlier, we must account for control traffic (as well as data traffic) that causes queueing delays; thus we define λ_i' to be the average arrival rate of all messages (data plus control) to the ith channel. Moreover, let us remove the assumption that nodal processing delay is the same for all nodes and also let us include the destination IMP-to-HOST transmission delay. Thus we obtain the following expression for the average delay T_{SP} experience by single-packet

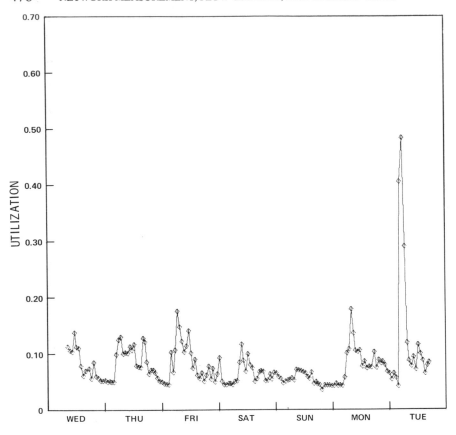

Figure 6.30(*b*) Utilization of the channel (HARVARD to ABERDEEN) with the highest hourly average (with overhead).

messages:

$$T_{SP} = \sum_{i=1}^{M} \frac{\lambda_i}{\gamma} \left[\frac{\lambda_i'/\mu' C_i}{\mu' C_i - \lambda_i'} + \frac{1}{\mu C_i} + P_i + K_l \right] + \sum_{j=1}^{N} \frac{\gamma_{\cdot j}}{\gamma} \left(K_j + \frac{1}{\mu_H C_{Hj}} \right) \quad (6.1)$$

where K_l is the packet processing time at node l, and l is the origin node of channel i, $\gamma_{\cdot j}$ is the mean departure rate of messages from the network to the HOST at site j, and $1/\mu_H C_{Hj}$ is the mean transmission time of messages to a HOST at site j.

The above formula assumes single-packet message traffic, while in the ARPANET, we segment messages into as many as eight packets. Our model may be extended to give the average multipacket message delay

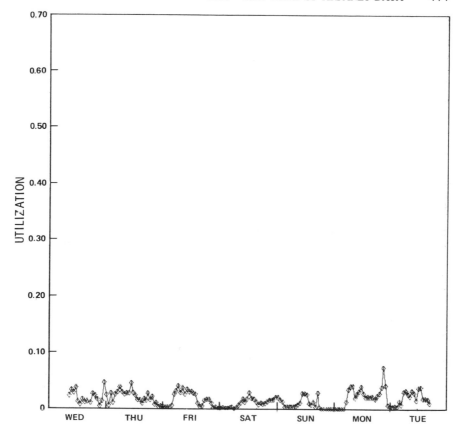

Figure 6.31(a)   Utilization of the channel (ISI to RML) with the highest weekly average (without overhead).

T_{MP} as follows [COLE 71, FULT 72, KLEI 74]. We first consider $\bar{\tau}_{jk}$, the mean interpacket arrival time (at the destination) for messages from source j to destination k. It is difficult to measure $\bar{\tau}_{jk}$ for each j-k pair in the network. We therefore introduce an approximation [COLE 71] that yields

$$E[\tau(n \text{ hops})] \cong \frac{\rho(1-\rho^{n-1})}{1-\rho} S_F(n)$$

The above expression gives the expected value of $\bar{\tau}_{jk}$ for nodes j and k that are n hops apart. It assumes that (i) the channel utilizations ρ_i for all channels in the path from j to k are constant and equal to ρ, (ii) that the path is unique and (iii) that $S_F(n)$ is the average time to transmit a full

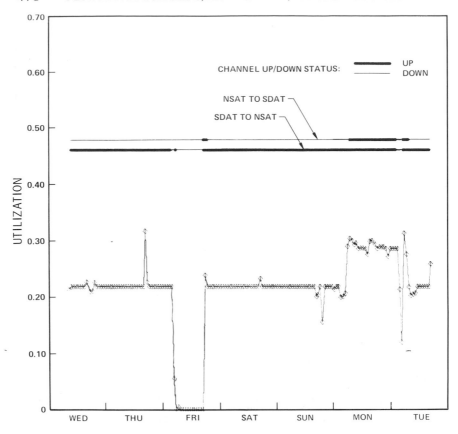

Figure 6.31(*b*)   Utilization of the channel (SDAC to NSAT) with the highest weekly average (with overhead).

packet on each of the *n* channels in the path. Let us now obtain an approximation to the network-wide mean interpacket gap $\bar{\tau}$. The average line utilization is $\bar{\rho} = \lambda/\mu CM$. The time it takes to transmit a full packet averaged over all channels in the network is

$$\bar{S}_F = \sum_{i=1}^{M} \frac{\lambda_i}{\lambda} \frac{1}{\mu_F C_i}$$

where $1/\mu_F$ is the length of a full packet. We now have the following approximation for $\bar{\tau}$:

$$\bar{\tau} \cong \bar{\rho} \left(\frac{1 - (\bar{\rho})^{\bar{n}-1}}{1 - \bar{\rho}} \right) \bar{S}_F$$

Now assuming that the last packets of multipacket messages have the

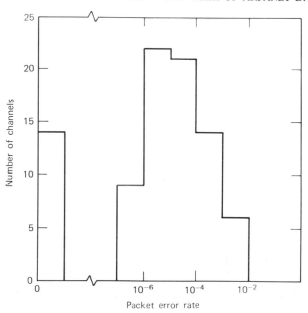

Figure 6.32   Packet error behavior of the channels (week-long measurement).

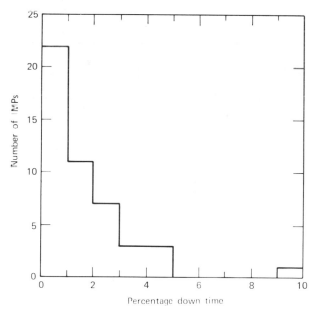

Figure 6.33   Nineteen-month reliability of IMPs (June 1972–December 1973).

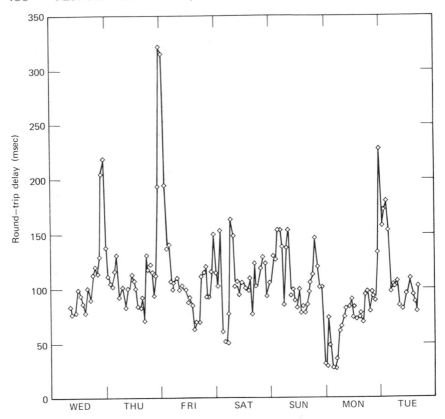

Figure 6.34   Measured average message delay.

same mean length as single-packet messages, we have the average message delay for multipacket messages:

$$
\begin{aligned}
T_{MP} = & \sum_{i=1}^{M} \frac{\lambda_i}{\gamma} \left[\frac{\lambda_i'/\mu'C_i}{\mu'C_i - \lambda_i'} + \frac{1}{\mu_F C_i} + P_i + K_l \right] \\
& + \sum_{i=1}^{M} \frac{\lambda_i}{\lambda} \left[(\bar{m}-2)\frac{1}{\mu_F C_i} + \frac{1}{\mu C_i} \right] \\
& + \sum_{j=1}^{M} \frac{\gamma_j}{\gamma} \left(K_j + \frac{1}{\mu_{FH} C_{Hj}} \right) + (\bar{m}-1)\bar{\tau}
\end{aligned}
\qquad (6.2)
$$

where $1/\mu_{FH} C_{Hj}$ is the transmission time of a full packet to a HOST at site j.

Let β be the fraction of the total number of messages that are single-packet messages. We then obtain the final expression for the average message delay in the network:

$$
T = \beta T_{SP} + (1-\beta) T_{MP}
\qquad (6.3)
$$

The measure of delay supplied by the IMPs is round-trip delay. Therefore, in order to compare the model with the measurements, we also need an expression for round-trip delay (that is, we must include the average RFNM delay, T_{RFNM}, in the model). A RFNM is simply another single-packet message traveling from destination to source. Thus it experiences the single-packet message delay T_{SP} with an appropriate value for μ and without the HOST transmission term, as follows:

$$T_{RFNM} = \sum_{i=1}^{M} \frac{\lambda_{Ri}}{\gamma} \left[\frac{\lambda_i'/\mu'C_i}{\mu'C_i - \lambda_i'} + \frac{1}{\mu_R C_i} + P_i + K_l \right] + \sum_{j=1}^{N} \frac{\gamma_{j.}}{\gamma} (K_j)$$

where λ_{Ri} is the mean arrival rate of RFNMs to channel i, $1/\mu_R$ is the length of a RFNM, and $\gamma_{j.}$ is the mean departure rate of RFNMs from the network to the HOSTs at site j (which equals the mean arrival rate of messages from the HOSTs at site j to the network).

The expression for mean round-trip delay T_R is therefore

$$T_R = T + T_{RFNM} \tag{6.4}$$

For the week-long measurement, we first calculated the *zero*-load value for T_R and obtained a value of $T_R = 69$ msec; the hourly variation of this quantity is shown in Figure 6.35. (The source of the time-variation is the shift in the origin-destination traffic mix.) This corresponds to forcing λ_i and γ to zero (keeping the same ratio as before for each i). This causes

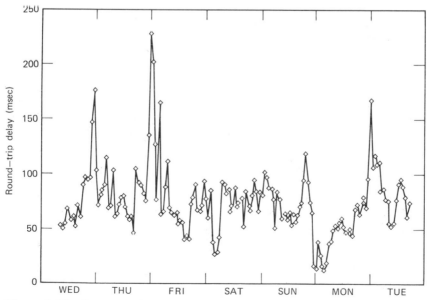

Figure 6.35   Computed (zero load) average message delay.

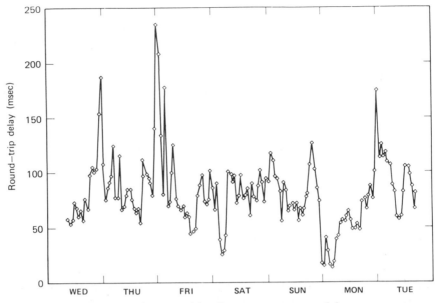

Figure 6.36   Computer (measured load) average message delay.

the average time in queue to go to zero. The zero load value must be less than the measured value, and is to be compared with the measurements displayed in Figure 6.34. Note that the network is introducing very small congestion effects. Furthermore, in Figure 6.36, we show the hourly variation of T_R (whose weekly average was $T_R = 73$ msec) for the *actual* load values as measured.

The model presented above is rather complex due mainly to the fact that not all channels (or IMPs) need have the same speed. In addition, the waiting-time terms complicate the expressions, and represent the part of the model that is most subject to question (namely, the Markovian assumptions). However, from Figures 6.35 and 6.36, we see that the zero-load and measured load calculations are nearly the same. This shows that the effect of the waiting time is quite negligible and so any improvement over Markovian assumptions will yield negligible changes to T for this lightly loaded subnet. This suggests a far simpler (no-load) model similar to the threshold model of Figure 5.24 for estimating T_R as follows. The expression for T_{SP} (and T_{RFNM}, which is similar in form) may be simplified by dropping the waiting time terms, and setting all $K_i = K$ (a constant), all $C_i = C$ (a constant at 50 KBPS), and $C_{Hj} = C_H$ (a constant at 100 KBPS). The result is

$$\hat{T}_{SP} = \bar{n}\left(\frac{1}{\mu C} + K\right) + K + \frac{1}{\mu_H C_H} + \sum_{i=1}^{M} \frac{\lambda_i}{\gamma} P_i \qquad \blacksquare \quad (6.5)$$

(and a similar expression for $\hat{T}_{RFNM}$). Except for the last summation, these parameters are easily computed. For the sum, one must estimate (or measure) the channel traffic λ_i and the network throughput γ. The propagation delays P_i are known constants. With these simplifications (and assuming $\beta = 1$, since the measured value of $\beta = 0.96$ was observed), we then have the approximation

$$\hat{T}_R = \hat{T}_{SP} + \hat{T}_{RFNM}$$

Our calculation gives $\hat{T}_R = 70$ msec,* which is an excellent approximation to the earlier stated value of $T_R = 69$ msec (at zero-load) and $T_R = 73$ msec (at measured load)!

On the other hand, the measured value of $T_R = 93$ msec is significantly larger than the measured load estimate of the model of $T_R = 73$ msec. This difference arises because the model does *not* include: any delay by the destination HOST in accepting the message; any delay due to the request for storage allocation at the destination IMP; any delay due to errors,

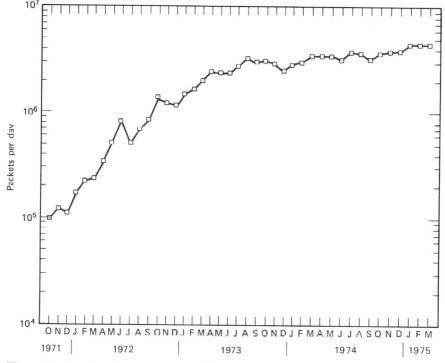

Figure 6.37   Growth in network traffic.

* The components for $\hat{T}_R$ are $\bar{n} = 3.31$, $1/\mu C = 8.2$ msec, $K = 0.75$ msec, $1/\mu_H C_H = 2.75$ msec, the propagation sum $= 11.4$ msec, and $1/\mu_R C = 3.36$ msec.

retransmissions, and looping; exact data on P_i; time variations in ρ finer than the hourly computations used; and non-Markovian assumptions. All the above omissions (except possibly the last) will increase the computed value of T_R.

Most of the measurement results reported in this section refer to only one particular week in the month of August 1973. The traffic patterns change as the network enlarges and so the names of the sites and their activity are not necessarily representative of the long-term state of affairs. Nevertheless, the *phenomena* we have discussed are present in the network (and change more slowly).

In order to provide a measure of how the traffic in the network has grown over the past $3\frac{1}{2}$ years, in Figure 6.37 we show the average number of packets per day entering the network as a function of time beginning with October, 1971 [MCKE 75]. We see that the traffic grew at a fairly regular rate for the first $1\frac{1}{2}$ years shown, leveled off thereafter, and experienced roughly $4\frac{1}{2}$ million packets per day passing through the network by March 1975. (This only includes the nonincestuous traffic; if we include incest, then March 1975 carried over 6 million packets per day.) As we approach the capacity of the subnetwork (which is currently at approximately ten million packets per day) both it and the HOST computing network will have to be expanded to provide both the communications and processing resources, respectively, that will be demanded through the network.

Let us now examine the details and effect of overhead on the ARPANET throughput.

6.7. LINE OVERHEAD IN THE ARPANET

We recognize that the choice of control characters, control messages, and control functions is one of the most important decisions in the design of computer-communication networks. More elaborate control procedures or protocols should tend to make the exchange of data smoother and safer in a well-designed system. However, as with sophisticated operating systems, there is a limit to the complexity of the control procedures; this limit is determined by the amount of overhead these procedures introduce. Beyond a certain point, the increase in overhead is too large to justify additional complexity for an intended improvement of service (and we have already seen cases where more sophisticated control led to degradations!). Therefore it is important to carefully analyze the overhead characteristics of a given system. The line overhead in communication networks depends critically on the characteristics of the exchanged messages [KLEI 76]. A large number of small messages involves more overhead than a small number of large messages to transfer the same number of data bits. The line capacity and protocol must therefore be selected with respect to the expected characteristics of the message exchange.

We define line overhead as all those characters transmitted that are not exchanged between user processes in the attached HOSTs. A user process is here defined to be any process that makes use of the system calls provided for in the Network Control Program (NCP). With this definition of a user process we exclude some higher-level overhead. For example, a process that controls the transfer of files will generate control messages that contribute to the total overhead. However, these higher-level control messages cannot be recognized as such in the subnet and so we exclude them. In what follows, then, we focus on the overhead introduced only by the subnet protocols and the HOST-to-HOST protocol. In this section we consider the version 2 system with the modification of an additional single (16-bit) IMP word that was later to be used in version 3.

The line overhead in the ARPANET may be classified according to the following four categories:

1. Level-0 overhead: Control of packet transmissions between adjacent IMPs.
2. Level-1 overhead: Message control in the subnet, that is, transmission control between source IMP and destination IMP.
3. Level-2 overhead: Message control between HOSTs.
4. Background traffic overhead: Routing messages, line status messages, status reports.

Table 6.1 gives a detailed explanation of the line overhead on all three levels of communication and for the background traffic. The format of a single-packet user message is further illustrated in Figure 6.38. As shown, there are nine hardware-generated characters for each transmitted packet. In addition, there are 16 bits for the IMP-to-IMP acknowledgement and 16 bits for the software checksum which contribute to the level-0 line overhead. Indeed, the overhead per transmitted data packet in version 3 has a value of 184 bits (and per stored packet it is 192 bits), whereas RFNMs are 168 bits; these values are 16 bits more than the version 2 values (see page 445).

The level-1 line overhead consists of two parts: 80 bits for the packet header of each user message and 64 bits for each subnet control message. The 80 bits of the packet header are not exclusively used for message control in the subnet. There is, for instance, a field for message identification that is passed unmodified from the source HOST to the destination HOST and might therefore be considered level-2 line overhead. For

Table 6.1
Line Overhead Classification

CATEGORY	NAME	NO. BITS	DESCRIPTION
Level-0:	SYN	16	Two hardware-generated SYN characters for clock synchronization
	DLE/STX DLE/ETX	32	Four hardware-generated control characters for message delimiting
	H-checksum	24	Hardware-generated checksum
	ACK-header	16	Software-generated control word carrying acknowledgement bits
	S-checksum	16	Software-generated checksum
Level-1:	Packet header	80	Five 16-bit words of packet header in each noncontrol message
	Subnet control	64	Four words for each subnet control message (see Table 6.2)
Level-2:	HOST-to-HOST protocol	40	Per message overhead specified by the HOST-to-HOST protocol
	HOST-to-HOST control	Average 93.5	Messages of different lengths for control of HOST-to-HOST traffic (see Table 6.3)
Background:	Routing messages	1160	Routing message sent every 640 msec (includes level-0 and level-1 overhead)
	IHY	152	I-heard-you message sent every 640 msec to determine up/down status of line (includes level-0 and level-1 overhead)
	Status reports	1728 (+336 for 2 RFNMs)	Status reports (2 packets) sent every 52.4 sec to the Network Control Center (NCC) (includes level-0 and level-1 overhead)

simplicity, however, all 80 bits of the packet header are counted here as level-1 overhead.

The level-2 line overhead also consists of two parts: 40 bits for the extended leader of HOST-to-HOST protocol* and an average number of

---

* These 40 bits are often considered part of the text as far as the subnet is concerned; thus the maximum "text" is $967 + 40 = 1007$ bits and the line overhead is 184 (nine hardware-generated characters plus 16 bits ACK plus 16 bits software checksum plus 80 bits packet header); see Figure 6.38.

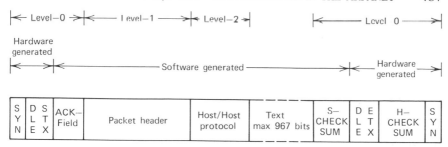

Figure 6.38   Format of single-packet user message.

93.5 bits for each HOST-to-HOST control message as observed in the subnet by the packet-tracing mechanism. (Padding bits are neglected.) We will assume that all messages adhere to the official HOST-to-HOST protocol (although it is known that other private protocols are in use, they represent only a very small fraction of the total traffic).

The background traffic consists mainly of routing messages that are exchanged between each pair of adjacent IMPs at least every 640 msec (for 50 KBPS lines). The I-heard-you messages that are sent to test the status of the phone lines and the status reports that are sent by each IMP every 52.4 sec to the Network Control Center represent a much smaller fraction of the background traffic. Since the background traffic is assumed to be independent of the network load, the level-0 and level-1 overhead of the background messages is included in the line overhead for the background traffic. This distribution of line overhead will facilitate our calculations below.

A week-long measurement experiment in May 1974 showed that very close to 50% of all packets transmitted in the ARPANET were subnet control messages. A list of all the subnet control messages, their frequency of occurrence, and their function is shown in Table 6.2. The relative frequency of these subnet control messages was determined by means of a measurement feature in the IMPs called packet tracing. Using this packet-tracing mechanism, about 75,000 subnet control messages were sampled from 35 different IMPs at different times of day and on different days of the week. The data of Table 6.2 represents the average over all these samples. Though not shown in the table, the deviation of the individual samples from the mean was remarkably small. RFNMs for single-packet messages represent by far the largest fraction of subnet control messages. This is not surprising since, as we have seen, some 96% of the messages entering the network are single-packet messages. It is interesting to note that we never observed a RFNM for a multipacket message that did not carry a "piggybacked" ALL. This means that there is

Table 6.2
Subnet Control Messages

NAME	PERCENT OF TOTAL	FUNCTION
RFNM-S	88.77	Sent from destination IMP to source IMP to signal the correct receipt of a *single-packet* message
RFNM-M	0.00	Sent from destination IMP to source IMP to signal the correct receipt of a *multipacket message*
ALL-S	3.98	Sent from destination IMP to source IMP to signal the *allocation* of one buffer for a *single-packet* message
REQALL	1.09	Sent from source IMP to destination IMP to *request the allocation* of eight buffers for a multipacket message
ALL-M	1.19	Sent from destination IMP to source IMP to signal the *allocation* of eight buffers for a *multipacket message*
GVB	1.04	Sent from source IMP to destination IMP to *give back* an unused buffer allocation that was received via a RFNM-ALL
RFNM-ALL	2.35	Combined effect of *RFNM-M* and *ALL-M*; the allocation for the next multi-packet message is piggybacked on the RFNM of the previous one
INCTRANS	1.13	An *incomplete transmission* message sent from destination IMP to source IMP for each message that could not be delivered correctly to its destination HOST
DESTDEAD	0.46	Sent from destination IMP to source IMP for each message that was sent to a *dead destination* HOST

so much reassembly buffer space available that in almost all cases eight packets can be allocated for the next transmission within the 1 sec timeout after the first packet of a multipacket message has been accepted by the destination HOST.

If our sampling technique were perfect, then the fraction of REQALL messages, the fraction of ALL-M messages, and the fraction of GVB messages would all be the same. This is because every request will sooner or later be granted and every allocation will, possibly after repeated use, be returned (see Figure 6.11). The amount by which these fractions differ gives an indication of the accuracy of our sampling method. Table 6.2 shows that slightly more than 50% of all multipacket messages that enter the ARPANET do not need to request a buffer allocation at the destination IMP since such an allocation is already waiting at the source IMP to be used. Phrased differently, we can say that slightly less than 50% of all piggybacked ALLs are not used by the source IMP and returned after a timeout of 250 msec to the destination IMP. This means that a much larger fraction of multipacket messages must wait for the necessary buffer allocation than one would have hoped in order to achieve a high throughput. There are two possible explanations for this behavior: (1) transfer of files that demand a long sequence of multipacket messages are relatively infrequent, (2) the timeout interval of 250 msec is too small compared to the HOST reaction time.

The large number of ALLs for single-packet messages is in agreement with the observations reported in [OPDE 74]; that is, since the IMPs obviously are not short of reassembly buffers, all these ALLs are due to single-packet messages that arrive out of order at their destination IMP. The fraction of ALLs for single-packet messages has decreased lately since the IMP program has recently been modified in such a way that continued retransmission of single-packet messages is no longer possible.

A surprisingly large fraction of subnet control messages are INCTRANS, which signal the source HOST that a message could not be delivered correctly to its destination HOST. The data of Table 6.2 indicates that, on the average, every hundredth message which enters the ARPANET will not reach its destination. The reason for this undesirable behavior is that many destination HOSTs are tardy in accepting messages. A HOST is declared down by its IMP if a message waits for more than 30 sec on the HOST output queue (from IMP to HOST) to be accepted. When this occurs, an INCTRANS control message is returned for every message that is waiting on the HOST output queue. (Future messages that reach the destination IMP after the HOST has already been declared down will generate a "destination dead" control message.) The frequent occurrence of incomplete transmissions is therefore not due to a failure of the subnet but to the unresponsiveness of some of the attached HOST computers.*

Compared with the number of incomplete transmissions, the number of destination dead control messages is rather small. These DESTDEAD messages appear to be generated mainly in cases where one HOST wants

---

* For example, the "software halts" of the TENEX operating system, which are unusual conditions from which the system cannot recover without operator intervention, are a major source of HOST unresponsiveness.

to find out which other HOSTs are currently responding to net traffic, and thereby send a "probe" message to dead HOSTs.

The packet-tracing mechanism allows one to distinguish between HOST-to-HOST control packets and data packets (as well as subnet control packets). From the examination of several thousand samples it was determined that some 41% of all HOST-to-HOST packets that traverse the network are NCP control commands! Let us examine the frequency with which each command type is sent, in order to determine what might be done to reduce this type of overhead.

In Table 6.3 we list the control commands, their length, and a short description together with bounds on their frequency of transmission. The frequency of the HOST-to-HOST control commands was derived from the length of the corresponding control messages as observed in the subnet. Since this method does not allow us to uniquely identify each control command, we can present only an upper and lower bound for the frequency of their occurrence. The basic HOST-to-HOST protocol involves the opening of connections (accomplished with the exchange of STR and RTS) and the closing of connections (i.e., the exchange of matching CLS commands). In addition, and of more relevance to this overhead discussion, is the allocation of buffer space at the destination HOST; this is not unlike the source-IMP-to-destination-IMP reservation procedure and is accomplished with the use of ALL, GVB, and RET commands. In Figure 6.39 we see the sequence of level 1 and level 2 control commands. At time a the STR is generated at the source HOST, it is passed as single packet message 1 from the source IMP to the destination IMP (which sends a RFNM back to the source IMP) and is received at the destination HOST at time b. The RTS is returned to the source HOST at time c via net message 2. Meanwhile, at time d, the destination HOST reserves some buffer space in its NCP store and signals this via a HOST-to-HOST ALL which becomes net message 3 and arrives at the source HOST at time e. The data flow now begins which causes a REQALL in the net, with a subnet ALL response back to the source IMP at time f. This triggers the data flow (message 4) which we assume to be a multipacket message of less than eight packets (because of a stingy ALL from the destination HOST). Luckily, the new HOST-to-HOST ALL "primes the pump" before the 250 msec source IMP timeout for the subnet ALL, and this triggers the rest of the data (message 6). After the second RFNM + ALL, the source HOST now times-out and the source IMP returns the last subnet ALL (i.e., the subnet GVB). Later, the source HOST closes the connection (via the CLS which journeys through the net as message 7) and the destination HOST finishes the dialogue with its matching CLS (message 8). In fact, the protocol is

Table 6.3
HOST to HOST Control Commands

NAME	LENGTH IN BITS	PERCENT OF TOTAL	FUNCTION
RTS	80	3.2–7.5	Sent from receiving HOST to sending HOST to set up a connection
STR	80		Sent from sending HOST to receiving HOST to set up a connection
CLS	72	3.2–7.5	Exchanged between receiving HOST and sending HOST to *close* a connection
ALL	64	63.8–79.0	Sent from receiving HOST to sending HOST to signal the *allocation* of message and bit space. The sending HOST is restricted from sending more messages or bits than have been allocated to him by the receiving HOST
GVB	32	0–10.0	Sent from receiving HOST as a request that the sending HOST *give back* all or part of its current allocation.
RET	64	0–10.0	Sent from sending HOST to receiving HOST to *return* all or part of its allocation (response to GVB)
INR	16		*Interrupt* command sent from the receiving HOST to the sending HOST
INS	16		*Interrupt* command sent from the sending HOST to the receiving HOST
ECO	16		*Echo* command to determine if some other HOST is ready for a network conversation
ERP	16	1.2–7.3	*Echo reply* command to return data from the echo command to its sender
RST	8		*Reset* command for the reinitialization of NCP tables
RRP	8		*Reset reply* command (response to reset command)
ERR	max 96		*Error* command
NOP	8		*No operation*

slightly more complex than this and involves a fuller exchange of RTS, STR, and CLS commands which permit logical communication link numbers to be assigned properly through the use of a shared link that acts as a "communication broker"; this process is referred to as the initial connection protocol (ICP) [POST 71].

If we examine the measurements reported in Table 6.3, the most striking among the bounds is the high frequency of the ALL (allocate) command. This is a network-wide characteristic. Let us consider the

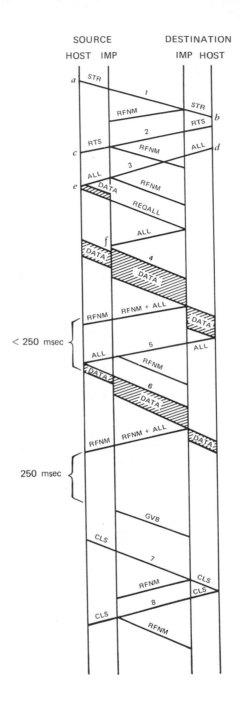

Figure 6.39   Level-1 and level-2 control commands.

492

impact of this phenomenon. The using up of network line bandwidth has little effect since, as we saw earlier, at the present time there is plenty to spare. Below, we do, however, show the effects of the size of the HOST allocation on the available capacity. If a user message is required to wait in the sending HOST until an ALL arrives from the receiving HOST, the effect would surely be noticeable. This may in fact be the case, and as such would contribute to some excessive delay as seen by users but not attributable to network delay alone. The fact that there are only about two data messages per ALL command indicates that the size of the allocation contained therein is very small. Note that the allocation size is a variable which depends on the NCP implementation.

Another consideration (possibly more important than the wasted network bandwidth and the extra waiting time for ALL control messages) is what portion of the HOST I/O and CPU bandwidth is spent in sending these overhead messages. The fewer messages sent and received by the NCP the smaller is the degradation to overall HOST performance.

An examination of the effect of allocation size would have been virtually impossible without the aid of an instrumented NCP. In the case of the ARPANET, there were several decisions to be made by NCP implementers. Among these were the number of buffers and buffer size per connection, fixed or dynamically allocated buffers, maximum number of allowable connections, and so on. In each case decisions such as these must be tested to ascertain their validity or to suggest improvement.

From Table 6.3 we conclude that most NCP control commands sent are of the ALL type. Therefore if one wants to reduce the overhead due to HOST-to-HOST control messages, the most effective first step (within the current protocol) is to reduce the number of ALL-type messages. Therefore, for those HOSTs that can afford to use larger and/or more buffers, the answer is simply to send larger allocations!

Let us now derive a simple formula for the line efficiency as a function of the traffic characteristics. As we have seen in Figure 6.39, the ARPANET HOST-to-HOST protocol provides for a connection-oriented message exchange involving opening and closing connections, and allocating buffers in the destination HOST.

Let N be a random variable representing the total number of bits that are to be transmitted and let A (also a random variable among HOSTs) be the number of bits that is allocated per ALL control message by the receiving HOST. Then the number of ALL control messages which must be sent from the receiver to the sender is*

$$a = \left\lceil \frac{N}{A} \right\rceil$$

* $\lceil s \rceil$ is the usual ceiling function and is equal to the smallest integer greater than or equal to s.

Define X to be the random number of data bits in a network data message. Note that X must be less than or equal to (a) N, the total number of transmitted bits (b) A, the number of allocated bits, and (c) 8023, the maximum number of data bits per message (we do not count the 40 bits of HOST-to-HOST protocol as pure data). Define Y to be the number of packets per message; we then have

$$Y = \left\lceil \frac{X+40}{1008} \right\rceil$$

Define m to be the total number of messages to be transmitted; we have

$$m = \left\lceil \frac{N}{X} \right\rceil$$

We denote the mean of these random variables by $\bar{N}$, $\bar{A}$, $\bar{a}$, $\bar{X}$, $\bar{Y}$, and $\bar{m}$.

Note that we have ignored all the overhead bits used for message padding. The number of padding bits depends on the word length of the HOST computer. The exclusion of padding has only a small effect on the computations below; as a consequence, the results may be viewed as being slightly optimistic.

Table 6.4 summarizes the line overhead involved in opening and closing a single connection and sending ALL control messages and data messages:

Table 6.4

Line Overhead Per Connection (in bits)

MESSAGE TYPE	NUMBER OF MESSAGES	OVERHEAD PER MESSAGE				TOTAL OVERHEAD PER MESSAGE
		LEVEL-0 OVERHEAD	LEVEL-1 OVERHEAD	LEVEL-2 OVERHEAD	RFNM OVERHEAD	
STR	1	104	80	40+80	168	472
RTS	1	104	80	40+80	168	472
ALL	$\bar{a}$	104	80	40+64	168	456
Data	$\bar{m}$	$104\bar{Y}$	$80\bar{Y}$	40	168	$184\bar{Y}+208$
CLS	2	104	80	40+72	168	464

We assume that no HOST-to-HOST control messages are piggybacked together as, for example, sending the first ALL together with the RTS (which is done by several HOSTs). Our measurement data shows that over 80% of all HOST-to-HOST control messages contain only one control command. If HOSTs maximize their message lengths (an assumption we shall make), we have

$$X = \min (N, A, 8023)$$

We define the average line efficiency η as the ratio of the total average number of data bits to the total average number of data plus overhead bits. Assuming that all the connections in the ARPANET can be described by the two variables $\bar{N}$ and $\bar{A}$, we make the following simple definition for the average line efficiency (see Table 6.4):

$$\eta = \frac{\bar{N}}{\bar{N} + 456\bar{a} + (184\bar{Y} + 208)\bar{m} + 1872} \qquad \blacksquare \quad (6.6)$$

Here, $456\bar{a}$ is the line overhead due to ALL commands, $(184\bar{Y} + 208)\bar{m}$ is the line overhead due to the overhead characters in data messages, and 1872 bits is the line overhead due to the opening and closing of a connection (one STR, one RTS, and two CLSs). Figure 6.40 shows the line efficiency η as a function of N for selected values of $\bar{A}$. The discontinuities in the curves are caused by message and packet boundaries. For $\bar{A} \geq \bar{N}$ only one ALL control message is necessary. Therefore the line efficiency is independent of $\bar{A}$ in this case. Note the low line efficiency for small values of $\bar{N}$. The line efficiency is only 0.32% if connections are used to transmit single characters ($\bar{N} = 8$). Even for large values of $\bar{N}$, the line efficiency is very low if the allocation size A is small. This shows what a drastic effect an NCP controlled parameter (A) can have on the efficiency of the communications subnet! A buffer shortage in the HOST computers can therefore directly lead to a decreased line utilization in the subnet. For the transfer of large quantities of data with a sufficiently large allocation size, the average line efficiency can be larger than 82% (in fact as $\bar{A}$, $\bar{N} \to \infty$, $\eta \to 0.827$).

Since our definition of average line efficiency does not include the background traffic, we must subtract the average bandwidth for the background traffic from the given physical bandwidth before applying the calculated percentages. In the ARPANET, the background traffic is 2.16 KBPS. The maximum possible bandwidth for process-to-process communication is therefore 82.7% of 47.84 KBPS or 39.56 KBPS. (Compare this value to the measured throughput of Figure 6.14.) This corresponds to 79% utilization of the 50 KBPS lines.

Let us now consider the case where a connection is used for a long interactive use of a HOST computer. In this case the overhead for opening and closing connections can be neglected. The line efficiency is not determined by N, the total number of transmitted bits, but by the average size $\bar{X}$ of each interactive message. The formula for the average line efficiency can now be simplified:

$$\eta_l = \frac{1}{1 + 456/\bar{A} + (184\bar{Y} + 208)/\bar{X}}$$

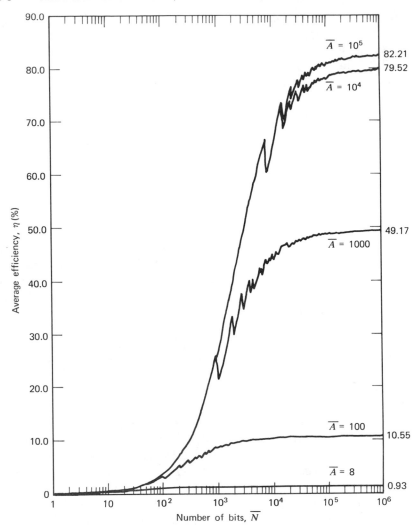

Figure 6.40   Average line efficiency as a function of data length.

Table 6.5 shows the average line efficiency, η_I, for interactive use as a function of $\bar{X}$ and $\bar{A}$. Part of this table is empty since the average message size $\bar{X}$ can never be larger than the average allocation size $\bar{A}$. We again notice the decreased line efficiency for small values of $\bar{A}$. However, even for $\bar{A} = \infty$ the line efficiency is only 2.00% if the messages are sent one character at a time.

We expect the line efficiency of the ARPANET as it is now being used to be rather low for the following two reasons: (1) the character-at-a-time

Table 6.5

Average Line Efficiency η_l in Percent

$\bar{X}_{(bits)}$ \ $\bar{A}_{(bits)}$	8	100	1000	∞
8	0.93	1.83	1.98	2.00
40		6.51	8.88	9.26
100		10.55	18.60	20.33
200			29.27	33.78
500			44.64	56.05
1000			49.21	63.45
2000				72.46
5000				81.59
8023				82.69

mode of communication with the TENEX system* (which represents a significant part of the total traffic) decreases the average number of data bits $\bar{X}$, and (2) the small buffer space in the TIPs decreases the average number of allocated bits $\bar{A}$.

As we have seen, the line efficiency in the ARPANET lies somewhere in the wide range between less than 1 and 79%. Let us now turn to measurement results that will allow us to calculate the current line efficiency. These results refer to the ARPANET as of May 1974 with 46 IMPs and 51 full duplex channels. To simplify matters, we make the following additional assumptions:

1. All lines have the same speed (50 KBPS).
2. All IMPs and lines are up.
3. The overhead for status reports can be equally allocated to all lines.

Table 6.6 gives a breakdown of all the bits transmitted per second in the measured ARPANET according to the line overhead classification of Table 6.1. The numbers represent an average over all 102 simplex lines. The contributions of the background traffic to the total traffic can be directly derived from Table 6.1. For the status reports we assumed that they were, on the average, sent over 6.25 hops before they reached the Network Control Center. (This number was computed for the topology of the ARPANET in May 1974.) The average number of packets per second per channel was measured to be 4.27 pkt/sec (excluding status reports). From this we easily derive the level-0 line overhead. The fact that roughly

---

* As we stated earlier, the TENEX system will be changed to improve this.

Table 6.6
Line Overhead in the ARPANET (May 1974)

CATEGORY	NAME	LINE BANDWIDTH		PERCENT OF LINE CAPACITY	
		BITS/SEC	SUM	%	SUM
Level-0:	SYN	68.32	444.08	0.14	0.89
	STX/ETX	136.64		0.27	
	H-Checksum	102.48		0.20	
	ACK-Header	68.32		0.14	
	S-Checksum	68.32		0.14	
Level-1:	Packet Header	173.60	308.00	0.35	0.62
	Subnet Control	134.40		0.27	
Level-2:	HOST-to-HOST Protocol	75.53	158.72	0.15	0.32
	HOST-to-HOST Control	83.19		0.17	
Background:	Routing messages	1812.50	2160.96	3.63	4.32
	IHY	237.50		0.48	
	Status Reports	110.96		0.22	
Data:	(Nonoverhead bits)	295.56	295.56	0.59	0.59
	Total sum:		3367.32 bits/sec		6.73%

50% of all transmitted packets represent subnet control messages allows us to determine the level-1 line overhead. The average number of bits per second per channel, excluding level-0 and level-1 line overhead and background traffic, was measured to be 454.28 bits/sec. Also, 87.02% of all packets are the first packet of a message and therefore carry the additional 40 bits of HOST-to-HOST protocol overhead. As previously stated, 41% of all packets exchanged between HOSTs are HOST-to-HOST control messages with an average length of 93.5 bits (excluding the 40 bits of HOST-to-HOST overhead). These numbers allow us to determine the level-2 line overhead, and from this we can determine the number of data bits exchanged between processes.

As can be seen from Table 6.6, about 64% of the traffic currently being carried by the ARPANET is background traffic. A large percentage of the background traffic is due to routing messages. The number of data bits per second is only about 0.6% of the line capacity. The line utilization including all types of overhead is 6.73%. (As mentioned before, this does not include the extra routing messages that are sent when the line utilization is low.)

Because of the low line utilization, some of these numbers might be misleading. Therefore let us try to assess the effect of increasing the load on the subnet. While the background traffic is held constant, we will assume that the level-0, level-1, and level-2 line overhead as well as the data bits are increased proportionally until the line utilization is 100%. This way we obtain an estimate for the overhead characteristics in a saturated net, *if the traffic characteristics are unchanged*. The result of this traffic projection is displayed in Table 6.7.

It is interesting to note that about 35% of all transmitted characters are now due to IMP-to-IMP (level-0) transmission control. The best line efficiency (i.e., percentage of data bits) one can hope to achieve is about 20% (a conservative estimate of the 23.44% shown; i.e., we must keep ρ below unity). This, of course, is an average number. In particular cases, one may get far better line utilization. However, if the overall traffic characteristics remain constant, not more than roughly 10 KBPS of the 50 KBPS will, on the average, be available for process-to-process communication. Note also that the background traffic, which currently represents more than 64% of all the traffic, becomes almost negligible as the net saturates.

It has been argued by some [METC 72] that there should be no difference between process communication within a HOST and process communication over a network. From a logical point of view this may be the right approach. As far as the efficient use of resources is concerned, such an approach may have disastrous results. In this sense the network is

Table 6.7
Projected Line Overhead

CATEGORY	NAME	LINE BANDWIDTH		PERCENT OF LINE CAPACITY	
		BITS/SEC	SUM	%	SUM
Level-0:	SYN	2709.28	17610.30	5.42	35.22
	STX/ETX	5418.55		10.84	
	H-Checksum	4063.92		8.13	
	ACK-Header	2709.28		5.42	
	S-Checksum	2709.28		5.42	
Level-1:	Packet Header	6884.23	12213.95	13.77	24.43
	Subnet Control	5329.72		10.66	
Level-2:	HOST-to-HOST Protocol	2995.19	6294.15	5.99	12.59
	HOST-to-HOST Control	3298.96		6.60	
Background:	Routing messages	1812.50	2160.96	3.63	4.32
	IHY	237.50		0.48	
	Status reports	110.96		0.22	
Data:	(Nonoverhead bits)	11720.64	11720.64	23.44	23.44
	Total sum:		50000.00 bits/sec		100.00%

not transparent to interprocess communication. Rather, the HOSTs must be aware of the fact that the allocation of network resources requires the same care as the allocation of any other resource. It appears that in some cases the freedom which the ARPANET protocols provide its implementers has been misused. In order to reduce the overhead, much more thought must be spent on the *efficient* implementation and use of network protocols, rather than only on their feasibility.

Although the overhead can be decreased, the designers of computer networks must realize that a significant percentage of the line utilization will always be needed for control information. The exact amount of the overhead depends critically on the type of traffic (or traffic mix) the network is intended to carry. Only a careful study will reveal what part of the physical bandwidth is actually available for user-process-to-user-process communication.

6.8. RECENT CHANGES TO THE FLOW CONTROL PROCEDURE

Some of the problems with the network control procedures prior to December 1974 (i.e., version 2) described above led to a revision of the message processing in the subnet and has resulted in version 3. In particular, message sequencing and limiting is now done on the basis of HOST-to-HOST pairs. The number of messages that can be transmitted simultaneously in parallel between a pair of HOSTs is eight. This is a significant departure from both version 1 (process-to-process limitation of one message per logical link) and version 2 (source-IMP-to-destination-IMP limitation of four messages per IMP-pair). Let us now describe the details of this new scheme.

Before a source HOST A at source IMP S can send a message to some destination HOST B at destination IMP D, a *message control block* must now be obtained in IMP S and IMP D. This message control block is used to control the transfer of messages. It is called a *transmit block* in IMP S and a *receive block* in IMP D. The creation of a transmit-block–receive-block pair is similar to establishing a (simplex) connection in the HOST-to-HOST (ICP) protocol. It requires an exchange of subnet control messages that is always initiated by the source IMP. The message control blocks contain, among other things, the set of message numbers in use and the set of available message numbers; it is through the use of a maximum of eight successive message numbers that the flow control is enforced on a HOST-pair basis.

After the first packet has been received from a HOST, the source IMP checks whether or not a transmit-block–receive-block pair exists for the transfer of messages from HOST A to HOST B. If HOST A has not sent any messages to HOST B for quite some time, it is likely that no such message control block pair even exists. Therefore, source IMP S creates a transmit block and sends a subnet control message to destination IMP D to request the creation of a receive block. When IMP D receives this control message, it creates the matching receive block and returns a subnet control message to IMP S to report this fact. When IMP S receives this control message the message control block pair is established.

A shortage of transmit and/or receive blocks will normally cause only an initial set-up delay. Currently there are 64 transmit and 64 receive blocks available in each IMP. This means, for example, that one HOST can transmit data to 64 different HOSTs simultaneously or that two HOSTs, attached to the same destination IMP, can each communicate with 32 different HOSTs simultaneously, and so on. Since 64 message blocks is a rather large number, it is unlikely that an initial set-up delay will be encountered.

The remaining resources are acquired in the following sequence as earlier: message number, reassembly buffers, and PLT entry. Since there are eight message sequence numbers that are allocated on a sending-HOST–receiving-HOST pair basis, a HOST is allowed to send up to eight messages to some receiving HOST without having received an acknowledgement for the first message. For multipacket traffic this is more than enough because there are currently still only six entries in the PLT. Therefore, the PLT has suddenly become a more prominent bottleneck than it used to be in the old message processing procedure when only four messages per IMP-pair could exist.

Note that a multipacket message tries to obtain the reassembly buffers *before* it asks for the PLT entry. (This sequence of resource allocation leads to difficulties as we describe below.) In case there is no reassembly buffer allocation waiting at the source IMP, the message number (and a PLT entry) is stolen by the REQALL as earlier (see Figure 6.11).

We now summarize some of the salient features of the three versions of flow control that we have described. Our main concern is with the location of control and with the details of the allocatable resources, namely, message number, PLT entries, capacity of the reassembly storage, subnet ALLs, HOST-to-HOST ALLs, and the scheme for IMP-to-IMP ACKs. In Figure 6.41, we show where the flow control throttle is applied; this is also described in the first two rows of Table 6.8. In particular, we see how version 1 limits the communication link between two HOST processes to only one message. Version 2 limits the IMP-pair communication to four messages; this constraint applies to the merged HOST data streams at the IMP-pair. Version 3 limits HOST-pairs to at

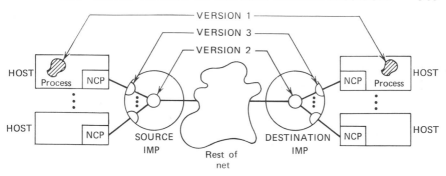

Figure 6.41   Points of control for the various flow control procedures.

most eight messages. Table 6.8 is drawn so that the columns are proportional to the period of time over which the particular versions have been in effect. The third row of the table gives the maximum number of PLT entries; rather than the PLT, version 1 used a link table in each IMP which permitted a total of 63 links from all HOSTs at a given IMP to the set of all other HOSTs in the net (with the restriction of one message number per link). The maximum number of reassembly buffers is a changing quantity, and as far as we know, it varied with time and version number roughly as shown in the fourth row of the table; these numbers are all for IMPs without a VDH (with a VDH, the system is restricted to at most *one* multipacket message in reassembly). We see the effect of upgrading the IMP core storage from 12 K to 16 K words late in 1972. The last three rows of the table are self-explanatory.

The throughput for this new control procedure (version 3) was measured in February 1975 and it was found to be significantly less than that shown in Figure 6.14 [KLEI 75]. One cause for this decrease is that the 316 IMPs are reaching their processing limit and are close to saturation. The second (and more subtle) cause for the throughput degradation, which we call *phasing*, is due to the sending of superfluous REQALLs! A REQALL is called superfluous if it is sent while a previous REQALL is still outstanding. This situation can arise if message i sends a REQALL but does not use the ALL returned by this REQALL because it obtained its reassembly buffer allocation piggybacked on a RFNM for an earlier message (which reached the source IMP before its requested ALL). The sending of superfluous REQALLs is undesirable because it unnecessarily uses up resources. In particular, each REQALL claims one PLT entry. Intuitively, it appears to be impossible that more than four eight-packet messages could be outstanding at any time since there is reassembly buffer space for only four such messages (34 reassembly

Table 6.8
Summary of the Various Flow Control Procedures

		1971	1972	1973	1974	1975
		VERSION 1		VERSION 2		VERSION 3
	WHERE	PROCESS-TO-PROCESS		SOURCE-IMP-TO-DESTINATION-IMP		HOST-TO-HOST
MESSAGE NUMBER CONTROL	MAXIMUM NUMBER	1		4		8
Maximum number of PLT entries		—		6		6
Maximum number of multipacket messages being reassembled		3	3	10	8	4
Subnet allocation required?		No		Yes (Piggyback)		Yes (Piggyback)
HOST-to-HOST allocation required?		Yes		Yes		Yes
Piggybacked IMP-to-IMP acknowledgements?		No		Yes		Yes

buffers). If, however, the buffer space that is freed when message i is reassembled causes an ALL to piggyback on a RFNM for message $i-j$ $(j \geq 1)$, then the RFNM for message i may queue up in the destination IMP behind $j-1$ other RFNMs. Thus only four messages really have buffer space allocated. Those messages that are outstanding in addition to these four have already reached the destination IMP and their RFNM's are waiting for buffer space (i.e., waiting for piggybacked ALLs). The sending of more than four eight-packet messages is initially caused by the sending of superfluous REQALLs. The PLT entries that were obtained by these REQALLs are later used by regular messages when the PLT is full; further input from the source HOST is stopped until a PLT entry becomes available (this results in the inefficient use of transmission facilities as we shall see). Thus we have a situation where a HOST may use all six entries in the PLT for the transmission to a destination HOST.

Figure 6.42 graphically depicts the kind of phasing we observed for almost all transmissions over more than four hops. Let us briefly explain the transmission of message i. At time a the last packet of message $i-1$ has been accepted and the input of the first packet of message i is initiated. This first packet is received by the source IMP at time b. Since there is a buffer allocation available (which came in piggybacked on the RFNM for message $i-7$), no REQALL is sent. However, let us assume that the PLT is full at time b. Therefore, message i must wait until time c when the RFNM for message $i-6$ frees a PLT entry and message i may then proceed. At time d all eight packets have been accepted by the source IMP. The first and eighth packets are received by the destination IMP at times e and f, respectively. The sending of the RFNM for message i is delayed until the RFNMs for messages $i-3$, $i-2$, and $i-1$ are sent. The buffer space that is freed when message $i+3$ reaches the destination at time g is piggybacked on the RFNM for message i which reaches the source IMP at time h. This effect may be seen in Figure 6.42 by observing the time slice picture while message i is in flight. Here we show messages as rectangles and RFNMs as ovals. Attached to RFNMs and messages are the ALL and PLT resources they own. We see the four ALLs owned by messages $i-1$, i, $i+1$ and by the RFNM for message $i-5$; we see the six PLTs owned by messages $i-1$, i and by the RFNMs for messages $i-5$, $i-4$, $i-3$, $i-2$. Message $i+1$ cannot leave the source IMP since it is missing a PLT; most of the PLTs are owned by RFNMs who are foolishly waiting for piggybacked ALLs which are *not* critical resources at the source IMP (message $i+1$ already has its ALL!). The trouble is clearly due to a poor phasing between PLTs and ALLs.

The phasing described above was observed for destination IMPs without the VDH software. For VDH IMPs, which can only reassemble *one*

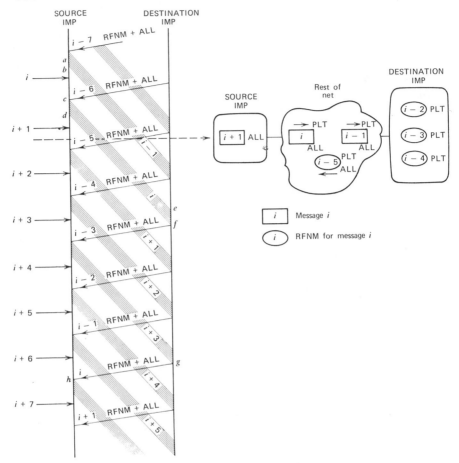

Figure 6.42   Phasing and its degradation to throughput.

message at a time (ten reassembly buffers), a different kind of phasing was observed which resulted in even more serious throughput degradations! In this case, a situation is created in which a REQALL control message is sent for every data message. The six PLTs are assigned to three RE-QALLS and three data messages. Figure 6.43 depicts this situation graphically. The first packet of message i is transmitted from the source HOST to the source IMP between time a and b. Since there is no buffer allocation available, the source IMP decides first to send a REQALL. However, all the PLTs are assigned and therefore the sending of the REQALL message is delayed until time c when the reply to an old REQALL (for message $i-3$) delivers an ALL and a PLT. At this time the PLT entry is immediately stolen by the delayed REQALL (generated

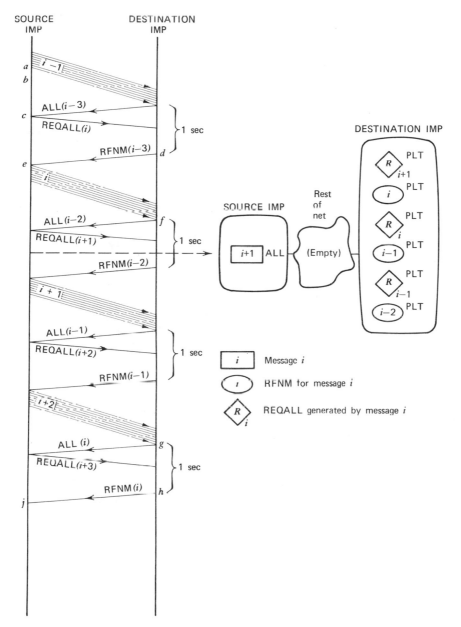

Figure 6.43 Phasing when only one multipacket message can be reassembled.

for message i). Note that at this point, message i gets the necessary buffer allocation but it cannot be sent to the destination because the PLT is once again full! Only when the RFNM for message $i-3$ times-out after 1 sec (time d) and is received by the source IMP (time e) without a piggy-backed ALL does a PLT entry become free for use by message i. At time f all eight packets have been received by the destination IMP. The sending of the RFNM for message i is now delayed by several seconds because the replies for messages $i-2$, $i-1$, and for two previous RE-QALLS must be sent first (see the snapshot in Figure 6.43). At time g the ALL control message responding to REQALL(i) is sent to the source IMP and 1 sec later at time h the RFNM for message i times-out. The RFNM is finally received for message i by the source IMP at time j.

The phasing in the case of destination IMPs with VDH software resulted in throughput degradations by a factor of three. This large decrease occurs because the system is stalled for almost 1 sec while the source IMP has the buffer allocation but no PLT entry; during this delay, the destination IMP, which can free a PLT entry by immediately sending a RFNM, is waiting for the buffer allocation (which cannot arrive) to use as a piggyback. There are two obvious ways to avoid this undesirable phasing of messages. First, one can avoid sending superfluous REQALLs that are the underlying cause for the phasing. Second, one can avoid the piggybacking of allocates on RFNMs as long as there are other replies to be sent. Here, as with all the other deadlocks and degradations we have seen, it is rather easy to find solutions once the fault has been uncovered; the challenge is to identify and remove these problems at the design stage.

6.9. THE CHALLENGE OF THE FUTURE

Our opening comments in the previous chapter expressed concern over the difficulty of wedding computers and communications. This apprehension was based on certain apparent incompatibilities between the computer and communication industries and between terminal and computer behavior. We have discussed one example of successful computer network operation, and this, among others, provides a strong basis for confidence in these networks. The conclusion one can draw from some of these analytic, simulation, and measurement studies is that the analytic modeling and design of computer networks has been rather successful.

The essence of the ARPANET technology is the introduction of "packet switching" as a new mode of communications. This point of view is discussed by Roberts [ROBE 70], who elaborates upon his original concept of network operation. Until such as the ARPANET, communication between two points was offered in the classical fashion, that is, a channel was set up (dialed, leased, purchased, etc.) between the two parties desiring communications service; this channel was then made available for their exclusive use until they voluntarily released it. As discussed earlier, such use is highly inefficient, and the notion of packet switching takes advantage of this fact. The packet-switching concept is that a very high bandwidth channel be made available to many users simultaneously (in the same spirit as ATDM). These users are characterized as being "bursty" in the sense that they have occasional need for service from the channel (occasional in terms of their average bandwidth needs as compared to the channel's total bandwidth capability). This collection of users then sees a high-speed channel available on a full-time basis (but which they each use at a very low utilization); the effect of the simultaneous demand for use either introduces occasional delays to transmission or occasional requests for repeated transmissions. The AR-PANET is an example of such a shared system. Also, the consideration of satellites as communication channels within such networks has unfolded a large number of interesting packet-switching problems.

However, there do remain several open questions in network analysis and design. For example, there are a number of phenomena that we have been unable to model, in particular those arising from routing and flow control procedures and many more for which only partial results exist. The major difficulty comes from the strongly coupled behavior among the various queues in a network of queues whenever we consider a realistic network model. For example, one would like to describe the behavior of the many packets of a multipacket message as they make their way through the network, and to evaluate the cost of sending multipackets (in terms of the buffer space and overhead costs incurred). Certainly the blocking effect and the lockup effects are still poorly understood; one approach is to design networks in which these effects disappear, and certainly this is an appropriate design philosophy [PRIC 73]; however, from an analytical point of view one would like not to ignore these problems just because they are difficult, but to study them in hopes of finding alternative solutions to their deleterious effects. In this chapter we have shown that message reassembly, message sequencing, and phasing are traps for the unwary system designer in that they can very easily lead to deadlock and degradation conditions. These and related issues are also treated in [CROW 75]. Other questions in network design come to mind. For example, what structure should a high-bandwidth IMP have? How can efficient use be made of a variety of high-bandwidth circuits? The entire question of large networks (thousands of nodes) poses numerous challenging questions: for example, how should these large networks be partitioned for effective design and what operational procedures should they follow [KAMO 76]?

Public packet-switching networks are springing up not only in the United States (ARPANET, Telenet's Intelligent Network), but also in Canada (Datapac), England (E.P.S.S.), Europe (E.I.N.), France (Transpac), Japan, Spain, Sweden, and elsewhere; this brings us face-to-face with the unusual, interesting, and very important internetting problems of interfacing separate networks [CERF 74, POUZ 73, ZIMM 73]; these are largely flow control and protocol problems again. The creation of the ARPANET has stimulated considerable research into how programs and operating systems should communicate with each other; this question is of interest even independent of the network operation. It is interesting to note that the present ARPANET is expanding rather quickly both in size and in traffic. As non-ARPA users gain access to the network, we must resolve the very nasty questions regarding charging mechanisms, privacy and security guarantees, guaranteed access and service, and so on. The Defense Communication Agency (DCA) took over responsibility for the ARPANET in July 1975; they are now faced with these nasty questions. Specialized carriers [GAIN 73] have been authorized to create packet-switched networks for commercial use. These are only some of the new and interesting problems in computer-communication networks [FOX 72, IEEE 72, DATA 73]. The analysis described in the early parts of Chapter 5 serves as a point of departure for some of the more recently discovered areas for investigation.

It is perhaps fair to say that whereas a variety of significant technical problems face us with regard to the growth of telecommunications and remote data processing, it is clear that these will not be the significant problem areas of the future. It does not take much thought to recognize that the major problems are social, economical, political, and even ecological in nature. Moreover, one must carefully examine the real goals that the customer of remote data processing may be measuring a proposal against. For example, the "underdeveloped" countries of this globe do not so rapidly take for granted the fact that more teleprocessing is necessarily better, but rather their point of view must be, "where can we best invest our finite resources for the largest and most needed gain?" Just because a service is good does not mean it is of highest priority. In addition, as our teleprocessing systems find their way deeper into the business applications, they will come under careful scrutiny as regards their efficiency, reliability, and robustness under heavy and demanding use. If they do not measure up, they will be discarded—"management is not known for its enduring patience!" Many of us have already had the pleasure (and the penalties) of remote access to data processing systems through portable consoles in our homes. We see the extension of these simple systems to networks that provide access to private files, the

Library of Congress, income tax aids, consumer shopping (imagine the deadly and unstable price wars), aids to education, and the like. It seems clear that the impact of these systems will not be limited by technology, but rather, by the socio-politico-economic, management, and ecological factors.

REFERENCES

BOBR 71   Bobrow, D. G., J. D. Burchfiel, D. L. Murphy, and R. S. Tomlinson, "TENEX, A Paged Time-Sharing System for the PDP-10," paper presented at the Third ACM Symposium on Operating System Principles, October 18–20, 1971. Published in *Communications of the Association for Computing Machinery,* **15,** 135–143. (1972).

CERF 74   Cerf, V. G., and R. E. Kahn, "A Protocol for Packet Network Communication," *IEEE Transactions on Communication,* **COM-22,** 637–648, (1974).

COLE 71   Cole, G. C., *Computer Network Measurements: Techniques and Experiments,* School of Engineering and Applied Science, University of California, Los Angeles, Engineering Report UCLA-ENG-7165, October 1971.

CROW 75   Crowther, W. R., F. E. Heart, A. A. McKenzie, J. M. McQuillan, and D. C. Walden, "Issues in Packet Switching Network Design," *AFIPS Conference Proceedings,* 1975 National Computer Conference, **44,** 161–175.

DATA 73   "Data Networks: Analysis and Design," *Proceedings of the Third IEEE Data Communications Symposium,* St. Petersburg, Florida, November 13–15, 1973.

DAVI 71   Davies, D. W., "The Control of Congestion in Packet Switching Networks," *Proceedings of the Second ACM IEEE Symposium in the Optimization of Data Communications Systems,* Palo Alto, California, 46–49, October 1971.

FORG 75   Forgie, J. W., "Speech Transmission in Packet-Switched Store-and-Forward Networks," *AFIPS Conference Proceedings,* 1975 National Computer Conference, **44,** 137–142.

FOX 72   Fox, J., ed., *Proceedings of the International Symposium on Computer-Communication Networks and Teletraffic,* Polytechnic Press, (Brooklyn, New York) April, 1972.

FRAN 72   Frank H., and W. Chou, "Topological Optimization of Computer Networks," *Proceedings of the IEEE,* **60** 1385–1397 (1972).

FULT 71   Fultz, G. L. and L. Kleinrock, "Adaptive Routing Techniques for Store-and-Forward Computer-Communication Networks," *Proceedings of the International Conference on Communications,* 39-1 to 39-8 June 14–16, 1971.

FULT 72   Fultz, G. L., "Adaptive Routing Techniques for Message Switching Computer-Communication Networks," School of Engineering and Applied Science, University of California, Los Angeles, Engineering Report UCLA-ENG-7252, July 1972.

GAIN 73   Gaines, E. C., "Specialized Common Carriers—Competition and Alternative," *Telecommunications*, **7**, No. 9, 15–26 (1973).

GERL 73   Gerla, M., "The Design of Store-and-Forward (S/F) Networks for Computer Communications," Computer Science Department, University of California, Los Angeles, Engineering Report UCLA-ENG-7319, 1973.

HAIG 63   Haight, F. A., *Mathematical Theories of Traffic Flow*, Academic Press (New York), 1963.

HEAR 70   Heart, F. W., R. E. Kahn, S. M. Ornstein, W. R. Crowther, and D. C. Walden, "The Interface Message Processor for the ARPA Computer Network," *AFIPS Conference Proceedings*, 1970 Spring Joint Computer Conference **36**, 551–567.

IEEE 72   *Proceedings of the IEEE* (Special Issue on Computer Communications), **60**, No. 11, 1243–1466 (1972).

KAHN 71   Kahn, R. E. and W. R. Crowther, "Flow Control in a Resource Sharing Computer Network," *Proceedings of the Second IEEE Symposium on Problems in the Optimization of Data Communications Systems*, Palo Alto, California, 108–116, October 1971.

KAMO 76   Kamoun, F., "Design Considerations for Large Computer-Communication Networks," Ph.D. Dissertation, Computer Science Department, School of Engineering and Applied Science, University of California, Los Angeles, 1976.

KLEI 74   Kleinrock, L. and W. E. Naylor, "On Measured Behavior of the ARPA Network," *AFIPS Conference Proceedings*, 1974 National Computer Conference **43**, 767–780, 1974.

KLEI 75   Kleinrock, L. and H. Opderbeck, "Throughput in the ARPANET—Protocols and Measurement," *Proceedings of the Fourth Data Communications Symposium*, Quebec, Canada, 6–1 to 6–11 October 1975.

KLEI 76   Kleinrock, L., W. E. Naylor, and H. Opderbeck, "A Study of Line Overhead in the ARPANET," *Communications of the Association for Computing Machinery* (1976).

MCKE 75   McKenzie, A. A., Letter to S. Walker, April 8, 1975.

MCQU 72   McQuillan, J. M., W. R. Crowther, B. P. Cossell, D. C. Walden, and F. E. Heart, "Improvements in the Design and Performance of the ARPA Network," *AFIPS Conference Proceedings*, 1972 Fall Joint Computer Conference, **41**, 741–754.

MCQU 73   McQuillan, J. M., "Throughput in the ARPA Network—Analysis and Measurement," Bolt Beranek and Newman, Inc. (Cambridge, Mass.), Report No. 2491.

MCQU 74a   McQuillan, J. M., "Design Considerations for Routing Algorithms in Computer Networks," *Proceedings of the Seventh Hawaii*

International Conference on System Sciences, University of Hawaii, Honolulu, Hawaii, January 8–10, 1974.

MCQU 74b   McQuillan, J. M., "Adaptive Routing Algorithms for Distributed Computer Networks," Bolt Beranek and Newman, Inc., (Cambridge, Mass.), Report No. 2831, May 1974 (available from the National Technical Information Service, AD 781467).

METC 72   Metcalfe, R. M., "Strategies for Operating Systems in Computer Networks," *Proceedings of the ACM Annual Conference,* 278–281, August 1972.

NAYL 75   Naylor, W. E., "A Loop-Free Adaptive Routing Algorithm for Packet Switched Networks," *Proceedings of the Fourth Data Communications Symposium,* Quebec, Canada, 7–9 to 7–14, October 1975.

OPDE 74   Opderbeck, H. and L. Kleinrock, "The Influence of Control Procedures on the Performance of Packet-Switched Networks," *Proceedings of the National Telecommunications Conference,* San Diego, California, December 1974.

POST 71   Postel, J., "Official Initial Connection Protocol," (NIC 7101), ARPA Network Information Center, Stanford Research Institute, Menlo Park, California, June 11, 1971.

POUZ 73   Pouzin, L., "Interconnection of Packet Switching Networks," INWG 42 (NIC 20792), ARPA Network Information Center, Stanford Research Institute, Menlo Park, California, October 1973 (also IFIP/TC6.1/42).

PRIC 73   Price, W. L., "Simulation of Packet-Switching Networks Controlled on Isarithmic Principles," *Proceedings of the Third IEEE Symposium on Data Networks Analysis and Design,* St. Petersburg, Florida, 44–49 November 13–15, 1973.

ROBE 70   Roberts, L. G. and B. D. Wessler, "Computer Network Development to Achieve Resource Sharing," *AFIPS Conference Proceedings,* 1970 Spring Joint Computer Conference **36,** 543–549.

ZEIG 71   Zeigler, J. F., "Nodal Blocking in Large Networks," School of Engineering and Applied Science, University of California, Los Angeles, Engineering Report UCLA-ENG-7167, 1971.

ZIMM 73   Zimmermann, H. and M. Elie, "Proposed Standard HOST-HOST Protocol for Heterogeneous Computer Networks: Transport Protocol," INWG 43 (NIC 20856) ARPA Network Information Center, Stanford Research Institute, Menlo Park, California, December 1973.

EXERCISES

6.1. Suppose that with each routing update message, each IMP also sends a "hop number" table (HNT) that contains one entry for each IMP in the network (including itself). Assume that all neighbors of an IMP (say, IMP n) send exactly one update to that IMP between successive updates generated by that IMP. Let the kth entry h_k in the next HNT to be generated by IMP n be calculated as follows:

$$h_k = \begin{cases} \min_{\{neighbors\}} h_k^{(i)} + 1 & k \neq n \\ 0 & k = n \end{cases}$$

where $h_k^{(i)}$ is the kth entry in the most recent HNT received by IMP n from his ith neighbor.

(a) For an N-IMP network of arbitrary (but connected) topology in which no IMP or line failures occur, determine how many update times are required before all HNTs cease to change if all HNTs are initialized to zero.

(b) What is the meaning of each entry after equilibrium is reached?

(c) After equilibrium is reached, suppose some IMP crashes. How many updates are required in an N-IMP net before all IMPs are certain that the crash occurred (and can identify the faulty IMP).

(d) For the net of Figure 5.20(b) (omitting LINC and HARV), assume all HNTs contain all zeros to begin. Show the HNT in UCLA at each update until it stops changing.

(e) From the equilibrium HNT at UCLA obtained in part (d), show the sequence of changing HNTs (until a new equilibrium or its functional "equivalent" is reached) if the RAND IMP crashes.

(f) After a long time, suppose RAND recovers. Now show the sequence (up to equilibrium) of UCLA HNTs.

6.2. The measured total throughput shown in Figure 6.9 was stated as being optimistic. Describe some upper bounds on this throughput that are violated by this fictitious picture.

6.3. Consider Figure 6.13.
(a) Assuming that each message shown is a full multipacket message, calculate the throughput from source IMP to destination IMP. Assume no other interfering traffic and that the time between message arrivals shown is 500 msec.

(b) Repeat part (a) assuming channels that are
(i) twice as fast.
(ii) twice as slow.

(c)   Calculate the idle time of the forward channels for the three cases above.

6.4.   Compare the curves in Figures 6.14 and 6.15 and reconcile the numbers shown there with the theoretical round-trip delays based on message length, line speed, overhead, and hop distance.

6.5.   [COLE 71] Consider messages whose lengths b are exponentially distributed with a mean length of $\bar{b}$ bits. Assume that any message is segmented into packets each of length L bits (maximum) to which is added a header (i.e., the software overhead) of H bits, giving a required buffer size of $L+H$ bits of memory. Let $\bar{b}_p -$ E[length of a packet] and $\eta = \bar{b}_p/(L+H)$ be the efficiency of the buffer utilization.

(a)   Show that

$$\bar{b}_p = \bar{b}[1 - e^{-L/\bar{b}}]$$

and therefore,

$$\eta = \frac{\bar{b}[1 - e^{-L/\bar{b}}]}{L+H}$$

(b)   Find the defining equation for the *optimum* packet length L_o.
(c)   Show that the optimum efficiency, η_o for any given $\bar{b}$ and H is simply

$$\eta_o - e^{-L_o/\bar{b}}$$

6.6.   Find the equation describing the optimum envelope for line efficiency as shown in Figure 6.40. Express this optimum in terms of $\bar{N}$ only.

6.7.   Consider Figure 6.42.
(a)   Draw the time slice picture shortly after time e.
(b)   Repeat for time h.

6.8.   Consider Figure 6.43. Draw the time slice picture shortly after time f.

Glossary of Notation*

(Only the notation used often in this book is included below.)

NOTATION†	DEFINITION	TYPICAL PAGE REFERENCE
ACK	Acknowledgment control message	440
ALL	Allocate control message	444
ARPA	Advanced Research Projects Agency	304
ARPANET	ARPA experimental computer network	304
ATDM	Asynchronous time-division multiplexing	244
$A(t)$	$P[\tilde{t} \le t]$	2
$A^*(s)$	Laplace transform of $a(t)$	3
a	One-way propagation time (normalized)	397
$a(t)$	$dA(t)/dt$	2
BTMA	Busy-tone multiple access	403
$B(x)$	$P[\tilde{x} \le x]$	2
$B^*(s)$	Laplace transform of $b(x)$	3
$\tilde{b}$	Message length (bits)	322
$b(x)$	$dB(x)/dx$	2
CA	Capacity-assignment problem	319
CBE	Concave branch elimination algorithm	353
CFA	Capacity and flow assignment problem	319
CPU	Control processing unit	157
CSMA	Carrier sense multiple access	394
C_i	Capacity of ith channel	315

* In those (few) cases where a symbol has more than one meaning, the context (or a specific statement) resolves the ambiguity.
† The use of the notation $y_n \rightarrow y$ is meant to indicate that $y = \lim_{n \to \infty} y_n$, whereas $y(t) \rightarrow y$ indicates that $y = \lim_{t \to \infty} y(t)$.

516

NOTATION	DEFINITION	TYPICAL PAGE REFERENCE
C_n	nth customer to enter the system	2
$C(u)$	$P[\tilde{u} \le u]$	23
$C^*(s)$	Laplace transform of $c(u)$	23
C_a^2	Coefficient of variation for interarrival time	288
C_b^2	Coefficient of variation for service time	288
C_p^2	Coefficient of variation for number of processing operations	288
$c(u)$	$dC(u)/du$	23
D	Denotes deterministic distribution	4
D	Dollar cost of a network	317
D_e	Excess dollars	331
d_i	Cost per unit of capacity on the ith channel	330
$d_i(C_i)$	Cost function for ith channel	316
d_k	$P[\tilde{q} = k]$	6
$E[X] = \bar{X}$	Expectation of the random variable X	2
E_k	System state k	6
E_r	Denotes r-stage Erlangian distribution	4
FA	Flow assignment problem	319
FB	Generalized foreground-background	172
FCFS	First-come–first-serve	107
FD	Flow deviation	342
FDMA	Frequency division multiple access	388
FET	First exit time	381
$\Gamma(w, t)$	$P[U(t) \le w]$	18
G	Denotes general distribution	4
G	Channel traffic for satellites and ground radio	364
$G(K)$	Constant for closed networks	14
$G(y)$	Busy-period distribution	112
$G^*(s)$	Laplace transform of $g(y)$	17
$G_b^*(s)$	Laplace transform of pdf for Y_b	112
$G_c^*(s)$	Laplace transform of pdf for Y_c	112
$G_0^*(s)$	Laplace transform of pdf for Y_0	112

NOTATION	DEFINITION	TYPICAL PAGE REFERENCE
$g(y)$	$dG(y)/dy$	11
h	Scaling factor on network throughput	344
HOL	Head of the line	119
HOST	Computer facility attached to a network	305
H_R	Denotes R-stage hyperexponential distribution	4
I	Duration of the idle period	11
IMP	Interface Message Processor	310
KBPS	Kilobits per second	310
L	Maximum packet length	459
LCFS	Last-come–first-serve	144
l_i	Length of ith channel	316
M	Denotes exponential distribution	4
M	Size of finite population	12
M	Number of channels in a network	315
ML	Multilevel	178
M^*	Saturation number	209
m	Number of servers	4
m	Stationary infinitesimal mean	73
$m(x, t)$	Infinitesimal mean	67
$\bar{n}$	Average path length	325
$n(x)$	Density of attained service	163
NCC	Network Control Center	305
NCP	Network Control Program	310
NIC	Network Information Center	305
NMC	Network Measurement Center	429
N_q	Number of customers in queue	6
$N(t) \rightarrow N$	Number of customers in system at time t	3
N	Number of nodes in a network	315
$o(x)$	$\lim_{x \rightarrow 0} o(x)/x = 0$	29
p	Index on priority groups ($p = 1, 2, \ldots, P$)	107
$\mathbf{P}$	Matrix of transition probabilities	7
PDF	Probability distribution function	2
PLT	Pending Leader Table	444

NOTATION	DEFINITION	TYPICAL PAGE REFERENCE
$P[A]$	Probability of the event A	2
$P[A\|B]$	Probability of the event A conditioned on the event B	7
$P_k(t)$	$P[N(t)=k]$	6
pdf	Probability density function	2
p_k	$P[k$ customers in system]	9
p_{ij}	$P[$next state is $E_j\|$current state is $E_i]$	7
$p_{ij}(t)$	$P[X(s+t)=j\|X(s)=i]$	8
$Q(z)$	z-transform of $P[N=k]$	3
$q_n \to \tilde{q}$	Number left behind by departure (of C_n)	15
$q'_n \to \tilde{q}'$	Number found by arrival (of C_n)	20
q_{ij}	Transition rates of Markov chain	8
R	Round-trip propagation time (normalized)	370
REQALL	Request for allocate	444
RFNM	Request for next message	430
RR	Round robin	166
r_{ij}	$P[$next node is $j\|$current node is $i]$	14
r_k	$P[\tilde{q}'=k]$	6
S	Throughput for satellite and ground radio packet switching	364
SEPT	Shortest-expected-processing-time-first	144
SERPT	Shortest-expected-remaining-processing-time-first	144
SIPT	Shortest-imminent-processing-time-first	144
SJF	Shortest-job-first	107
SPT	Shortest-processing-time-first	144
SRPT	Shortest-remaining-processing-time-first	144
SSA	Selfish scheduling algorithms	188
$S(y)$	$P[\tilde{s}\le y]$	3
$S(y\|x)$	$P[\tilde{s}\le y\|\tilde{x}=x]$	161
$S^*(s)$	Laplace transform of $s(y)$	3
$S^*(s\|x)$	Laplace transform of $dS(y\|x)/dy$	165
s	Laplace transform variable	3
$s_n \to \tilde{s}$	Time in system (for C_n)	3

NOTATION	DEFINITION	TYPICAL PAGE REFERENCE
$s(y)$	$dS(y)/dy$	3
$\bar{s}_n \rightarrow \bar{s} = T$	Average time in system (for C_n)	3
T	Average time in system	3
T_0	No-load delay	325
$T(x)$	Average response time for job requiring x sec of service	161
TCFA	Topology, capacity, and flow assignment problem	319
TDMA	Time division multiple access	386
TELNET	Telephone network program	310
TENEX	Time-sharing operating system for PDP-10 computer	305
TIP	Terminal IMP	313
T_p	Average time in system for priority group p	108
T_i	Average time waiting for and using ith channel	320
$t_n \rightarrow \tilde{t}$	Interarrival time (between C_{n-1} and C_n)	2
$\bar{t} = 1/\lambda$	Average interarrival time	2
$\overline{t^k}$	kth moment of $a(t)$	2
t_0	Natural unit of time	90
U_n	$U_n = u_0 + u_1 + \cdots + u_{n-1}$	24
$U(t)$	Unfinished work in system at time t	18
$\bar{U}$	Average unfinished work	117
$\bar{U}_0$	Natural unit of backlog	91
$u_0(t)$	Unit impulse function	11
$u_n \rightarrow \tilde{u}$	$u_n = x_n - t_{n+1} \rightarrow \tilde{u} = \tilde{x} - \tilde{t}$	22
VDH	Very distant HOST	451
$v_n \rightarrow \tilde{v}$	Number of arrivals during service time (of C_n)	15
$v'_n \rightarrow \tilde{v}'$	Number of customers served during $(\tau_n, \tau_n + t_n)$	20
W	Average time in queue	3
W_p	Average wait for priority group p	108
W_0	Average remaining service time	109

NOTATION	DEFINITION	TYPICAL PAGE REFERENCE
$W(x)$	Average waiting (wasted) time for job requiring x sec of service	161
$W(y)$	$P[\tilde{w} \leq y]$	3
$W^*(s/x)$	Laplace transform of $dW(y/x)/dy$	191
$W^*(s)$	Laplace transform of $w(y)$	3
$w_n \to \tilde{w}$	Waiting time (for C_n) in queue	3
$w(y)$	$dW(y)/dy$	3
$w_n \to \tilde{w} = W$	Average waiting time (for C_n)	3
$X(t)$	Stochastic process	8
$x_n \to \tilde{x}$	Service time (of C_n)	2
$\bar{x} = 1/\mu$	Average service time	2
$\overline{x^k}$	kth moment of $b(x)$	2
$\overline{x_p}$	Average service time for pth priority group	108
$\overline{x_p^2}$	Second moment of pth priority group's service time	109
Y	Busy-period duration	11
Y_b	Delay busy-period duration in delay cycle	111
Y_c	Delay cycle duration	111
Y_0	Initial delay in delay cycle	111
z	z-transform variable	3
$\alpha(t)$	Number of arrivals in $(0, t)$	327
γ_i	(External) input rate to node i	14
γ_{jk}	(External) input message rate with origin j, destination k	315
γ	$\sum_j \sum_k \gamma_{jk}$	315
$\delta(t)$	Number of departures in $(0, t)$	327
λ	Average arrival rate	2
λ_i	Average arrival rate to channel i in network	316
λ_k	Birth (arrival) rate when $N = k$	9
λ_p	Average arrival rate for pth priority group	108
μ	Average service rate	2

NOTATION	DEFINITION	TYPICAL PAGE REFERENCE
μ_i	Average service rate at node i	14
μ_k	Death (service) rate when $N = k$	9
$\pi^{(n)} \rightarrow \pi$	Vector of state probabilities $\pi_k^{(n)}$	7
$\pi_k^{(n)} \rightarrow \pi_k$	$P[\text{system state (at } n\text{th step) is } E_k]$	7
$\prod_{i=1}^k a_i$	$a_1 a_2 \cdots a_k$ (product notation)	9
ρ	Utilization factor	5
ρ_p	Utilization factor for priority group p	108
σ	Root for G/M/m	20
σ^2	Stationary infinitesimal variance	73
σ_a^2	Variance of interarrival time	25
σ_b^2	Variance of service time	18
$\sigma^2(x, t)$	Infinitesimal variance	67
σ_p	$\sum_{i=p}^P \rho_i$	121
τ_n	Arrival time (of C_n)	2
$\Phi_+(s)$	Laplace transform of $W(y)$	24
$\triangleq$	Equals by definition	32
$\cong$	Approximately equals	30
$(0, t)$	The interval from 0 to t	57
$\bar{X} = E[X]$	Expectation of the random variable X	2
$(y)^+$	$\max[0, y]$	23
$(y)^-$	$-\min[0, y]$	24
A/B/m	m-server queue with $A(t)$ and $B(x)$ identified by A and B, respectively	4
$f^{(k)}(t_0)$	$d^k f(t)/dt^k \vert_{t=t_0}$	3
⊛	Convolution operator	23

Summary of Important Results

Following is a collection of the basic results from this text in the form of a list of equations. To the right of each equation is the page number where it first appears in a meaningful way; this is to aid the reader in locating the descriptive text and theory relevant to that equation.

GENERAL

$\rho = \lambda \bar{x}$ (G/G/l) — 5

$\rho = \lambda \bar{x}/m$ (G/G/m) — 5

$T = \bar{x} + W$ — 5

$\bar{N} = \lambda T$ (Little's result) — 6

$\bar{N}_q = \lambda W$ — 6

$\bar{N}_q = \bar{N} - m\rho/(G/G/m)$ — 6

$dP_k(t)/dt$ — flow rate into E_k — flow rate out of E_k — 6

$p_k = r_k$ (for Poisson arrivals) — 6

$r_k = d_k$ [$N(t)$ makes unit changes] — 6

POISSON PROCESSES

$$P_k(t) = \frac{(\lambda t)^k}{k!}\, e^{-\lambda t}$$
9

$\bar{N}(t) = \lambda t$ — 9

$\sigma^2_{N(t)} = \lambda t$ — 9

$E[z^{N(t)}] = e^{\lambda t(z-1)}$ — 9

BIRTH–DEATH SYSTEMS

$$p_k = p_0 \prod_{i=0}^{k-1} \frac{\lambda_i}{\mu_{i+1}}$$
(equilibrium solution to birth–death system) — 9

$$p_0 = \cfrac{1}{1 + \sum_{k=1}^{\infty} \prod_{i=0}^{k-1} \lambda_i / \mu_{i+1}}$$ 9

M/M/1

$$p_k = (1 - \rho)\rho^k$$ 10

$$\bar{N} = \rho/(1 - \rho)$$ 10

$$\sigma_N^2 = \rho/(1 - \rho)^2$$ 10

$$W = \frac{\rho/\mu}{1 - \rho}$$ 10

$$T = \frac{1/\mu}{1 - \rho}$$ 10

$$s(y) = \mu(1 - \rho)e^{-\mu(1-\rho)y} \qquad y \geq 0$$ 11

$$S(y) = 1 - e^{-\mu(1-\rho)y} \qquad y \geq 0$$ 11

$$w(y) = (1 - \rho)u_0(y) + \lambda(1 - \rho)e^{-\mu(1-\rho)y} \qquad y \geq 0$$ 11

$$W(y) = 1 - \rho e^{-\mu(1-\rho)y} \qquad y \geq 0$$ 11

$$P[\text{interdeparture time} \leq t] = P[\text{idle period duration} \leq t] = 1 - e^{-\lambda t}$$ 11

$$g(y) = \frac{1}{y\sqrt{\rho}} e^{-(\lambda+\mu)y} I_1(2y\sqrt{\lambda\mu})$$ 11

$$f_n = \frac{1}{n}\binom{2n-2}{n-1}\rho^{n-1}(1+\rho)^{1-2n}$$ 12

$$p_k = \cfrac{[M!/(M-k)!](\lambda/\mu)^k}{\sum_{i=0}^{M} [M!/(M-1)!](\lambda/\mu)^i} \qquad \text{(M/M/1 finite population)}$$ 12

$$Q(z) = \frac{\mu(1-\rho)(1-z)}{\mu(1-z) - \lambda z[1 - G(z)]} \qquad \text{(M/M/1 bulk arrival)}$$ 12

$$p_k = \left(1 - \frac{1}{z_0}\right)\left(\frac{1}{z_0}\right)^k \qquad k = 0, 1, 2, \ldots \qquad \text{(M/M/1 bulk service)}$$ 12

M/M/m

$$p_k = \begin{cases} p_0 \dfrac{(m\rho)^k}{k!} & k \leq m \\[3mm] p_0 \dfrac{(\rho)^k m^m}{m!} & k \geq m \end{cases}$$ 13

$$p_0 = \left[\sum_{k=0}^{m-1} \frac{(m\rho)^k}{k!} + \frac{(m\rho)^m}{m!}\left(\frac{1}{1-1\rho}\right)\right]^{-1}$$ 13

$$p_k = \frac{(\lambda/\mu)^k/k!}{\sum_{i=0}^{m} (\lambda/\mu)^i/i!} \qquad \text{M/M/m (loss system)}$$

13

$$T = \frac{m}{\mu C} + \frac{P_m}{\mu C(1-\rho)}$$

280

$$P_m = \frac{p_0(m\rho)^m}{(1-\rho)m!}$$

280

$$T(1, \lambda, C) \le T(m, \lambda, C)$$

283

$$T(m, a\lambda, aC) = \frac{1}{a} T(m, \lambda, C)$$

283

$$W(1, \lambda, C) \ge W(m, \lambda, C)$$

285

$$W(m, a\lambda, aC) = \frac{1}{a} W(m, \lambda, C)$$

284

$$\bar{N}(m, a\lambda, aC) = \bar{N}(m, \lambda, C)$$

284

$$\bar{N}_q(m, a\lambda, aC) = \bar{N}_q(m, \lambda, C)$$

284

LIFE AND RESIDUAL LIFE

$$f_X(x) = \frac{xf(x)}{m_1} \qquad \text{(lifetime density of sampled interval)}$$

16

$$\hat{f}(y) = \frac{1-F(y)}{m_1} \qquad \text{(residual life density)}$$

16

$$\hat{F}^*(s) = \frac{1-F^*(s)}{sm_1} \qquad \text{(residual life transform)}$$

16

$$r_1 = \frac{m_2}{2m_1} \qquad \text{(mean residual life)}$$

16

$$r(x) = \frac{f(x)}{1-F(x)} \qquad \text{(failure rate)}$$

16

M/G/1

$$r_k = p_k = d_k$$

15

$$\bar{N} = \rho + \frac{\lambda \overline{x^2}/2}{(1-\rho)} \qquad \text{(P-K mean value formula)}$$

16

$$W = \frac{W_0}{1-\rho} \qquad \text{(P-K mean value formula)}$$

16

$$W_0 = \frac{\lambda \overline{x^2}}{2}$$

16

$$Q(z) = B^*(\lambda - \lambda z) \frac{(1-\rho)(1-z)}{B^*(\lambda - \lambda z) - z} \qquad \text{(P-K transform equation)} \qquad 17$$

$$W^*(s) = \frac{s(1-\rho)}{s - \lambda + \lambda B^*(s)} \qquad \text{(P-K transform equation)} \qquad 17$$

$$S^*(s) = B^*(s) \frac{s(1-\rho)}{s - \lambda + \lambda B^*(s)} \qquad 17$$

$$\sigma_{\bar{w}}^2 = W^2 + \frac{\lambda \overline{x^3}}{3(1-\rho)} \qquad 17$$

$$P[I \le y] = 1 - e^{-\lambda y} \qquad 17$$

$$G^*(s) = B^*(s + \lambda - \lambda G^*(s)) \qquad 17$$

$$g_1 = \frac{\bar{x}}{1-\rho} \qquad 18$$

$$\sigma_g^2 = \frac{\sigma_b^2 + \rho(\bar{x})^2}{(1-\rho)^3} \qquad 18$$

$$F(z) = zB^*[\lambda - \lambda F(z)] \qquad 18$$

$$h_1 = \frac{1}{1-\rho} \qquad 18$$

$$\sigma_h^2 = \frac{\rho(1-\rho) + \lambda^2 \overline{x^2}}{(1-\rho)^3} \qquad 18$$

$$\frac{\partial F(w, t)}{\partial t} = \frac{\partial F(w, t)}{\partial w} - \lambda F(w, t) + \lambda \int_{x=0}^{w} B(w-x) \, d_x F(x, t) \qquad 18$$

$$\text{(Takács integrodifferential equation)}$$

$$F^{**}(r, s) = \frac{(r/\eta)e^{-\eta w_0} - e^{-r w_0}}{\lambda B^*(r) - \lambda + r - s} \qquad 19$$

$$Q(z) = \frac{(1-\rho)(1-z)B^*(\lambda - \lambda G(z))}{B^*(\lambda - \lambda G(z)) - z} \qquad \text{(bulk arrival)} \qquad 17$$

$$W^*(s) = 1 - \rho + \frac{\lambda(1 - G^*(s))}{s + \lambda - \lambda G^*(s)} \qquad \text{(LCFS)} \qquad 119$$

$$\sigma_{\text{FCFS}}^2 = (1-\rho)\sigma_{\text{LCFS}}^2 - \rho W^2 \qquad 150$$

$$W_0 = \sum_{i=1}^{P} \frac{\lambda_i \overline{x_i^2}}{2} \qquad \text{(P priorities)} \qquad 109$$

$$G_b^*(s) = G_0^*(\lambda - \lambda G^*(s)) \qquad 112$$

$$G_c^*(s) = G_0^*(s + \lambda - \lambda G^*(s)) \qquad 113$$

$$\sum_{p=1}^{P} \rho_p W_p = \begin{cases} \dfrac{\rho W_0}{1-\rho} & \rho < 1 \\ \infty & \rho \geq 1 \end{cases} \qquad \text{(M/G/l priority conservation law)} \qquad 114$$

$$W_p = \frac{W_0}{(1-\sigma_p)(1-\sigma_{p+1})} \qquad p = 1, 2, \ldots, P \qquad\qquad 121$$

$$\text{(HOL nonpreemptive priorities)}$$

$$T_p = \frac{\bar{x}_p(1-\sigma_p) + \sum_{i=p}^{P} \lambda_i \overline{x_i^2}/2}{(1-\sigma_p)(1-\sigma_{p+1})} \qquad \text{(HOL preemptive priorities)} \qquad 125$$

$$W(x) = \frac{W_0}{\left[1 - \lambda \int_0^{x^-} yb(y)\,dy\right]\left[1 - \lambda \int_0^{x^+} yb(y)\,dy\right]} \qquad \text{(SJF)} \qquad 124$$

$$W_p = \frac{[W_0/(1-\rho)] - \sum_{i=1}^{p-1} \rho_i W_i[1 - (b_i/b_p)]}{1 - \sum_{i=p+1}^{P} \rho_i[1 - (b_p/b_i)]} \qquad p = 1, 2, \ldots, P \qquad 131$$

$$\text{(delay-dependent priorities)}$$

$$W(y) = \frac{W_0}{[1 - \rho + \rho\beta(y^+)][1 - \rho + \rho\beta(y^-)]} \qquad\qquad 138$$

$$\text{(bribing for queue position)}$$

M/G/∞

$$p_k = \frac{\rho^k}{k!}\, e^{-\rho} \qquad\qquad 19$$

$$T = \bar{x} \qquad\qquad 19$$

$$s(y) = b(y) \qquad\qquad 19$$

G/M/1

$$r_k = (1-\sigma)\sigma^k \qquad k = 0, 1, 2, \ldots \qquad\qquad 20$$

$$\sigma = A^*(\mu - \mu\sigma) \qquad\qquad 20$$

$$W(y) = 1 - \sigma e^{\mu(1-\sigma)y} \qquad y \geq 0 \qquad\qquad 20$$

$$W = \frac{\sigma}{\mu(1-\sigma)} \qquad\qquad 20$$

G/M/m

$$\sigma = A^*(m\mu - m\mu\sigma) \tag{21}$$

$$P[\text{queue size} = n \,|\, \text{arrival queues}] = (1-\sigma)\sigma^n \qquad n \geq 0 \tag{21}$$

$$R_{k-1} = \frac{R_k - \sum_{i=k}^{m-2} R_i p_{ik} - \sum_{i=m-1}^{\infty} \sigma^{i+1-m} p_{ik}}{p_{k-1,k}} \tag{21}$$

$$p_{ij} = 0 \qquad \text{for} \quad j > i+1 \tag{21}$$

$$p_{ij} = \int_0^\infty \binom{i+1}{j} [1 - e^{-\mu t}]^{i+1-j} e^{-\mu t j}\, dA(t) \qquad j \leq i+1 \leq m \tag{21}$$

$$\beta_n = p_{i,i+1-n} = \int_{t=0}^{\infty} \frac{(m\mu t)^n}{n!} e^{-m\mu t}\, dA(t) \tag{21}$$

$$0 \leq n \leq i+1-m, \qquad m \leq i$$

$$p_{ij} = \int_0^\infty \binom{m}{j} e^{-j\mu t} \left[\int_0^t \frac{(m\mu y)^{i-m}}{(i-m)!} (e^{-\mu y} - e^{-\mu t})^{m-j} m\mu\, dy \right] dA(t) \tag{21}$$

$$j < m < i+1$$

$$J = \frac{1}{[1/(1-\sigma)] + \sum_{k=0}^{m-2} R_k} \tag{21}$$

$$W = \frac{J\sigma}{m\mu(1-\sigma)^2} \tag{21}$$

$$w(y \,|\, \text{arrival queues}) = (1-\sigma) m\mu e^{-m\mu(1-\sigma)y} \qquad y \geq 0 \tag{22}$$

$$W(y) = 1 - \frac{\sigma e^{-m\mu(1-\sigma)y}}{1 + (1-\sigma) \sum_{k=0}^{m-2} R_k} \qquad y \geq 0 \tag{22}$$

$$W \cong \frac{\sigma_a^2 + (1/m^2)\sigma_b^2}{2\bar{t}(1-\rho)} \qquad \text{(heavy traffic approximation)} \tag{47}$$

$$W \leq \hat{W} \leq \frac{\sigma_a^2 + (1/m^2)\sigma_b^2}{2\bar{t}(1-\rho)} \tag{50}$$

$$\hat{W} - \left(\frac{m-1}{m}\right)\bar{x} \leq W \leq \hat{W} \tag{50}$$

$$\bar{\bar{N}} - \rho(m-1) \leq \bar{N} \leq \bar{\bar{N}} \tag{50}$$

G/G/1

$$w_{n+1} = (w_n + u_n)^+ \tag{23}$$

$$c(u) = a(-u) \circledast b(u) \tag{23}$$

$$\bar{u} = \bar{t}(\rho - 1) \tag{23}$$

$$W(y) = \begin{cases} \displaystyle\int_{-\infty}^{y} W(y-u)\, dC(u) & y \geq 0 \\ 0 & y < 0 \end{cases}$$

24

(Lindley's integral equation)

$$A^*(-s)B^*(s) - 1 = \frac{\Psi_+(s)}{\Psi_-(s)}$$

24

$$\Phi_+(s) = \frac{1}{\Psi_+(s)} \lim_{s \to 0} \frac{\Psi_+(s)}{s} = \frac{W(0^+)}{\Psi_+(s)}$$

25

$$W = \frac{\sigma_a^2 + \sigma_b^2 + (\bar{t})^2(1-\rho)^2}{2\bar{t}(1-\rho)} - \frac{\overline{I^2}}{2\overline{I}}$$

25

$$\tilde{w} - \sup_{n \geq 0} U_n$$

24

$$w(y) = \pi(w(y) \circledast c(y))$$

24

$$W(y) \cong 1 - \exp\left[-\frac{2\bar{t}(1-\rho)}{\sigma_a^2 + \sigma_b^2}\, y \right] \qquad \text{(heavy traffic approximation)}$$

31

$$W \cong \frac{(\sigma_a^2 + \sigma_b^2)}{2\bar{t}(1-\rho)} \qquad \text{(heavy traffic approximation)}$$

31

$$W \leq W_U = \frac{\sigma_a^2 + \sigma_b^2}{2\bar{t}(1-\rho)} \qquad \text{(upper bound)}$$

34

$$\frac{\overline{[(\tilde{u})^+]^2}}{2\bar{t}(1-\rho)} = W_K \leq W \qquad \text{(lower bound)}$$

36

$$W_M \leq W \qquad \text{where } W_M \text{ solves}$$

40

$$W_M = \int_{-W_M}^{\infty} [1 - C(u)]\, du \qquad \text{(lower bound)}$$

$$\frac{\rho^2 C_b^2 - \rho(2-\rho)}{2\lambda(1-\rho)} = W_U - \frac{\rho(2-\rho) + C_a^2}{2\lambda(1-\rho)} \leq W \qquad \text{(lower bound)}$$

43

$$W_U - \frac{1}{2}\bar{t}(1+\rho) \leq W \leq W_U \qquad (\bar{t} - \text{MRLA/G/l})$$

42

$$\lambda W_U - \frac{1+\rho}{2} \leq \bar{N}_q \leq \lambda W_U \qquad (\bar{t} - \text{MRLA/G/l})$$

42

$$W_U - \frac{1}{2}\bar{t}(C_a^2 + \rho) \leq W \leq W_U \qquad (\text{IFR/G/l})$$

42

$$\lambda W_U - \frac{C_a^2 + \rho}{2} \leq \bar{N}_q \leq \lambda W_U \qquad (\text{IFR/G/l})$$

43

$$T_U(1, a\lambda, aC) = \frac{1}{a}\, T_U(1, \lambda, C)$$

288

$$\sigma_b^{\ 2} \leq \sigma_{\tilde{w}}^{\ 2} \leq \sigma_a^{\ 2} + \sigma_b^{\ 2} - 2 W_M \bar{t}(1-\rho) \qquad\qquad 40$$

$$\gamma e^{-s_0 y} \leq 1 - W(y) \leq e^{-s_0 y} \qquad\qquad 45$$

$$\frac{\partial f}{\partial t} = -\frac{\partial}{\partial w}[m(w,t)f] + \frac{1}{2}\frac{\partial^2}{\partial w^2}[\sigma^2(w,t)f] \qquad \text{(diffusion equation)} \qquad 71$$

$$\frac{\partial F(w,t)}{\partial t} = (1-\rho)\frac{\partial F(w,t)}{\partial w} \qquad \text{(fluid approximation)} \qquad 72$$

$$\frac{\partial F}{\partial t} = -m\frac{\partial F}{\partial w} + \frac{1}{2}\sigma^2\frac{\partial^2 F}{\partial w^2} \qquad \text{(diffusion approximation: D/A)} \qquad 73$$

$$F(w) = 1 - e^{2mw/\sigma^2} \qquad w \geq 0 \quad \text{(equilibrium D/A solution)} \qquad 74$$

$$F(w,t) = \Phi\left(\frac{w - w_0 - mt}{\sigma\sqrt{t}}\right) - e^{2mw/\sigma^2}\Phi\left(\frac{-w - w_0 - mt}{\sigma\sqrt{t}}\right) \qquad 77$$

$$\text{(transient D/A solution)}$$

$$\frac{\partial F}{\partial t'} = \frac{\partial F}{\partial w'} + \frac{1}{2}\frac{\partial^2 F}{\partial w'^2} \qquad \text{(scaled D/A)} \qquad 76$$

$$\text{Relaxation time} \approx \frac{\sigma^2}{m^2} = \frac{\lambda \overline{x^2}}{(1-\rho)^2} \qquad \text{(D/A)} \qquad 77$$

$$E[U_d(t)\,|\,U_d(0) = w_0] \rightarrow (\rho - 1)t + w_0 + \frac{\lambda \overline{x^2} \exp[-2(\rho-1)w_0/\lambda \overline{x^2}]}{2(\rho - 1)} \qquad 85$$

$$\text{(asymptotic D/A solution for time-dependent wait for } \rho > 1)$$

$$E[U_d'(t')\,|\,U_d'(0) = 0] = (1 + t'/2)[2\Phi(\sqrt{t'}) - 1] - \frac{t'}{2} - \frac{1}{2}P\left(\frac{3}{2}, \frac{t'}{2}\right) \qquad 87$$

$$\bar{U} = \rho W + \frac{\overline{x^2}}{2\bar{t}} \qquad 117$$

$$\sum_{p=1}^{P} \rho_p W_p = \bar{U} - W_0 \qquad \text{(G/G/l priority conservation law)} \qquad 117$$

G/G/m

$$W(y) \cong 1 - \exp\left(-\frac{2\bar{t}(1-\rho)}{\sigma_a^{\ 2} + (\sigma_b^{\ 2}/m^2)}\,y\right) \qquad \text{(heavy traffic approximation)} \qquad 47$$

$$W \cong \frac{\sigma_a^{\ 2} + \sigma_b^{\ 2}/m^2}{2\bar{t}(1-\rho)} \qquad \text{(heavy traffic approximation)} \qquad 47$$

$$\hat{W} - \frac{[(m-1)/m]\overline{x^2}}{2\bar{x}} \leq W \leq \frac{\sigma_a^{\ 2} + (1/m)\sigma_b^{\ 2} + [(m-1)/m^2](\bar{x})^2}{2\bar{t}(1-\rho)} \qquad 49$$

$$\frac{\rho^2 C_b^2 - \rho(2-\rho)}{2\lambda(1-\rho)} - \frac{[(m-1)/m]\overline{x^2}}{2\overline{x}} \le W \qquad \text{(lower bound)}$$
50

$$T_U(m, a\lambda, aC) = \frac{1}{a} T_U(m, \lambda, C)$$
289

$$T(1, \lambda, C) \le T(m, \lambda, C) \qquad \text{for} \quad C_p^2 \le 1$$
289

TIME SHARING: SINGLE RESOURCE M/G/1

$$T(x) = x + W(x)$$
161

$$n_q(x) = \lambda[1 - B(x)]\frac{dW(x)}{dx}$$
164

$$n(x) = \lambda[1 - B(x)]\frac{dT(x)}{dx}$$
164

$$S^*(s|x) = \frac{s(1-\rho)e^{-sx}}{s - \lambda + \lambda B^*(s)} \qquad \text{(batch processing)}$$
165

$$T(x) = \frac{W_0}{1-\rho} + x \qquad \text{(batch processing)}$$
165

$$T(x) = \frac{x}{1-\rho} \qquad \text{(RR and LCFS)}$$
168

$$T(x) = \frac{W_x + x}{1 - \rho_x} \qquad \text{(FB)}$$
174

$$S^*(s|x) = W_x^*(s + \lambda - \lambda G_x^*(s)) \exp[-x(s + \lambda - \lambda G_x^*(s))]$$
175
$$\text{(FB, and ML with FB)}$$

$$T(x) = \frac{W_{a_i} + x}{1 - \rho_{a_{i-1}}} \qquad \text{(ML with FCFS)}$$
181

$$W^*(s|x) = \left(\frac{1-\rho}{1-\rho'}\right)\left(\frac{s - \lambda' + \lambda' B^*(s)}{s - \lambda + \lambda B^*(s)}\right)\hat{W}_\lambda^*(s|x) \qquad \text{(SSA)}$$
194

$$T(x) - \frac{\lambda \overline{x^2}}{2(1-\rho)} - \frac{\lambda' \overline{x^2}}{2(1-\rho')} + \hat{T}_\lambda(x) \qquad \text{(SSA)}$$
194

$$\int_{0^-}^{\infty} T(x)[1 - B(x)]\, dx = \frac{\overline{x^2}}{2(1-\rho)} \qquad \text{(conservation law)}$$
199

$$\int_{0^-}^{\infty} W(x)[1 - B(x)]\, dx = \frac{\rho \overline{x^2}}{2(1-\rho)} \qquad \text{(conservation law)}$$
199

$$\frac{dW(x)}{dx} \ge 0$$
201

$$\frac{\lambda \overline{x_x^2}}{2(1-\rho_x)} \le W(x) \le \frac{\lambda \overline{x^2}}{2(1-\rho_x)(1-\rho)} + \frac{x\rho_x}{1-\rho_x} \qquad \text{(tight bounds)} \qquad\qquad 204$$

$$\frac{\partial}{\partial W}[C(W,x)W] = k[1-B(x)] \qquad \text{(optimum algorithm)} \qquad\qquad 261$$

TIME-SHARING: MULTIPLE RESOURCES (QUEUEING NETWORKS)

$$T = \frac{M/\mu}{1-p_0} - \frac{1}{\lambda} \qquad\qquad 208$$

$$M^* = \frac{1/\mu + 1/\lambda}{1/\mu} \qquad \text{(saturation number)} \qquad\qquad 209$$

$$T(x) \cong \mu T x \qquad\qquad 210$$

$$T = T_a \cong \frac{Mx_s}{m_s \mu_N x_N} - \frac{1}{\lambda} \qquad \text{(closed network, } M \gg M^*) \qquad\qquad 220$$

$$M^* = \frac{m_s}{x_s} \sum_{i=1}^{N} x_i \qquad \text{(closed network)} \qquad\qquad 221$$

$$f_i(1) = \frac{x_i/m_i}{x_1 + x_2 + \cdots + x_N} \qquad \text{(closed network)} \qquad\qquad 223$$

$$f_i(M) \le \begin{cases} Mf_i(1) & M \le M^* \\ M^* f_i(1) = \dfrac{x_i/m_i}{x_s/m_s} & M \ge M^* \end{cases} \qquad \text{(closed network)} \qquad\qquad 224$$

$$\lambda_i = \gamma_i + \sum_{j=1}^{N} \lambda_j r_{ji} \qquad \text{(open or closed network)} \qquad\qquad 214$$

$$p(k_1, k_2, \ldots, k_N) = p_1(k_1)p_2(k_2) \cdots p_N(k_N) \qquad \text{(open network)} \qquad\qquad 215$$

where $p_i(k_i)$ is solution to an isolated M/M/m_i system

$$p(k_1, k_2, \ldots, k_N) = \frac{1}{G(K)} \prod_{i=1}^{N} \frac{x_i^{k_i}}{\beta_i(k_i)} \qquad \text{(closed network)} \qquad\qquad 14$$

$$p(k_1, k_2, \ldots k_N) = \frac{1}{G(K)} \prod_{i=1}^{N} x_i^{k_i} \qquad \text{(closed network, } m_i = 1) \qquad\qquad 216$$

$$\mu_i x_i = \sum_{j=1}^{N} \mu_j x_j r_{ji} \qquad i = 1, 2, \ldots, N \qquad\qquad 15$$

$$G(K) = \sum_{\mathbf{k} \in A} \prod_{i=1}^{N} \frac{x_i^{k_i}}{\beta_i(k_i)} \qquad\qquad 15$$

$$P[\tilde{k}_i = k_i] = \frac{x_i^{k_i}}{G(K)}[G(K-k_i) - x_i G(K-k_i-1)] \qquad\qquad 217$$

$$E[\tilde{k}_i] = \sum_{k_i=1}^{K} x_i^{k_i} \frac{G(K-k_i)}{G(K)} \qquad\qquad 217$$

$$\beta_i(k_i) = \begin{cases} k_i! & k \le m_i \\ m_i! \, m_i^{k_i - m_i} & k \ge m_i \end{cases}$$

15

$$e_i(l) = \gamma_i(l) + \sum_{j=1}^{N} e_j(l) r_{ji}(l) \qquad \text{(open and closed networks)}$$

227

$$p(k_1, k_2, \ldots, k_N) = p_1(k_1) p_2(k_2) \cdots p_N(k_N) \qquad \text{(open network)}$$

228

where

$$p_i(k_i) = \begin{cases} (1 - \rho_i)\rho_i^{k_i} & \text{if node type is FCFS } (\cdot/M/1), \\ & \text{RR } (\cdot/G/1) \text{ or LCFS } (\cdot/G/1) \\ \dfrac{\rho_i^{k_i}}{k_i!} e^{-\rho_i} & \text{if node type is } \cdot/G/\infty \end{cases}$$

228

and where

$$\rho_i = \begin{cases} \sum_l \dfrac{e_i(l)}{\mu_i} & \text{if node type is FCFS } (\cdot/M/1) \text{ and } \mu_{ik_i} = \mu_i \\ \sum_l \dfrac{e_i(l)}{\mu_{il}} & \text{if node type is RR } (\cdot/G/1), \\ & (\cdot/G/\infty) \text{ or LCFS } (\cdot/G/1) \end{cases}$$

229

$$P(\alpha_1, \alpha_2, \ldots, \alpha_N) = C g_1(\alpha_1) g_2(\alpha_2) \cdots g_N(\alpha_N)$$

227

$$p(k_1, k_2, \ldots, k_N) = C h_1(k_1) h_2(k_2) \cdots h_N(k_N) \qquad \text{(closed network)}$$

228

where

$$h_i(k_i) = \begin{cases} \left(\sum_l \dfrac{e_i(l)}{\mu_{ik_i}} \right)^{k_i} & \text{if the node is of type FCFS } (\cdot/M/1) \\ \left(\sum_l \dfrac{e_i(l)}{\mu_{il}} \right)^{k_i} & \text{if the node is of type RR } (\cdot/G/1) \text{ or LCFS } (\cdot/G/1) \\ \dfrac{1}{k_i!} \left(\sum_l \dfrac{e_i(l)}{\mu_{il}} \right)^{k_i} & \text{if the node is of type } (\cdot/G/\infty) \end{cases}$$

228

$$p(k_1, k_2, \ldots, k_N) = \frac{1}{G(K)} \prod_{i=2}^{N} \left(\frac{\mu_1 \rho_i}{\mu_i} \right)^{k_i} \qquad \text{(central server network)}$$

232

COMPUTER-COMMUNICATION NETWORKS

$$T = \sum_{i=1}^{M} \frac{\lambda_i}{\gamma} T_i$$

321

$$p(b) = \mu e^{-\mu b} \qquad b \ge 0 \qquad \text{(independence assumption)}$$

322

$$T = \sum_{i=1}^{M} \frac{\lambda_i}{\gamma} \left(\frac{1}{\mu C_i - \lambda_i} \right)$$

322

$$T = K + \sum_{i=1}^{M} \frac{\lambda_i}{\gamma} \left[\frac{\lambda_i / \mu' C_i}{\mu' C_i - \lambda_i} + \frac{1}{\mu C_i} + P_i + K \right]$$

323

$$\bar{n} = \frac{\lambda}{\gamma} \qquad \text{(average path length)} \qquad\qquad 327$$

$$T_0 = \bar{n} \sum_{i=1}^{M} \frac{\lambda_i/\lambda}{\mu C_i} \qquad \text{(no-load delay)} \qquad\qquad 327$$

$$\hat{T}_{SP} = \bar{n}\left(\frac{1}{\mu C} + K\right) + K + \frac{1}{\mu_H C_H} + \sum_{i=1}^{M} \frac{\lambda_i}{\lambda} P_i \qquad\qquad 482$$

$$\text{(no-load approximation)}$$

$$\eta = \frac{\bar{N}}{\bar{N} + 456\bar{a} + (184\bar{Y} + 208)\bar{m} + 1872} \qquad \text{(ARPANET line efficiency)} \quad 495$$

$$C_i = \frac{\lambda_i}{\mu} + \left(\frac{D_e}{d_i}\right)\frac{\sqrt{\lambda_i d_i}}{\displaystyle\sum_{j=1}^{M} \sqrt{\lambda_j d_j}} \qquad i = 1, 2, \ldots, M \qquad\qquad 331$$

$$\text{(optimum assignment for linear costs)}$$

$$T = \frac{\bar{n}}{\mu D_e}\left(\sum_{i=1}^{M} \sqrt{\frac{\lambda_i d_i}{\lambda}}\right)^2 \qquad \text{(minimum delay for linear costs)} \qquad 332$$

$$C_i = \frac{\lambda_i}{\mu} + \mu C(1 - \bar{n}\rho)\frac{\sqrt{\lambda_i}}{\displaystyle\sum_{j=1}^{M} \sqrt{\lambda_j}} \qquad i = 1, 2, \ldots, M \qquad\qquad 332$$

$$\text{(optimum assignment for } d_i = 1)$$

$$T = \frac{\bar{n}\left(\displaystyle\sum_{i=1}^{M} \sqrt{\lambda_i/\lambda}\right)^2}{\mu C(1 - \bar{n}\rho)} \qquad \text{(minimum delay for } d_i = 1) \qquad 332$$

$$l_i \triangleq \frac{\partial T}{\partial(\lambda_i/\mu)} = \frac{C_i}{\gamma\left(C_i - \dfrac{\lambda_i}{\mu}\right)^2} \qquad\qquad 343$$

$$\text{(length definition for flow deviation algorithm)}$$

$$l_i \triangleq \frac{\partial T}{\partial(\lambda_i/\mu)} = \frac{\bar{n}\displaystyle\sum_{j=1}^{M} \sqrt{\lambda_j d_j/\lambda}}{D_e}\left[\sqrt{\frac{d_i}{\lambda\lambda_i}} + \frac{d_i}{\mu D_e}\sum_{j=1}^{M} \sqrt{\frac{\lambda_j d_j}{\lambda}}\right] \qquad 349$$

$$\text{(lengths for optimum linear costs)}$$

$$C_i = \frac{\lambda_i}{\mu} + \left(\frac{\lambda_i}{\mu\gamma T_{MAX}}\right)\frac{\displaystyle\sum_{j=1}^{M} \sqrt{\lambda_j d_j}}{\sqrt{\lambda_i d_i}} \qquad \text{(optimum assignment for dual)} \qquad 350$$

$$D = \sum_{i=1}^{M} \frac{\lambda_i d_i}{\mu} + \frac{1}{\gamma T_{MAX}}\left[\sum_{i=1}^{M} \left(\frac{\lambda_i d_i}{\mu}\right)^{1/2}\right]^2 \qquad \text{(minimum cost for dual)} \qquad 350$$

$$l_i \triangleq \frac{\partial D}{\partial(\lambda_i/\mu)} = d_i \left(1 + \frac{\sum\limits_{j=1}^{M} \sqrt{\lambda_j d_j}}{\gamma T_{MAX} \sqrt{(\lambda_i d_i)}}\right) \qquad \text{(lengths for optimum linear costs in dual problem)} \qquad 351$$

$$\gamma_0 \cong mD + b \qquad\qquad 360$$

SATELLITE PACKET SWITCHING

$$S = Gp_0 \qquad\qquad 364$$

$$S = Ge^{-2G} \qquad \text{(pure ALOHA)} \qquad\qquad 364$$

$$S = Ge^{-G} \qquad \text{(slotted ALOHA)} \qquad\qquad 365$$

$$S = G\left(1 - \frac{G}{M}\right)^{M-1} \qquad \text{(finite population equal rates)} \qquad 366$$

$$G = 1 \qquad \text{(optimum throughput)} \qquad\qquad 367$$

$$T = R + 1 + \frac{1-q}{q_t}\left[R + 1 + \frac{K-1}{2}\right] \qquad \text{(slotted ALOHA)} \qquad 372$$

GROUND RADIO PACKET SWITCHING

$$S = \frac{Ge^{-aG}}{G(1+2a) + e^{-aG}} \qquad \text{(nonpersistent CSMA)} \qquad\qquad 398$$

$$S = \frac{aGe^{-aG}}{(1 - e^{-aG} + a} \qquad \text{(slotted nonpersistent CSMA)} \qquad 399$$

$$S = \frac{G[1 + G + aG(1 + G + aG/2)]e^{-G(1+2a)}}{G(1+2a) - (1 - e^{-aG}) + (1 + aG)e^{-G(1+a)}} \qquad\qquad 399$$

$$\text{(1-persistent CSMA)}$$

$$S = \frac{Ge^{-G(1+a)}[1 + a - e^{-aG}]}{1 - e^{-aG} + ae^{-G(1+a)}} \qquad \text{(slotted 1-persistent CSMA)} \qquad 399$$

$$S(G, p, a) = \frac{(1 - e^{-aG})[P'_s\pi_0 + P_s(1 - \pi_0)]}{(1 - e^{-aG})[a\bar{t}'\pi_0 + a\bar{t}(1 - \pi_0) + 1 + a] + a\pi_0} \qquad\qquad 399$$

$$\text{(p-persistent CSMA)}$$

$$T = \left(\frac{G}{S} - 1\right)[2a + 1 + \beta + \bar{X}] + 1 + a \qquad \text{(nonpersistent CSMA)} \qquad 401$$

Index